Air Pollution Modeling and Its Application III

NATO • Challenges of Modern Society

A series of edited volumes comprising multifaceted studies of contemporary problems facing our society, assembled in cooperation with NATO Committee on the Challenges of Modern Society.

Air Pollution Modeling and Its Application III

Edited by
C. De Wispelaere

Prime Minister's Office for Science Policy
Brussels, Belgium

Published in cooperation with
NATO Committee on the Challenges of Modern Society

PLENUM PRESS • NEW YORK AND LONDON

Library of Congress Cataloging in Publication Data

International Technical Meeting on Air Pollution Modeling and Its Application (13th: 1982: Ile des Embiez, France)
Air pollution modeling and its application III.

(NATO Challenges of modern society; v. 5)
Includes bibliographical references and index.
1. Atmospheric diffusion—Mathematical models—Congresses. 2. Air—Pollution—Meteorological aspects—Mathematical models—Congress. I. Wispelaere, C. de. II. Title. III. Series.
QC880.4.D44I57 1982   628.5′3′0724   83-17800

ISBN-13: 978-1-4612-9673-7   e-ISBN-13: 978-1-4613-2691-5
DOI: 10.1007/978-1-4613-2691-5

Proceedings of the Thirteenth International Technical Meeting on
Air Pollution Modeling and Its Application, held September 14–17, 1982,
at Ile des Embiez, France

© 1984 Plenum Press, New York
Softcover reprint of the hardcover 1st edition 1984

A Division of Plenum Publishing Corporation
233 Spring Street, New York, N.Y. 10013

PREFACE

In 1969 the North Atlantic Treaty Organization established
the Committee on the Challenges of Modern Society. Air Pollution
was from the start one of the priority problems under study within
the framework of the pilot studies undertaken by this Committee.
The organization of a yearly symposium dealing with air pollution
modeling and its application is one of the main activities within
the pilot study in relation to air pollution.

After being organized for five years by the United States
and for five years by the Federal Republic of Germany, Belgium,
represented by the Prime Minister's Office for Science Policy
Programming, became responsible in 1980 for the organization of
this symposium.

This volume contains the papers presented at the 13th International Technical Meeting on Air Pollution Modeling and its Application held at Ile des Embiez, France, from 14th to 17th September
1982. This meeting was jointly organized by the Prime Minister's
Office for Science Policy Programming, Belgium, and the Ministère
de l'Environnement, France. The conference was attended by 120
participants and 45 papers have been presented. The closing session of the 13th I.T.M. has been attended by Mr. Alain Bombard,
French Minister of the Environment. The members of the selection
committee of the 13th I.T.M. were A. Berger (Chairman, Belgium),
W. Klug (Federal Republic of Germany), K. Demerjian (United States
of America), L. Santomauro (Italy), M.L. Williams (United Kingdom),
H. Van Dop (The Netherlands), H.E. Turner (Canada), C. De Wispelaere
(Coordinator, Belgium).

The main topic of this 13th I.T.M. was Langrangian Modeling.
On this topic a review paper was presented by Anton Eliassen (Norwegian Meteorological Institute). Other topics of the conference
were : Modeling cooling tower and power plant plumes, Modeling the
dispersion of heavy gases, Remote sensing as a tool for air pollution
modeling, Dispersion modeling including photochemistry. Dr. J. Knox
(Lawrence Livermore National Laboratory) presented a review paper
as an introduction to the topic Modeling the dispersion of heavy gases.

On behalf of the selection committee and as organizer and editor I should like to record my gratitude to all participants who made the meeting so stimulating and the book possible. Among them I particulary mention the chairmen and rapporteurs of the different sessions. Thanks also to the local organizing committee, especially Dr. J.C. Oppenau and Miss J. Maréchal, who was the Conference Secretary. Finally it is a pleasure to record my thanks to Mrs Desees, Mrs Van Saen and Miss De Corte, for preparing and typing these papers.

C. De Wispelaere

CONTENTS

LAGRANGIAN MODELING

MODELING COOLING TOWER AND POWER PLANT PLUMES

MODELING THE DISPERSION OF HEAVY GASES

CONTENTS

REMOTE SENSING AS A TOOL FOR AIR POLLUTION MODELING

DISPERSION MODELING INCLUDING PHOTOCHEMISTRY

EVALUATION OF MODEL PERFORMANCES IN PRACTICAL APPLICATIONS

CONTENTS

1: LAGRANGIAN MODELING

Chairmen: K. Klug
     J. Schreffler
     A. Eliassen

Rapporteurs: P. Zannetti
      C. M. Bhumralker
      R. Berkowicz

ASPECTS OF LAGRANGIAN AIR POLLUTION MODELLING

Anton Eliassen

The Norwegian Meteorological Institute
P.O. Box 320, Blindern
Oslo 3, Norway

INTRODUCTION

Lagrangian models are models in which parcels of air are followed as they blow with the wind. The models keep track of the pollutant content of the parcels. This is in contrast to Eulerian models, where the integration of the mass-balance equation is performed in a geographically fixed grid. Lagrangian models are popular, because their basic principle is easy to grasp also for non-professionals. In addition some numerical problems associated with the advection terms in the Eulerian mass-balance equation are avoided by Lagrangian models.

Considering the number of papers that have accumulated on Lagrangian air pollution models, a complete review of the subject represents a formidable task. This contribution is limited to a discussion of the basic principles of such models, and looks at typical examples of the main model types. The sensitivity of model calculations to some of the physical parameters is also discussed.

Lagrangian models for long range transport of air pollution

Consider a well-mixed parcel of air, with base area $A(t)$ and height $h(t)$, that moves with a certain height-independent advection wind. Assume further that the pollutant considered is depleted by dry deposition, and by a general first-order chemical decay. The mass balance equation for the pollutant within the air-parcel is

$$\frac{d}{dt}(Ahq) = v_d Aq - kAhq + \dot{Q}A \qquad (1)$$

3

where q is the pollutant concentration (mass per unit volume), v_d is the deposition velocity, k is the rate of the first-order decay, and Q is the pollutant emission per unit area and time. The time derivative is the total (Lagrangian) derivative. The equation above is the basic equation for Lagrangian models of air pollution.

The Lagrangian models fall into two main types : source-oriented and receptor-oriented. In a source-oriented model, the positions of puffs, consecutively emitted from each source, are traced as a function ot time. At any instant, the concentration field is given as the sum of concentrations due to each puff. For each puff, the source term in (1) acts only to give the puff a certain mass or concentration at the time of emission. In a receptor-oriented model on the other hand, the pollutant content of an air parcel is followed until the air parcel arrives at one of the selected receptor points. During its travel, the air parcel receives emitted material from the sources it passes over, i.e. the source term in (1) acts continuously.

Source-oriented Models   For a source-oriented model, (1) takes the form

$$\frac{d}{dt} (Ahq) = - v_d Aq - kAhq, \qquad \text{for } t>0 \qquad (2)$$

Introducing the pollutant mass M of the puff

$$M = Ahq \qquad (3)$$

we obtain simply

$$\frac{dM}{dt} = - \frac{v_d}{h} M - kM \qquad (4)$$

where, in principle, v_d, h and k can be functions of t. If (4) is integrated to give M(t), then the concentration q can be found using (3), if the horizontal area A(t) and the height h(t) of the puff are prescribed. Note that if the three-dimensional divergence of the velocity field is assumed to be zero, then A(t) and h(t) can both change either by turbulence or by systematic horizontal divergence. However, the volume A(t)h(t) will only change due to turbulence.

A good example of a model of this type is the EURMAP/ENAMAP model (Johnson et al. 1978, Bhumralkar et al. 1979). This model has been used to calculate the transport and deposition of airborne sulphur pollution over Europe and over Eastern North America.

The original EURMAP-1 model was designed to predict long-term
($\sim$ annual) deposition and concentration pattern. The radius r of
the cylindrical puffs was specified as a function of travel time t
as

$$r = (r_o^2 + t.10^4 m^2 s^{-1})^{1/2} \qquad\qquad (5)$$

where r_o is an initial radius of $\sim$ 30 km. In the vertical,
instantaneous mixing up to a constant mixing height of 1000 m was
assumed. A refined scheme for vertical mixing has been adopted in
EURMAP 2, designed for episode studies. Here, the mixing height is
estimated as a function of time and space, from 00 and 12 GMT
radiosonde data. The vertical extent of the puff is prescribed as
a function of travel distance and stability. When the puff penetrates
the mixing height, its vertical growth rate is reduced by means of
an assumed stability increase.

The MESOS model (ApSimon and Goddard 1976, Wrighley et al.,
1979), is designed to estimate the dispersion of radioactive
material over Europe. This model has now been used together with
a meteorological data base covering the year 1976, to obtain dispersion statistics for releases from a single source of duration
3 hours, 6 hours, 12 hours, 1 day, 3 days and 7 days. In the MESOS
model, a well-mixed turbulent boundary layer is assumed to be sur-
mounted by a series of stably stratified layers, each 100 m thick,
and with relatively little mixing between these layers. Vertical
dispersion of a puff in the well-mixed layer is based on local
Pasquill dispersion parameters and stability categories. Eventually
the material fills the mixing layer completely. If then the mixing
layer becomes deeper, the pollutant is diluted and mixed up to the
new height. If the mixing layer becomes shallower, some of the
material becomes isolated in the stable layers aloft and cannot
be depleted by dry deposition. This vertical stratification of the
puff represents a refinement compared to other operative Lagrangian
models, since it allows for the isolation of pollutants from the
surface by stable stratification. The isolated pollutants are not
subject to dry deposition.

The next logical step is to allow for wind shear, so that each
of the stably stratified layers is advected by a different horizontal
wind. The puff model discussed by Pack et al. (1978) has been
modified (Draxler and Taylor, 1982) to include this effect. The
model has been tested against data for the dispersion of Kr-85,
released from the Idaho National Engineering Laboratory in 1974
(Draxler, 1982). In this model, no vertical mixing is assumed
during the night, when a puff splits into five equal sublayers,
each of 300 m height. These sub-layers are followed as separate
trajectories for all subsequent calculations. During the next

daytime phase the elevated layers are permitted to mix downward and upward at a rate of 300 m per 3 hours. This procedure
results in an increase in the number of trajectories by a factor
of 5 each day.

The wind shear/vertical diffusion mechanism causes an
instantaneous puff release to be gradually elongated. The
mechanism has been studied theoretically by Taylor (1982). He shows
that at long travel times ($\sim 10^5$s) the spread of material in the
long direction of the puff is dominated by the shear effect, except
in very light winds or very large values of the horizontal
diffusivity. Across the puff, however, the effect of horizontal
diffusion on the spread of material is always comparable to that
of shear, and eventually dominates it for very long travel times.

Clearly, the wind shear effect is important for
instantaneous releases. However, model predictions are often compared with measurements that are averages over a certain sampling
time. During the sampling time trajectories consecutively emitted
from a source will take different paths, and cause a horizontal
spread of the plume. This effect is further discussed in Section 3.

Table 1. The average spread of an instantaneous puff release
σ_h as a function of travel distance d, compared to
the mode Θ of the angular spread and the width of
the plume Θd for a sampling time of 24 hours.
Values for σ_h given by the theory of Taylor (1982),
with a geostrophic wind speed of 11 ms^{-1} and horizontal and vertical diffusivities of 10^4 m^{2}s^{-1} and
10^2 m^{2}s^{-1}. The material presented by Gifford (1982)
gives similar values for σ_h. Values of Θ taken
from Figure 2, reproduced from Smith (1979).

d (km)	σ_h(km)	Θ(radians)	Θd (km)
100	10	0.63	63
250	20	0.58	144
500	30	0.52	260
1000	60	0.49	490
2000	110	0.43	860

Here we draw attention to Table 1, where instantaneous puff diffusion is compared to the diffusion due to sampling time. We see that for a sampling time of 24 hours, the "synoptic swinging" of trajectories is the dominating factor for plume spread. This means that the wind shear effect, and in general the rate of spread assumed for instantaneous puff releases is important only for very short-term averages in connection with episode studies, for example.

Receptor-oriented Models   For a receptor-oriented model (1) may be written as

$$Ah \frac{dq}{dt} = - q\frac{d}{dt} (Ah)_{turb} - v_dAq - kAhq + \dot{Q}A \qquad (6)$$

where $\frac{d}{dt}$ $(Ah)_{turb}$ is due only to turbulence. In such models the air parcels usually pick up pollution from the area-sources of an emission grid. The size of the emission grid elements is typically 80-150 km, so that the pollutant has dispersed horizontally up to this scale already. A diffusing puff reaches this scale only after a travel time of order 10^5s, so that with such emission data, the part of the diffusive term in (6) due to a change in A drops out for shorter transport times. Anyway, for sampling times of a day or more, the synoptic swinging of trajectories will dominate, in which case the "A-part" of the diffusive term also can be ignored.

Furthermore, for such sampling times, the effect of the diurnal cycle of the mixing layer is to some extent averaged out. It is tempting, then, to ignore the term $-q\frac{d}{dt}$ $(AH)_{turb}$ altogether, and to use climatological values for h. Under this assumption (6) simplifies to

$$\frac{dq}{dt} = - \frac{vd}{h}q - kq + \frac{\dot{Q}}{h} \qquad (7)$$

in which v_d, h and k may be functions of time. However, the change of h with time may only be due to systematic con- or divergence, and not to turbulence. This means that the top of the mixing layer behaves as a material surface, implying strictly that the exchange of air between the mixing layer and the free troposphere is zero. Equation (7) is similar to (4), but determines q instead of M. We recall, however, that to obtain (4) it was unnecessary to make any simplyfying assumptions about the effect of turbulence.

Equation (7) forms the basis of source-oriented models such as the Canadian AES model (Voldner et al., 1980, Olson and Voldner, 1981), and the model used in the OECD programme on long range transport of air pollutants (OECD, 1977 ; Eliassen, 1978).

A modified version of the latter model is used by the so-called
Meteorological Synthesizing Centre, West (MSC-W) in connection with
the European programme on long range transport (EMEP, see Eliassen
and Saltbones, 1982)

A receptor-oriented trajectory model based on eq. (7) might be
termed a "moving box" model. It has the agreeable property that
non-linear chemistry can be included fairly easily. This is not so
in source-oriented models, where puffs emitted from different sources
at different times may overlap. The chemical species in the over-
lapping puffs should then be allowed to interact. Because of the
nonlinearity, however, the distribution of pollutants back into
the individual puffs after the chemical interaction is
indeterminate. This advantage of receptor-oriented models disappears
as soon as the horizontal diffusion is made a function of travel
time. As discussed above, the latter refinement is relevant only
in case studies or when sampling times are short.

Trajectories

Pollutant concentration, wind speed and direction will
generally change with height. Although theory and observations
give some information on the distribution of these quantities within
the boundary layer, the choice of a wind for pollutant advection
to some extent remains arbitrary.

As a result, the existing Lagrangian models use different
advection winds. For example, Heffter (1980) has chosen to use
vertically averaged boundary-layer winds, ApSimon and Goddard (1976)
use a surface geostrophic wind modified according to surface type
and time of day, whereas Eliassen (1978) uses 850 mb isobaric
trajectories. Winds from a lower level than 850 mb would probably
be a better choice, especially over land in winter.

Trajectories based on different choices of advection wind
can be compared in various ways. Figure 1 shows a comparison
between surface geostrophic trajectories and isobaric 850 mb
trajectories based on a direct wind analysis (OECD, 1977). The
trajectories were compared by measuring the distance between the
respective intersections of a circle of radius 1000 km draw around
their common endpoint. Excluding one out of 180 cases, the mean
distance between two corresponding trajectories was 246 km, and
the median 140 km.

This illustrates that the determination of true trajectories
of boundary-layer air involves considerable uncertainty. Whereas
one part of this uncertainty arises from the limited amount of

meteorological information available from the boundary layer,
another part is due to the computational method. Walmsley and
Mailhot (1982) have investigated the magnitude of the errors arising from the different steps in normal computation procedures for
trajectory calculation. They find that the horizontal interpolation
of wind data between gridpoints represents an important source of
error. If linear interpolation is used, the average position error
is estimated to about 350 km for a transport time of 42 hours. The
use of a cubic scheme reduces this to about 90 km. A linear time
interpolation of winds between the six-hourly synoptic observations
appears to be sufficiently accurate, considering the errors arising
from observations and horizontal interpolation. The effect of such
position errors will decrease with increasing sampling time, since
the errors are masked by so-called"synoptic swinging" of the
trajectories. Smith (1979) has calculated the angular spread of a
sequence of trajectories released from a common origin as a function
of sampling time and travel distance. Figure 2 shows the mode Θ
of the angular spread as a function of these two quantities. Θ
appears to decrease slightly with distance r in such a way that

$$\Theta \sim r^{-0.125} \tag{8}$$

For a sampling time of one day and 1000 km transport, Θ is about
25°, corresponding to a plume width of about 430 km. This is
comparable to the above-mentioned average crosswind difference of
246 km between surface geostrophic and 850-mb trajectories after
1000 km of transport, and also to the numerical errors estimated
by Walmsley and Mailhot (1982).

Obviously then, the average pollutant concentration over long
time periods depends on the geographical distribution of trajectory
paths. A correct determination of each individual trajectory is
not essential. Assume for the moment that the task is to calculate
the contribution of each emission grid element to the pollutant
concentration or deposition at a selected receptor point. As a
first approximation, the contribution from the i-th element will
be proportional to

$$n_i Q_i f(r_i) \tag{9}$$

where n_i is the frequency of trajectories crossing the i-th element
and ending at the receptor point, Q_i is the emission strength of
the i-th grid element, and $f(r_i)$ is a function accounting for
dilution and removal of the pollutant as a function of distance r.
If a map of n is used together with a map of the distance-corrected
emission $Q f(r_i)$, one can quickly estimate the relative long-term
contributions of the various emission grid elements.

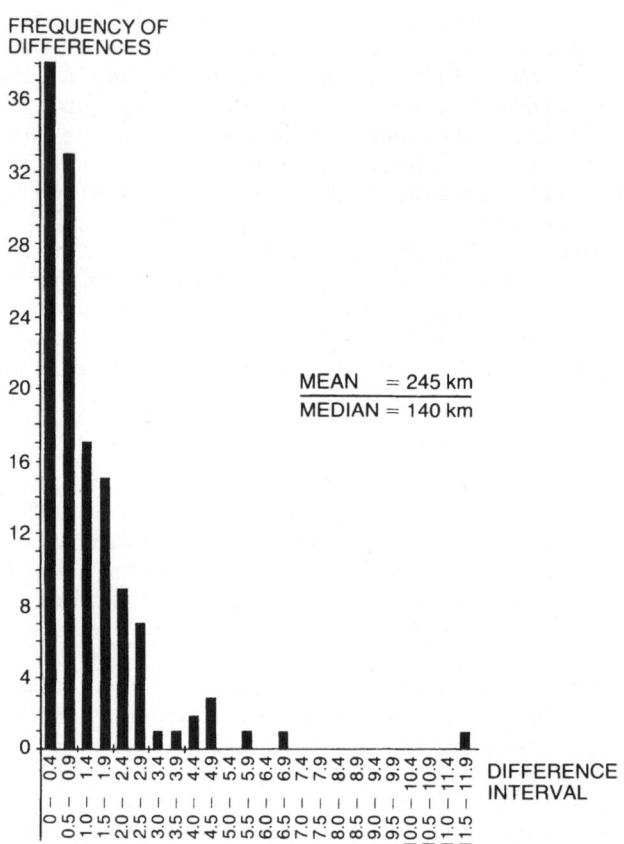

Fig. 1. Distance between trajectories crossing a circle of
         radius 1000 km around a common endpoint in southern
         Norway. January to March 1974, trajectories arriving
         at 00 and 12 GMT. 1 unit of difference interval
         =200 km. Surface geostrophic trajectories versus
         850 mb trajectories. The "rogue" value at 11.7 was
         excludes when calculating the mean. (From OECD, 1977).

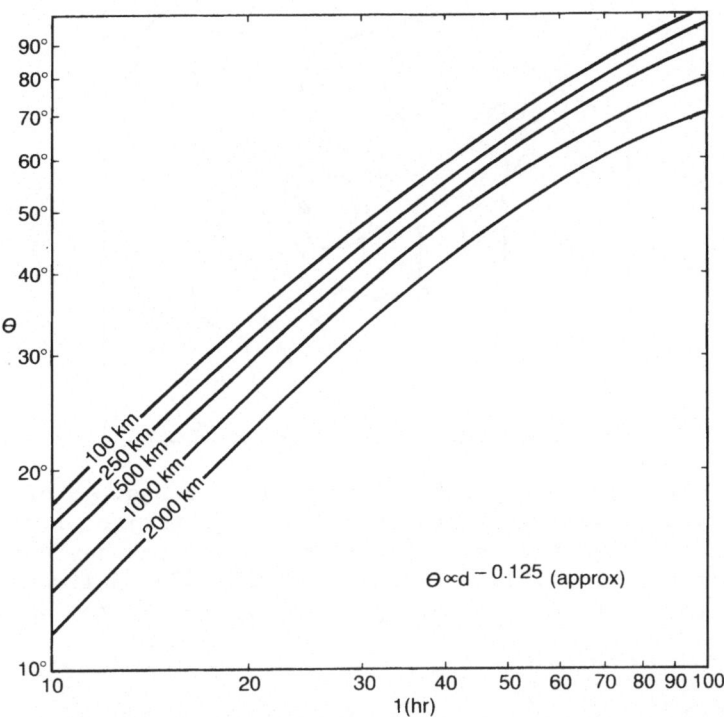

Fig. 2.   Variation of the mode of the angular spread Θ of
          geostrophic trajectories as a function of sampling
          time t and range d. (From Smith, 1979).

A. ELIASSEN

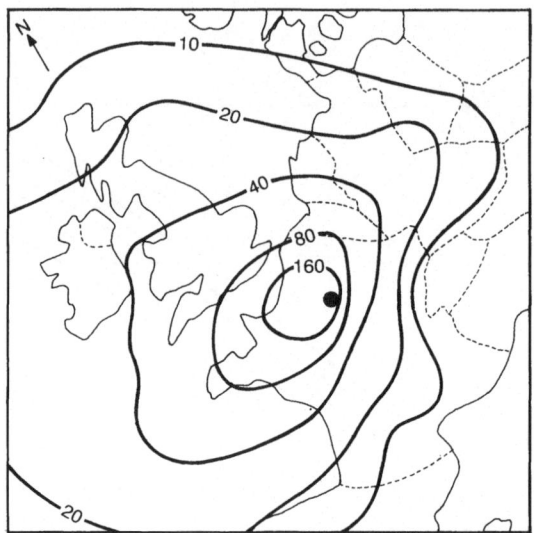

Fig. 3.   The frequency of 850 mb trajectory crossings of
           127-km grid elements, based on 4016 trajectories
           arriving at Vert-le-Petit four times per day during
           the period 1 July 1972 – 31 March 1975. Units are
           scaled to a total of 730 trajectories. (From OECD, 1977)

        Figure 3, taken from OECD (1977) shows the frequency of
127 km grid-element crossings for 850-mb trajectories arriving
Vert-le-Petit just south of Paris. We see that the most probable
path for a trajectory arriving at Paris is from the West accross
Bretagne, and the least probable is over the Alps from the East.
Other choices of advection wind will produce similar maps, in spite
of the large differences between different trajectory types that
can occur from case to case. The long-term contribution of an
emission region to a receptor point is therefore considerably less
sensitive to the choice of advection wind than the short-term contribution.

Calculation of Pollutant Deposition

Wet Deposition   A pollutant may be incorporated into a raindrop in many ways. Pollution particles may act as condensation nuclei. Gases may diffuse into liquid droplets, and be subject to chemical reactions in the liquid phase allowing for inclusion of additional gas. The rates involved vary widely from one precipitation event to the next, depending on cloud characteristics. A reflection of this complex situation is the variety of wet deposition rates that can be inferred from field experiments. For sulphur, however, there seems to be a broad consensus that its removal during precipitation is rapid with a characteristic time scale of the order of 1 hour.

The long-term average wet deposition pattern depends of course not only on the rate of removal during precipitation, but also on how often precipitation occurs. It may be useful to consider this important point in some detail. Following Rodhe and Grandell (1972, 1981), we assume a constant scavenging coefficient λ_p to apply during precipitation. If we think of the sequence of dry and wet periods experienced by a pollutant-carrying particle or trajectory, the rate of wet removal that applies may be taken as a stochastic process (t)defined by

$$\lambda(t) = \begin{cases} 0 \text{ if dry period at } t \\ \lambda_p \text{ if precipitation occurs at } t \end{cases} \qquad (10)$$

Assuming now that $\lambda(t)$ is a Markov process with stationary transition probabilities, the lengths of the dry and wet periods are exponentially distributed. If τ_d and τ_p are the average lengths of dry and wet periods experienced by the moving particle, it can be shown that the average residence time T_p for the pollutant with respect to wet deposition is

$$T_p = \tau_d P_d + \frac{\tau_d + \tau_p}{\tau_p} \quad \frac{1}{\lambda p} \qquad (11)$$

where p_d is the probability of a dry period at the starting point of the moving particle. Given the transport winds then T_p, together with the corresponding residence time T_d for dry deposition, determine the deposition of pollutant as a function of travel distance.

Relationships similar to (11) may be derived with slightly other assumptions, see Rodhe and Grandell (1981). The quantities τ_d and τ_p have been estimated by Hamrud et al. (1981) from trajectories and precipitation data. τ_d was found to be 51 hours in summer

and 53 hours in winter, whereas τ_p was 11 hours in both cases. If
we take τ_d = 50 hours, τ_p = 10 hours, λ^{-1} = 1 hour and approximate
the Eulerian probability from the Lagrangian data, i.e.

$$P_d \simeq 1 - \frac{\tau_p}{\tau_d + \tau_p} = 0.83$$

then (11) yields

$$T_p = 50 \text{ hours} \times 0.83 + 6 \times 1 \text{ hour} = 47.5 \text{ hours}$$

This shows that for pollutants that are relatively efficiently
removed during rain, such as sulphur, the average length of a dry
period is a key quantity for the determination of T_p, and hence
for the wet deposition pattern. T_p is less sensitive to λ: if λ
is changed by a factor of two in either direction, T_p varies between
44.5 and 53.5 hours.

It is therefore important that the precipitation data used
by any model imply a correct value of T_p. This may be difficult
to achieve, due to the time and space resolution of precipitation
data available from the WMO station network. Precipitation amounts
are generally measured in periods of 12 or 6 hours. In Europe the
spacing of the stations available on the WMO telecommunication
network is such that a 150 km grid element over land will contain
two to three reporting stations on the average. If the calculations
are based on gridded precipitation amounts, these are averaged out
in time and space, with overestimated duration and underestimated
intensity. For a trajectory moving through such a precipitation
field, τ_d will be too short, perhaps leading to an incorrect value
for T_p.

The modellers, however, have to base their approach on the
available data. In the MESOS model, for example (ApSimon and Goddard,
1976; ApSimon et al., 1980) the calculation of wet deposition is
based on observations of present weather rather than on precipitation amounts. These observations are available every 3rd hour.
The amounts are roughly estimated from these data in which
precipitation is classified into continuous or showery, and slight,
moderate or heavy. If the accurate precipitation amounts are needed,
however, these are generally not available on a time resolution
better than 6 hours. Such data were used in the OECD programme on
long range transport (OECD, 1977; Eliassen, 1978), and currently
in the European Programme on long range transport (EMEP), see
Eliassen and Saltbones (1982). The EURMAP model of Bhumralkar
et al. (1980) also employs 6-hourly precipitation amounts.

Dry Deposition   In general, the dry deposition of a contaminant can be accounted for in models by applying the condition

$$v_d q(z) = K \frac{\partial q}{\partial z} \tag{12}$$

at a height z_1, at the surface of just above it. Lagrangian models are usually one-layer models with no vertical structure. The flux due to dry deposition is then simply proportional to the concentration. The proportionality factor v_d, usually termed the deposition velocity, is different for different species. Furthermore, it depends on z_1, and on the state of the surface itself. For SO_2, v_d is of the order 1 cms^{-1}.

The sensitivity of estimated dry deposition to v_d can be conveniently investigated using a simple statistical model (Rodhe, 1972). Consider dispersion of sulphur dioxide from a point source of strength Q within a sector spanning an angle θ. The long-term average flux to the surface D(r) due to dry deposition is

$$D(r) = \frac{Q v_d t}{\theta h u} \cdot \frac{1}{r} \exp\left(-\frac{r}{u}\left(\frac{v_d}{h} + k_t + k_w\right)\right) \tag{13}$$

where r is the distance from the source, t is the time in which the wind at the source lays within the sector, u is an average wind-speed within the sector, h is an average mixing height, and k_t and k_w are average rates for transformation to sulphate and dry deposition, respectively. Differentiation of (13) with respect to v_d gives

$$\frac{dD}{D} = (1 - \frac{r v_d}{h u}) \frac{d v_d}{v_d} \tag{14}$$

relating the relative change in the deposition velocity with the relative change in dry deposition.

Figure 4 shows $(1 - \frac{r v_d}{h u})$ as a function of r and v_d with $h = 10^3$ m and u = 10 m/s. For $r v_d$ = hu = $10^4 m^2 s^{-1}$, a small change in v_d gives no change in D. This is because an increase in v_d is compensated by a resulting decrease in the concentration. If v_d is $\approx$1cms^{-1}, as for sulphur dioxide, then $\left|\frac{dD}{D}\right| < 0.5 \left|\frac{d v_d}{v_d}\right|$ from r $\approx$ 500 km to r $\approx$ 1500 km. On the regional scale, therefore, the dry deposition pattern is only moderately sensitive to the deposition velocity assumed for sulphur dioxide. In a way this is fortunate, since v_d for sulphur dioxide by no means is constant, but appears to vary between a few mms^{-1} up to about two cms^{-1}, depending on atmospheric turbulence and surface conditions.

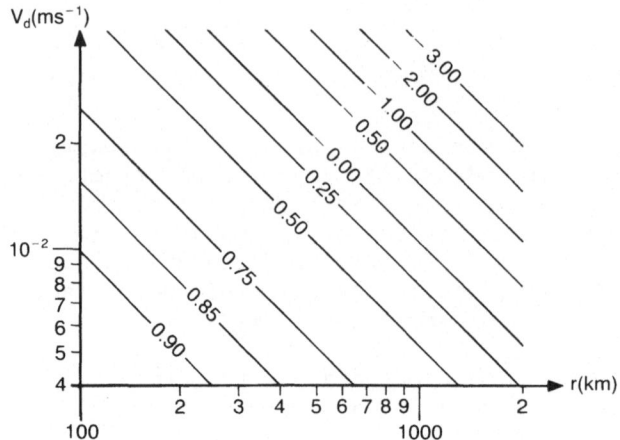

Fig. 4.   Atmospheric turbulence and surface conditions.
            $(1 - \frac{rv_d}{hu})$ as a function of v_d and r, for $h = 10^3$ m
            and $u = 10$ m/s. For further explanation see text.

Discussion of Model Results

In the previous sections it has been attempted to show how
the long-term ($\sim$ annual) pollutant deposition pattern over a
region essentially depends on the value of a few parameters,
such as the height of the mixing layer, the deposition velocity
and the average length of a dry period experienced by a trajectory.
The errors in the model calculations from case to case will have
a tendency to cancel in the long run. This is why all long-term
model calculations of the sulphur deposition over Europe, for
example, broadly give the same results, in reasonable agreement
with measurements (Eliassen, 1980). A recent example of a calculated
sulphur deposition pattern for Europe is shown in Figure 5.

In case studies, one would expect that the more advanced
models would give better results than the simpler ones. The advanced
version of the EURMAP model, outlined in Section (Bhumralkar et al.
1981), was tested in several case studies against the simpler

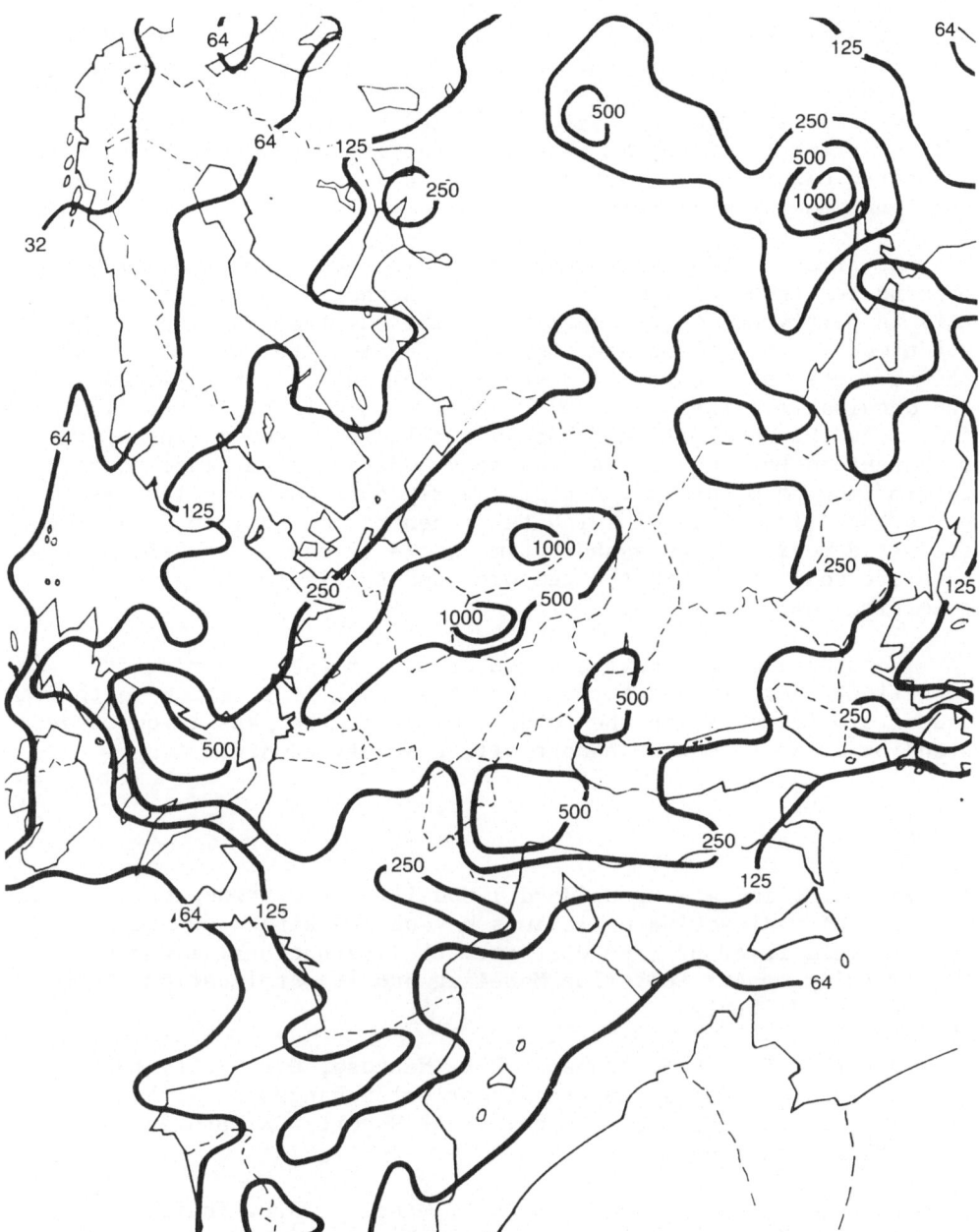

Fig. 5.   Calculated average monthly dry plus wet sulphur
           deposition over Europe, for the two-year period
           starting 1 October 1978. Unit : mg/m^2 as S.
           (From Eliassen and Saltbones, 1982)

trajectory model used in the OECD programme, without performing
significantly better. Maul (1980) has developed a model based on
analytical solutions for the vertical concentration distribution,
and in which the pollution in the upper and lower layers can
follow different trajectories. Maul used his model to study the
sulphur transport from the UK to Norway in wet deposition episodes.
He tested his model against a model with a constant mixing height,
and found no large differences.

At present, the more advanced models seem unable to perform
much better than the simpler ones. The author's feeling is that
this is partly due to the quality of the chemical measurements with
which model predictions are compared. The measurements often contain a number of errors (Eliassen and Saltbones, 1982), which limit
the obtainable correlation with model predictions. Another limiting
factor is the information actually available on the motion of the
pollutant-containing air. An advanced model requires more correct
information to be useful. A particularly difficult problem seems
to be the calculation of sulphate concentration in rain, and hence
the wet deposition, on a short time scale (1 day). The difficulties
here are connected with the temporal and spatial patchiness of
precipitation events.

Considering the above arguments, the prospects for swift
modelling progress do not seem the very best. If we stick to the
use of simple models in the future, however, the gaps in our understanding of pollutant transport will probably remain.

REFERENCES

ApSimon, H.M. and A.J.H. Goddard : Modelling the atmospheric disper-
        sal of radioactive pollutants beyond the first few hours of
        travel. Paper presented at the 7th International Technical
        Meeting on Air Pollution Modeling and its Application, Sep-
        tember 7-10, 1976.

Bhumralkar, M.B. ; W.B. Johnson, R.L. Mancuso, D.E. Wolf, 1979.
        Regional patterns and transfrontier exchanges of airborne
        sulphur pollution in Europe. Final Report, SRI project 4797,
        SRI International.

Bhumralkar, C.M. : W.B. Johnson, R.L. Mancuso, R.H. Thuillier,
        D.E. Wolf, 1980. Interregional exchanges of airborne sulphur
        pollution and deposition in Eastern North America. Paper
        presented at Second Joint Conference on Applications of Air
        Pollution Meteorology, New Orleans. American Meteorological
        Society.

Draxler, R.R. and A.D. Taylor, 1982. Horizontal dispersion parameters for long-range transport modelling. Journ. of Appl. Meteor. 21, 367-372.

Eliassen, A., 1978. The OECD study of long range transport of air pollutants : Long range transport modelling. Atmos. Environ 12, 479-487.

Eliassen, A., 1980. A review of long-range transport modelling. Journ. of Appl. Meteor. 19, 231-240.

Eliassen, A. and J. Saltbones, 1982. Modelling of long range transport of sulphur over Europe : A two-year model run and some model experiments. EMEP/MSCW Report 1/82, Norwegian Meteorological Institute. To appear in Atmospheric Environment.

Gifford, F.A., 1982. Horizontal diffusion in the atmosphere : A Lagrangian-dynamical theory. Atmos. Environ. 16, 505-512.

Hamrud, M., H. Rodhe and J. Grandell, 1981. A numerical comparison between Lagrangian and Eulerian rainfall statistics. Tellus 33, 235-241.

Heffter, J.L.,1980. Air Resources Laboratories atmospheric transport and dispersion model. NOAA Technical Memorandum ERL ARL-81.

Johnson, W.B., D.E. Wolf and R.L. Mancuso, 1978. Long term regional patterns and transfrontier exchanges of airborne sulphur pollution in Europe. Atmos. Environ. 12, 511-527.

Maul, P.R.,1980. Atmospheric transport of sulphur compound pollutants. CEGB Midlands Region Report No. MID/SSD/80/0026/R.

OECD, 1977. The OECD programme on long range transport of air pollutants. Measurements and findings. OECD, Paris.

Olson, M.P. and E.C. Voldner, 1981. Documentation of the Atmospheric Environment Service long-range transport of air pollutants model. Work Group 2 Report 2-5 AES, Toronto.

Pack, D.H. ; G.J. Ferber, J.L. Heffter, K. Telegadas, J.K. Angell, W.H. Hoeker and L. Machta, 1978. Meteorology of long-range transport. Atmos. Environ. 12, 425-454.

Rodhe, H. and J. Grandell, 1972. On the removal time of aerosol particles from the atmosphere by precipitation scavenging. Tellus 24, 442-454.

Rodhe, H. and J. Grandell, 1981. Estimates of characteristic times
    for precipitation scavenging J. of the Atmos. Sci. 38, 370–386.

Smith, F.B., 1979. The character and importance of plume lateral
    spread affecting the concentration downwind of isolated sources
    of hazardous airborne material. WMO symposium on the long-range
    transport of pollutants and its relation to general circulation
    including stratospheric/tropospheric exchange processes.
    Sofia, 1–5 October 1979. WMO No. 538,241–251.

Taylor, A.D., 1982. Puff growth in an Ekman layer. Journ. of the
    Atmos. Sci. 39, 837–850.

DISCUSSION

F.B. SMITH   In your paper you quote some
            results of A.D. Taylor on the downwind spread of an
            instantaneous puff and relate these values to the
            effect of wind shear. The variation of with down-
            wind distance d in the Table is like d^a where a $\approx$ 0.8
            Wind shear within a layer of finite depth can only
            produce a variation close to $d^{1/2}$, once the layer
            is filled, provided stabilisation does not set in,
            inhibiting vertical mixing. However as Gifford has
            shown large mesoscale and synoptic scale turbulence is
            strong enough to maintain a nearly linear rate of
            growth. Do you have any comments on this ?

A. ELIASSEN   To obtain the numbers in Table
            1 of my paper I have read off the curve given for $\sigma_h(t)$
            in Draxler and Taylor (1982). This curve is supposed
            to represent the analytic solution for puff spread in
            an Ekman layer. According to that curve, $\sigma_h \sim t^a$ where a
            $\approx$ 0.8, which contradicts your statement that in this
            case a should not be much larger than 0.5. I have no
            good explanation for this situation.

W. KLUG   Could you please comment on the
            virtues and drawbacks of Lagrangian versus Eulerian
            modeling from your point of view ?

A. ELIASSEN   I think the main virtue of
            Lagrangian models is that one avoids the numerical
            problems associated with the advective terms in the
            Eulerian mass-balance equation. Furthermore, Lagrangian
            models are economic since one can perform the
            integration of this equation only along chosen trajec-
            tories istead of in a complete Eulerian grid. The vir-
            tue of Eulerian models is that in principle, all rele-
            vant physical and chemical processes, like for example

wind shear and nonlinear chemistry, can be handled. At present the usefulness of this virtue is limited by the information content of available meteorological data.

THE INTEGRAL EQUATION OF DIFFUSION

F.B. Smith

Meteorological Office
Bracknell,
Berkshire RG12   2SZ   U.K.

INTRODUCTION

Various theories are now available to describe the dispersion of passive material released from a source into the atmosphere. These theories include the eddy-diffusivity equation of diffusion, statistical theory, similarity theory, higher-order closure solutions of the basic equation of motion and continuity, and rondom walk modelling. All these methods explicitly or implicitly require some knowledge of the Lagrangian character of the turbulence field. Some methods only apply to rather idealised states of flow (e.g. statistical theory only applies to uniform flow with homogeneous turbulence) whereas others are much more versatile - (e.g. random walk modelling). Some are simple in concept and application, others are much more complex. In recent years the virtues of random walk modelling have been recognized and the technique widely exploited. The only obvious disadvantages are:

(i)   In non-homogeneous turbulence bias--velocities are required to maintain zero-mass-fluxes and Wilson et al (1981) and Thomson & Ley (1982) have made progress in prescribing these.

(ii)   The linearity of the simple correlation equation, basic to the method, has only been verified experimentally in quasi-homogeneous turbulence. (Hanna, 1979)

(iii) The method, although very simple to code and run on a computer, requires many thousands of trajectories before a reliable concentration profile can be established.

An alternative method with many of the virtues of random walk modelling is given in this paper. Whilst being marginally harder to grasp than random walk modelling, especially in complex flow situations, it does have the advantage of being very quick to run on a computer. As far as the author is aware this is the first time this approach has been applied to atmospheric dispersion.

THE INTEGRAL EQUATION OF DIFFUSION APPLIED TO HOMOGENEOUS TURBULENCE

The approach has its origins both in random walk modelling and in the classical basic equation of diffusion :

$$C(\underline{r}) = \int S(\underline{r}') \ P(\underline{r} - \underline{r}') \ d\underline{r}' \qquad (1)$$

where C is the concentration at point $\underline{r}$, S is the source strength at $\underline{r}'$ and P is the probability transfer-function between $\underline{r}'$ and $\underline{r}$. To give this equation teeth the nature of P has to be defined, and therein lies the problem.

The following approach is suggested : Consider a single continuous source in a turbulent stationary airstream. Let the particles leaving the source move in a random walk, a series of straight paths separated by "collisions" in which all Lagrangian memory is lost. The length of the paths varies from one to another and is determined in each case by a random process equivalent to assuming that during each time-interval δt along the path the probability of collision is $\delta t / \tau$. τ is the local Lagrangian time-scale of the real turbulent field being simulated and may thus vary with position.

In practice, the concentration field may be built up within a grid, whose elemental gridlengths are small compared to the length-scale of turbulence, starting at the source and working progressively downstream. The concentration at any new receptor gridsquare can be calculated from the number of particles being advected through the square which experienced their last collision within upwind squares, intermediate between the source and the receptor square, where the concentration is already determined. Thus for a specified number of particles emitted from the source, the number N that pass through (X,Z) is given by :

$$N(X,Z) = N(x,z) \text{ x (Probability P' of collision at (x,z))}$$
$$\text{x (Probability P'' of no further collision}$$
$$\text{between (x,z) \& (X,Z))}$$
$$\text{x (Probability P of the velocity w =}$$
$$\text{u(Z-z)/(X-x) occurring, required to take}$$
$$\text{the particle from (x,z) to (X,Z))...(2)}$$

For the sake of clarity, it will be assumed first of all that the turbulence is homogeneous, the downstream velocity u is constant and there are no boundaries. The gridlengths are defined as

$$\Delta x = \frac{u\tau}{L}$$

$$\Delta x = \frac{w_m\tau}{L}$$

(3)

where L is an arbitrary large constant (10 can be shown to be a good choice, and is used in all calculations), τ is the Lagrangian time-scale and w_m can be identified either with the root-mean-square w-velocity σ_w or, if w has extreme values $\pm w_m$, with this extreme.

Then $P' = 1/L$ and $P'' = (1 - \frac{1}{L})^{n-1}$ (4)

Where $X - x = n\Delta x$, $Z - z = -m\Delta z$

and to ensure no loss of particles :

$$P = \Delta P(\frac{m}{n}) = \int_{\frac{m-1/2}{n}}^{\frac{m+1/2}{n}} p(w)\,dw$$

(5)

where p(w) is the probability density function of the turbulent velocity w. The particles which originate from the source and undergo no collision until reaching (X,Z) make a contribution to N like all other squares except that P' is put equal to 1 and $N(0,0)$ is put equal to $uQ/w_m \Delta x$ (where Q is the source strength, or the total number of particles released).

Expressed in terms of concentrations :

$$C(N,M) = \frac{Q}{w_m \Delta x}(1 - \frac{1}{L})^{N-1} \Delta P(\frac{M}{N}) + \sum_{m,n} \frac{1}{L}(1 - \frac{1}{L})^{n-1} \Delta P(\frac{m}{n}) C(n,m)$$ (6)

where $M = Z/\Delta z$, $N = X/\Delta x$.

In the homogeneous field this is exceedingly easy to code for a computer and quick to run because the coefficients :

$$q(n,m) = (1 - \frac{1}{L})^{n-1} \Delta P(\frac{m}{n})$$

(7)

are independent of N and M and need to be calculated only once. In terms of q :

$$C(N,M) = Aq(N,M) + \sum_{m,n} \frac{1}{L} q(n,m) C(n,m)$$

(8)

where A is a large constant representing the source strength.

Equation (6) may be solved with $A = 10^4$, $L = 10$ and with

$$p(w) = \frac{15}{16w_m} (1 - (\frac{w}{w_m})^2)^2 \qquad for \ -w_m \leqslant w \leqslant w_m \qquad (9)$$

$$= 0 \qquad for \ |w| > w_m$$

This distribution has a standard deviation $\sigma_w = 0.378 \ w_m$, a zero mean, and a shape very similar to the Gaussian distribution without the rather unrealistic "tails" at large w characteristic of the latter distribution. The resulting concentration distribution is very quickly obtained on a computer, and has moments in very close agreement with analytically-derived moments. Thus the second moment σ^2 is in complete agreement with Taylor's statistical theory result with an expontial correlogram with the same Lagrangian time-scale τ .

$$\sigma^2 (T) = 2\tau\sigma_w^2 (T + \tau e^{-T/\tau} - \tau) \qquad (10)$$

Higher moments can be obtained analytically from the integral form of equation (6) :

$$C (X,Z) = \frac{Q}{uT} exp(- \frac{T}{\tau}) \ P(w) + \frac{1}{\tau_o} \int_0^T exp \ (- \frac{T - t}{\tau}) \ \int_{-\infty}^{\infty} C(x,z)P(w) \ dw \ dt \qquad (11)$$

where $X = uT$, $Z = wT$ in the first term on the right-hand-side, and $x = ut$ and $w = u(Z-z)/CX-x)$ in the second term. Of particular interest is the fourth moment

$$\mu_4 = \frac{u}{Q} \int_{-\infty}^{\infty} Z^4 C (X,Z) \ dZ \qquad (12)$$

which gives the kurtosis (or flatness factor) of the distribution :
$\gamma = \frac{\mu_4}{\sigma^4} - 3$ If $\gamma = 0$ the distribution has the same kurtosis as a Gaussian distribution. If however $\gamma > 0$ the distribution is more strongly peaked than a Gaussian distribution with the same σ and is called "leptokurtic", whereas if $\gamma < 0$ it is less strongly peaked and is called "platykurtic". For equation (11) the concentration distribution can be integrated to give a kurtosis :

$$\gamma = 3(J \gamma_w + (K-1)) \qquad (13)$$

where γ_w is the kurtosis of the w-distribution,

$$aJ = 2v - 6 + (6 + 4v + v^2) exp(-v)$$
$$aK = v^2 - 6 + (6 + 6v + 2v^2) exp(-v)$$

$$a = v^2 + 1 - 2v + \exp(-2v) - 2(1-v)\exp(-v)$$

$$v = T/\tau$$

J is very close to 1 until v exceeds about 2, and then falls slowly to zero. K always just exceeds 1, reaching a maximum of 1.224 near v = 5. Thus at small T, γ is determined largely by the kurtosis of the w-distribution γ_w but at larger T (near v = 4) the particle distribution becomes somewhat leptokurtic due to the combined effect of the two terms. (Note that the sum of two perfectly Gaussian distributions with different σ's is also leptokurtic). Eventually the second term in (13) becomes dominant and, regardless of γ_w, the value of γ approaches zero. The concentrated distribution therefore becomes Gaussian at large T/τ in accord with random walk theory.

THE EQUATION APPLIED TO PARTICLES WITH SPECIFIED INITIAL TURBULENT VELOCITIES

The conditional particle-motion theories developed by Smith (1968) and more recently by Gifford (1981) draw interest on the sub-ensemble of particles that leave the source with a given value of w. Since the width of such sub-ensemble plumes does not depend on the magnitude of w, we may arbitrarily put w = 0 for all particles in the first term on the right-hand side of equation (8), i.e. $\Delta\rho = 0$ unless m = 0 when $\Delta\rho = 1$. The resulting solutions of (8) enable the width of the sub-ensemble plume to be compared with the width of the full plume (unrestricted w at the source). Table (1) shows this comparison :

Table (1) The normalised width of two plumes are compared at different distances downwind from the source. The first plume contains all particles released whereas the second contains only the sub-ensemble with zero initial w.

$x/u\tau$	0.1	0.2	0.5	1	2	5	10
$\sigma/w_m\tau$: whole ensemble	.040	.075	.175	.324	.567	1.052	1.541
$\sigma/w_m\tau$: w(0) = 0	.005	.016	.065	.168	.389	0.918	1.455

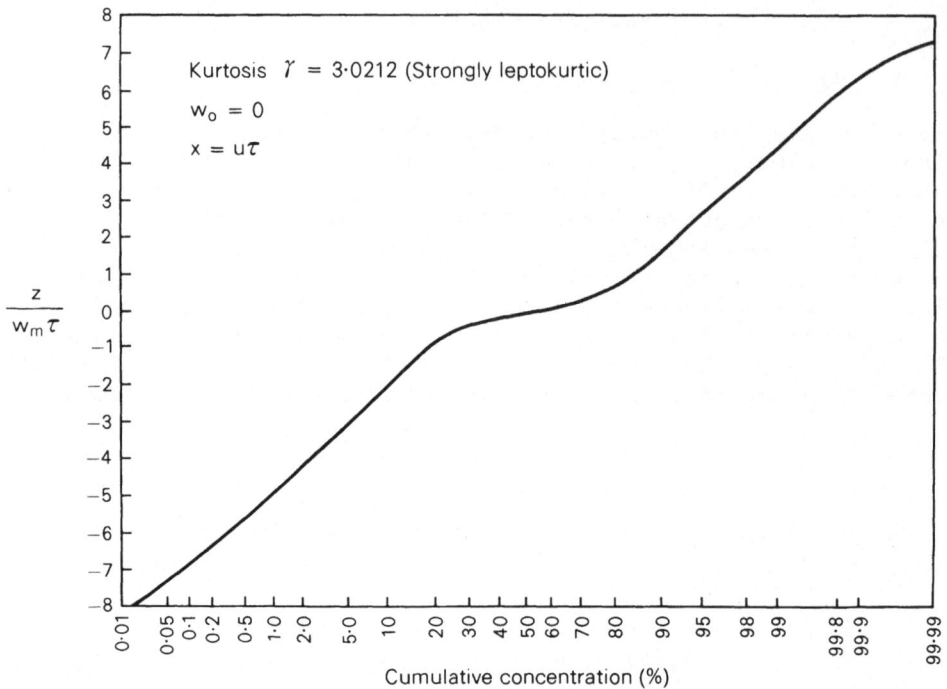

Figure 1. Cumulative concentration profile at x=uτ for initial w=0

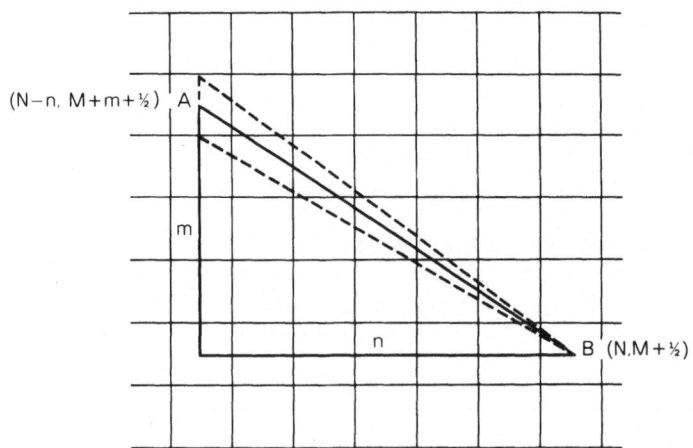

Figure 2. Integration grid with definition of notation.

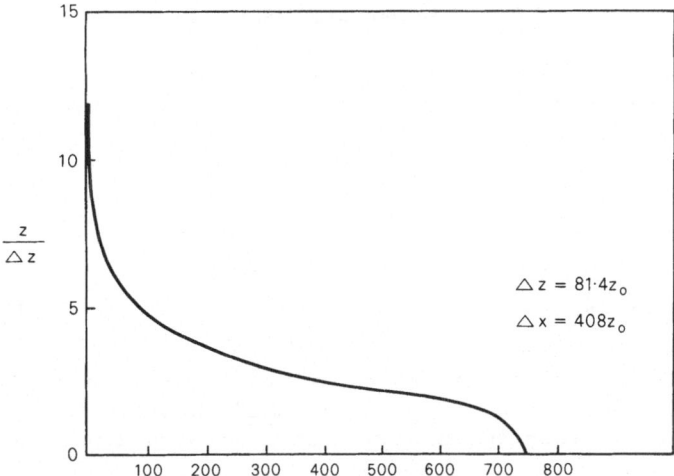

Figure 3. Concentration profile at x=10Δx for "Ground level" release
(h=1/2 Δ z).

Similarly a comparison may be made between the respective values
of the kurtosis (see Table (2)). For the w-distribution given in
equation (9), the kurtosis $\gamma_w = -\frac{2}{3}$ showing the distribution to be
platykurtic (i.e. slightly flattened compared to a Gaussian distribution).

Table (2). A comparison of the kurtosis of two plumes at different
downwind distances. The whole-ensemble plume demonstrates decreasing
platykurticity reflecting a decreasing influence of the shape of the
w-distribution, whereas the sub-ensemble plume, in which the initial
w is always zero, demonstrates decreasing peakiness or leptokurticity.
The peakiness in the second plume is clearly very strong at small
$x/u\tau$.

$x/u\tau$	0.1	0.5	1	2.5
δ whole ensemble	-.632	-.500	-.358	-.065
δ sub-ensemble w(0)=0	-	7.292	3.021	0.820

An example of the concentration profile at x = uτ for particles emitted with zero w is provided in Figure (1). This shows the
cumulative concentration as the plume is traversed. Note that 80%
of the plume is contained between z = $\pm$ 2$w_m\tau$ compared with 60 %
between z = $\pm$0.3$w_m\tau$ showing the highly peaked nature of the distribution. These results are in complete agreement with Smith's
(1968) theory.

APPLICATION TO THE NEUTRAL SURFACE-STRESS LAYER

The technique can be extended easily to treat diffusion in the neutral surface stress layer in which the wind is varying logarithmically with height :

$$u = \frac{u_*}{k} \ln \frac{z}{z_0} \tag{14}$$

and the Lagrangian time-scale varies linearly

$$\tau = k u_* z / \sigma_w^2 \tag{15}$$

as used by Reid (1979) and others in random walk modelling. The vertical velocity variance σ_w^2 is assumed constant with height, and the probability distribution $p(w)$ remains as in equation (9).

A height $z = z^*$ is chosen (typically z^* may be $1000\, z_0$ within the surface layer and the value of L there is put equal to 10 (as in the homogeneous case). Assuming $\sigma_w = 1.3 u_*$, Δx and Δz, defined as before, are equal to

$$\left. \begin{aligned} \Delta x &= 408.7\ z_0 \\ \Delta x &= \ \ 81.4\ z_0 \end{aligned} \right\} \quad \text{calculated at } z = z^*$$

These gridlenghts are then held fixed at these values throughout the grid. This implies that $L = u\tau/\Delta x$ varies with z, since both u and τ increase with height, and is less than 1 in the lowest two gridsquares. Referring back to the description of the method in the homogeneous case, the equivalent of equation (4), (5) and (6) are:

$$P' = 1/L_n, \quad P'' = \prod_{i=1}^{n-1} (1 - \frac{1}{L_i}) \text{ where } L_i = \frac{\Delta x}{u_i \tau_i} = (0.11785\ r_i\ \ln 81.4 r_i)^{-1}$$

and $r_i = M + \frac{1}{2} + mi/n$. M,N,m and n have the same meaning as before except that M = 0 is at ground-level, and the source is at a height $(M_s + \frac{1}{2})$.

$$u(r_0)\ C(N,M) = \frac{Q u (M + \frac{1}{2})}{w_m\ \Delta x} \prod_{i=1}^{N-1} (1 - \frac{1}{L_i})\ \Delta P\ (\frac{M - M_s}{N})\ +$$

$$\sum_{m,n} \frac{1}{L} \prod_{i=1}^{n-1} (1 - \frac{1}{L_i})\ \Delta P (\frac{m}{n}) u(r_n)\ C(N-n, M+m) \tag{16}$$

The factor $(1 - L_i^{-1})$ is put zero whenever $L_i \leqslant 1$. The $\Delta \rho$ are defined as follows (see figure (2)):

$$\Delta z = \frac{w_m \tau^*}{L^*} , \qquad \Delta x = \frac{u^* \tau^*}{L^*}$$

where u^*, τ^* and $L^* = 10$ are defined at $z = z^*$. The particles

which collide at A and move to B are assumed to move in straight
lines retaining their u and w velocities unchanged and as defined
at A.

Thus $\dfrac{w}{u} = -\dfrac{m\Delta z}{n\Delta x} = -\dfrac{m}{n}\ \dfrac{w_m\ \tau^*}{L^*}\ \dfrac{L^*}{u^*\tau^*} = -\dfrac{m}{n}\ \dfrac{w_m}{u^*}$

i.e. $\dfrac{w}{w_m} = -\dfrac{m}{n}\dfrac{u}{u^*} = -\dfrac{m}{n}\ \dfrac{\ln z/z_0}{\ln z^*/z_0}$ \hfill (17)

where $z = M + m + \dfrac{1}{2} = r_n$ and z^*/z_0 is specified. ΔP is now defined
as before in equation (5). Equation (16) can now be solved very
easily. Computer time consumed is still very short although clearly
longer than for the homogeneous case since the $q(n,m)$ defined in
equation (8) are now functions of N and M as well as n and m.

Figure (3) shows as an example the concentration profile at
$x = 10\Delta x$ when the source is close to the ground at $z = \Delta z/2$
If $z_0 = 1$ cm this would mean the profile at $x = 8$ metres with the
source at about 0.4 metres. The profile conforms very closely with
the expected exponential form:

$C = C_0 \exp (-az^{3/2})$ \hfill (18)

for a neutral surface layer.

Figure (4) shows concentration contours for a crosswind
elevated line source at $z = 4.5\Delta z$.

CONCLUSIONS

In summary the integral equation of diffusion approach has
the following advantages :

(a)   it is basically simple to understand and formulate
in a variety of situations.

(b)   it has the merits of random walk modelling but is
much faster in terms of computer time.

(c)   in some simple applications, analytical solutions
can be obtained.

(d)   unlike in classical statistical theory and in
similarity theory, concentrations are readily obtained.

However like all theories of diffusion some knowledge of the
Lagrangian parameters, in particular the time-scale, is essential
and these parameters cannot be directly measured and have to be
inferred from Eulerian measurements with all the uncertainties this
implies.

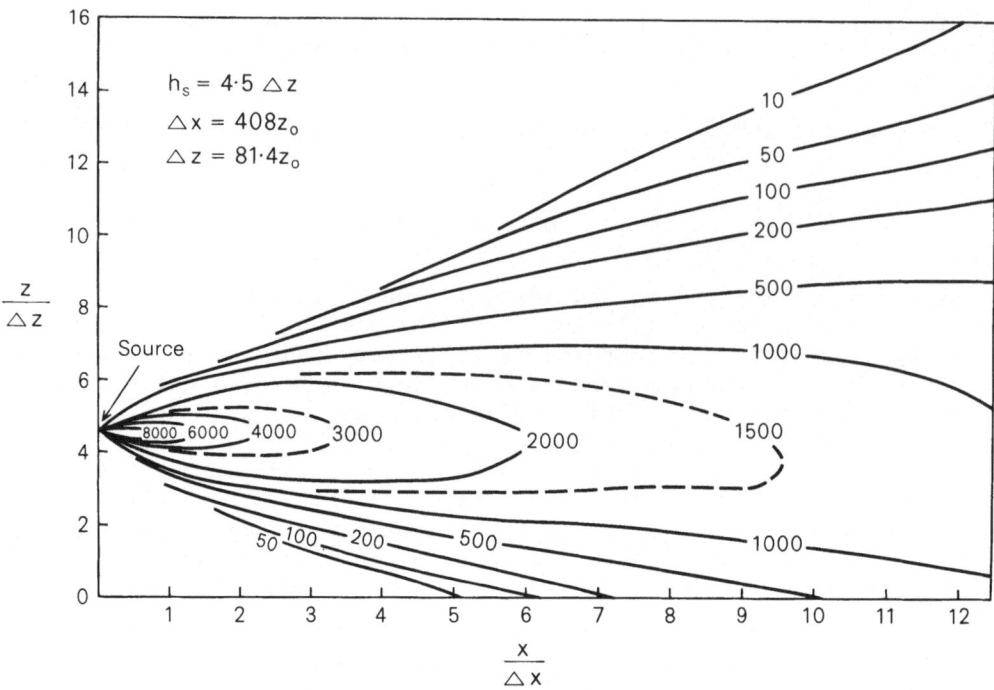

Fig.4. Contours of concentration downwind from a source at a height
of 4.5 Δz. Units depend on wind speed and source strength.

REFERENCES

Hanna, S.R.,   1979 Some statistics of Lagrangian and Eulerian wind
       fluctuations. J.Appl.Met., 18,4.

Reid, J.D.,   1979 Markov-chain simulations of vertical dispersion
       in the neutral surface layer for surface and elevated releases.
       B.L. Met., 16, pp 3-22.

Smith, F.B., 1968 Conditioned particle motion in a homogeneous tur-
       bulent field : Atmos. Environ., 2, pp 491-508.

Taylor, G.I., 1921 Diffusion by continuous movements. Proc. London
       Math. Soc., Ser.2., 20 p 196.

Thomson D., and Ley A., 1982 A random walk dispersion model applicable
       to diabatic conditions. Internal Met Office Note T.D.N. 138.

Wilson, J.D., Thurtell, G.W. and Kidd, G.E., 1981 Numerical
       simulations of particle trajectories in inhomogeneous tur-
       bulence. Part II : systems with variable turbulent velocity
       scales. B.L. Met., 21, pp 423-441.

DISCUSSION

R. BERKOWICZ   Can the sub-ensembles of particles
with the same initial velocity be interpreted as
puffs ?

F.B.SMITH   No, not strictly, because we assume
the sub-ensemble is made up of effectively independent particles. Nevertheless the behaviour of the
sub-ensemble plume (i.e. the way its width varies
with downwind distance) is qualitatively very
similar to the behaviour of an expanding puff, but
there the similarity ends : a puff is a "multi-par-
ticle" diffusion problem, the sub-ensemble is a
"one-particle" diffusion problem.

E. RUNCA   How is the p.d.f. of w at source
height taken into account by your model ?

F.B. SMITH   The p.d.f. of w is accounted for
explicitly in the model through the probability
$P = \Delta P(m/n)$ defined in equation (5) and appearing
in the equations for the concentration (equation
(6) and elsewhere). The nice thing about this
model is that it can do this very easily for any
p.d.f. which may or may not vary with height or
position.

W.KLUG   You did not say anything about the
boundary conditions, but U understand that this
can be easily taken care of?

F.B.SMITH   Yes, boundaries present no
particular problem because all aspects of the
turbulence respond to the presence of the boundary
and in particular the Lagrangian timescale
approaches zero there. Consequently both L and
ΔP change as the boudary is approached in such a
way that "particles" do not penetrate the boundary
and no special conditions have to be imposed
except that $(1- 1/L_i)$ in not allowed to become
negative). This is made clear in the written text.

A.VENKATRAM   Could you please amplify on the
concept of "collision" used in your theory.

F.B.SMITH   When "particles" undergo collision
in my theory they completely lose correlation with
the velocity immediately before collision. It is
thus a "drunkards walk". Collisions occur along the

straight-line paths of the particles with
probability $\frac{\delta t}{\tau(z)}$ in any time interval δ_t, where
$\tau(\underline{z})$ is a function of position $\underline{z}$. In spite of this
very simple representation of the true continuous
Lagrangian behaviour of the particles motion, the
width of the ensemble-averaged plume is fully consistent with, for example, the plume given by
Taylor's statistical theory in homogeneous turbulence and with the predictions of similarity
theory in a neutral surface-stress layer.

LAGRANGIAN LONG RANGE AIR POLLUTION

MODEL FOR EASTERN NORTH AMERICA

C.M. Bhumralkar[1][*], R.M. Endlich[1], K.C. Nitz[1],
R. Brodzinsky[1], and P. Mayerhofer[2]

[1]SRI International, Menlo Park, CA., USA;
[2]Meteorology Applications Inc., Menlo Park, CA., USA

INTRODUCTION

The transport of airborne sulfur and nitrogen compounds and
their deposition on the earth's surface by dry and wet processes
have been simulated using Lagrangian numerical models developed at
SRI over the past several years. The EURMAP model (Johnson et al.,
1978; Bhumralkar et al., 1981, 1982) developed under sponsorship of
Umweltbundesamt of the Federal Republic of Germany, applies to
Europe; and the ENAMAP model (Bhumralkar et al., 1980; Mayerhofer
et al., 1981; Endlich et al., 1982), developed under sponsorship
of EPA, applies to eastern North America. Emissions data are obtained for the areas of interest and are represented on a grid mesh
with spacing of 50 km for Europe and 70 km for North America. The
computations are based on grid-point analyses of winds, stability,
and precipitation made from standard meteorological reports, and
thereby incorporate the effects of changing weather patterns. The
most recent versions of the models use three layers in the vertical
and correct the winds for the influences of smoothed terrain
(Bhumralkar et al., 1982; Endlich et al., 1982).

The model version for sulfur compounds, ENAMAP-2S, has been used
to compute monthly average maps of airborne concentrations and dry
and wet deposition for both SO_2 and sulfates. The puff (Lagrangian)
methodology also identifies source-receptor relationships, i.e.,
how much of the pollution in a region was emitted locally, and how
much originated in other specified areas. The model version for
nitrogen compounds, ENAMAP-2N, treats the five species NO, NO_2, PAN,
HNO_3 and nitrates and makes computations of the concentrations and
depositions of each.

[*]On temporary assignment to NOAA, Rockville, MD, USA.

FUNCTIONAL PARTS OF ENAMAP

POLLUTION EMISSIONS
TRANSPORT BY WINDS
VERTICAL MIXING
CHEMISTRY —— Transformation, Dry Deposition, Wet Deposition
RESULTS —— Maps, Interstate Pollution Exchange Tables
VERIFICATION —— Comparison to Air Quality Measurements

Fig. 1. The parts of the ENAMAP-2 pollution transport model.

Fig. 2. The eastern North American model domain.

OUTLINE OF THE MODEL

The functional parts of ENAMAP are shown in Fig. 1. These different parts are typical of those used in a Lagrangian model, which tracks pollution puffs emitted over a grid at specified intervals (such as 12 hours). Each puff is normally tracked for 2 to 3 days using 3-hour time steps, until the puff becomes very diffuse or passes out of the domain. As a puff travels with the winds, it loses matter by dry and wet deposition processes. Also, chemical changes are simulated, and are discussed later. The model domain (which covers eastern North America) is shown in Fig. 2. The area is represented by a 41 by 46 grid mesh with 70 km spacing.

The model is divided into 3 vertical layers and the layers are formulated in terrain-following coordinates. Winds within the layers are made nondivergent to account for the effects of smoothed terrain on the flow. The layers also are used in the initial insertion of the pollution. In the daytime, the pollution puffs are distributed

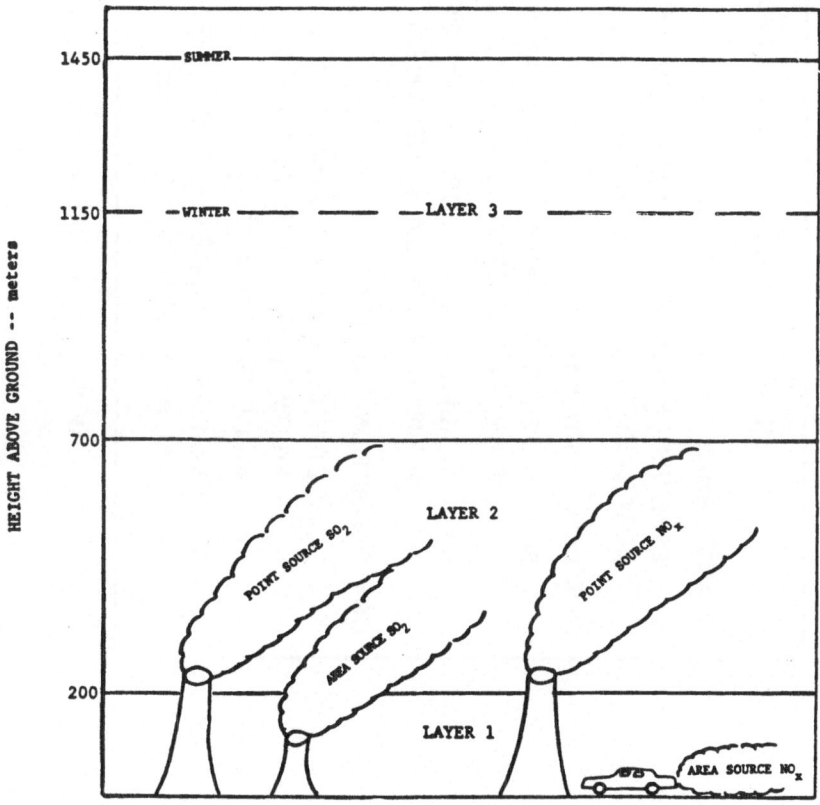

Fig. 3. The nighttime injection of emissions in layers.

CHEMICAL PROCESSES FOR SULFUR COMPOUNDS

TRANSFORMATION OF SO_2 TO $SO_4^=$

    DEPENDS ON SOLAR INSOLATION.   VARIES FROM 0.01 h^{-1} (1% h^{-1}) IN WINTER TO 0.04 h^{-1} IN SUMMER.

WET DEPOSITION RATE

    SO_2: DEPENDS ON RAINFALL TYPE AND RATE.   FOR 5 mm h^{-1} OF RAIN, VARIES FROM 0.03 h^{-1} (WINTER) TO 0.17 h^{-1} (SUMMER).

    $SO_4^=$: SIMILAR TO SO_2.   FOR 5 mm h^{-1} OF RAIN, VARIES FROM 0.06 h^{-1} (WINTER) TO 0.43 h^{-1} (SUMMER).

DRY DEPOSITION VELOCITY

    SO_2: DEPENDS ON LAND CHARACTERISTICS AND VEGETATION, STABILITY, SEASON AND TIME OF DAY.   VALUES VARY FROM 0.05 cm s^{-1} TO 0.85 cm s^{-1}.

    $SO_4^=$: SIMILAR TO SO_2 EXCEPT NO DEPENDENCE ON TIME OF DAY. VALUES VARY FROM 0.25 cm s^{-1} TO 0.95 cm s^{-1}.

Fig. 4. ENAMAP-2 transformation and deposition rates for sulfur compounds.

throughout the layers so that initial concentrations in the layers are uniform. At night, mixing is limited to stable lapse rates. For point sources of either SO_2 or NO_x, the typical plume rise is several hundred meters; therefore, point sources initially enter layer 2 (see Fig. 3). The 1977 SO_2 emissions were predominantly (92 percent) from point sources (1977 National Emissions Report, 1980); therefore at night all SO_2 emissions are placed initially in layer 2. NO_x pollutants were divided (approximately) evenly between point and area sources. Point source NO_x is placed initially in layer 2 for the same reasons given above. On the other hand, approximately 90 percent of NO_x area emissions come from vehicles; therefore, as shown in Fig. 3, all NO_x area emissions are placed initially in layer 1.

Vertical mixing is based on a finite-difference form of the expression

$$\partial C/\partial t = \partial/\partial z \left[D_z (\partial C/\partial z) \right] \ .$$

The vertical diffusion coefficients D_z are computed as a function of wind shear, stability, and the height of the mixing depth. The values used in the computations reported in this paper fall in the range from 1 to 10 $m^2 s^{-1}$. The simulataneous solution for new concentration values in a column uses the Crank-Nicolson method (see Crank, 1967).

THE SULFUR MODEL AND APPLICATIONS

The chemical processes simulated in ENAMAP-2S are summarized in Fig. 4. The transformation of SO_2 and $SO_4^=$ depends on solar insolation and, therefore, varies with season and latitude. The dry deposition velocity is based on the surface characteristics (water or land-use type) as well as on thermal stability (Sheih et al., 1979). The stability class is computed from wind speed, cloud cover, and time of day. The dry deposition velocities vary over a wide range. For SO_2 at night, a relatively low dry deposition rate of 0.07 cm s^{-1} is used everywhere to account for low absorption by plant surfaces. Wet deposition rates used in ENAMAP-2S are the largest for summertime (convective) rain (Scott, 1978). In our earlier work, we used larger wet depositon rates, especially in winter. All these chemical rates are somewhat uncertain because they are based to a large extent on laboratory measurements.

The model was run for the months of January and August 1977; and monthly average values of wet and dry deposition and airborne concentration were computed for SO_2 and $SO_4^=$. Maps showing typical results for January 1977 are given in Figs. 5 and 6. The patterns of concentration (and also deposition) naturally are closely related

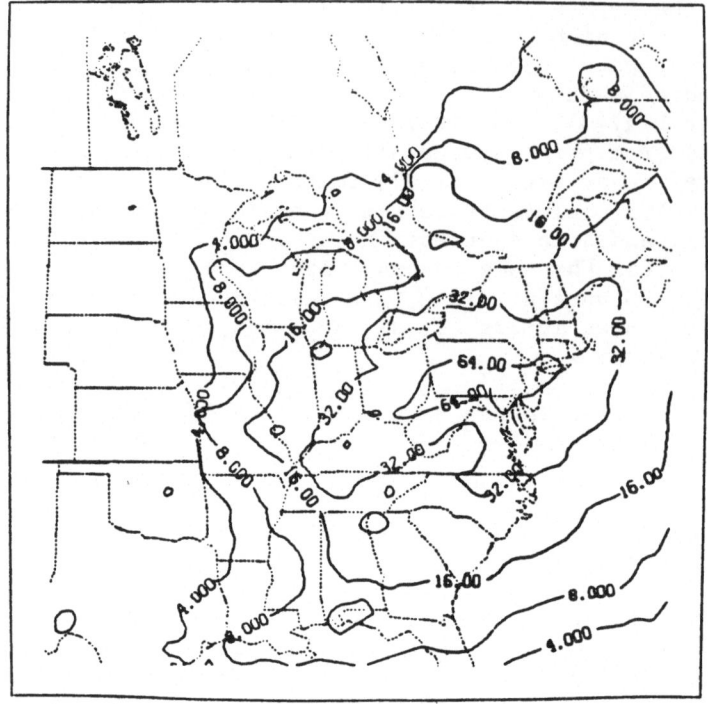

(Upper)

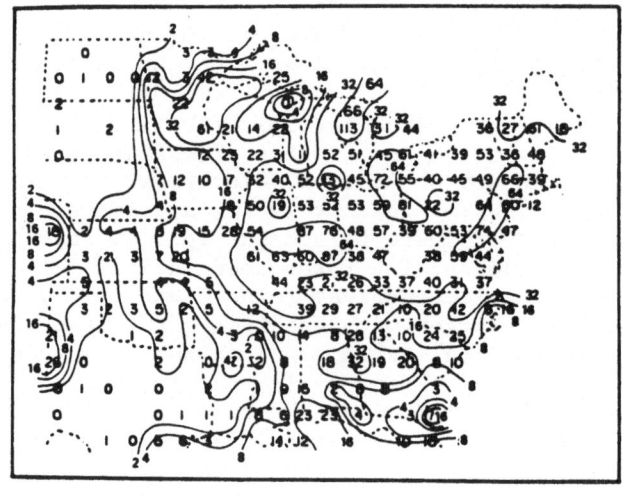

(Lower)

Fig. 5. Calculated (upper) and observed (lower)
        airborne SO_2 concentrations ($\mu g\ m^{-3}$) for
        January 1977.

(Upper)

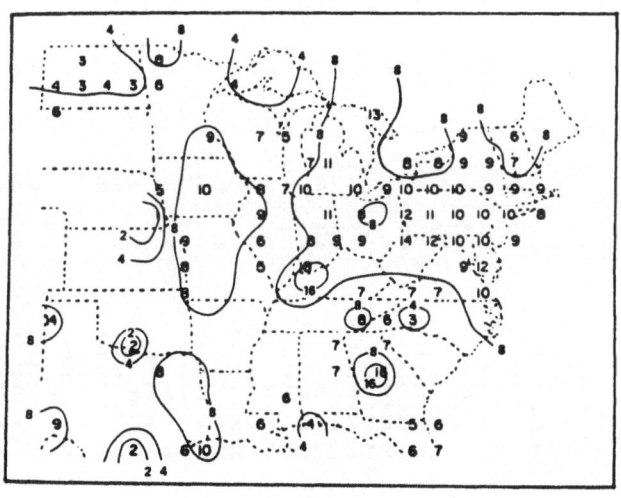

(Lower)

Fig. 6. Calculated (upper) and observed (lower)
airborne $SO_4^=$ concentrations (μg m^{-3}).

Table 1. Example of computed sulfur deposition due to interregional ex-- changes of airborne sulfur in January 1977 (Units are KTON Sulfur).

EMITTER REGION	TOTAL CONTRIBUTIONS TO				S DEPOSITIONS WITHIN RECEPTOR REGIONS										
	16	17	18	19	20	21	22	23	24	25	26	27	28	29	30
1 ALABAMA															
2 ARKANSAS															
3 COLORADO															
4 DEL/MD/DC															
5 FLORIDA															
6 GEORGIA															
7 ILLINOIS															
8 INDIANA															
9 IOWA															
10 KANSAS															
11 KENTUCKY															
12 LOUISIANA															
13 MAINE															
14 MICHIGAN															
15 MINNESOTA															
16 MISS.															
17 MISSOURI															
18 MONTANA															
19 NEBRASKA															
20 NH/VT															
21 NEW JERSEY															
22 NEW MEXICO															
23 NEW YORK															
24 N CAROLINA															
25 N DAKOTA															
26 OHIO															
27 OKLAHOMA															
28 PENN.															
29 RI/MA/CT															
30 S CAROLINA															
31 S DAKOTA															
32 TENNESSEE															
33 TEXAS															
34 VIRGINIA															
35 W VIRGINIA															
36 WISCONSIN															
37 MANITOBA															
38 N BRUNS															
39 ONTARIO															
40 QUEBEC															
41 NEWFOUND															
TOTAL (KTON S)	8.8	8.7	.0	.1	9.5	9.0	.0	27.7	23.5	.2	28.7	1.9	34.5	11.9	12.0

NITROGEN CHEMISTRY

$$NO \rightleftharpoons NO_2; \quad K = \frac{[NO_2]}{[NO]};$$ $\quad K = 2 \text{ (Day)}, \quad K = 50 \text{ (Night)}$

$$NO_2 \overset{k_n}{\Rightarrow} PAN + HNO_3 + NO_3^-;$$ $\quad k_n = 0.1 \text{ (Day)}, \quad k_n = 0.02 \text{ (Night)}$

$$PAN \overset{k_d}{\Rightarrow} PAN + HNO_3 + NO_3^-;$$ $\quad k_d = 0.1 \text{ (Day)}, \quad k_d = 0.0 \text{ (Night)}$

$$PAN \overset{k_p}{\Rightarrow} NO_2;$$ $\quad k_p = 0 \text{(Day)}, \quad k_p = 0.02 \text{ (Night)}$

$$\frac{d[NO_2]}{dt} = -k_n[NO_2] + k_p[PAN] - (k'_{DRY} + k'_{WET})[NO_2]$$

$$\frac{d[NO]}{dt} = -(k'_{DRY} + k'_{WET})[NO]$$

$$\frac{d[PAN]}{dt} = 0.5(k_n[NO_2] + k_d[PAN]) - (k_d + k_p)[PAN]$$
$$- (k''_{DRY} + k''_{WET})[PAN]$$

$$\frac{d[HNO_3]}{dt} = 0.4(k_N[NO_2] + k_d[PAN]) - (k'''_{DRY} + k'''_{WET})[HNO_3]$$

$$\frac{d[NO_3^-]}{dt} = 0.1(k_N[NO_2] + k_d[PAN]) - (k''''_{DRY} + k''''_{WET})[NO_3^-]$$

Fig. 7. Summary of the reactions used in ENAMAP-2 for nitrogen compounds.

**DRY DEPOSITION VELOCITIES, v_d,
FOR NITROGEN COMPOUNDS IN THE ATMOSPHERE**

COMPOUND	v_d (cm s^{-1})
NO_2, NO	0.2
HNO_3	1.0
PAN	0.25
NO_3^- (aerosol)	0.6

**WET DEPOSITION RATES FOR ATMOSPHERIC NITROGEN COMPOUNDS
EXPRESSED AS FRACTION OF RATES FOR SO$_x$**

COMPOUND	RELATIVE RATE
NO_2, NO	$0.25 \times a_{SO_2}$
HNO_3	$0.50 \times a_{H_2SO_4}$
PAN	$0.50 \times a_{SO_2}$
NO_3^- (aerosol)	$a_{H_2SO_4}$

Fig. 8. Dry and wet deposition rates used in the model.

to the emission sources; and the SO_2 and $SO_4^=$ concentrations have largest values downwind from the Ohio River valley. The calculated and observed concentrations of SO_2 are in general agreement; however, there are significant local differences. The calculated $SO_4^=$ concentrations generally are somewhat higher than the observed values. The deposition maps are not shown here; however, the wet deposition patterns reflect the rainfall patterns for the month. The dry deposition amounts in January are much larger and more windspread than the wet deposition amounts. Similar patterns were computed for August 1977.

The model also gives source-receptor relationships as illustrated in Table 1. The columns of Table 1 show the contribution of each individual area to the total sulfur depositon (SO_2 and $SO_4^=$ combined) within that area. For example, column 23 (for New York) shows that 4.1 KTON S originated in Michigan (row 14), 2.7 KTON originated in New York itself, and 4.5 KTON originated in Ohio (row 26). The total deposition in New York in January 1977 was 27.7 KTON S. The model also gives all of these entries as percentages of totals received. From this information, the influence of pollution sources in one area on the surrounding areas can be determined.

The verification of computations of this type is a difficult task for several reasons. These include problems in measuring emissions, concentrations, and depositions, and the need to avoid local, unrepresentative effects. For the January and August 1977 computations the principal validation data are SO_2 and $SO_4^=$ concentrations from the SURE and SAROAD inventories. As shown by Mayerhofer et al (1981) and Endlich et al (1982), the simulated airborne monthly concentrations of SO_2 and sulfate generally agree reasonably well with measured air quality values (linear correlation coefficients are in the range from 0.2 to 0.7), and are generally within a factor of 2 of the measured values.

THE NITROGEN MODEL AND APPLICATIONS

The features of the NO_x model are very similar to those used for the sulfur compounds; however, the chemistry for nitrogen compounds is considerably more complex. The algorithm uses five chemical species: NO, NO_2, PAN, HNO_3, and NO_3^-. The reaction scheme is shown in Fig. 7. The reaction rates shown are best estimates based on available information, and are subject to revision. The dry and wet deposition rates (based on Garland and Penkett, 1976; McMahon and Denison, 1979; and Sehmel, 1980) are shown in Fig. 8. The latter are given in relation to wet deposition rates for either SO_2 or sulfates. In each time step the computations are made as shown in Fig. 9.

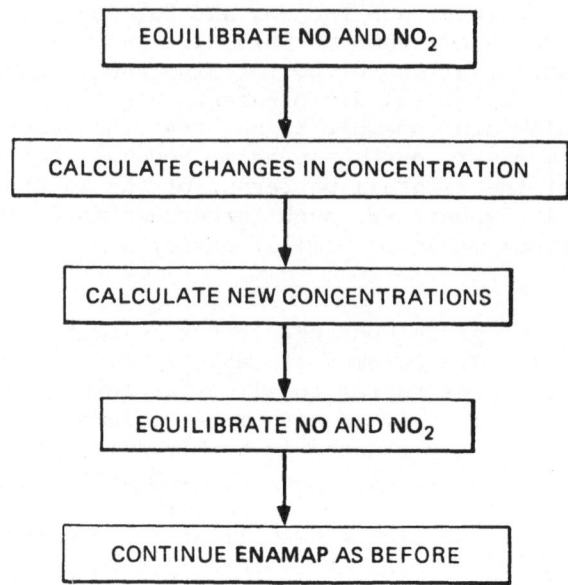

Fig. 9. The steps used in calculating changes
in nitrogen compounds.

To illustrate the computations, we have applied the algorithm
(without deposition processes) to a hypothetical puff having an
initial concentration of 100 parts of NO and NO_2. The amount of
each of the five compounds versus time is shown by the curves in
Fig. 10. (The time scale has 12 hours of day followed by 12 hours
of night, etc.) The relative amounts of NO and NO_2 change markedly
at the transition from day to night, or night to day. The overall
pattern is a conversion of NO and NO_2 to HNO_3 and nitrates. These
curves are a reasonably good approximation to laboratory measurements
made by Spicer (private communication).

The ENAMAP model was run for the month of January 1977. The
calculated monthly average concentrations of each of the five compounds are shown in Fig. 11. The concentrations of NO and NO_2 have
largest values around metropolitan areas. Concentrations of the
products PAN, HNO_3 and nitrates are smoother and more widespread due
to transport and diffusion processes. (The deposition patterns and
interregional exchange tables were also computed, but are not shown.)

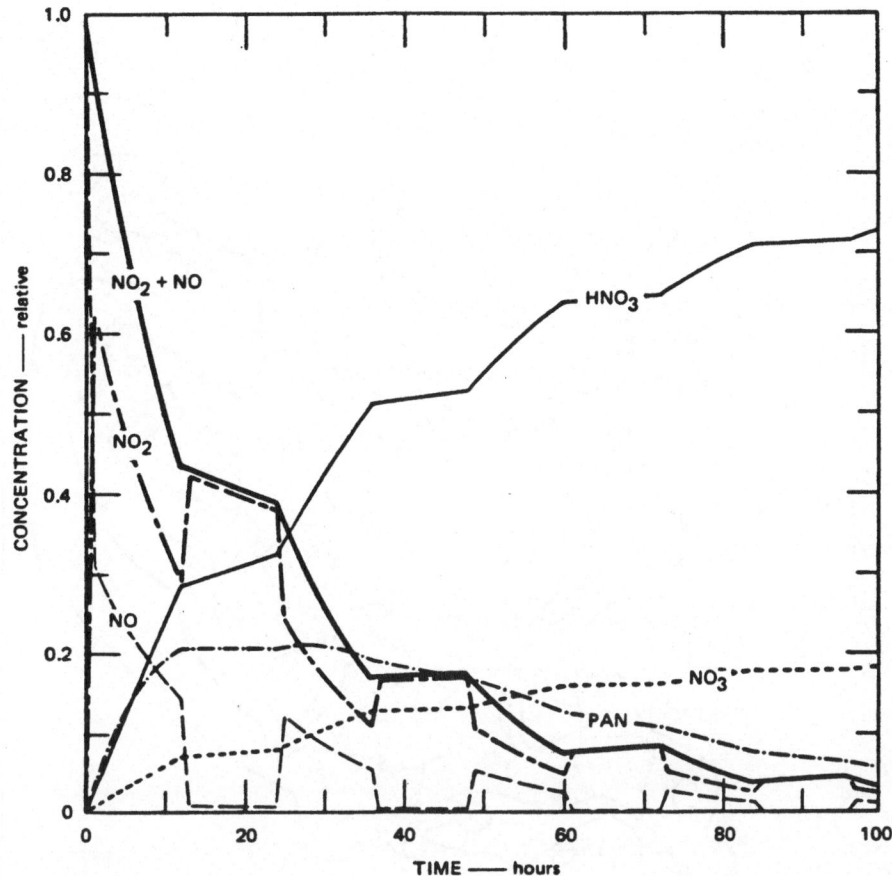

Fig. 10. Illustration of the "five compound" algorithm for nitrogen compounds for a hypothetical puff.

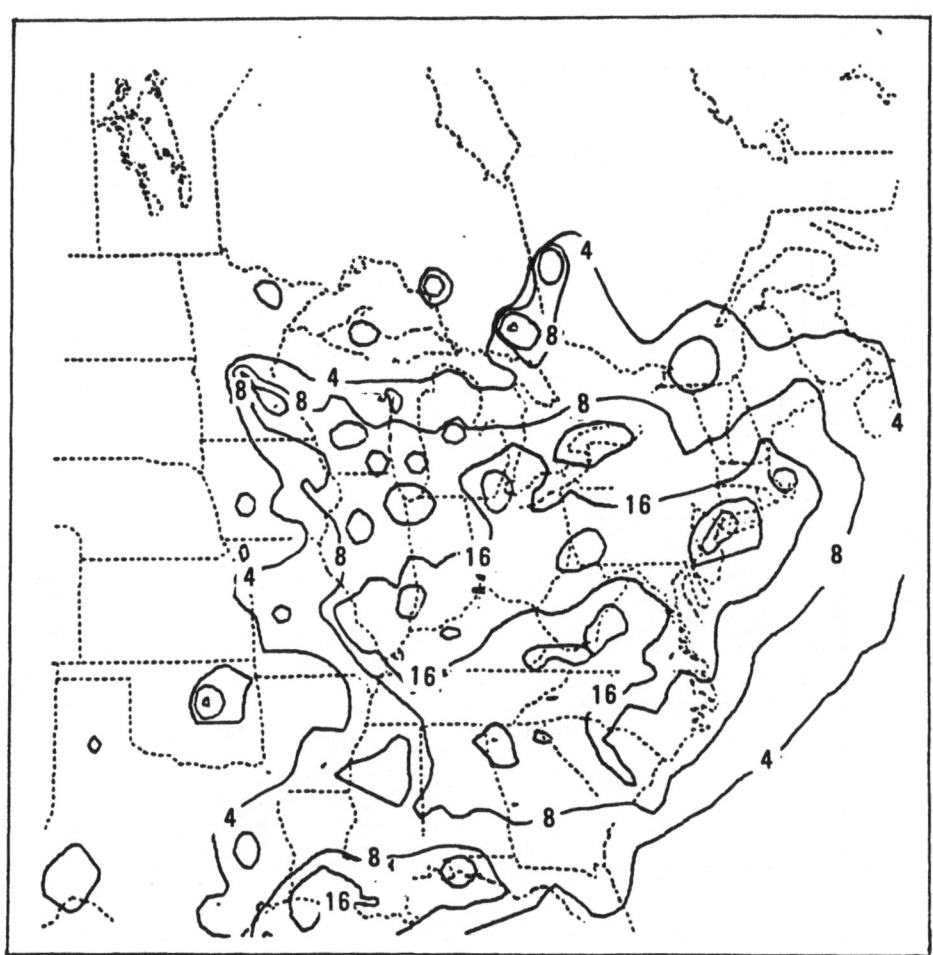

Fig. 11. Calculated concentrations of nitrogen compounds
        (μg m^{-3}) for January 1977.
        a. NO$_2$ concentration

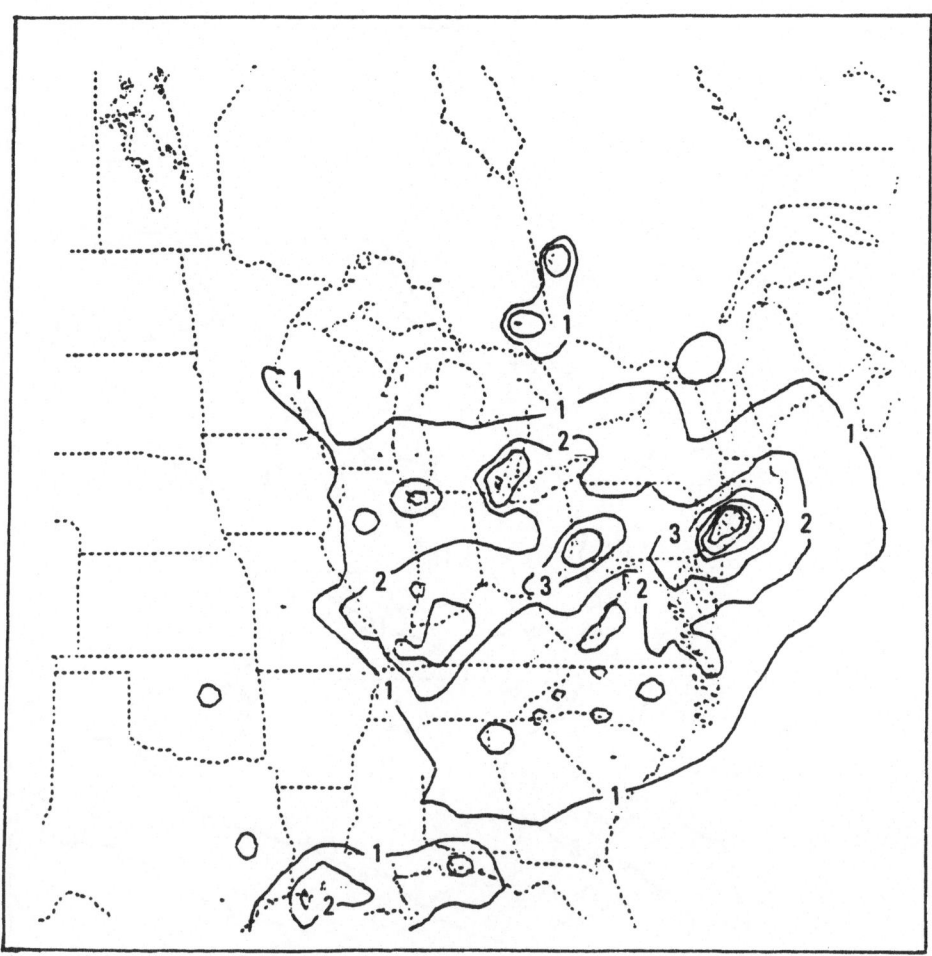

Fig. 11. Calculated concentrations of nitrogen compounds
(μg m^{-3}) for January 1977.
b. NO concentration

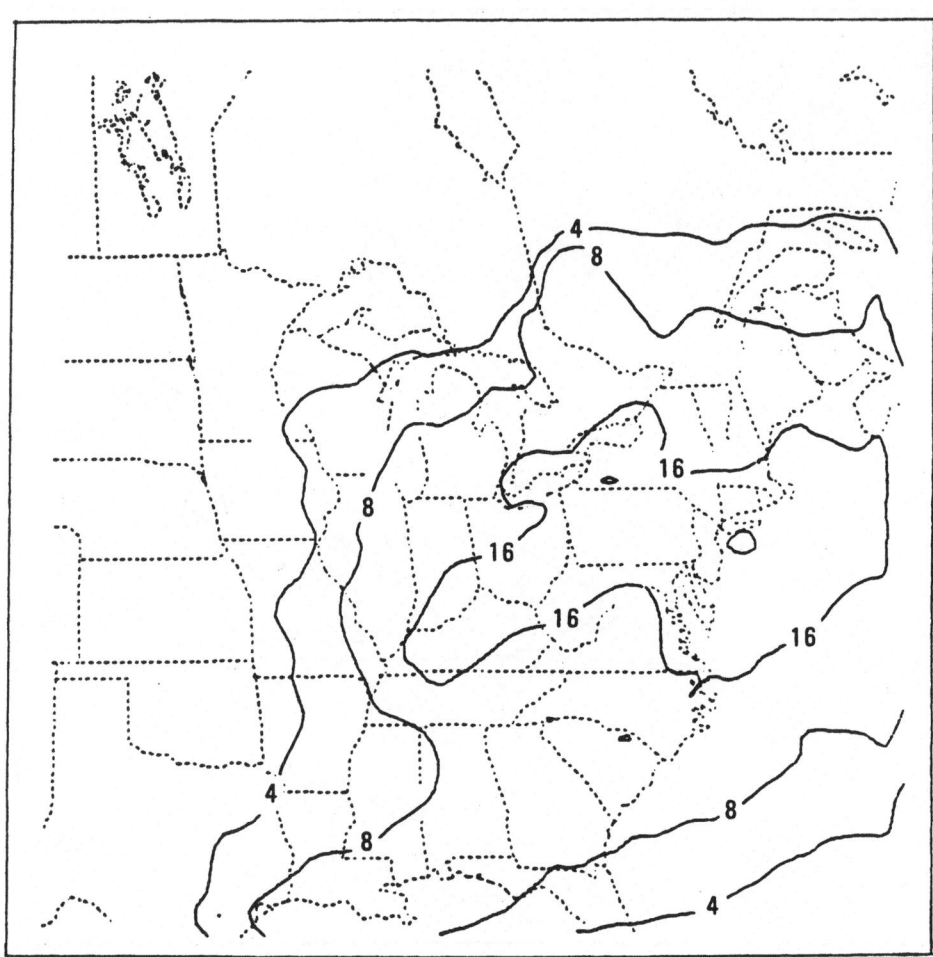

Fig. 11. Calculated concentrations of nitrogen compounds
($\mu g\ m^{-3}$) for January 1977.
c. PAN concentration

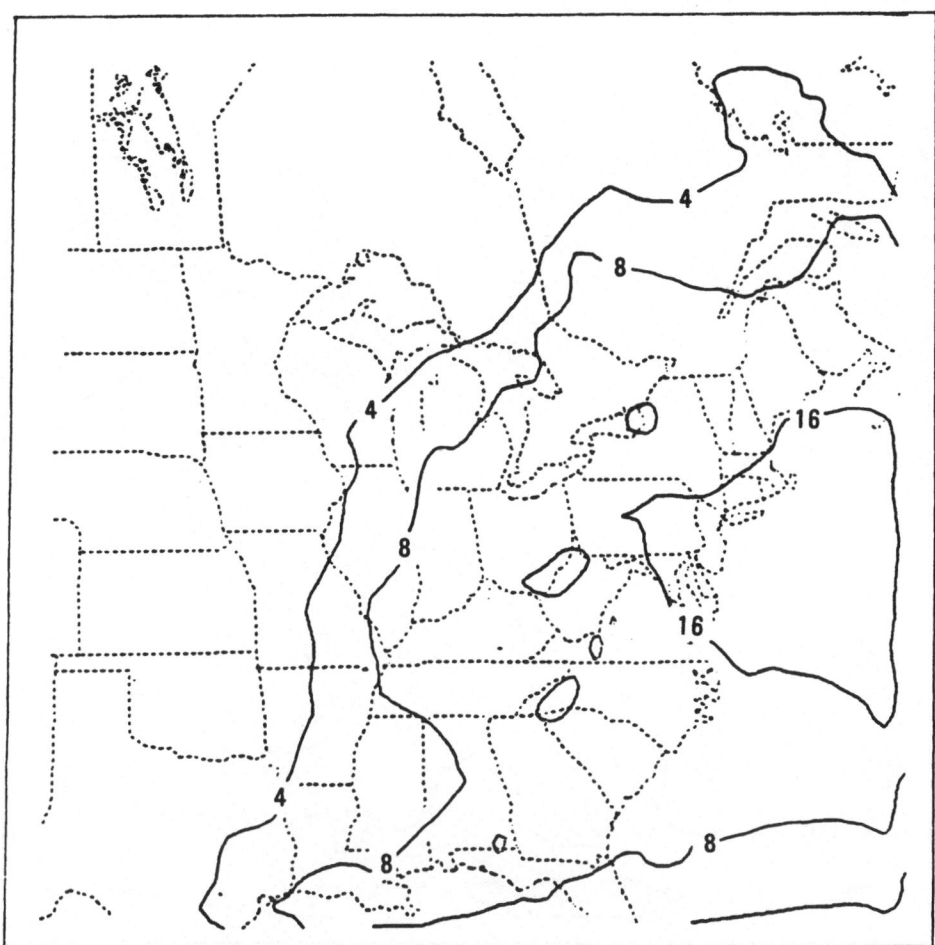

Fig. 11. Calcualted concentrations of nitrogen compounds
        (μg m^{-3}) for January 1977.
        d. HNO$_3$ concentration

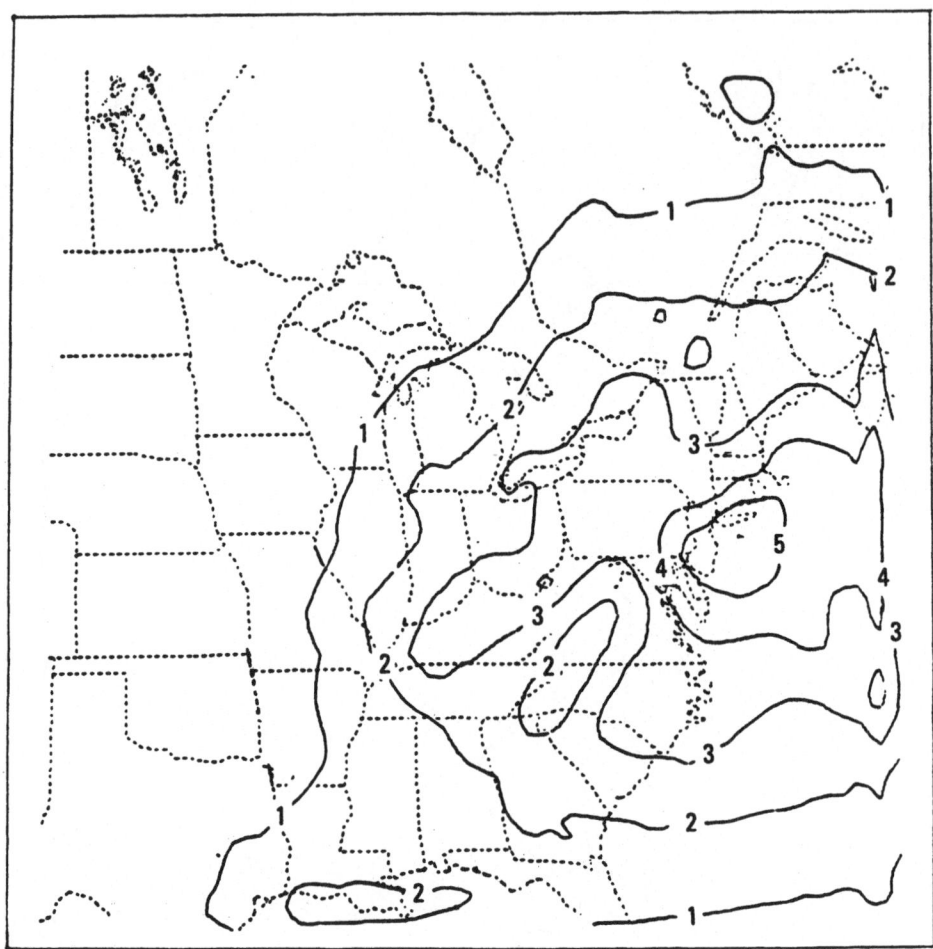

Fig. 11. Calculated concentrations of nitrogen compounds
        (μg m^{-3}) for January 1977.
        e. NO$_3^-$ concentration

Table 2. Budget of the enanox model for illinois nitrogen emissions for January 1977 (Units Kilotons)

	Wet Deposition	Dry Deposition	Total Deposition	Total Deposition in N units	Flux out of grid region
NO	0.0	3.0	3.0	1.5	0.3
NO_2	0.2	14.1	14.3	5.0	2.2
PAN	0.6	39.4	40.0	5.8	10.8
HNO_3	1.9	92.8	94.7	28.0	31.3
NO_3^-	0.8	23.5	24.3	7.0	6.7

The budget of the model can be seen from the computations for Illinois (Table 2). (These computations involved 1612 puffs that were tracked for a total of 61,702 time steps, illustrating the large amount of computation involved.) Table 2 shows that the principal product of the simulation is HNO_3, followed by nitrates and PAN. The relative proportions appear to be realistic.

SUMMARY

Experience with ENAMAP-2 shows that it has the following advantages:

1) The puff-tracking methdology identifies the source-receptor relationships among different areas, i.e., the amount of pollution in an area that originates in all areas of the model domain.

2) Emissions can be introduced in layers in an appropriate manner.

3) Variable parameters can be adjusted as new information becomes available.

4) The model is sufficiently economical so that computations can be made for long time periods (such as a month).

5) The features of the model are sufficiently realistic so that we believe it can be applied to time periods as short as one day.

6) For SO_x pollutants the accuracy of the results is reasonably good. Improved results are expected by tuning the model with available observations.

7) The initial results of the NO_x simulations appear to be reasonable. This indicates that NO_x pollution transport and deposition can be computed in a manner similar to that used previously for sulfur compounds.

The model is subject to the following limitations:

1) The results are dependent on rates for several chemical processes; these rates are rather uncertain and are under continuing investigation by various groups.

2) The pollution patterns from single sources of emissions are not identified.

3) It is not possible to include nonlinear chemistry in this type of model, as chemical processes due to mixing of puffs are not simulated.

A complete report concerning the model, including its application to NO_x emissions, is currently being prepared.

ACKNOWLEDGMENT

We appreciate the technical contributions to model development made by Mr. Terry L. Clark of the Environmental Sciences Research Laboratory, Environmental Protection Agency, Research Triangle Park, NC, who is Project Officer for the contract.

DISCLAIMER

Although the research described in this article has been funded wholly or in part by the United States Environmental Protection Agency under Contract 68-02-3424 to SRI International, it has not been subjected to Agency review. Therefore, it does not necessarily reflect the views of the Agency, and no official endorsement should be inferred.

REFERENCES

Bhumralkar, C.M., R.L. Mancuso, D.E. Wolf, K.C. Nitz, and W.B. Johnson, 1980, "Adaptation and application of the ENAMAP-1 model to eastern North America--Phase II," Final Report, Contract 68-02-2959, SRI International, Menlo Park, CA.

Bhumralkar, C.M., R.L. Mancuso, D.E. Wolf, and W.B. Johnson, 1981, Regional air pollution model for calculating short-term (daily) patterns and transfrontier exchanges of airborne sulfur in Europe, Tellus, 33:142.

Bhumralkar, C.M., R.M. Endlich, R. Brodzinsky, K.C. Nitz, and W.B. Johnson, 1982, "Further studies to develop and apply long- and short-term models to calculate regional patterns and transfrontier exchanges of airborne pollution in Europe," final report to Umweltbundesamt, Federal Republic of Germany, Project 8365, SRI International, Menlo Park, CA.

Crank, J., 1967, "The Mathematics of Diffusion," Oxford University Press, London.

Endlich, R.M., C.M. Bhumralkar, R. Brodzinsky, K. Nitz, B. Cantrell, and P. Mayerhofer, 1982, "ENAMAP long term air pollution model," paper presented at 75th Annual Meeting, Air Pollution Control Assoc., June 22, 1982.

Garland, J.A., and S.A. Penkett, 1976, Absorption of peroxy acetyl nitrate and ozone and natural surface, Atmos. Environ.,10:1127.

Johnson, W.B., D.E. Wolf, and R.L. Mancuso, 1978, Long term regional
    patterns and transfrontier exchanges of airborne sulfur pollu-
    tion in Europe, Atmos. Environ., 12:511.
Mayerhofer, P.M., R.M. Endlich, B.K. Cantrell, R. Brodzinsky, and
    C.M. Bhumralkar, 1981, "ENAMAP-1A long term SO_2 sulfate air
    pollution model: refinement of transformation and deposition
    mechanisms," final report, contract 68-02-3424, SRI Interna-
    tional, Menlo Park, CA.
McMahon, T.A., and P.J. Denison, 1979, Review paper; empirical at-
    mospheric deposition parameters--a survey, Atmos. Environ.,
    10:571.
Scott, B.C., 1978, Parameterization of sulfate removal by precipita-
    tion, J. Appl. Meteor., 17:1375.
Sehmel, G.A., 1980, Particle and gas dry deposition: a review,
    Atmos. Environ., 14:983.
Sheih, C.M., M.L. Wesely, and B.B. Hicks, 1979, Estimated dry depo-
    sition velocities of sulfur over the eastern United States and
    surrounding regions, Atmos. Environ., 13:1361.
Spicer, C.W., 1981, The distribution, fate and reaction rate of NO_x
    in the atmosphere, private communication.

DISCUSSION

N.D. VAN EGMOND   Did you consider the effect of
            the limited availability of oxidant (= NO_2 + O_3) in
            the oxidation of NO to NO_2? This strongly affects the
            NO_2 concentration and thus the formation rate of HNO_3
            form NO_2.

C.M. BHUMRALKAR   No this is our first attempt
            to incorporate NO treatment in the Lagrangian.

F. FANAKI   In determining the vertical lo-
            cation of the sources do you use the physical height
            of the exit or do you use or specific plume rise for-
            mula ?

C.M. BHUMRALKAR   The model is rather coarse both
            in time and space. So we do not use any specific plume
            rise formula.
            Also, the emission data used do not provide all the
            detailed information. We only know the point and area
            source emissions from grids each of which has a size
            of 70 km x 70 km.

SOME COMPARISONS BETWEEN LAGRANGIAN AND SIMPLE SOURCE-ORIENTED
MODELS FOR LONG RANGE ATMOSPHERIC TRANSPORT AND DISPERSAL
OF RADIOACTIVE RELEASES

H.M. ApSimon, A. Davison and A.J.H. Goddard

Nuclear Power Section
Imperial College of Science & Technology
London   SW7 2BX

INTRODUCTION

In this paper results obtained with a more complex Lagrangian
model, MESOS, developed to study long-range effects of atmospheric
releases of radioactivity from nuclear installations, will be
compared with results from simpler models based on extrapolating
Gaussian plume models out to longer distances using only meteorological conditions at the source.

MESOS is a fully Lagrangian model, using extensive databases
of meteorological information to deduce the trajectories of a
sequence of puffs across Europe and to simulate dilution and
depletion according to local conditions in the atmosphere along
these trajectories at the time of passage. It has been described
at previous NATO/CCMS conferences and elsewhere[1,2,3]. The model
may be used to calculate time-integrated atmospheric concentrations,
dry deposition, and wet deposition at a large number of points for
sequential 3 hour releases of several pollutants simultaneously from
a point source. Results for longer release periods are obtained by
summing over component 3 hour releases. The model has been applied
to several source locations in Europe using databases spanning 2
years, 1973 and 1976, mainly for radioactive releases (but also for
some studies of stack releases of sulphur from coal). Predicted
long-term average levels of contamination have been used to estimate
collective doses to the population of EEC countries arising from
routine continuous releases from nuclear power plants, and results
for hypothetical short-term releases used for statistical surveys of
possible contamination resulting from accidental releases.

The comparisons of MESOS results with those from simpler models
in this paper apply to point sources, and time integrated atmospheric
concentrations arising from releases of fixed duration, rather than
average concentrations over fixed periods of time due to all
contributing release periods. Nevertheless the conclusions are
probably relevant to other long-range atmospheric transport
problems such as SO_2 or NO_x in which area sources are widespread,
and air concentrations over specified measurement intervals are of
interest.

Initially comparison will be confined to long-term average
contamination from prolonged continuous releases to discuss possible
systematic differences which may not average out over time. This
is followed by a discussion of trajectory effects relevant to
modelling dispersal of short-period releases in the contexts of
obtaining statistical information on the probability distributions
of various levels of contamination at a remote point for accidental
releases, and assessing critical episodes.

LONG TERM AVERAGE CONCENTRATIONS FROM CONTINUOUS RELEASES

The traditional methods of estimating long-term average
concentrations within a few tens of kilometres from the source
usually involve considering a series of direction sectors, and
calculating the average concentration by summing concentrations for
different weather categories weighted according to the frequency
with which the wind blows into each sector in each category.
Vertical dispersion in each weather category is based on Gaussian
plume models, and at longer distances the plume may be multiply
reflected between the ground and an inversion at a typical height,
eventually leading to a uniform concentration profile over this
depth. Depletion by dry deposition, wet deposition and radioactive
decay are incorporated by simple source depletion methods. Thus
the concentration at distance r in sector i may be written

$$\overline{C_i}(r) = Q \sum_j f_{i,j} \frac{X_j(r).R_j(r)}{r\alpha u_j} \tag{1}$$

where Q is the release rate, $f_{i,j}$ is the frequency of the wind in
category j into sector i, α is the angle of the sector in radians,
$1/u_j$ is an average inverse windspeed for category j conditions,
$X_j(r)$ is a function representing the dependence of concentration on
distance r in category j as a result of vertical dilution, and $R_j(r)$
is the fraction remaining in the plume at distance r allowing for
depletion.

A model named SIMPLE[4] has been developed based on equation (1)
extrapolated out to longer distances, and the results compared with
those from MESOS. In order to obtain good general agreement with the

Lagrangian model for a range of nuclides with different depletion characteristics it was found necessary to allow for the following systematic effects.

The Finite Persistence of Stable Conditions

The restricted vertical dispersion of material during stable nocturnal conditions is likely to break up with increased mixing after dawn. When this is allowed for the most frequent neutral conditions dominate in equation (1) at longer distances instead of the less frequent stable ones. This is fortunate for simple models because the stronger winds common in these neutral conditions tend to lead to straighter more direct trajectories. It also means that the average concentrations are less sensitive to the frequency of occurrence of stable conditions, which can vary considerably between for example a source at the coast and a source a few miles inland. In SIMPLE the break up of stable conditions has been allowed for by subdividing each stable category into four subcategories with different durations ($1\frac{1}{2}$ hrs, $4\frac{1}{2}$ hrs, $7\frac{1}{2}$ hrs and $10\frac{1}{2}$ hrs) of travel before dawn and a transition to neutral conditions; at the transition the Gaussian vertical dispersion parameter, is continuous, but then grows at a rate appropriate to the neutral conditions; an inversion ceiling appropriate to neutral conditions is imposed throughout.

An Increase in the Mean Speed of Advection u_j

The values of u_j used in equation (1) for low level releases are frequently based on winds at a height of 10 metres. As material spreads vertically the mean speed of advection increases due to the increase of windspeed with height over the wind profile. This increase in mean wind speed is not compensated for by trajectory meander; if it is not allowed for travel times may be overestimated, so that fluxes surviving $R_j(r)$ may be considerably underestimated for depleting nuclides. As an approximate guide the geostrophic windspeed is roughly twice the 10 metre wind speed, and windspeeds in SIMPLE were progressively increased over the first 12 hours of travel to 1.8 times the 10 metre windspeed as an estimate of the mean windspeed over the mixing layer.

Frequency Distributions $f_{i,j}$ Based on Geostrophic Windroses

It is preferable to use geostrophic windroses rather than observed low level winds to assess the frequency with which the wind blows into each sector. Comparison at several sites has shown that although there is good agreement between the two for some flat inland sites with no topographical features when the backing of wind direction with height is allowed for in simple fashion, the observed low level winds are frequently influenced by purely local effects such as funelling along a valley or land-sea breeze circulations, and are therefore not a good indicator of the longer-range behaviour. The persistence of trajectories and the correlation with geostrophic wind is discussed later in this paper.

Spasmodic Depletion of Wet Deposition

It is important at longer distances to allow for the spasmodic nature of wet deposition. Wet deposition is frequently allowed for close to the source in one of two extreme ways. Either a fraction of the emission corresponding to the fraction of the time it is raining is depleted continuously using a wash-out coefficient at full strength, in which case only this finite fraction can ever be deposited by rain; or the entire emission may be depleted with a wash out rate of only a fraction of the full value in which case all the material could ultimately be deposited. At longer distances neither of these extremes is adequate. In SIMPLE the spasmodic nature of rainfall is taken into account in an approximate way by choosing an average value for the persistence of rainy and dry periods T_R and T_D, and assuming that after every T_R hours of travel time the particular portion of the emission constituting the rain-scavenged fraction changes with an appropriate probability of transition from dry to wet conditions and vice-versa. T_R and T_D were taken as 6 and 60 hours respectively but the results are not sensitive to the exact values. An investigation of these quantities from puff histories generated with MESOS using the meteorological databases, showed no significant difference between Eulerian estimates based on conditions at a fixed source, and Lagrangian estimates based on estimated puff trajectories, at least for source locations not subject to high orographic rain.

In this context it is perhaps worth mentioning some of the difficulties of prescribing rainfall and assessing wet deposition with a Lagrangian model. As far as observational data are concerned there is a choice between direct measurements of rainfall, and reports from synoptic stations under the "present weather" code which classify precipitation according to type, for example "rain showers, slight". Direct measurements have the virtue of being fairly precise for moderate or heavy rainfall, whereas reports of rain under the present weather code are qualitative, each category encompassing a range of rainfall rates. (This problem has been apparent in some studies of high sulphate episodes over Norway[5] using the MESOS databases, for which rainfall is estimated from the present weather observations.) On the other hand the synoptic stations give far better spatial and temporal resolution than direct rain measurements, and also have the great advantage that they are reported by ships over the sea where direct measurements are not available.

COMPARISONS OF RESULTS FROM SIMPLE AND MESOS

The agreement between predictions of longterm average concentrations using straightforward models like SIMPLE and more complex Lagrangian models like MESOS is illustrated in Figure 1 for

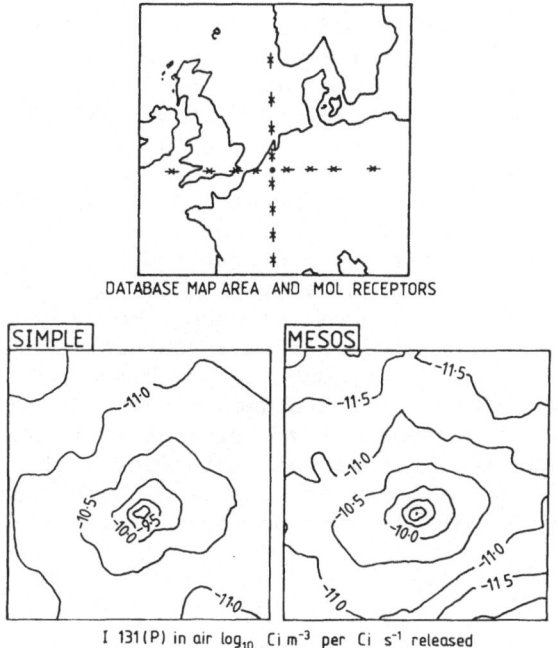

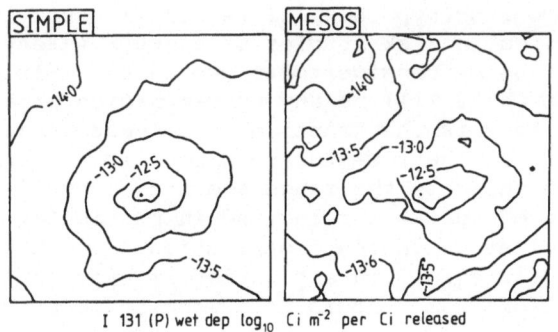

Figure 1 Comparison of isopleths of concentration in air estimated with models SIMPLE and MESOS for a hypothetical source at Mol in 1973

Figure 2 Comparison of isopleths of wet deposition calculated with the models SIMPLE and MESOS for a hypothetical source at Mol in 1973

a release of I^{131} from a hypothetical source at Mol in Belgium,
normalised to a unit release rate of 1 Ci.s^{-1}. The top part of the
figure shows the map area considered and the source location, and the
lower part isopleths of concentration at logarithmic intervals of $\sqrt{10}$
calculated according to both models. I^{131} has a half-life of 8.1
days; and a basic deposition velocity of 0.3 cm.s^{-1}, and a wash-out
coefficient of $5.10^{-5}.J^{0.8}$s^{1}, (J being the rainfall rate, in mm/hr)
have been used, in both cases. The general agreement is quite good
although it deteriorates in the outer regions where systematic
synoptic scale effects on the trajectories are not allowed for in
SIMPLE, as discussed later. The SIMPLE model has certainly proved
adequate for such applications as collective dose calculations
arising from routine continuous releases from nuclear installations.
However prediction of wet deposition is far less satisfactory - see
Figure 2. This is partly because the occurrence of rain and wind
direction are correlated, and the geostrophic windrose in wet
conditions is quite substantially different from the overall
geostrophic windrose. The spatial variation of orographic rain also
leads to significant differences, the Lagrangian model MESOS showing
the effect of high rainfall over Norway for example.

SHORT TERM RELEASES

   The simulation of the dispersal of short-term releases from a
discrete source introduces extra complexities, and simple
models based on source meteorology are unlikely to produce good
results in individual situations. Application of MESOS to simulate
dispersal of the Windscale release of 1957 generally gave agreement
within a factor of 3 of observations, but illustrated the sensitivity
of contamination at the observation points to uncertainties in the
trajectories[3]. However it is perhaps still meaningful to compare
predictions from simple and Lagrangian models in the form of
statistical distributions of contamination at remote receptor points
over a large number of hypothetical releases; such information is
frequently required in risk studies of reactor safety for example.
This comparison can be considered in two parts. First there is the
probability that there will be any contamination at all at the
receptor point; that is the fraction of releases that are transported
sufficiently near to the receptor to expose it. Secondly when the
point is exposed there is the range and frequency distribution of
different levels of contamination, and in particular the highest
levels of contamination in the distribution. In this context, MESOS
has been used for a full investigation of hypothetical releases
varying in duration from 3 hours to a week at 16 points round each
of 5 different source locations in Western Europe using the 1973 and
1976 databases (see figure 1 for the 16 selected points round Mol).

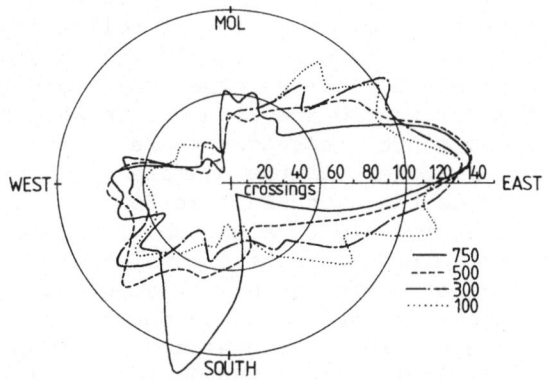

Figure 3   Trajectory rose for Mol 1973

TRAJECTORY ANALYSIS AND THE PROBABILITY OF EXPOSURE AT A POINT

The probability that material will be transported towards points at different distances and in different directions from the source can be examined using trajectory roses. These show the frequency with which trajectories originating at the source cross different ten degree sectors of circles centred on the source. Agreement between sim ple source oriented and Lagrangian models will depend partly on the per sistence of wind in the same direction and the degree to which turning of trajectories out of a sector is compensated for by turning of trajectories into a sector. An example of a trajectory rose is given in Figure 3 for the case of Mol for distances of 100, 300, 500 and 750 km from the source, as deduced from trajectories calculated with MESOS for the year 1973. With MESOS, trajectories are estimated from geostrophic winds deduced from 3 hourly surface pressure data, backed and reduced in strength according to appropriate vertical wind profiles over the vertical distribution of pollutants. It is clear that the wind roses are far from isotropic. At distances of 100 km there is a close correlation between the trajectory rose and the geostrophic windrose at the source as might be expected. However as the distance from the source increases there is increasing deforma- tion of the trajectory rose. In the example given relatively few trajectories persist to the South East at 750 km; this coincides with trajectories over the Alps. Although with strong orographic influences there are several reasons for which trajectories deduced from surface level geostrophic pressure fields could be inaccurate, in this case the effect is due to the synoptic scale low pressure regions which form in the lee of the Alps with winds from the North West, diverting trajectories round the cyclonic region. Another effect illustrated in Figure 3 is the accumulation of trajectories to the North at 750 km. This was apparent for several source locations and for different years, and may be due to quite frequent high- pressure areas over Scandinavia. Thus there do seem to be system- atic effects on trajectories due to common synoptic weather patterns, which cannot be allowed for in simple fashion in Eulerian models. Such effects may be quite significant within a few hundred kilometres from the source. The effects mentioned above are evident in the annual average levels of contamination in Figure 1 where concentra- tions fall off in the South East corner, but remain relatively high over Norway.

In order to simulate trajectory meander some authors have calculated puff trajectories as though each puff's velocity changed in synchronisation with the wind at the source. This will not be successful in representing the synoptic scale effects either. More- over because of the greater variability in surface winds due to small scale local effects and changes in stability, trajectory meander may be considerably overestimated. Agreement is likely to be best with stronger winds.

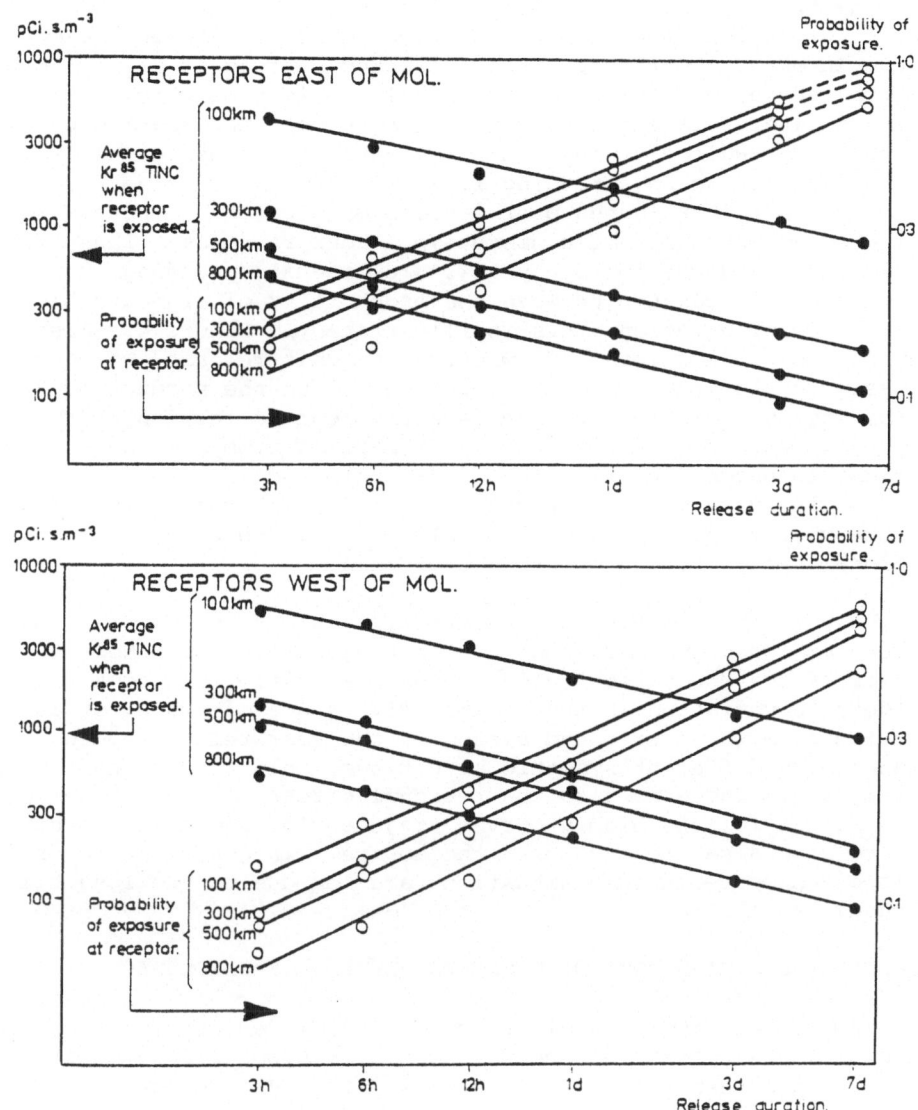

Figure 4   Variation of probability of exposure and average time
         integrated atmospheric concentrations with release duration

The wind roses are not the only factor in determining the probability of exposure of a remote receptor point; the probability also depends on the lateral spreading of material and the area exposed along the trajectories. For longer distances there is likely to be a large contribution from synoptic scale divergence between traject- ories. In MESOS discrete puffs are tracked setting of every 3 hours from the source; a continuous 3 hour release is considered as a series of intermediate puffs fanning out between the tracked puffs at the beginning and end of the 3 hour release and exposing the area in between; smaller scale spreading is allowed for by growth of the individual puffs. Releases over periods greater than 3 hours are considered as assembles of consecutive 3 hour releases. Thus there is specific allowance for synoptic scale lateral spreading, which affects the relationship between the probability of exposure and the overall release duration; this is illustrated in Figure 4. Also given is the average level of exposure for unit release, which varies inversely as the probability of exposure, since the product of probability and average exposure is equivalent to the long-term average exposure from a continuous release. A simple power low relationship between probability of exposure and release duration seems to fit quite well. For wet deposition the probability of exposure varies more sharply with release duration because of the additional factor of the occurrence of rain.

The variation of trajectory spreading as a function of distance has been analysed by Smith[6] using back trajectories from Norway. Analysis of the MESOS results indicates that there is also a considerable variation in the lateral divergence of trajectories with direction from the source and season of the year related to the dependence on different synoptic situations; this is currently being investigated. There is also the residual effect of indirect exposures when material follows highly curved trajectories and may even circle round to give a second pass over the source. In general however the levels of exposure in such situations are low due to the long time of travel.

PROBABILITY DISTRIBUTIONS OF DIFFERENT LEVELS OF EXPOSURE

Examples of frequency distributions of different levels of contamination at a remote receptor point for hypothetical short term releases, have been given at a previous NATO/CCMS conference. These show certain characteristic features. They are generally skew with a strong peak corresponding to the most probable levels of contamina- tion at levels usually between one and two orders of magnitude less than the highest levels predicted, and an extended tail of the distribution towards lower exposures. The peak corresponds mostly to the fairly direct trajectories in moderate or strong winds, and the tail to slow meandering trajectories. The conditions leading to the highest contamination vary with the nuclide depletion processes, and distance from the source as discussed below. The range of levels of contamination also varies with the nuclide; depletion processes of decay and dry or wet deposition introduce extra variable factors tending to reduce concentrations in air and spreading the distributions out to lower levels of contamination. The spread of the distributions generally decreases with distance from the source because the maximum and most probable levels of contamination decrease with distance, but the lower exposures from meandering trajectories vary little with source distance. The shape of the distributions for wet deposition is a little different, but those for time integrated air concentrations and dry deposition are very similar. It has been found that the cumulative probability distributions approximate well to Weibull distributions, which are currently being used to analyse the MESOS results, relating the values of the Weibull parameters to such factors as distance and direction from the source, release duration, and nuclide depletion parameters in an attempt to provide a simple statistical model for use in reactor accident studies.

It seems difficult to obtain similar distributions from simple Eulerian models. For example considering the simple case of the time-integrated atmospheric concentration as a release passes over a remote receptor point, the level of contamination will depend on the flux surviving depletion and decay, the lateral, and the vertical dispersion, and the speed of transit overhead. To obtain realistic distributions the variability of all these factors needs to be allowed for; however they are not independent but are correlated with different synoptic situations. The problems are even greater for releases of more than a few hours duration. Comparison has been made for the simple case of a non-depositing nuclide between results for 3 hour releases from MESOS and results based on an extrapolation of the Gaussian plume model, similar to that in SIMPLE, with Gaussian plumes radiating from the source into each sector (that is based on a wind rose and allowing for the finite persistence of stable conditions after leaving the source). The range of contamination is much narrower from the simple model than from MESOS and there is no tail of the distribution towards the lower exposures, but the most probable levels of contamination correspond quite well. The differences increase when decay and depletion are involved as well. Similar differences occur when results from MESOS are compared with some trajectory models (e.g. TALD[8]) which consider a simple constant Gaussian plume along each trajectory calculated with dispersion depending purely on travel time irrespective of the meteorological history. Such models do not allow for the important effect of release duration either.

SITUATIONS LEADING TO HIGH LEVELS OF CONTAMINATION

The fact that the most probable levels of contamination arise largely from fairly direct trajectories, and lie towards the upper end of the distribution, implies that simple Eulerian models may

be used to make fairly pessimistic estimates of the higher levels
of contamination likely to arise from an accidental release; conservative estimates of lateral dispersion and straight line
trajectories may be incorporated (see for example Jones[9] or
Cagnetti[10]).

For simple cases with non-depleting nuclides comparison of
predictions from these models with results from MESOS shows that
they do indeed give values close to the highest few per cent of
values from the MESOS distributions.

A survey has been made of the highest two values of contamination predicted using MESOS for 16 receptor points with 3 hour
releases from Mol to see what meteorological situations are associated with them. An interesting aspect of this survey was that high
exposures tended to occur clustered in episodes of a few days
duration. At distances of the order of 100 km from the source high
values of all nuclides considered tended to occur simultaneously.
Many of the critical releases are associated with slow transport in
anticyclonic conditions and evening and night time releases are
prevalent; for the low level releases (50 to 100 m) considered,
material released after dusk can travel this far in stable night
time conditions with low vertical dispersion before dawn and the
break-up of these conditions. For higher level releases high exposures are likely over a relatively small area when fumigation
takes place after dawn following transport above a nocturnal inversion.

Time of release is not so crucial for exposures at longer
distances because stable conditions will have terminated, although
the flux surviving to such longer distances may be affected[11];
different situations become critical for different nuclides.
Relatively slow transport episodes still dominate for inert non-
depositing nuclides like Kr^{85}, but for short-lived nuclides faster
trajectories become important; in fact similar travel times lead
to the highest contamination for Xe^{135} (half-life 9.1 hours) at a
large range of distances. For nuclides subject to wet deposition
such as I^{131} dry anticyclonic conditions maintain high fluxes;
different situations would of course be important for high levels
of wet deposition. Vertical dispersion is usually of the order of
1000 metres or more by the time material has travelled a few hundred
kilometres since either travel times involve at least one diurnal
cycle of mixing, or high velocities give considerable mechanical
mixing. Some interesting exceptions did occur however. For example
night-time releases from an inland source could reach the coast
before dawn and then be transported a long distance over a cold sea
with a surface inversion. Very high levels were predicted in
Norway for a hypothetical release from Mol in such a situation,
with the added factor that the wind dropped as the release approached
the Norwegian coast. This deceleration was a common feature in the

severest situations with winds dropping to about 3 m.s^{-1} during
transit over receptor points giving relatively long exposure
times, after more rapid passage between source and receptor.
Another important feature of many of the releases giving high exposures at all distances is that the synoptic divergence between
trajectories is very small giving minimal lateral dispersion.
Persistence of trajectories in stationary synoptic situations becomes
increasingly important for releases of longer duraction; for releases over a matter of days the worst conditions correspond to
blocking situations irrespective of the nuclide properties[12].

CONCLUSIONS

It has been shown that suitably extrapolated simple Gaussian
plume models based on source meteorology can give comparable results
with more complex Lagrangian models for atmospheric concentrations
and dry deposition for the purpose of collective dose calculations
for continuous routine releases from nuclear installations, although
systematic effects of frequent synoptic situations are not represented. Prediction of patterns of wet deposition is more difficult.
For full statistical predictions of risk from short term accidental
releases Lagrangian models are necessary. However simple models
based on pessimistic assumptions about dispersion appear to give
estimates of levels of contamination falling close to the top few
per cent of the cumulative frequency distributions generated with
the full Lagrangian model; even though analysis of the situations
giving the highest predicted levels using MESOS revealed significant
spatial and temporal changes along the trajectories. Accurate prediction in a particular situation for a point source release is
very difficult at long range, since contamination at a set of
observation points can be very sensitive to the exact trajectories,
but Lagrangian models may help by giving a qualitative picture of
the transport and dispersal of the release.

REFERENCES

1. H.M. ApSimon, A.J.H. Goddard. "Modelling the atmospheric dis-
    persal of radioactive pollutants beyond the first few houses
    of travel". Proc.7th NATO/CCMS Conference, Airlie, Virginia,
    September 1976.
2. J. Wrigley et al. "Meteorological data and the MESOS model for
    the long-range transport of pollutants". Proc.WMO.Symposium,
    Bulgaria, 1979.
3. H. ApSimon et al. "Estimating the transfrontier consequences
    of accidental releases: the MESOS model for long-range atmos-
    pheric dispersal". Proc.CEC.Seminar 22-25 April 1980. Risø,
    Denmark, Vol.II, p.819-842.
4. S. Crompton. "Modelling long range exposure from continuous
    releases". Internal report Nuclear Power Section, Imperial
    College, London, SAF21, November 1979.

5. P. Maul. "Atmospheric transport of sulphur compound pollutants".
    CEGB Report MID/SSD/0026, March (1980).

6. F.B. Smith. "The influence of meteorological factors on radio-
    active dosages and depositions following an accidental release".
    Proc.CEC Seminar on Radioactive Releases and their Dispersion.
    Risø, April(1980),Vol.I, pp.223-245.

7. H. Bultynck et al. "Application of the meso-scale diffusion
    models MESOS and TALD to accidental releases of radioactivity".
    Proc.10th NATO/CCMS Conf., Rome, Italy, October(1979).

8. J. Le Grand, A. Depres, A. Doury. "Le program TALD pour
    l'evaluation des transports atmospheriques a longue distance
    dans les basses couches de l'atmosphere". Proc.CEC. Seminar,
    April(1980),Risø, Denmark, Vol.II, pp.743-776.

9. J.A. Jones. "A model for long-range atmospheric dispersion of
    radionuclides released over a short period". NRPB Report R124,
    (1981).

10. P. Cagnetti and V. Ferrara. "Le transport et la dispersion de
    rejets radioactifs dans l'atmosphere a moyennes et grandes
    distances: evaluation des concentrations dans l'air et des
    depots an sol". Vol.II, CEC Report, June(1980).

11. H. ApSimon et al. "On the importance of the vertical structure
    of the boundary layer in modelling the long-range transport
    following an accidental release of pollutant". WMO Symp. on
    Boundary Layer Physics. Norköping, Sweden, June (1978).

12. P. Cagnetti et al. "Conditions meteorologiques defavorables a
    lar diffusion sur grandes distances de rejets aeriformes
    prolonges dans le temps: les situations de blocage".
    Proc.CEC Seminar, April (1980), Risø, Denmark, Vol.II,
    pp.777-796.

DISCUSSION

F. FANAKI   Will any attempt be made or
                 has been made to compare your models with actual
                 observations ?

H. APSIMON   We have applied the MESOS mo-
                 del to simulation of the dispersion of I^{131} from
                 the Windscale release of 1957. The agreement with
                 observed values was generally within a factor 3; but
                 I think we were lucky because the meteorological si-
                 tuation was complicated, and predicted values at ob-
                 servation points were very sensitive to the exact
                 trajectories, apart form uncertainties in the re-
                 lease itself.

F. LUDWIG   Have you compared the performance of your simple model using straight trajectories with its performance using curved trajectories whose curvature is defined by the streamline curvature at the source ?

H. APSIMON   No. The only other variation we have considered at all in ranging puff velocities along trajectories in synchronisation with changing winds at the source.

E. RUNCA   What is meant by exact trajectory ? Pollutant particles follow different trajectories depending on the height where they are. Can you comment on that ?

H. APSIMON   I should not perhaps have used the word "exact". By puff trajectory I meant the mean motion of the pollutant material in the puff averaged over the vertical distribution at any time. I should have said that in trying to validate models in specific situations it should be recognized that predicted values of contamination at selected locations may be substantially altered by small perturbations to the trajectories, for example small changes in backing angle for the vertical wind profile.

APPLICATION OF PERSONAL COMPUTERS

TO LAGRANGIAN MODELING IN COMPLEX TERRAIN

F.L. Ludwig

SRI International

Menlo Park, California

INTRODUCTION

Personal computers are now capable of solving complex problems related to the transport and diffusion of materials released into the atmosphere. Making such calculations rapidly and "on the spot" would be of great use in planning evacuations and cleanup activities when toxic materials are spilled as the result of some accident. It would be desirable if such calculations were not limited to the simple, steady-state assumptions usually required for such applications. A more realistic simulation, accounting for wind fields which change in space and time and variations in the meteorological conditions would be better. As personal micro-computers become ever more readily available, it is no longer unreasonable to believe that these capabilities could be supplied relatively cheaply. Mass-produced personal computers are already relatively inexpensive. In fact, their cost is, on an equivalent basis, less than that of the mechanical desk calculator of 20 years ago. Personal computers are becoming more powerful every year. However, they have limitations in the amount of core storage that is available and in their speed, so their use requires careful programming and some simplifications of the theory.

More or less coincident with the development of the personal micro-computer, atmospheric scientists have been working on ways of simplifying point-source, non-steady-state diffusion models (Ludwig, Gasiorek, and Ruff, 1977) and methods for calculating mass-consistent flow fields in complex terrain (Ludwig and Byrd, 1980; Bhumralkar, Ludwig, and Mancuso, 1978). This paper discusses how these modeling techniques can be combined and implemented on a small computer. At the time that this paper is being written, a

model like that described here has been installed and operated on
a personal computer. Although the debugging of the program was not
completed in time to include results, the program was complete enough
to provide information on the amount of core storage and the time
required for typical calculations.

MODELING APPROACH

     As noted earlier, the use of time-varying meteorological condi-
tions (e.g., stability, mixing height, and so forth) and time- and
space-varying winds is very desirable. Although it does not appear
possible at this time to implement a full boundary layer model (including thermodynamic effects) on a personal computer, there are
techniques by which mass-conserving winds in complex terrain can be
calculated quickly and efficiently from a few scattered wind observations in the area. The technique requires that some prior calculations be made using a larger computer. In order to take full
advantage of time- and space-varying winds, the atmospheric transport
and diffusion model that is used must be of comparable sophistication. In the application described here, we have used a Gaussian
puff model that relies on some "tricks" to conserve core memory and
simplify calculations, while still retaining the ability to account
for varying winds and hourly changes of stability and mixing depth.

The Wind Interpolation Module

     The approach used to generate mass-consistent wind fields from
observed wind data is based on the technique described by Ludwig
and Byrd (1980). In general, schemes for deriving mass conserving
wind fields from a limited number of observations begin by defining
a trial wind field that is derived from the observations by ordinary
interpolation. If we suppose that the nondivergent wind field that
is sought is $\underset{\sim}{V}$, and that $\underset{\sim}{V}$ is related to $\underset{\sim}{V}^O$ (the original trial wind
field) as follows:

$$\underset{\sim}{V} = \underset{\sim}{V}^O + \frac{1}{2\alpha^2} \nabla\Phi$$

the resulting solutions will minimize the differences between the
trial wind field and the resulting nondivergent field (see e.g.,
Sherman, 1978). Because $\underset{\sim}{V}$ is nondivergent,

$$\underset{\sim}{\nabla} \cdot \underset{\sim}{V} = 0 \, ,$$

so a set of partial differential equations of the following form
is to be solved:

$$\nabla^2 \Phi = F(x,y,z)$$

where $F(x,y,z)$ is proportional to the divergence in the original interpolated wind field. The approach that was adopted takes advantage of the linear properties of the above relationship. If Φ_i is a solution for F_i, then $\Sigma a_i \Phi_i$ will be a solution for $\Sigma a_i F_i$.

Figure 1 schematically represents the formation of solutions from standardized input data sets. The column vectors shown at the top of Figure 1 represent an arbitrary set of observed winds that have been decomposed into a linear combination of independent data sets. In this case, the observed wind components are shown as the sum of their corresponding means (represented by the column vector at the right) and multiples of a set of normalized eigenvectors. The normalized eigenvectors can be derived from the covariance matrix for wind component data from an extended time period. Any set of linearly independent column vectors of appropriate number and dimension could be used, but the eigenvectors provide some advantages in that they can be used to reduce the required number of inputs and to introduce some smoothing of the data.

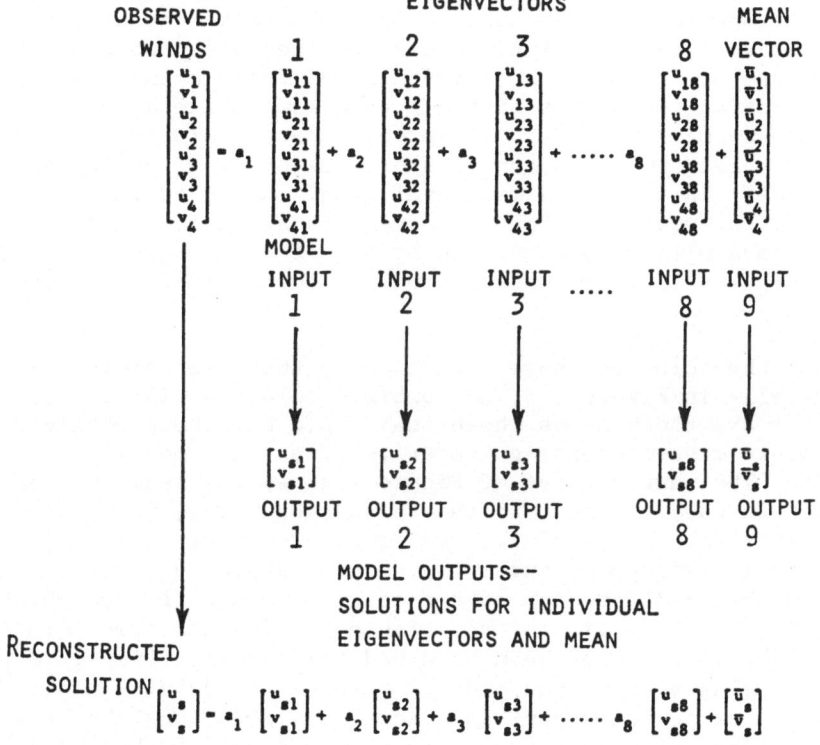

Fig. 1. Schematic representation of the method by which solutions are reconstructed.

The coefficients used in the linear combination shown in Figure 1 are symbolized by a_1 through a_8. For a set of linearly independent column vectors there will be a set of coefficients that satisfy the equation, which will be given by the inner products of the input data vector and each of the independent column vectors (eigenvectors), i.e.

$$A_j = \sum_{k=1}^{N} \left[(U_k - \bar{U}_k) \cdot u_{jk} + (V_k - \bar{V}_k) \cdot v_{jk} \right] ,$$

where a_j is the coefficient for the jth column vector, U_k and V_k are the U and V component for the kth station in the input data set. The overbars indicate averages of the U or V components at the kth observation site over the data sets from which the eigenvectors were derived. The terms u_{jk} and v_{jk} are kth pair of the jth column vector, and N is the number of sites from which data are available. There will be 2N components in each eigenvector.

Solutions can be obtained for each of the eigenvectors using one of the many models available (e.g., Bhumralkar et al., 1980; Sherman, 1978). These solutions are obtained ahead of time on a larger computer, because the computational requirements of the iterative relaxation techniques that are used would take a long time to execute on a small machine. The precalculated solutions are stored on a disk. Each solution consists of sets of u, v, and w values for each cell in a three-dimensional grid. A set of values is required for the mean and input eigenvector data sets. Inasmuch as the solutions will change with changes in the mixing depth, a different set of solutions must be provided for each of several mixing depth categories.

Once the solutions have been stored, they are combined as shown schematically in Figure 1. The combined solution will provide results that are the same as those that would have been obtained had the complete mass-conserving iterative relaxation model been used directly. The central part of Figure 1 shows that each column vector, when used as input to the relaxation model, provides a solution for each grid point, the figure indicates a solution for only a single grid point and only two components (u_{sj}, v_{sj}). The solutions obtained for each of the column vectors are recombined as shown in the last line of the figure to give the same solutions (U_s, V_s) that would have been obtained for the specified grid point using the relaxation method and the given input data. Thus, if one selects linearly independent data sets, all solutions can be expressed as linear combinations of the solutions for those data sets, as described by Ludwig and Byrd (1980).

The advantages of using linear combinations of solutions are obvious. The wind fields can be generated simply by reading solution matrices from disk storage, calculating the inner products of the observed input winds with eigenvector winds to obtain the appropriate coefficients. These coefficients are then used to form the linear combinations of solution wind fields that are used by the puff model described in the next section.

The Lagrangian, Gaussian Puff Transport and Diffusion Module

Transport and diffusion is simulated as a series of puffs, within which the concentration has a Gaussian distribution, characterized by a standard deviation σ_y in the horizontal, and σ_z in the vertical. In order to keep the computations and the memory requirements to a minimum, several simplifications have been introduced, following the example of Ludwig, Gasiorek, and Ruff (1977). The most important of the simplifications are:

● The number of puffs used to simulate the plume is kept to a minimum.

● Integer variables were used wherever possible.

● Computations of concentration contributions from a given puff were limited to those receptors where the contributions were significantly large.

● Some approximations were used for exponential quantities.

The basic equations used for this model are the same as those given by Ludwig, Gasiorek, and Ruff (1977). The variations of σ_y and σ_z with travel distance have been approximated with exponential functions of distance and the concept of virtual travel distance (Ludwig, 1982) has been invoked to account for changes in puff behavior with changes in atmospheric stability. The treatment of the effects of restricted mixing is as described by Ludwig, Gasiorek and Ruff (1977); the puff first undergoes unrestricted growth. Then, as it is affected by the top of the mixing layer, it is assumed to be "reflected", and, finally, when mixing is complete, a "box" model is used. A puff remains in this final state even if the mixing depth subsequently lowers (i.e., it is not compressed). A rising mixing depth will allow the puff to expand further in the vertical.

The description of each puff requires seven parameters, which are summarized in Table 1. An eighth parameter, the age of the puff, has also been included so that the model can easily be adapted to describe the behavior of pollutants that undergo exponential decay. Five of the variables are described with integers. This reduces memory requirements, but imposes some restrictions. For example, using the basic unit of one meter for length means that the coordinates of the puff must stay within the range ±32,767m for applications with most personal computers. Of course, if this were a severe problem, the distances could be scaled using 5- or 10-m

Table 1. Parameters Used To Describe Each Puff

Parameter	Unit	Variable Type	Remarks
x,y,z coordinate	m	Integer	
Puff strength	g	Floating point	
Virtual travel distances (two values)	m	Floating point	One value is used to calculate σ_y, the other σ_z
Age (not used)	min	Integer	Could be used to estimate decay for radioactive materials or 1st order chemistry
Maximum mixing height	m	Integer	The largest mixing depth that has influenced growth of this puff is retained and used to estimate the vertical puff dimension when mixing height decreases

units. Table 1 shows that we chose to use floating point numbers for virtual travel distance and puff strength. It would be possible to introduce automatic scaling techniques and use integers to reduce memory requirements, but it is not necessary for applications on at least one widely available personal computer.

As noted earlier, the number of puffs is minimized in order to reduce calculation time and memory requirements. Two approaches are used. First, we purge puffs from further consideration after they have traveled outside a 30-km square centered on the source. Within that square, the spacing between puffs is checked every six minutes (simulation time), and closely spaced puffs are merged so that one puff will replace two; the remaining puff will contain all the material in the original two puffs. The other parameters listed in Table 1 will be assigned the average of the values for the original two puffs. Figure 2 schematically illustrates the criteria used to decide whether two puffs should be merged. That figure shows the normalized concentrations arising from a series of identical, equally spaced puffs. The figure shows the effects of puff spacing, ranging from $0.5\sigma_y$ to $4\sigma_y$. Puff separations less than about $2\sigma_y$ cause variabilities of only a few percent along a line parallel to the plume. Thus, if separations are kept to $2\sigma_y$ or less, the model will produce reasonably continuous results. Puff merging also is done every six minutes.

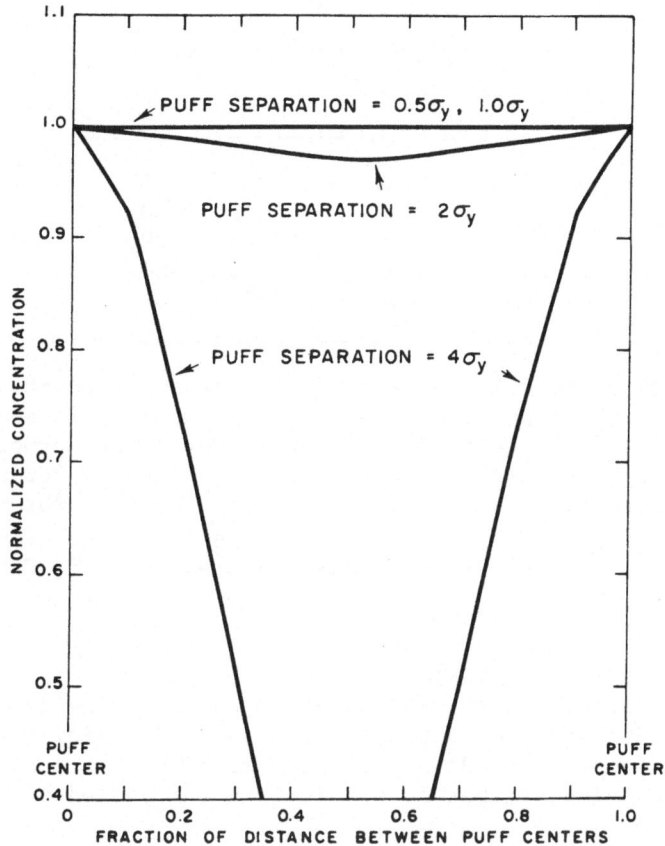

Fig. 2. Normalized concentration between two puffs in a string of puffs equal in size and spacing.

If the exact equation of σ_y were used, it would require frequent and time consuming evaluation of the expression $x^{0.9}$, because the variation of σ_y with travel distance is given by an equation of the following form:

$$\sigma_y = ax^{0.9} .$$

The value of a depends on the atmospheric stability. An approximation is used to determine whether or not to merge puffs. We have assumed that $\sigma_y \simeq 0.4ax$. Table 2 compares values obtained by this approximation to the values obtained from the above equation for a wide range of x. It can be seen that the approximation is quite conservative and that puffs separated by less than 0.4ax will always be within $1.5\sigma_y$ of each other. The exact equation is used when calculating concentrations, or recalculating σ_y for a change in stability class (Ludwig, 1982).

Table 2.   Comparisons of σ_y, $1.5\sigma_y$, and $0.4ax$

x	$\sigma_y = ax^{0.9}$	$1.5\sigma_y$	$0.4ax$
100	63a	95a	40a
1000	500a	750a	400a
5000	2100a	3200a	2000a
10000	4000a	6000a	4000a
50000	17000a	25000a	20000a

Not all puffs will contribute significantly to the concentrations at every receptor, so the computations are limited to those
puffs that do fall within $3\sigma_y$ of a receptor. Furthermore, the values
of the exponential terms appearing in the Gaussian formulae have
been approximated. Values of the function exp(-x) for x are stored
for integers between x=0 and x=9. This allows the approximation of
concentrations contributed by puffs within about $4\sigma_y$ of the surface.

IMPLEMENTATION ON A PERSONAL COMPUTER

The techniques described above have been used in a computer
program written in BASIC and installed on an Apple-II personal computer with 48K words of available storage and a mini-disk drive.
This is a standard, readily available machine that costs about $2,000
with a video display (but no printer) in the United States.

Figure 3 is a flowchart showing the general structure of the
program. From that figure it can be seen that the atmospheric stability, the mixing depth, and the source strength are input each
hour. Next, the winds at the beginning and end of the hour are input
and the inner products (the values of the a's in Figure 1) are calculated. Values of these inner products are linearly interpolated
at ten-minute intervals for intermediate times. Inside the time
loop, the wind fields are updated every ten minutes. Within these
ten-minute time periods, the advection is calculated for two-minute
steps. As the flowchart in Figure 3 shows, the winds are only constructed from the input solutions (for the appropriate mixing depth
class) in those cells that contain puffs at some time during the
ten minutes. Cells for which the wind has not yet been calculated
are identified by a large negative vertical velocity value. Inasmuch as puffs will tend to follow similar paths, usually it will not
be necessary to determine the winds for most of the cells in the
array. This appreciably reduces the necessary computations.

At the end of each ten-minute time period, the entire list of
remaining puffs is scanned, and those that have passed outside the
region of interest are removed from the list. Adjacent puffs that

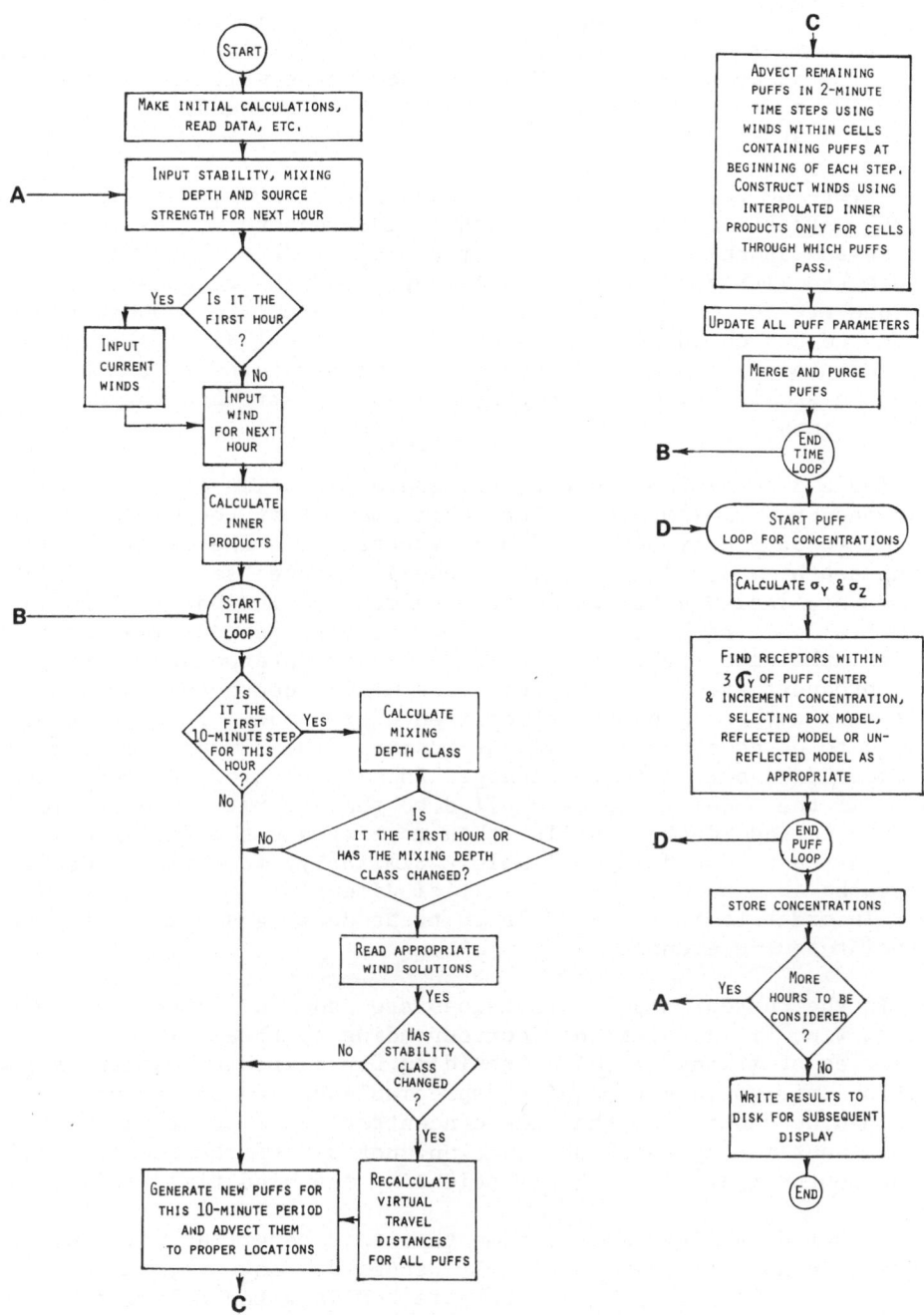

Fig. 3. Simplified flowchart.

are sufficiently close together are combined. Preliminary test calculations, using reasonable wind speeds and directions, indicate that the number of required puffs tends to stabilize at values somewhere between about 50 and 100. The program as written has provision for as many as 200 puffs, which should be sufficiently large to cover most situations.

After the puffs have been advected for an hour, calculations for concentrations are made at each of 100 receptors. As noted earlier and as shown in Figure 3, the program determines which receptors are within $3\sigma_y$ of the puff center and the receptor concentrations are incremented accordingly. The calculated concentrations are stored and could be used later with graphical display. If more hours are to be considered, then the program returns to the point where the stability, mixing depth, and source strength for the next hour are input.

Table 3 summarizes some of the characteristics of the model as it has been implemented. The table shows that the winds are determined on a 6-by-6 grid of 5-km squares. In the vertical, the mixing depth is divided into three equally spaced segments. Provisions are included for fully three-dimensional winds, but puffs are not allowed to have centers below the surface. As was noted earlier, the wind field is changed by linear interpolation of the inner products every ten minutes. The mixing depth and the stability are not changed more often than every hour. They are also assumed to be constant through the region. The current version of the model does not allow for changes in source strength except every hour. If the model is to be applied to the determination of cloud travel after accidental spills, then more frequent updating of source strength would be desirable. Similarly, it would be desirable to be able to specify the initial dimensions of the emitted cloud in order to better be able to treat dense gas spills or other non-point-source events.

In the current model, provisions are made for three different sets of wind field solutions (corresponding to three different categories of mixing depth). Providing for a greater number of such solution sets would not require appreciable change in the program. It should be understood that the concentration calculations use the actual input mixing depth observation, not one of the standardized mixing depths that are used for selecting the wind field solutions.

As noted earlier, at the time that this paper was written, the program had not yet been completely debugged. However, the process was sufficiently far along to indicate how long the calculations take, and how much memory is required. As described here, the model uses only about 75 percent of the available core memory of a typical personal computer. Thus, there is sufficient spare memory to allow further refinement, or expansion of the wind field array, an

Table 3. Summary of Characteristics of the Implemented Model

Charactertistics	Description
Horizontal and Vertical wind resolution	A 6x6x3 grid of 5-km squares; the vertical extent of each is 1/3 the local mixing depth
Temporal wind resolution	Wind field changed (by linear interpolation of inner products) every 10 minutes
Spatial resolution for mixing depth and stability	Uniform throughout the region
Temporal resolution of source strength, mixing depth, and stability	Changed hourly
Receptors for calculating concentration	Regular 10x10 grid, centered on the source; spacing can be specified (0.5 km is used in current version)
Number of mixing depth categories for wind field colutions	Three categories: <400 m, 400 to 800 m, and >800 m
Number of allowable wind observation sites	Five, could be four surface stations and a geostrophic wind

increase in the number of puffs used, and so forth. It also suggests that the model could be implemented on a machine with less memory than the Apple-II, especially if the solution wind sets were read from a disk at each ten-minute interval rather than stored in memory. Reading these data sets requires about one or two minutes, which would appreciably add to the time required for the calculations.

As noted before, the program is written in BASIC and normally uses an interpreter rather than a compiler to execute the instructions. When used with the standard "Apple" interpreter, the calculations for one hour require nearly 15 minutes. This includes the time required to read data from the disk and input wind and stability data from the keyboard. If a compiler is used, an appreciable time saving should be possible. It should be understood that the time required for the calculations will vary with the meteorological conditions, but the above examples are believed to be reasonably typical.

CONCLUSIONS AND RECOMMENDATIONS

It has been demonstrated that it is feasible to use relatively inexpensive personal computers to simulate the behavior or emissions from point sources in a relatively complicated time- and space-varying wind field. These first attempts are quite encouraging, but obviously many refinements could be introduced. The following would be among the more useful of the possible modifications;

- Provision for the use of output from numerical weather forecast models as input on a continuous, real-time basis.

- Provisions for more frequent updating of source strength, dimensions, and so forth.

- More frequent calculation of concentrations.

- Graphical output that could be interpreted easily by personnel responding to an emergency.

None of the above refinements represents an insurmountable extension beyond what has already been demonstrated. It is quite evident that it is possible to provide inexpensive computing capability to many agencies (such as fire departments, companies that ship toxic materials, hospitals, and so forth) for estimating the concentrations of gases and smokes more realistically than has been possible in the past.

ACKNOWLEDGMENT

Mr. Gerald Pierce, Research Engineer at SRI International, allowed the author to use his personal computer and contributed many valuable suggestions. Without his support, advice, and cooperation, this work would not have been possible.

REFERENCES

Bhumralkar, C.M., R.L. Mancuso, F.L. Ludwig, and D.S. Renne, 1980, A practical and economical method for estimating wind characteristics at potential wind energy conversion sites, Solar Energy, 25:55-65.

Ludwig, F.L., 1982, Effect of a change of atmospheric stability on the growth rate of puffs used in plume simulation models, J. Appl. Meteorol., 21 (in press).

Ludwig, F.L., and G. Byrd, 1980, An efficient method for deriving mass-consistent flow fields from wind observations in rough terrain, Atmos. Environ., 14:585-587

Ludwig, F.L., L.S. Gasiorek, and R.E. Ruff, 1977, Simplification of a Gaussian puff model for real-time minicomputer use, Atmos. Environ., 11:431-436

Sherman, C.E., 1978, A mass-consistent model for wind fields over complex terrain, J. Appl. Meteorol., 17:312-319

DISCUSSION

E. FANAKI   This is more for clarification than a question.
1. Do u,v represent same notation as in meteorology ?
2. What do you mean by mixing depth class.

F. LUDWIG   1. Yes, we use a 3-dimensional wind field and u and v are the two horizontal components of the wind; w, the vertical component, is obtained from mass-consistency considerations and the shape of the terrain.

2. This has been introduced to make the problem tractable because the wind field solutions vary according to the height of the top layer. These heights are broken into categories to reduce the number of solution sets that have to be considered.

ON DISPERSION IN THE CONVECTIVE BOUNDARY LAYER

A. Venkatram

Environmental Research & Technology, Inc.
696 Virginia Road
Concord, MA 01742

ABSTRACT

By assuming that the dispersion time scale in the convective boundary layer is "infinite," we have derived a simple relationship between the ground-level concentration and the probability density function of vertical velocities at source height. This formulation yields results that compare favorably with the water-tank experiments of Willis and Deardorff (1978).

INTRODUCTION

Recent studies (Lamb 1981, Willis and Deardorff 1978) show that dispersion in the convective boundary layer is strongly influenced by the relative longevity of convective downdrafts and updrafts. The majority of particles released in downdrafts travel continuously downward until they reach the vicinity of the ground. Numerical experiments (Lamb 1981) indicate that to a useful degree of approximation, the velocity of these particles can be taken to be constant at the value at the release point.

Weil and Furth (1981) have exploited this approximation in constructing a simple Monte Carlo model. They find that the use of a plausible probability density function for vertical velocities yields results very similar to those obtained by Lamb (1978) from an elaborate numerical model.

In this paper, we derive results that go beyond those discussed by Weil and Furth (1981). We show how the assumption of

an "infinite" Lagrangian time scale allows us to formulate an
explicit relationship between ground-level concentration and the
vertical velocity probability density function at source height.
With recently available (Lamb 1981) information on vertical
velocity frequencies, this expression is used to compute
ground-level concentrations that compare well with laboratory
experimental results (Willis and Deardorff 1978).

The analysis in this paper is similar to that proposed by
Misra (1982). However, the reader will notice that our assumptions regarding the dispersion time scale and the probability
density function of the vertical velocities are quite different
from his.

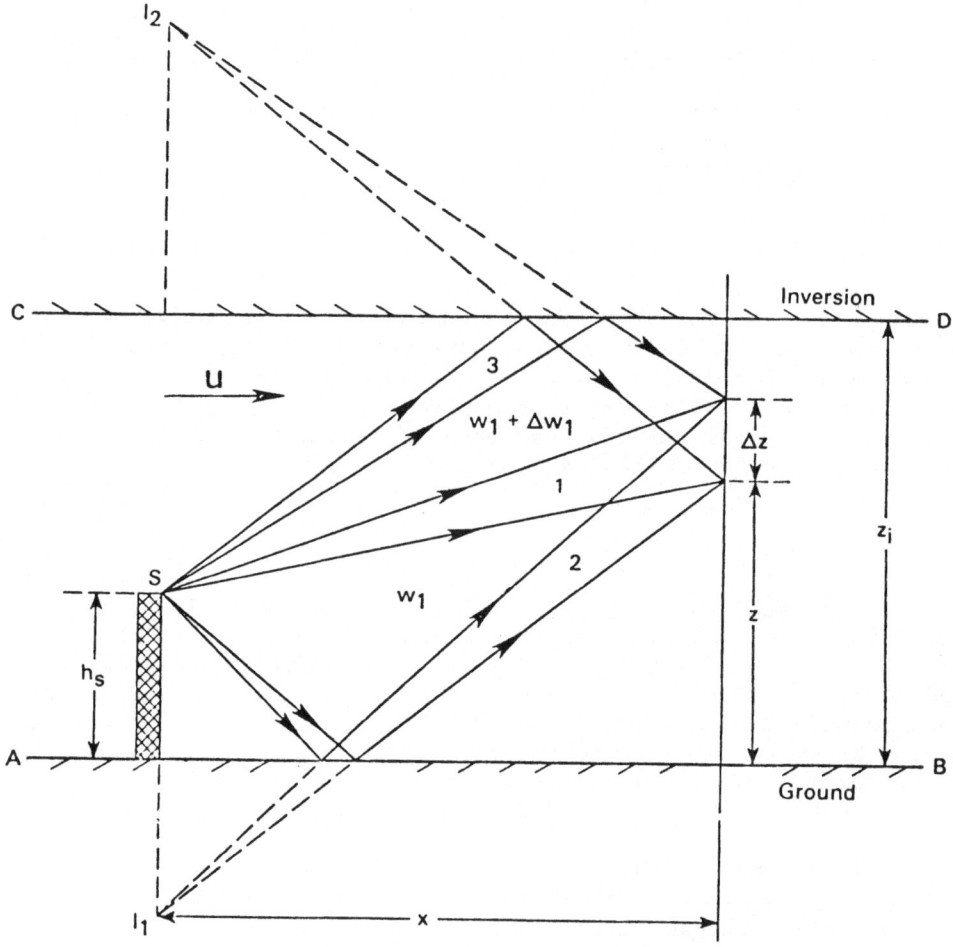

Figure 1   Geometry of Dispersion Along Straight Lines

The Model

The main assumption of the following analysis is that particles remember their velocities at release for an infinite time. This implies that particles travel along straight lines with velocities they acquire at the release point. Figure 1 shows the geometry of this dispersion problem. For the time being, we will restrict our analysis to the cross-wind integrated concentration, which we denote by $\overline{C}y(x,z)$.

Particles released from the source S can pass through the element Δz at height z by an infinite number of paths. Path 1 is the most direct way particles reach Δz from S. Particles can also reach Δz by being reflected off the ground. This is shown as path 2. The process of reflection at the ground will depend on the rate at which particles are swept into convective updrafts. Rather than model this process, we will simply assume that particles are reflected off the ground at the same velocity as they hit it. This then allows us to replace path 2 by that from the image source I_1 to Δz. Similarly, we can replace path 3 by that from the image of S reflected about CD to Δz. The infinitely large number of paths from S to Δz can be represented by the infinite reflected images of S about AB and CD.

To calculate the concentration at z, let us consider the particles passing through Δz along path 1. To make progress, we will assume that the horizontal flow velocity is a constant u. This is equivalent to assuming that there is no along-wind dispersion. Then, the fraction of particles released from S traveling along path 1 is simply $QP(w_1|h_s)\Delta w_1$ where Q is the emission rate and $P(w|z)$ is the probability density of vertical velocity w at height z. It is seen from the figure that $w_1 = u(z-h_s)/x$ and $\Delta w_1 = u\,\Delta z/x$. Denote the concentration associated with path 1 by $\overline{C}^y_1(x,z)$. Then a mass balance on the element Δz results in:

$$\overline{C}^y_1\,\Delta z u = QP(w_1|h_s)u\,\Delta z/x \tag{1a}$$

$$\text{or}\quad \overline{C}^y_1 = QP(w_1|h_s)/x \tag{1b}$$

where

$$w_1 = u(z-h_s)/x \tag{1c}$$

Using similar reasoning, the concentration $\overline{C}y(x,z)$ can be written as:

$$\overline{C}^y(x,z) = Q \sum_{n=1}^{\infty} P(w_n|\,h_s)/x \tag{2}$$

where the subscript 'n' refers to the path. For our purposes, we
will consider only four paths, three of which are shown in the
figure. Path 4 refers to particles reaching Δz after being
reflected off the inversion CD and the ground AB. This can be
represented by the reflection of I_2 about AB. Then the expression
for $\overline{C}^y(x,z)$ becomes

$$\overline{C}^y(x,z) \simeq Q \sum_{n=1}^{4} P(w_n|h_s)/x \tag{3}$$

where

$$w_1 = u(z-h_s)/x \tag{4a}$$

$$w_2 = -u(z+h_s)/x \tag{4b}$$

$$w_3 = u(2z_i-h_s-z)/x \tag{4c}$$

$$w_4 = u(2z_i-h_s+z)/x \tag{4d}$$

The ground-level concentration can be written as

$$\overline{C}^y(x,o) = \frac{2\,Q}{x}\left[P(w_1|h_s) + P(w_3|h_s)\right] \tag{5}$$

where

$$w_1 = -uh_s/x \tag{6a}$$

and $$w_3 = u(2-h_s/z_i)/x \tag{6b}$$

For most practical applications, we are primarily interested
in ground-level concentrations. Therefore, our analysis will
concentrate on Equation 5. Note that the term $P(w_3|h_s)$ will
disappear if the dispersion is not limited by the inversion at the
top of the mixed layer.

Application of Equation 5

Equation 5 can be used to estimate the ground-level
concentration if $P(w|z)$ is known. The simplest assumption about
this probability density function is that it is normal:

$$P(w|z) = \frac{1}{\sqrt{2\pi}\sigma_w(z)}\exp\left[-\frac{w^2}{2\sigma_w^2}\right] \tag{7}$$

In Equation 7, the standard deviation of the vertical velocity σ_w is a function of z. To conserve mass the mean vertical velocity has to be taken to be zero. If we substitute (7) into (5), the expression for $\bar{C}^y$ reduces to the commonly used Gaussian equation

$$\bar{C}^y = \sqrt{\frac{2}{\pi}} \frac{Q}{u\sigma_z} \quad \exp\left[-\frac{h_s^2}{2\sigma_z^2}\right] \tag{8a}$$

where

$$\sigma_z = \sigma_w x/u \tag{8b}$$

(In writing Equation 8, we have neglected the term $P(w_3|h_s)$ for convenience.) This exercise suggests the intuitively obvious conclusion that the Gaussian/dispersion equation is appropriate if $P(w|z)$ is Gaussian.

The main reason that Gaussian formulations do not work is that $P(w|z)$ in the convective boundary is positively skewed. This property leads to the descent of the plume centerline when the release is elevated. Examples of $P(w|z)$ for $z = 0.24z_i$ and $0.49z_i$ are shown in Figure 2. These were derived by Lamb (1981) from vertical velocity data generated from Deardorff's (1973) simulations of the convective boundary layer. Note that the mode for both probability density functions is close to $-0.5w_*$. Lamb (1981) shows that the mode moves to $-w_*/4$ when $z = 0.75z_i$ and approaches zero for $z \simeq z_i$.

To use this information on $P(w|z)$ to estimate ground-level concentrations, let us recast Equation 5 in terms of the nondimensional variables $X = w_* x/z_i u$ and $\bar{C}^Y = \bar{C}^y z_i u/Q$

$$\bar{C}^Y = \frac{2}{X}\left[P(\bar{w}_1|\bar{h}_s) + P(\bar{w}_3|\bar{h}_s)\right] \tag{9}$$

where

$$\bar{h}_s = h_s/z_i \tag{10a}$$

$$\bar{w}_1 = -\bar{h}_s/X \tag{10b}$$

$$\bar{w}_3 = (2-\bar{h}_s)/X \tag{10c}$$

Table 1 presents estimates of $\bar{C}^Y$ from Equation 9 for $h_s/z_i = 0.24$. The observations of the last column are those made by Willis and Deardorff (1978) in their water-tank experiments. It is seen that the correspondence between model estimates and measurements is reasonable. Although the predicted maximum

compares well with the observed, the predictions increase more
rapidly than the observations close to X = 0. Note that particles
reflected off the top of the boundary layer do not affect the
ground-level concentration until X = 1.0. Furthermore, the
maximum concentration around X ≃ 0.4 is determined almost
entirely by particles caught in downdrafts.

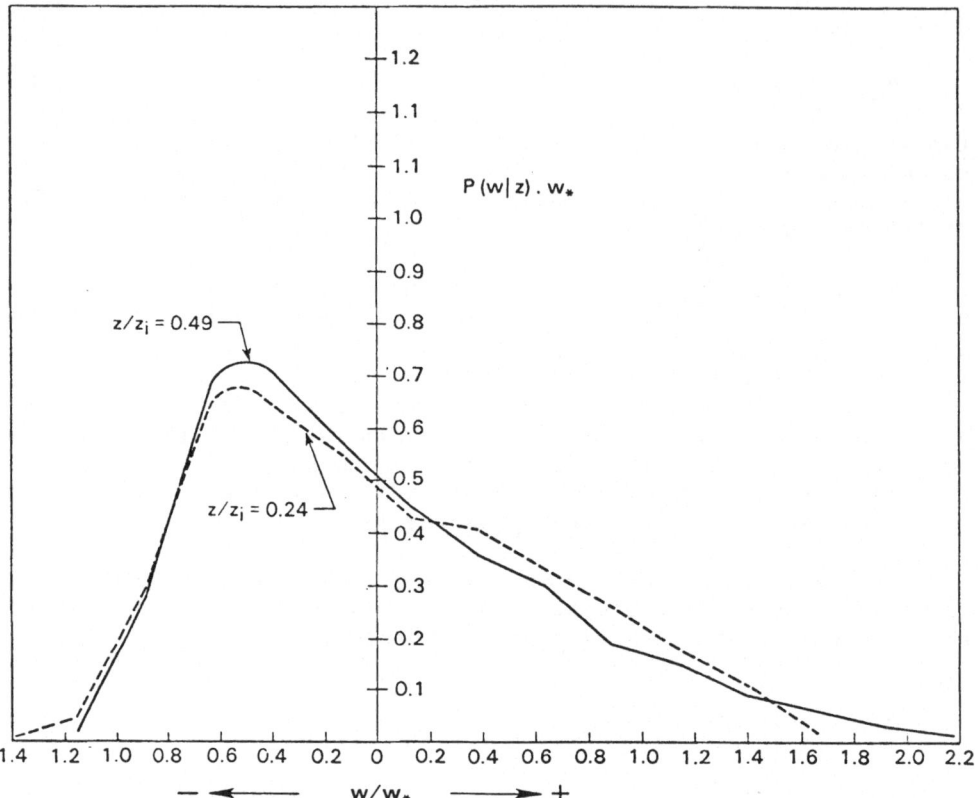

Figure 2   Probability Density of Vertical Velocity at Two Levels
in the Convective Boundary Layer (Derived by Lamb from
Deardorff 1973)

Table 2 presents similar calculations for h_s/z_i = 0.49.
Model predictions compare very favorably with measurements. Note
that the maximum concentration $\overline{C}^Y$ around X ≃ 1.0 is affected
by particles reflected off the top of the mixed layer.

Table 1. Predictions of $\overline{C}^Y$ for $h_s/z_i = 0.24$

X	$\overline{w}_1$	$\overline{w}_3$	$P(\overline{w}_1) \cdot w_*$	$P(\overline{w}_3) \cdot w_*$	$\overline{C}^Y$	$\overline{C}^Y$ (Observed)
0.2	−1.20	8.80	0.04	0	0.4	0.4
0.3	−0.80	5.90	0.42	0	2.8	1.6
0.4	−0.60	4.40	0.66	0	3.3	2.8
0.54	−0.44	3.26	0.66	0	2.4	3.0
0.6	−0.40	2.93	0.65	0	2.2	2.8
0.7	−0.34	2.51	0.62	0	1.8	2.4
0.8	−0.30	2.20	0.61	0	1.5	2.0
1.0	−0.24	1.76	0.59	0.05	1.3	1.5
1.5	−0.16	1.17	0.56	0.17	1.0	0.9

Table 2. Predictions of $\overline{C}^Y$ for $h_s/z_i = 0.49$

X	$\overline{w}_1$	$\overline{w}_3$	$P(\overline{w}_1) \cdot w_*$	$P(\overline{w}_3) \cdot w_*$	$\overline{C}^Y$	$\overline{C}^Y$ (Observed)
0.5	−1.00	3.00	0.17	0	0.68	0.73
0.6	−0.83	2.50	0.38	0	1.27	1.22
0.7	−0.71	2.14	0.57	0.01	1.66	1.60
0.8	−0.63	1.88	0.69	0.03	1.80	1.80
0.9	−0.56	1.67	0.72	0.06	1.73	1.82
1.0	−0.50	1.50	0.72	0.08	1.60	1.75
1.5	−0.33	1.00	0.67	0.17	1.12	1.22
2.0	−0.25	0.75	0.63	0.25	0.88	1.00

The probability density function $P(w|z)$ computed by Lamb neglects subgrid-scale energy in Deardorff's numerical model. This subgrid contribution is around 20% of the total vertical velocity variance between $\overline{h}_s = 0.25$ to 0.5 (Deardorff 1973). This suggests that the disagreement between observations and model predictions is expected. As the subgrid energy is expected to become Gaussian or even negatively skewed as z approaches zero, it would tend to shift the mode of $P(w|z)$ (without subgrid energy) towards $w = 0$ as $z \to 0.0$. This could explain why the predicted concentration maximum for $\bar{h}_s = 0.24$ occurs closer to X = 0 than it should. This effect should be less important for $\bar{h}_s = 0.49$. The other possible explanation for the unsatisfactory predictions for $\bar{h}_s = 0.24$ is that the assumption of an "infinite" dispersion time scale is poor for low release heights.

The Position of the Maximum Concentration

The formulation of the concentration distribution in terms of the probability density function of vertical velocities allows us to investigate some special features of dispersion in the convective boundary layer. Close to the source, the expression for the concentration $\bar{C}^y(x,z)$ can be written as

$$\bar{C}^y(x,z) = \frac{Q}{x} P(w_1|h_s) \qquad \text{(neglecting all the reflection terms)} \qquad (11a)$$

where

$$w_1 = -u(h_s-z)/x \qquad\qquad (11b)$$

The position of the centerline concentration is given by that of the maximum concentration at a fixed x. Therefore, the locus of the centerline concentration is given by:

$$\frac{\partial \bar{C}^Y}{\partial z} = \frac{dP}{dw_1} \cdot \frac{\partial w_1}{\partial z} = 0 \qquad\qquad (12a)$$

$$\text{or} \quad dP/dw_1 = 0 \qquad\qquad (12b)$$

By definition, the mode of $P(w|z)$ satisfies Equation 12b. Therefore, the motion of the centerline is given by:

$$z = h_s + w_m x/u \qquad\qquad (13)$$

Since the mode w_m is negative, the centerline of an elevated plume descends.

When $h_s/z_i < 0.5$, the ground-level concentration $C^y(z,0)$ is approximately given by

$$\bar{C}^y = \frac{2 Q}{x} P(w|h_s) \qquad \text{(reflection off } z_i \text{ is neglected)} \qquad (14a)$$

where

$$w = -uh_s/x \qquad\qquad (14b)$$

The ground reflection term is included in Equation 14a. The position of the maximum concentration is given by the solution of

$$\frac{\partial \bar{c}^y}{\partial x} = 0.0 \tag{15a}$$

or $\quad 1 = \frac{-1}{P} \frac{dP}{dw} w \tag{15b}$

To make further progress, we notice that the solution of Equation 15b yields the position of $\bar{C}^y_{max}$ in the region where $P(w|h_s)$ is an increasing function of w. From Figure 2, we see that $P(w|h_s)$ can be approximated by

$$P(w|h_s) = m(h_s)[w + w_o] \tag{16}$$

when $-w < w_m$. Note that $P(w|h_s) = 0$ when $w = -w_o$. Substituting (16) into (15b), we find that the maximum concentration corresponds to the vertical velocity w:

$$w = -w_o/2 \tag{17a}$$

or $\quad x_{max} = 2uh_s/w_o \tag{17b}$

or $\quad X_{max} = 2(h_s/z_i)/(w_o/w*) \tag{17c}$

Figure 2 indicates that $w_o/w* \simeq 1.2$ so that $X_{max} \simeq 1.7 \ (h_s/z_i)$. When $h_s/z_i = 0.24$, $X_{max} = 0.4$. The maximum $\bar{C}y$ is given by

$$\bar{c}^y_{max} = \frac{Q}{uh_s} P(-w_o/2|h_s)w_o \tag{18}$$

With nondimensional variables, Equation 18 becomes

$$\bar{c}^Y_{max} = \frac{2}{X_{max}} P(-w_o/2w_* \mid \bar{h}_s). w_* \tag{19}$$

Using $w_o/w_* \simeq 1.2$ we find $\bar{C}^Y_{max} = 3.3$ for $\bar{h}_s = 0.24$. Similar calculations for $h_s/z_i = 0.49$ yield $X_{max} = 0.82$ and $\bar{C}^Y_{max} = 1.71$, which agree better with experiments than those corresponding to $\bar{h}_s = 0.24$. This exercise indicates that information on the gross features of $P(w|z)$ allows us to make useful estimates of $\bar{C}^Y_{max}$. Note that $\bar{C}^Y_{max}$ will correspond to w_m rather than $-w_o/2$ if $w_m < -w_o/2$.

The preceding calculations emphasize the relationship between maximum concentrations and $P(w|z)$. Since the form of $P(w|z)$

varies with z, it is necessary to have information on this variation in order to make reliable estimates of $\overline{C}^y_{max}$.

Concluding Remarks

The long dispersion time scales of the convective boundary
layer allow us to relate concentrations to the probability density
function of vertical velocities at source height. This relationship yields concentration estimates that compare favorably with
the results of the water tank experiments conducted by Willis and
Deardorff (1978).

By expressing concentrations in terms of $P(w|z)$ we can under-
stand important features of dispersion in the convective boundary
layer. For example, we find that the observed descent of the
centerline of an elevated plume is a result of the mode of $P(w|z)$
being negative for $h_s/z_i > 0.1$. Our analysis also indicates that
the location of the maximum ground-level concentration can be
estimated with information on the gross features of $P(w|z)$.

REFERENCES

Deardorff, J.W. 1973. Three-dimensional numerical study of
     turbulence. Boundary Layer Meteorology 1: 169-196.
Lamb, R.G. 1981. Diffusion in the convective boundary layer. In:
     A Short Course on Atmospheric Turbulence and Air Pollution
     Modelling, 21-25 September, 1981, The Hague, The
     Netherlands. Proceedings available from D. Reidel Publishing
     Company.
Misra, P.K. 1982. Dispersion of non-buoyant particles inside a
     convective boundary layer. Atmospheric Environment
     16: 239-244.
Weil, J.C. and W.F. Furth 1981. A simplified numerical model of
     dispersion from elevated sources in the convective boundary
     layer. Fifth Symposium on Turbulence, Diffusion and Air
     Pollution, March 9-13, 1981, Atlanta, GA, American
     Meteorological Society, Boston.
Willis, G.E. and J.W. Deardorff 1978. A laboratory study of
     dispersion from an elevated source within a modeled
     convective planetary boundary layer. Atmospheric Environment
     12: 1305-1311.

DISCUSSION

E. RUNCA   In what way is your model different from other models of this type ?

A. VENKATRAM   It is different in the following ways :
1. The solution to the dispersion problem is analytical. No monte-carlo numerical technique is used.
2. The p.d.f. used for w corresponds to observations.
3. The analytical solution allows for the examination
   of several useful qualitative features of dispersion in the convective boundary layer.

F.B. SMITH   I see some inconsistency in your
bouncing ball concept - in the use of a single probability density function to estimate the effect of particles reflected from the top of the boundary layer.

A. VENKATRAM   I do not think so. I use the
positive axis of the p.d.f. for w to calculate the
concentrations caused by reflection off the top of
the mixed layer.

A NEW MONTE-CARLO SCHEME FOR SIMULATING LAGRANGIAN PARTICLE DIFFUSION WITH WIND SHEAR EFFECTS

Paolo Zannetti (x)

Kuwait Institute for Scientific Research
Environmental Sciences Department
P.O. Box 24885, Safat, Kuwait

INTRODUCTION

Simulation modeling is a problem of numerical discretization of a physical system. Such discretization, performed through computer experiments, is particularly important in those cases where physical theories need to be investigated or verified but laboratory experiments are unable to reproduce the complexities of the real world (e.g. stellar evolution).

In the last few decades, under the strong influence of a huge development of computational capabilities, discretization methods have been a major subject of investigation and development. Four major computational techniques have been developed and applied up to now :

- finite difference methods (Richtmyer and Morton, 1967)
- finite element methods (Strang and Fix, 1973)
- boundary element methods (Brebbia, 1978)
- particle methods (Hockney and Eastwood, 1981)

The last technique probably seems today the most advanced available numerical algorithm and, even more important, the most promising tool for numerical simulations with future generation of computer systems.

(x) On leave of absence from AeroVironment, Inc., 145 Vista Ave., Pasadena, CA 91107, U.S.A.

Using particle models, the temporal evolution of a physical
system is described by the dynamics of a finite number of interacting particles. Therefore, such models are typically Lagrangian
ones, while, for example, finite difference methods are purely
Eulerian representations of physical systems.

Three main types of particle models can be defined (Hockney
and Eastwood, 1981) :

- particle-particle (PP) models, where all interaction forces
  between particles are computed at each time step
- particle-mesh (PM) models, where forces are computed using
  a field equation (on a grid) for the potential
- PP-PM (or P^3M) models, a hybrid approach where interparticle
  forces are splitted into a short range component (computed
  using the PP method) and a slowly varying one (represented
  in mesh system by the PM method).

Particle models can be purely deterministic or (partially)
based on statistical methods. In the first case the simulation
of particle time evolution is unique. In the second case, Monte
Carlo techniques are used to produce semi-random "perturbations",
and, therefore, model outputs represent just a realization from
an infinite set of possible solutions.

Length and time scales (as in all discretization systems) play
an important role in particle models. In particular, the relation
between the actual physical particles (or elements) and the computer
model simulation particles is an important factor for the interpretation of the simulation results. In general, three possible
cases can be found (Hockney and Eastwood, 1981) :

- a one-to-one correspondence between actual and simulated
  particles, as, for example, in molecular dynamics simulation
- a description of fluid elements (position, vorticity) as
  particles, as, for example, in vortex fluid simulations,
  where the correspondence to physical particles (molecules)
  is totally lost
- the use of "superparticles", i.e., simulation particles
  representing a cloud of physical particles having similar
  characteristics.

Particle models have been mostly applied for simulating (and
understanding) the spiral structure of the galaxies, for plasma
dynamics simulation and for obtaining realistic representations of
turbulence in fluid. Air pollution dispersion by particle methods
is at its infancy, even though interesting studies have been
published in the last few years (e.g., Watson and Barr, 1976;
Hanna, 1979a; Lamb, et al. 1979a; Lange, 1978; Patterson et al.,
1981).

This paper, after a description of the discretization problems related to air pollution diffusion simulation by particles, presents a new Monte-Carlo scheme especially designed for handling wind shear effects in the atmospheric boundary layer. Finally, a brief discussion is provided concerning the suitability of new advanced meteorological instrumentation (e.g., the Doppler acoustic sounder) in providing data input for air pollution particle simulation models.

AIR POLLUTION SIMULATION BY LAGRANGIAN PARTICLES

"The diffusion of a substance released into a turbulent flow cannot be described by any model or theory" (Hunt, 1981). A short look at all available theories and numerical techniques can easily convince us that the above statement is basically true :

- Gaussian models strongly depend on sigma parameters generally computed through a questionable discretization (stability classes) of atmospheric turbulence status, or by using semi-empirical formulas.
- Grid models have a high operating cost, and moreover, are affected by numerical problems (instabilities, artificial numerical diffusion errors) and physical limitations (e.g., the K-theory is not appropriate for describing large turbulent eddies).
- Particle models are still under development and require meteorological Lagrangian measurements which are generally unavailable, and, however, extremely difficult to make.

Nevertheless, Lagrangian particle methods are very appealing. Emitted gaseous material is characterized by particles and each particle is "moved" at each time step by pseudo-velocities, which take into account the three basic dispersion components : 1) the transport due to the mean fluid velocity, 2) the (seemingly) random turbulent fluctuations of wind components (both horizontal and vertical), and 3) the molecular diffusion (if not negligible). With existing computers (the next generation will be even better), enough particles can be stored in memory (say, more than a few thousand) to accurately describe the characteristics of a single plume, or, better, an instantaneous puff release. Particle resolution plays, in fact, an important role. A one-hour simulation of an industrial emission of 1 kg/s of SO_2 from a stack emitting 20 m^3/s of gases will require, for example, 3600 "particles", each representing, in reality, a growing puff containing 1 kg of SO_2 and having an initial size of 20 m^3. The behaviour of this "superparticle" is certainly different from that of a single SO_2 molecule. (x)

(x) Puff Modeling (Zannetti, 1981a) is a powerful Lagrangian technique for treating transport and diffusion of such superparticles.

Nevertheless, we must keep in mind that, in air pollution modeling, we do not need to follow precisely each molecule in the atmospheric turbulent flow, but just to define an algorithm for particle displacement computation which gives an accurate particle density distribution. In mathematical notation, if a particle is located in $\underset{\sim}{x}(t_1)$ at t_1, its position at t_2 will be

$$\underset{\sim}{x}(t_2) = \underset{\sim}{x}(t_1) + \int_{t_1}^{t_2} \underset{\sim}{u}\left[\underset{\sim}{x}(t), t\right] dt \tag{1}$$

where $\underset{\sim}{u}$ is the "instantaneous" wind vector in each point $\underset{\sim}{x}(t)$ of the particle trajectory between t_1 and t_2.

Atmospheric turbulent properties make $\underset{\sim}{u}$ practically impossible to know, especially due to its semi-random components caused by atmospheric eddies. But the "equivalent" wind vector $\underset{\sim}{u}_e$ can be considered

$$\underset{\sim}{u}_e = \int_{t_1}^{t_2} \underset{\sim}{u}\left[\underset{\sim}{x}(t), t\right] dt/(t_2 - t_1) \tag{2}$$

which moves the particle directly from $\underset{\sim}{x}(t_1)$ to $\underset{\sim}{x}(t_2)$ in the interval (t_1, t_2). The problem is then to estimate $\underset{\sim}{u}_e$ from $\underset{\sim}{u}$ measurements, keeping in mind that $\underset{\sim}{u}_e$ must approximate the integral term in Eq. (2) only on a particle <u>ensemble</u> basis. For example, we can define

$$\underset{\sim}{u}_e = \bar{\underset{\sim}{u}}_e + \underset{\sim}{u}'_e \tag{3}$$

where $\bar{\underset{\sim}{u}}_e$ is our best estimate of the average Eulerian wind vector (transport) at $\underset{\sim}{x}(t_1)$, and $\underset{\sim}{u}'_e$ is a "diffusivity velocity". In other words, $\bar{\underset{\sim}{u}}_e$ (a smoothly variable term) represents our deterministic understanding of the average transport process, based on Eulerian wind measurement interpolation or a meteorological model output, while $\underset{\sim}{u}'_e$ is a single artificial numerical perturbation.

Since, in Eq. (3) $\bar{\underset{\sim}{u}}_e$ is supposed to be known, computing $\underset{\sim}{u}'_e$ is the key problem of Lagrangian particle modeling. Two fundamental approaches can be followed : the deterministic and the statistical ones.

<u>The deterministic computation of $\underset{\sim}{u}'_e$</u>

A typical example of the deterministic approach is given by the particle-in-cell method of Lange (1978), where, after some manipulation of the K-theory diffusion equation, we obtain

$$\underset{\sim}{u}'_e = (-\frac{K}{C}) \nabla C \tag{4}$$

where K is the usual eddy diffusion coefficient and C the concentration, computed as the number of particles in the cell containing $x(t_1)$. This method requires partitioning the computational domain into cells and it is able to duplicate K-theory dispersion with the important feature of removing the numerical advection errors associated with finite-difference solutions.

Using this method, the motion of a single particle will be affected by the time-varying concentration field, i.e., by the positions of the other particles (PM model).

The statistical computation of u'_e

The statistical approach (Monte Carlo-type models) certainly seems to be more flexible and appealing. According to the statistical approach, u'_e is a semi-random component computed by manipulating computer-generated random numbers. To perform this computation, it has been generally assumed that Eulerian measurements of u can provide statistical information on u'_e. However, these two parameters are not the same and further investigation is required to fully assess this point.

As a first approximation, however, we can accept the above assumption and use, for the diffusivity velocity u'_e a statistical generation scheme based on our understanding (and Eulerian measurements) of u. In particular, Hanna (1979b) has shown that it is a plausible assumption to describe both Eulerian and Lagrangian wind vector fluctuations by a simple Markov process (autocorrelation process of the first order).

If we extend this assumption to u'_e, we have (x)

$$u'_e(t_2) = R_e(t_2-t_1) \, u'_e(t_1) + u''_e(t_2) \tag{5}$$

where $R_e(t_2-t_1)$ contains the autocorrelations with lag $\Delta t = t_2-t_1$ of the u'_e components, and u''_e is a purely random vector.

Equation (5) is the key formula for statistically computing u'_e which will simply be a recursive sum of two terms---the first function of "previous" u'_e of the same particle, and the second purely randomly generated. Since Eq. (5) will be computed independently for each particle, two eventually coincident particles at t_1 will have, in general, different displacements, even if their past "history" is the same. Using this approach the motion of a particle is not affected by the position of the other particles and, therefore, this numerical algorithm is extremely fast since no interacting forces need to be computed.

(x) In this formula (and the following ones) each component of the vector on the left side is computed using only the corresponding component of each vector in the right side (scalar computations).

To apply Eq. (5) we need the initial $\underset{\sim e}{u}'(t_0)$ for each particle at its generation time t_0 (often assumed to be a zero vector) and the computation of $\underset{\sim e}{R}$ and $\underset{\sim e}{u}''$.

Due to the Lagrangian nature of $\underset{\sim e}{u}$, $\underset{\sim e}{R}$ has been often identified with $\underset{\sim L}{R}$, the autocorrelations of the Lagrangian wind vector $\underset{\sim L}{u}$. $\underset{\sim L}{R}$ can be related to Lagrangian turbulence time scales, for example, by

$$\underset{\sim L}{R} = \exp\left[-(t_2-t_1)/\underset{\sim L}{T}\right] \tag{6}$$

where $\underset{\sim L}{T}$ contains the two horizontal and the one vertical Lagrangian time scales. Generally, Lagrangian measurements of $\underset{\sim L}{T}$ (or $\underset{\sim L}{R}$ directly) are not available, but empirical relations have been proposed (e.g., Hanna, 1981) to estimate $\underset{\sim L}{T}$ from Eulerian measurement.

Assuming $\underset{\sim e}{u}''$ a purely random vector with zero-mean, normally-distributed independent components, we have that $\underset{\sim e}{u}''$ is completely characterized by $\underset{\sim u''}{\sigma}$, i.e. the standard deviations of its components. In this case, taking the variances of Eq. (5), we obtain

$$\underset{\sim u''}{\sigma} = \underset{\sim u'}{\sigma}\left[1 - \underset{\sim e}{R}^2(t_2-t_1)\right]^{1/2} \tag{7}$$

requiring the knowledge of $\underset{\sim u'}{\sigma}$, the standard deviations of $\underset{\sim e}{u}'$ which, again, can be approximated by the standard deviations of available Eulerian wind measurements.

From the standard deviations $\underset{\sim u''}{\sigma}$ of Eq. (7) it is easy, using commonly available computer programs, to generate each particle's $\underset{\sim e}{u}''$ term to be used in Eq. (5).

$\underset{\sim e}{R}$ and $\underset{\sim u''}{\sigma}$ are, in general, time-dependent (but constant between t_1 and t_2) and space-dependent. Therefore, they can fully utilize a three-dimensional meteorological input (Eulerian values) and can, at least theoretically, simulate extremely complex atmospheric diffusion conditions, otherwise impossible to treat with other numerical schemes.

Conclusion

Dispersion simulation by Lagrangian particles has been called "natural" modeling. These models do not need inputs of artificial stability classes, empirical sigma curves, or diffusion coefficients that are practically impossible to measure. Instead, diffusion characteristics are simulated by attributing a certain degree of "fluctuation" to each particle, using, for example, the computer's capability to generate semi-random numbers.

The basic advantages of this approach (e.g., see Lamb et al., 1979a; Lange, 1978) are :

- Compared with grid models, this method avoids the artificial initial diffusion of a point source in the corresponding cell and the advection numerical errors.
- This method is practically free of restricting physical assumptions, since all uncertainties are combined into the correct determination of pseudo-velocities.
- Each particle can be tagged with its coordinates, source indicator, mass, activity, species and size, allowing computation of wet and dry deposition, decay, and particle size distribution.
- If chemistry is required, a grid can be superimposed and concentrations in each cell computed by counting the particles of each species, allowing the use of any reaction scheme at each time step (x).
- The meteorological input required should be very close to actual measured data. The primary information needed seems to be (Lamb et al., 1979a) the variance of wind velocity fluctuations and the Lagrangian autocorrelation function, which can be related to Eulerian measurements (e.g., Hanna, 1981).

Potentially, the method is superior in both numerical accuracy and physical representativeness. However, much research is still needed to extract, from meteorological measurements and our limited theoretical understanding of turbulence processes, the meteorological input required to run this model (i.e., the pseudo-velocities to move each particle at each time step).

THE TREATMENT OF WIND SHEAR EFFECTS

Shear flow effects in the atmospheric boundary layer are characterized by the following three factors :

- Vertical variation of the average wind vector (changes in both speed and direction).

(x) A rigorous concentration computation should not just add up the number of particles in a given cell at a given time. In fact, concentrations should be computed using the total time spent by each particle in the receptor volume during each time step (as in Lamb et al., 1979b). Moreover, nonlinear chemistry computations (at least for fast reactions) should take into account the effects on the reaction rates of concentration turbulent fluctuations, which seems extremely complicated using Lagrangian methods.

  - Vertical variation of the intensity of wind fluctuations
    (especially vertical fluctuations).
  - Zero correlation between the along wind and the cross-wind
    fluctuations.
  - Negative correlation between the along wind and the vertical
    wind fluctuations.

According to the above considerations, derived from the analysis of Eulerian wind measurements, we can expect that a correct computation of u_e in Eq. (3) will incorporate the first factor in $\bar{u}_e$ and the other three in u'_e. Therefore, a need exists to generate cross correlated random velocity components for u'_e, and not just independent components as in Eq. (5). To this end, a new Monte-Carlo method has been anticipated in Zannetti (1981b) which is fully discussed below.

Let us consider a special reference system (SRS) where the x axis, for each specified altitude z, is chosen to coincide with the horizontal component of $\bar{u}_e$. Then, with the notation

$$u'_e(t) = \left[u'(t), \ v'(t), \ w'(t) \right] \tag{8}$$

$$u''_e(t) = \left[u''(t), \ v''(t), \ w''(t) \right] \tag{9}$$

the following scheme can be used for u'_e

$$u'(t_2) = \phi_1 \ u'(t_1) + u''(t_2) \tag{10}$$

$$v'(t_2) = \phi_2 \ v'(t_1) + v''(t_2) \tag{11}$$

$$w'(t_2) = \phi_3 \ w'(t_1) + \phi_4 \ u'(t_2) + w''(t_2) \tag{12}$$

which is similar to Eq. (5), but with the new Eq. (12) for handling the correlation $r_{u'w'}$ between u' and w'.

If we multiply Eq. (10) by $u'(t_1)$ and then take the average <> we obtain (u" is a purely random component)

$$\phi_1 = \frac{\langle u'(t_1) \ u'(t_2)\rangle}{\langle u'^2 (t_1)\rangle} = \frac{\langle u'(t_1) \ u'(t_2)\rangle}{\sigma^2_{u'}} = r_{u'}(\Delta t) \tag{13}$$

where $r_{u'}(\Delta t)$ is the autocorrelation of u' with time lag $\Delta t = t_2 - t_1$.

In an analogous way, working on Eq. (11), we obtain

$$\phi_2 = \frac{<v'(t_1)\, v'(t_2)>}{\sigma_{v'}^2} = r_{v'}(\Delta t) \tag{14}$$

where r_{v}' is the autocorrelation of v'.

Multiplying Eq. (12) by $w'(t_1)$ and taking the average $<>$ we have

$$r_{w'}(\Delta t)\, \sigma_{w'}^2 = \phi_3\, \sigma_{w'}^2 + \phi_1\, \phi_4\, r_{u'w'}(0)\, \sigma_{u'}\, \sigma_{w'} \tag{15}$$

where $r_{w'}$ is the autocorrelation of w' and $r_{u'w'}(0)$ is the cross correlation (with no time lag) between u' and w'.

Multiplying the corresponding members of Eqs. (10) and (12) and taking the average $<>$ we have

$$r_{u'w'}(0)\, \sigma_{u'}\, \sigma_{w'} = \phi_1\, \phi_3\, r_{u'w'}(0)\, \sigma_{u'}\, \sigma_{w'} + \phi_4\, \sigma_{u'}^2 \tag{16}$$

Eqs. (15) and (16) represent a system of two equations in two unknowns, which gives the solution

$$\phi_3 = \frac{r_{w'}(\Delta t) - \phi_1\, r_{u'w'}^2(0)}{1 - \phi_1^2\, r_{u'w'}^2(0)} \tag{17}$$

$$\phi_4 = \frac{r_{u'w'}(0)\, \sigma_{w'}\, \left[1 - \phi_1 r_{w'}(\Delta t)\right]}{\sigma_{u'}\, \left[1 - \phi_1^2\, r_{u'w'}^2(0)\right]} \tag{18}$$

Finally, taking the variances of Eqs. (10) – (12) we obtain the variances of the purely random fluctuations to be generated at each time step

$$\sigma_{u''}^2 = \sigma_{u'}^2\, (1 - \phi_1^2) \tag{19}$$

$$\sigma_{v''}^2 = \sigma_{v'}^2\, (1 - \phi_2^2) \tag{20}$$

$$\sigma_{w''}^2 = \sigma_{w'}^2(1 - \phi_3^2) - \phi_4^2\, \sigma_{u'}^2 - 2\, \phi_1\, \phi_3\, \phi_4\, r_{u'w'}(0)\sigma_{u'}\sigma_{w'} \tag{21}$$

Eqs. (10)-(12) can then be applied using Eqs. (13), (14), (17)-(21) if the proper meteorological input $r_{u'}$, $r_{v'}$, $r_{w'}$, $r_{u'w'}$, $\sigma_{u'}$, $\sigma_{v'}$, $\sigma_{w'}$ is available at each altitude.
In such case, where the scheme of Eqs. (10)-(12) is used instead of Eq. (5), $r_{u'w'}$ is the only additional input required.

The entire scheme above has been numerically tested and it has been verified its capability of correctly producing $\underset{\sim}{u}'_e$ components with any acceptable degree of autocorrelation $\underset{\sim}{R}_e$

$$\underset{\sim}{R}_e = (r_{u'}, r_{v'}, r_{w'}) \tag{22}$$

and cross correlation $r_{u'w'}$.

METEOROLOGICAL INPUT

As previously discussed, Monte-Carlo techniques require a suitable meteorological input. Meteorological tower and Doppler acoustic sounder instrumentations are the best sources of such data. They provide, at different altitudes and during a selected time interval (e.g., 20 minutes), the average wind vector and the standard deviations of the wind fluctuations with respect to a fixed orthogonal system (x,y,z). Therefore, in such coordinate system, at a certain altitude z, the average measured wind vector will be

$$\underset{\sim}{\bar{u}} = (\bar{u}_x, \bar{u}_y, \bar{u}_z) \tag{23}$$

and σ_{u_x}, σ_{u_y}, σ_{u_z} will be the measured standard deviations of the wind fluctuations. If we assume that such Eulerian measurements are a good representation of the statistical properties of $\underset{\sim}{u}_e$, then we can firstly say that

$$\underset{\sim}{\bar{u}}_e = \underset{\sim}{\bar{u}} \tag{24}$$

$$\sigma_{w'} = \sigma_{u_z} \tag{25}$$

However, as introduced in the previous chapter, we often need the horizontal wind fluctuations in a special reference system (SRS). Therefore, wind fluctuations in $(x,y,z,)$ must be projected into such system through a horizontal rotation of an angle θ

$$\theta = \text{arctg} \; \frac{\bar{u}_y}{\bar{u}_z} \tag{26}$$

Taking the variances of such projected wind fluctuations and remembering the important property that, in the SRS, the two horizontal wind fluctuations are non-correlated, we obtain

$$\sigma^2_{u'} = \frac{\bar{u}^2_x \sigma^2_{u_x} - \bar{u}^2_y \sigma^2_{u_y}}{\bar{u}^2_x - \bar{u}^2_y} \tag{27}$$

$$\sigma^2_{v'} = \frac{\bar{u}^2_x \sigma^2_{u_y} - \bar{u}^2_y \sigma^2_{u_x}}{\bar{u}^2_x - \bar{u}^2_y} \tag{28}$$

that together with Eq. (25) can be used in Eqs. (19)-(21) to provide the random fluctuations of each particle at each altitude.

In flat terrain conditions, Doppler or tower measurements can be horizontally extrapolated, therefore providing an appropriate, fully three-dimensional input for particle modelling using statistical Monte-Carlo techniques.

CONCLUSIONS AND POSSIBLE DEVELOPMENTS

Many aspects of the above described techniques need further investigation. As already discussed, a major problem is the utilization of available Eulerian measurements for inferring $\underset{\sim}{u}_e$ properties.

Moreover, the assumption of no correlation between the two horizontal components in the SRS, is acceptable only for small time intervals not affected by systematic horizontal wind direction meandering across the average value $\bar{\underset{\sim}{u}}_e$. If these conditions are not met, persistent, highly correlated, values between the two components are often measured, unless the meandering factor is preliminary removed from wind measurement data.

To solve all these problems, ad-hoc tracer experiments should be designed specifically for evaluating and validating particle diffusion theories.

It is true, nevertheless, that particle models can easily reproduce, under specific simplifying assumptions, the diffusion results obtainable by other modeling techniques, as the Gaussian model and the K-theory. Particle modeling "natural" approach to air pollution simulation seems, therefore, the most suitable technique for future air pollution modeling research and development.

REFERENCES

Brebbia, C.A., 1978, The Boundary Element Method for Engineers,
    Pentech Press, London.
Hanna, S.R., 1979a, A Statistical Diffusion Model for use with va-
    riable wind fields. 4th Symposium on Turbulence, Diffusion
    and Air Pollution, AMS, Jan. 15-18, Reno, N.V.
Hanna, S.R., 1979b, Some Statistics of Lagrangian and Eulerian wind
    fluctuations, J. of Appl. Meteor., 18, 518.
Hanna, S.R., 1981, Turbulent Energy and the Lagrangian time scales
    in the planetary boundary layer. 5th Symposium on Turbulence,
    Diffusion and Air Pollution, AMS, March 9-13, Atlanta, Georgia.
Hockney, R.W., and Eastwood, J.W., 1981, Computer Simulation Using
    Particles, McGraw-Hill, Inc.
Hunt, J.C.R., 1981, Diffusion in the stable boundary layer. A short
    course on Atmospheric Turbulence and Air Pollution Modelling
    Sept. 21-25, The Hague, Netherlands.
Lamb, R.G., Hogo, H., and Reid, L.E., 1979a, A Lagrangian Monte
    Carlo Model of air pollutant transport, diffusion and removal
    processes. 4th Symposium on Turbulence, Diffusion and Air
    Pollution, AMS, Jan. 15-18, Reno, N.V.
Lamb, R.G., Hogo, H. and Reid, L.E., 1979b, A Lagrangian Approach to
    Modeling Air Pollutant Dispersion-Development and Testing in
    the Vicinity of a Roadway, EPA-600/4-79-023 Research Report.
Lange, R., 1978, ADPIC-- A Three Dimensional Particle-in-cell Model
    for the Dispersal of Atmospheric Pollutants and its Comparison
    to Regional Tracer Studies. J. of Appl. Meteor., 17, 320.
Patterson, D.E., Husar, R.A., Wilson, W.E., and Smith, C.F., 1981,
    Monte Carlo Simulation of Daily Regional Sulfur Distribution:
    Comparison with SURE Sulfate Data and Visual Range Observations
    during August 1977. J. of Appl. Meteor., 20, 404.
Richtmyer, R.D., and Morton, K.W., 1967, Difference Methods for
    Initial-Value Problems, Interscience Publishers, John Wiley
    and Sons, Inc., New-York.
Strang, G., and Fix, G., 1973, An Analysis of the Finite Element
    Method, Prentice-Hall, Inc., Englewood Cliffs, N.J.
Watson, C.W., and Barr, S., 1976, Monte Carlo Simulation of the
    Turbulent Transport of Airborne Contaminants. Los Alamos
    Scientific Laboratory, Technical Report LA-6103.
Zannetti, P., 1981a, An Improved Puff Algorithm for Plume Dispersion
    Simulation, J. of Appl. Meteor., 20, 1203.
Zannetti, P., 1981b, Some Aspects of Monte Carlo Type Modeling
    of Atmospheric Turbulent Diffusion. 7th Conference on Proba-
    bility and Statistics in Atmospheric Sciences, AMS, Nov. 2-6,
    1981, Monterey, CA.

DISCUSSION

J. KNOX   1. How do you calculate concen-
           trations ?
                                               2. Do you satisfy standard cou-
           rant conditions for large Δt, how does model behave.

P. ZANNETTI   1. By counting the number of
           particles in volume Δv during time Δt. A more ri-
           gorous way is to calculate the total time spent by
           all particles in ΔV during Δt. This total time is
           proportional to the concentration.
                                               2. It does present problems if
           there are not enough particles for large Δt. Courant
           condition need not to be satisfied. We only need
           enough particles for a good resolution of the concen-
           tration field.

PRELIMINARY TESTS OF THE INTRODUCTION OF LAGRANGIAN DISPERSION

MODELLING IN A 3-D ATMOSPERIC MESO-SCALE NUMERICAL MODEL

Christian Blondin and Gérard Therry

Etablissement d'Etudes et de Recherches Météorologiques
73 - 77, Rue de Sèvres
92106 BOULOGNE-BILLANCOURT CEDEX

INTRODUCTION

One has often opposed Eulerian and Lagrangian numerical mo-
dellings as more or less irreconciliable technics to treat atmospheric pollution problems. In fact, the use and improvement of
both methods by specific scientific teams are strongly connected
to objective factors such as : computing ressources, available
routine data (concerning pollution and meteorology) scientific
skills.

Nevertheless, from a basic point of view, it appears clearly
that the transport and diffusion of atmospheric pollutants is a
Lagrangian problem, that can be solved efficiently by Lagrangian
numerical modelling, whereas an Eulerian formalism seems more
appropriate to take into account the physico-chemical aspects
(transformation of species, rain and wash - out, ...). (For a
review, see Eliassen 1980).

Since most of actual pollution problems (high concentration
episodes, accidental releases, ...) cannot or more exactly must
not, be investigated using simple analytic solutions (Gaussian
plume model, for example), and since turbulent diffusion is a
fundamental phenomenum to deal with, the current "state of the
art" of the turbulent planetary boundary layer meteorology and
the fact that pollution and meteorological measurements provide
information under Eulerian form induces one to choose Eulerian
technics as the more adaptedone in a wide range of applications,
in spite of the nature of the transport - diffusion problem.

The main obstacle to applicability of Lagrangian modelling
is the lack of experimental data concerning Lagrangian properties
in the turbulent planetary boundary layer.

Following for example, the basic contents of the famous
Taylor's formula, the key parameters for numerical computation of
pollutants diffusion are the turbulent wind components variances.
The practical use of these variances in Lagrangian schemes changes
from one author to the other (Mc Nider et alii, 1980 ; Runca et
alii, 1981), even if the main idea remains the same, namely :
independent particles move according to a velocity depending on
three factors : the mean wind speed, a fluctuating component
related to Reynold's tensor, and a purely random component, so
that the total variances of these particles velocity components
computed at each time step equal prescribed values.

The scope of this paper is to present a method of estimation
of wind components variances at the meso-scale (characteristic
length scale : 10 km) and to show examples of the influence of the
thermodynamic structure of the atmosphere upon its turbulent behaviour and pollutants diffusion capability.

A MODEL OF THE TURBULENCE STRUCTURE OF THE ATMOSPHERE

The basic principle is to model the spatio-temporal evolution
of the turbulent kinetic energy $\bar{e}$. One shall find in C. Blondin
and G. Therry (1981) a complete description of the way to close
the turbulence equations system in this framework. In this paper,
we mentioned that several questions were still under investigation
which we want to discuss now.

Turbulent energy flux

With usual assumptions and notations, the one dimensional
(refering to the vertical axis) version of the $\bar{e}$ equation is :

$$\frac{\partial \bar{e}}{\partial t} = -\frac{\partial (\overline{w'e})}{\partial z} + K\left\{ \left(\frac{\partial \bar{u}}{\partial z}\right) + \left(\frac{\partial \bar{v}}{\partial z}\right) - \alpha_t \beta \left(\frac{\partial \bar{\theta}}{\partial z} - \gamma\right) \right\} - C_\varepsilon \frac{\bar{e}^{3/2}}{1_\varepsilon}$$

with

$$K = C_k 1_k \bar{e}^{1/2}$$

1_k, 1_ε mixing lengths for diffusion and dissipation. A classical approach leads to the following expression for the turbulent
energy flux :

$$\overline{w'e} = - K_e \frac{\partial \bar{e}}{\partial z}$$

But it is now well-know from experimental data and numerical simulation (André 1980) that this flux is oriented upward throughout the whole depth of the planetary boundary layer (P.B.L.) in convective situation. The previous expression fails to reproduce this feature close to the ground where $\bar{e}$, positive quantity, decreases with height. According to the $\overline{w'e}$ rate equation, one can achieve the following diagnostic expression in stationary equilibrium condition :

$$\overline{w'e} = -\frac{1}{c_\varepsilon c_8} \cdot \frac{1\varepsilon}{\bar{e}^{1/2}} \cdot \left\{ 3 \, \bar{w}'^2 \, \frac{\partial \bar{e}}{\partial z} - \frac{3}{2} \, \beta \overline{w'^2\theta'} + \overline{u'w'} \cdot \frac{\partial \overline{u'w'}}{\partial z} + \overline{v'w'} \cdot \frac{\partial \overline{v'w'}}{\partial z} \right\}$$

which is a refined expression of the classical one since   may be written

$$\overline{w'e} = - Ke \, \frac{\partial \bar{e}}{\partial z} + \frac{\alpha e \, Ce \, Ck}{3} \cdot \frac{1\varepsilon}{\bar{e}^{1/2}} \cdot \left\{ \frac{3}{2} \, \beta \overline{w'^2\theta'} - \overline{u'w'} \cdot \frac{\partial \overline{u'w'}}{\partial z} - \overline{v'w'} \cdot \frac{\partial \overline{v'w'}}{\partial z} \right\}$$

The additionnal term helps to avoid the mentioned defect, mostly because the buoyancy production is positive in the lower part of a convective P.B.L.

Since $\overline{w'^2\theta'}$ is not available directly, it must be parameterized using the convective velocity $w_* = (\beta Q_0 \, Zi)^{1/2}$ (Deardorff, 1970), this final equation can be used :

$$\overline{w'e} = -\alpha_e \, C_k \left\{ 1_k C^{1/2} \, \frac{\partial \bar{e}}{\partial z} - 0.2 \, C_e \frac{1_\varepsilon}{\bar{e}^{1/2}} \, \beta w_* \, \overline{w'\theta'} \right\}$$

Mixing length problem

In C. Blondin and G. Therry (1981), we insisted on the fact that two mixing lengths are necessary to close the turbulence equations system, and we proposed expressions for them specially adapted for convective evolution simulations.

In fact, the behaviour of these lengths in stable stratification cases was poorly represented, in particular the systematic decrease at the upper stable part of a convective P.B.L.

In this respect, the guiding idea is that, in stable stratified layers, both turbulent diffusion and dissipation are due to small eddies, so that a determination by local parameters is to be expected, leading to :

$$1\alpha \quad (\bar{e} \, / \, \beta \, \frac{\overline{\partial\theta}}{\partial z})^{1/2}$$

Keeping in mind that the ratio l_k/l_ϵ is related to $\overline{w'^2/e}$, which is smaller in stable than in neutral stratification, the mixing lengths are computed according to :

$$\frac{1}{l} = \frac{1}{kz} + \frac{C_1}{zi} - \left(\frac{1}{kz} + \frac{C_2}{zi} \right) \cdot m_1 \cdot m_2 + \frac{C_5}{ls}$$

with

$$m_1 = \frac{1}{1 + C_3 \, zi/kz} \quad ; \quad m_2 = \begin{cases} 0 & \text{if } L_{MO} \leq 0 \\ 1 \\ \dfrac{1}{1 - C_4 \, L_{MO}/zi} & \text{if } L_{MO} > 0 \end{cases}$$

$$\frac{1}{ls} = \begin{array}{l} 0 \ \text{if} \ \partial\overline{\theta}/\partial z < 0 \\[4pt] \left(\beta \dfrac{\partial\overline{\theta}}{\partial z} \Big/ e \right)^{1/2} \ \text{if} \ \dfrac{\partial\overline{\theta}}{\partial z} > 0 \end{array}$$

and for l_ϵ $c_1 = 15$, $c_2 = 5$, $c_3 = 5 \times 10^{-3}$, $c_4 = 1$, $c_5 = 1.5$

for l_k $c_1 = 15$, $c_2 = 11$, $c_3 = 2.5 \times 10^{-3}$, $c_4 = 1$, $c_5 = 3$

The complete equation of $\overline{e}$, with advection and horizontal diffusion terms, is added to the classical equations set of a mesoscale numerical model. A realistic information on the turbulent kinetic energy field in various contexts (convection situation, thermal and orographic forcings) is then available for pollutants diffusion studies.

ESTIMATION OF WIND COMPONENTS VARIANCES

The Lagrangian numerical modelling implies the knowledge of each wind component variance, and the previous model provides finally as information on their sum. It is well known that the contribution of horizontal and vertical wind components variances to the turbulent kinetic energy depends on the vertical stability. The larger the buoyancy effects, the greater the importance of $\overline{w'^2}$. This behaviour is clearly explained when examining the w'^2 rate equation as it appears in André et alii (1978) :

$$\frac{\partial \overline{w'}^2}{\partial t} = \frac{\partial \overline{w'}^3}{\partial z} + 2 \, \overline{\beta w'\theta'} - C_4 C_\epsilon \frac{\overline{e}^{1/2}}{l_\epsilon} \left(\overline{w'^2} - \frac{2}{3} \overline{e} \right) - \frac{2}{3} c_\epsilon \frac{\overline{e}^{3/2}}{l_\epsilon}$$

where C_4 is Rotta's (1951) constant.

By assuming stationarity and neglecting diffusion terms, the previous equation reads :

$$\frac{\overline{w'^2}}{\overline{e}} = \frac{2}{3} \frac{C_4 - 1}{C_4} - 2 \; 1_\varepsilon \frac{\beta \overline{w'\theta'}}{C_4 C_\varepsilon \overline{e}} \; 3/2$$

The first term of the sum in the r.h.s. corresponds to the value in neutral stratification, which is thought to be approximately 0,4; 0,5 from experimental evidences, and gives $C_4 \sim 4$.

To rewrite the $\overline{e}$ rate equation assuming also stationarity and neglecting diffusion terms leads to the formal expression :

$$0 = P_w + P_\theta - \varepsilon$$

where P_w stands for wind shear production ; P_θ buoyancy production and ε dissipation.

Remembering that the Richardson number can be then expressed by

$$Ri = - P_\theta / P_w$$

the previous relationship gives

$$\overline{w'^2} = 0.5 \; (1 + P_\theta/\varepsilon)$$

And so

$$\frac{\overline{w'^2}}{\overline{e}} = 0.5 \; ((1-2 \; Ri)/(1-Ri))$$

Obviously, this formula is valid for $- \infty < R_i < 0.5$ which is, however, compatible with the idea of critical Richardson number. Nevertheless, estimation of Richardson number from experimental data can lead to large positive values, especially in well mixed boundary layer, due to very small values of both wind speed and temperature vertical gradients.

It must be kept in mind that this approximation of $(\overline{w'^2}/\overline{e})$ is a very crude one, which only aims to distinguish between unstable, neutral and stable stratification influences upon the various components of the turbulent kinetic energy. Following this idea and extending in R_i the domain of evaluation of $(\overline{w'^2}/\overline{e})$, we use practically

$$\frac{\overline{w'^2}}{\overline{e}} = 0.5 \; (1 + \tanh \; (-Ri))$$

Considering the partition between $\overline{u'^2}$ and $\overline{v'^2}$, it is far more difficult to precise a general policy. Because of lack of experimental (in situ or laboratory) pertinent data, we shall simply assume that $\overline{u'^2} = \overline{v'^2}$.

NUMERICAL TESTS OF LAGRANGIAN MODELLING IN MESO-SCALE MODELS

One is invited to refer to Blondin (1980) for a brief review of atmospheric meso-scale models used at the Etablissement d'Etudes et de Recherches Météorologiques. The main goal of numerical simulations of pollutants dispersion carried out in this Institute is to emphazise the prominent role of atmospheric thermodynamics in the behaviour of plumes. In this respect, complex terrain wind patterns (channeling effects, Venturi effects, kabatic forcings, upward motion forcing) and thermal local circulations (breezes,...) combined with atmospheric turbulence behaviour govern the diffusion-transport of pollutants over medium distances. The following described experiments aim to show that a simplified Lagrangian numerical scheme allows to simulate satisfactoringly the print of atmospheric properties upon plumes structures.

Pollutants emission is simulated by the injection of 1000 inert particles in the atmosphere (we treat here a simple point source).

Each particle moves independently, and its position is computed by :

$$\overline{X_i^{n+1}} = \overline{X_i^n} + (\overline{U_i^n} - \overline{U'_i^n}) . \Delta t$$

where

$\overline{X_i^{n+1}}$ is the position at time step (n+1) along the i axis

$\overline{X_i^n}$ is the position at time step n along the i axis

Δt is the time step

$\overline{U_i^n}$ is the mean wind component at time step n

$\overline{U'_i^n}$.   is a random velocity perturbation of the i wind component at time step n

$\overline{U_i^n}$ is computed by linear interpolation between the velocity values at the grid modes of the meso-scale model.

U'^n_i is generated by a random numerical processus, so that the distribution of a lot of realisations would be a centered Gaussian distribution of variance U'^{2n}_i, estimated from the turbulent kinetic energy field.

The time step is chosen to be systematically greater than the Lagrangian time scale (here we adopt 30 s).

This procedure leads to a method far from accuracy since it does not take into account the complete Reynold's tensor and presents systematic shortcomings when strong inhomogeneous turbulence is experienced.

We carried out the following numerical experiments : using a 2-D numerical meso-scale model, we studied the air flow motion over a hill range (500 m high) under various vertical stratification conditions. For each case, after one hour real time simulation a quasi steady state is achieved, for wind speed, temperature and turbulent kinetic energy field which are exhibited in the pictures referenced respectively .a, .b,.c. Then, in this stationnary flow, we used the previous Lagrangian scheme to simulate the displacements of 1000 independent particles emitted at 350 m above the ground. Assuming that a continuous emission can be simulated by injection of particles at each time step behaving exactly as the first 1000 emitted particles, concentrations for continuous emission point source can be computed. The mean trajectory of the particles and the concentration in a continuous plume are drawn on figures .d, with in the upper part of the figure, the number of particles among the 1000 original ones deposited on the ground.

In all these experiments, the wind profile at the enter lateral boundary (left hand side) is the same on all levels that is 10 m/s.

The series of figures 1 refers to an air flow motion in a purely neutral atmosphere. One can notice the quasi-symmetric structure for both horizontal and vertical wind components, except close to the ground, where vertical wind shear is more pronounced on the downwind side of the hill. This feature explains the maximum energy area in this region. The plume structure shows the influence at the vertical air motion forcing, and a continuous dilution, increasing the wideness of plume during its travel. Very few deposition occurs in the domain, due to initial emission height.

Stable atmosphere air flow is studied in the series of figures 2. A strong wind speed up occurs over the top of the hill range, while the vertical velocity field is more complex. The potential temperature field exhibits a dynamic foehn effect and a smooth lee wave pattern after the hill range downwind. This subsidence effect is strong enough to create an inversion layer

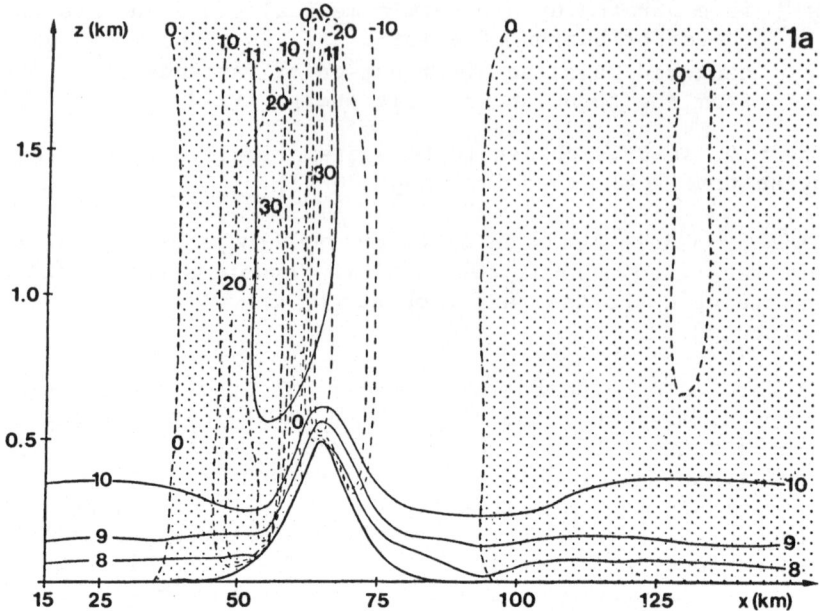

Fig. 1 a : Neutral atmosphere. Vertical cross-section of horizontal wind speed in m/s (full lines) and vertical
wind speed in cm/s (broken lines). Shadowed areas
corresponds to upward notion.

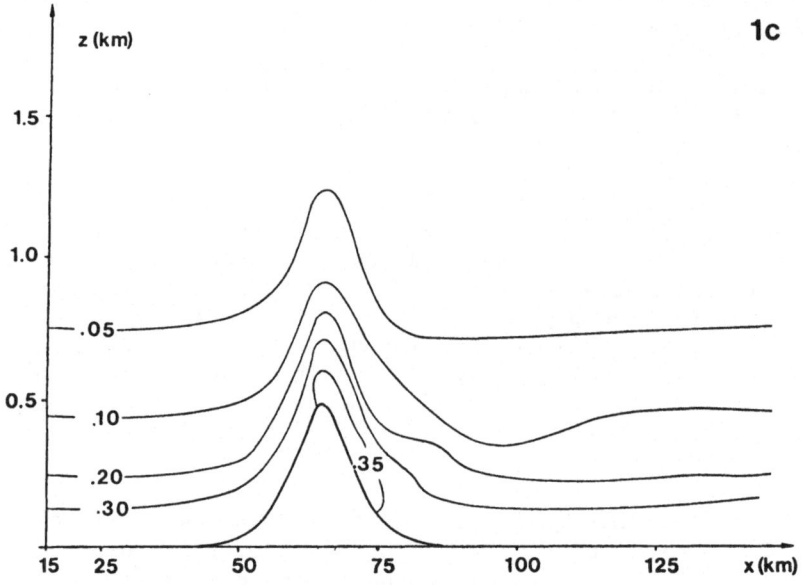

Fig. 1 c : Neutral atmosphere. Vertical cross-section of turbulent kinetic energy in m^2/s^2.

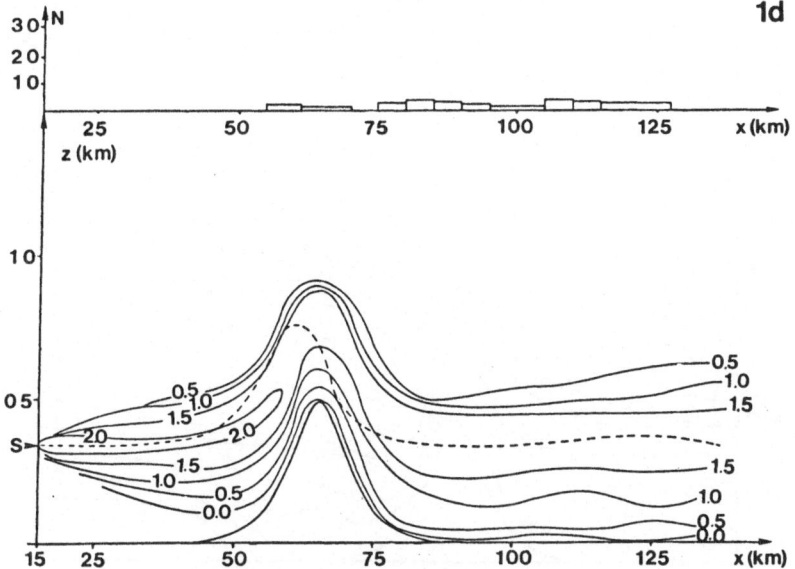

Fig. 1 d : Neutral atmosphere. Vertical cross-section of the
logarithme of concentration (arbitrary units). The
broken line corresponds to the mean trajectory.
In the upper part, number of particles deposited
on the ground.

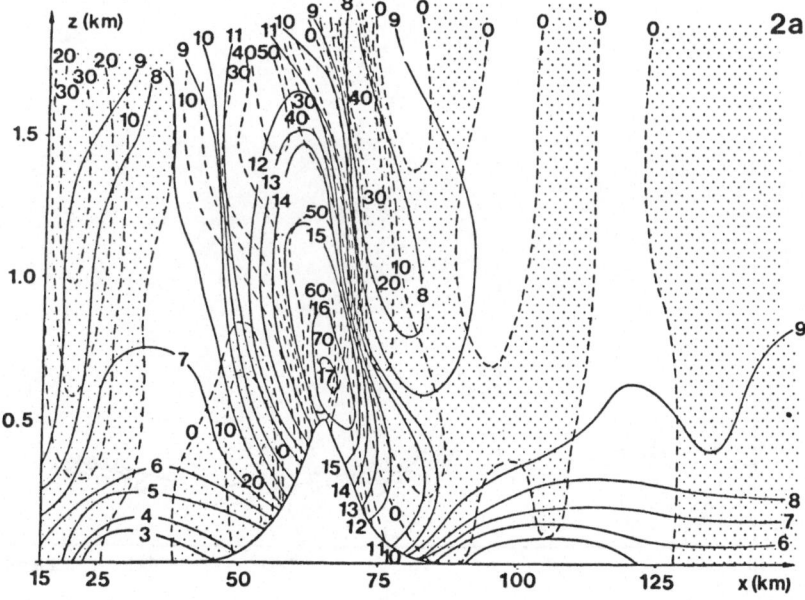

Fig. 2 a : Stable atmosphere. Horizontal and vertical wind speed.

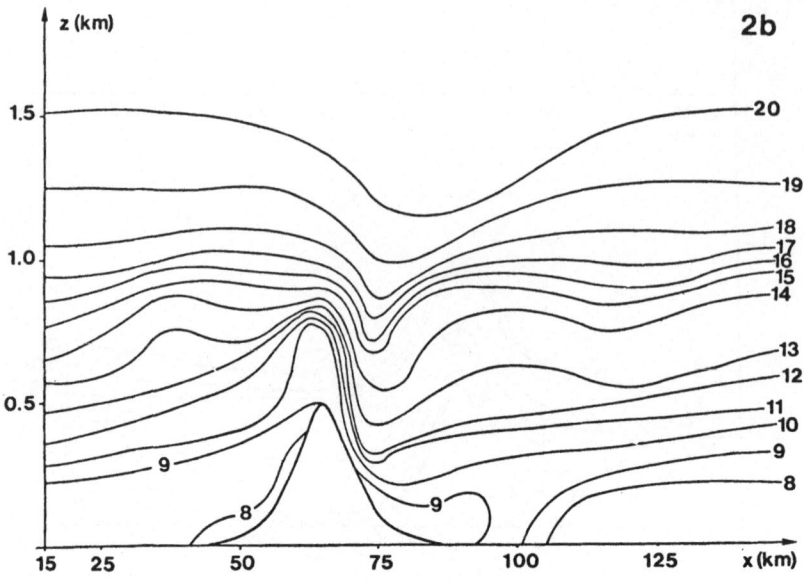

Fig. 2 b : Stable atmosphere. Vertical cross-section of poten-
         tial temperature (θ – 273°K).

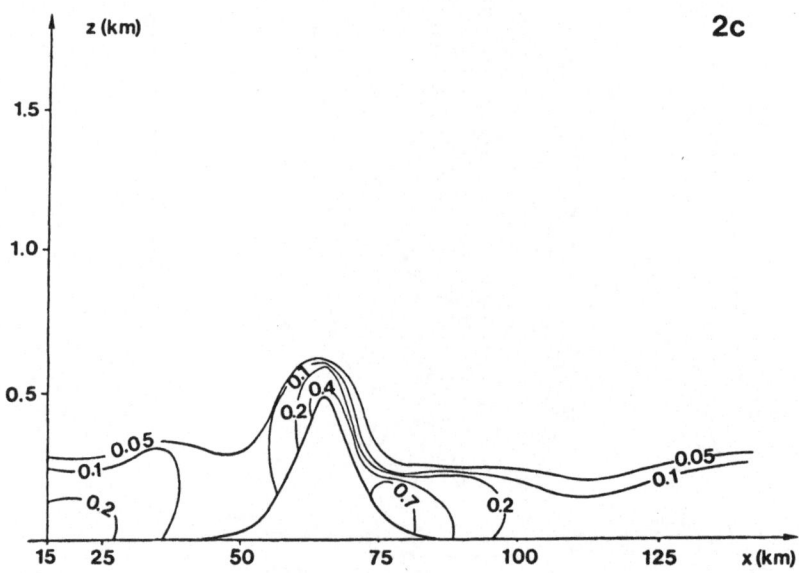

Fig. 2 c : Stable atmosphere. Turbulent kinetic energy.

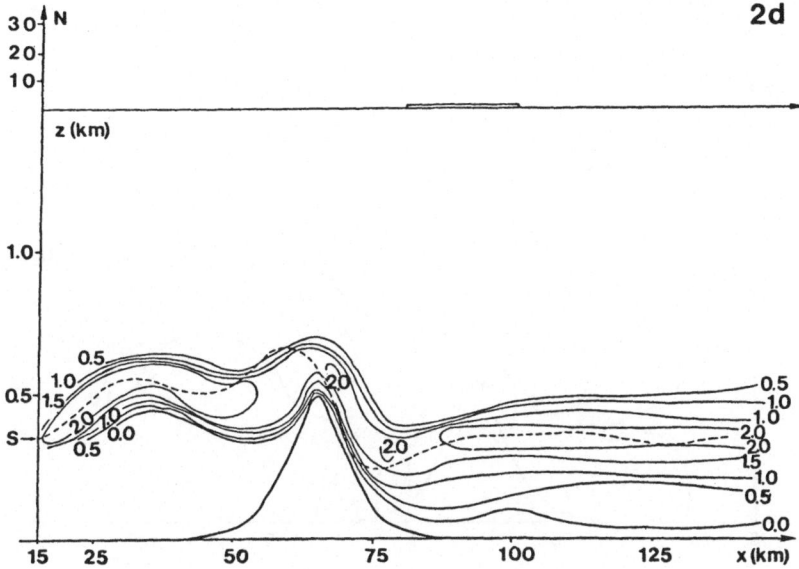

Fig. 2 d : Stable atmosphere. Concentration and number of
particles deposited on the ground.

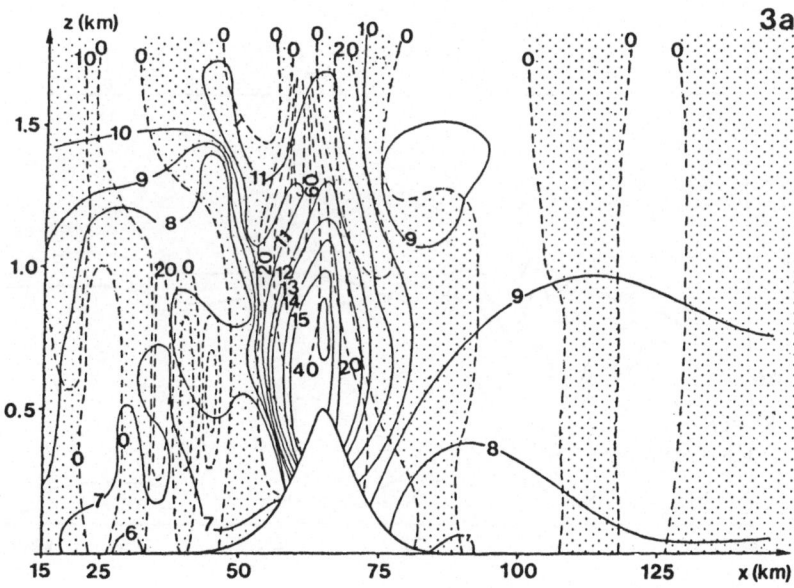

Fig. 3 a : Unstable atmosphere. Horizontal and vertical wind speed.

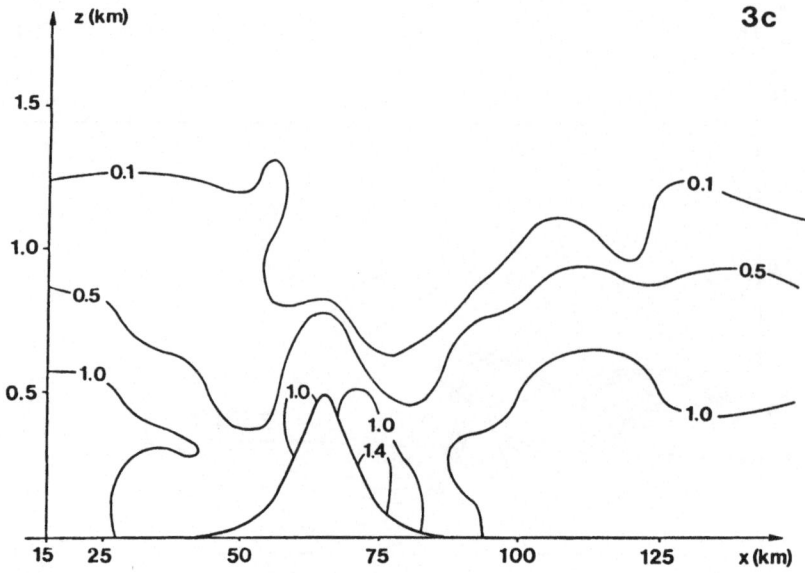

Fig. 3 c : Unstable atmosphere. Turbulent kinetic energy.

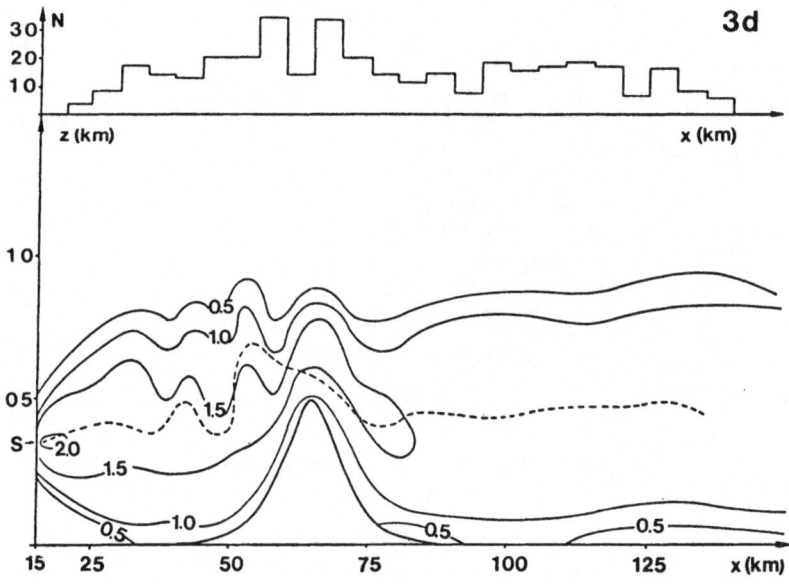

Fig. 3 d : Unstable atmosphere. Concentration and number of
           particles deposited on the ground.

over the downwind side of the range so that the turbulent kinetic
energy is not important in this region in spite of wind shear production. All the turbulent kinetic energy is so concentrated near
the ground. The plume behaviour is coherent with these features.
The plume remains narrow, till it encounters significant turbulent
energy, when arriving near the range. Then particles are speeded
up on the downwind side but are trapped under the inversion. The
plume begins to extend on the vertical after, but is less diluted
than in the neutral case and presented concentrations as high as
close to the emission point.

The air flow for lapse conditions is also simulated (cf. figures 3). The instability is generated by a positive heat flux at
the ground but is limited by a quasi horizontal inversion layer
1500 m high. So, the acceleration of the wind over the hill range
is mainly due to a kind of Venturi effect. As in the stable case,
downward velocities occur on both sides of the hill range. Obviously, the turbulent kinetic energy presents stronger values than in
the previous experiments, and consequently, the plume is far more
diluted. In this case, the deposition on the ground is significant
and begins only 7 km of the source downwind. After the hill range,
the plume becomes a well mixed one as could be expected.

CONCLUSION

The influence of thermodynamics structure of the atmosphere
upon plumes behaviour has been clearly identified using a simple
Lagrangian scheme to simulate transport diffusion of inert particles. The relative success of these experiments lies on a correct estimation of both horizontal and vertical wind velocity
components variances.

It can be reasonably hoped that, incorporated in a 3-D nume-
riacl meso-scale model, this Lagrangian scheme will be a useful
tool to study medium range pollutants transport problems.

REFERENCES

André J.C.n De Moor G., Lacarrère P., Therry G., du Vachat R.
    1978, Modelling the 24-Hour evolution of the mean and tur-
    bulent structure of the planetary boundary layer, J. Atmos.
    Sci. 35 : 1861 - 1882

André J.C., Lacarrère P., 1980, Simulation numérique détaillée de
    la couche limite atmosphérique. Comparaison avec la situa-
    tion des 2 et 3 juillet 1977 à Voves, La Météorologie, 6 :
    5 - 49

Blondin C., 1980, Interest of an atmospheric meso-scale model for
    air pollution transport studies over medium distances, in :
    Proc. 11th I.T.M. on Air Pollution modeling and its appli-
    cations, Amsterdam, the Netherlands.

Blondin C., Therry G., 1981, Analysis of particles sea breeze
    cycle using txo dimensionnal numerical meso-scale models,
    in : Proc. 12th I.T.M. on Air Pollution Modeling and its
    applications ; Palo-Alto, United States.

Deadorff J.W., 1970, Convective velocity and temperature scales
    for the instable planetary boundary layer and for Rayleigh
    convection, J. Atmos. Sci., 27 : 1211 - 1213.

Eliassen A., 1980, A review of long range transport modeling,
    J. Atmos. Sci., 19 : 231 - 240.

Mc Nider R.T., Hanna S.R., Pielke R.A., 1980, Sub-grid scale
    plume dispersion in coarse resolution meso-scale models,
    in : Proc. 2nd joint conference on Application of air
    pollution meteorology, New-Orleans, United States.

Runca E., Bonino G., Posch M., 1981, Lagrangian Modelling of
    air pollutants dispersion from a point source, in : Proc.
    12th I.T.M. on Air Pollution Modeling and its applications
    Palo-Alto, United States.

DISCUSSION

F. FANAKI   1) Are you modeling a ridge or hill ?
                2) Does the plume impinge on the hill ?

C. BLONDIN   1) I am modeling a ridge (2-D model).

                2) If the computed altitude of a particle is lower
                   than the ground elevation, this particle is assumed
                   to be deposited on the ground and to remain there.
                   Refering to lapse rate conditions, the plume does
                   impinge on the ridge.

F.B. SMITH   In the U.K. Meteorological Office,
                we have been making measurements of wind speed
                and turbulence around rather simply-shaped hills.
                Although I cannot quote numbers from memory, I am
                certain that the speed-up factors and turbulence
                levels near the hill crests greatly exceed the values
                you have given in your paper for neutral stability
                conditions. Have you tested your results against
                such field data ? Your numbers are hard to swallow.
                They do not appear to be consistent with my expe-
                rience in U.K.

C. BLONDIN

In our simulation, the stability is neutral in the whole domain. In such conditions, there is no reason for great speed up factors and turbulence levels.

In the field, "neutral stability conditions" correspond to a more or less thick neutral layer bounded by a stable layer above. I think our model, running with such data, will give results which compare favorabily with field data.

A first idea of these possible results may be given by the figures 3a and 3c.

STUDIES OF PLUME TRANSPORT AND DISPERSION OVER

DISTANCES OF TRAVEL UP TO SEVERAL HUNDRED KILOMETRES

John Crabtree

Meteorological Office
Bracknell
England

INTRODUCTION

In the United Kingdom the Meteorological Office is collaborating with the Central Electricity Research Laboratories (CERL) in a study of the long-range transport of pollutants over the North Sea. For these experiments the plume from Eggborough power station in South Yorkshire is "labelled" with sulphur hexafluoride so that it can be uniquely identified. The movement of the plume is predicted by a numerical trajectory model, and the Hercules aircraft of the Meteorological Research Flight, Farnborough is used to intercept and sample the gases at selected distances downwind. The usual meteorological, turbulence and cloud physics instrumentation on the aircraft (Nicholls, 1978) has been augmented by chemical sampling equipment (Crabtree and Marsh, 1981) in order to measure, on a time-scale short enough to reveal the detailed structure of the plume, the concentrations of the tracer gas, sulphur and nitrogen compounds, and ozone. Several flights have been carried out so far, including one experiment in which the same portion of the plume was sampled on two successive days. Plumes have been detected at distances up to more than 600 kilometres from the source.

PLUME TRANSPORT

The Trajectory Model

The position of the plume from Eggborough power station (53° 42'N, 1° 10'W) is predicted at six-hour intervals up to three days in advance using the output of the Meteorological Office 10-level numerical forecast model (Burridge and Gadd, 1975).

129

In this model, data from most of the northern hemisphere are analysed every twelve hours, and prognoses are produced at six-hour intervals throughout the forecast period of 72 hours. During trajectory computation linear interpolation is carried out over the six-hour interval between successive fields, and between grid-points (the grid-points form a square network on a polar stereographic projection: the separation varies with latitude, being 300 km at 60°N). Each six-hour element of a trajectory is calculated by a method of successive approximations not very different from that followed by a subjective analyst. The first estimate assumes that the wind at the position and time of the start of the element applies throughout the whole six hours. The second estimate halves the time-step, taking the wind as constant over each three-hour period. The process is continued until the vector difference between two successive estimates is less than a specified small value.

The 10-level model is based on the primitive equations of motion so the calculated wind takes into account not only the geostrophic balance, but also isallobaric, curvature and inertial terms. The main problem arises in the estimation of the wind at plume height. In reality, this varies with stability and wind speed, but for present purposes has been taken to be 400 metres. Winds at this height are not directly available as output from the forecast model, but are estimated by inverting the relationships used by the model to derive surface winds (see Findlater et al., 1966) and applying a simple boundary-layer model to those winds. (Smith, 1975).

Errors of the trajectory model

The accuracy of the trajectory predictions depends on a number of factors, including the performance of the 10-level model: of particular importance are the availability of meteorological data for the area of interest, and the relative size of the grid resolution compared with the scale of surface inhomogeneities affecting the plume transport. Also relevant are the way the 400m wind is derived from the model, how accurately the 400m wind represents plume transport etc. Although the forecasts are used to indicate suitable conditions for the experiment up to three days ahead, their main purpose is to facilitate the planning of sampling flights shortly before take-off. The analyses presented here therefore concentrate on the errors of trajectories computed on the day of the flight, from analyses and forecasts based on meteorological data for 0000 GMT that day.

Crosswind error and the influence of sub-grid-scale motions

Lateral errors, taken as the crosswind distance between the actual plume and its forecast position, are shown in Fig. 1: the sign convention is such that a positive error occurs when the

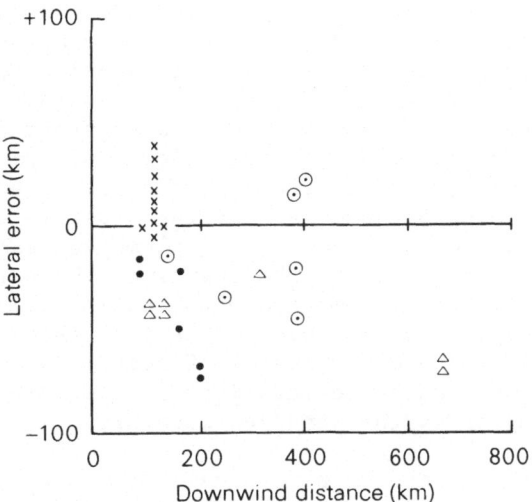

Fig. 1. Trajectory errors of computer forecasts based on data
        for 0000 GMT on day of flight.

         •   18 June   1980
         ⊙   22 October   1980
         x   11 December   1980
         △   28-29 January 1981

plume is to the left of the forecast position looking downwind
(i.e. the forecast wind is veered from the true wind). Overall
there is a wide scatter of points with a preponderance of negative
errors, but interestingly no apparent indication of any systematic
increase in error as distance of travel increases. An examination
of aircraft winds, however, reveals an explanation for at least
some of these errors. For example the first flight (18 June 1980)
was carried out in a weak ridge well ahead of a warm front. During
the period when the plume was being advected towards the sampling
area, a mesoscale trough formed which remained almost stationary
for a time over or just off the east coast of England. This feature was too small to be predicted by the coarse-mesh numerical
model, so for much of the time the plume was carried well to the
south of the forecast position. By the time of the last sampling
run the trough had almost disappeared so the error was reduced to
about -20km. The second flight (22 October 1980) took place in a

cyclonic circulation around a complex low over and to the west of
Britain.   The plume was broader than expected, again suggesting an
influence from mesoscale eddies, the errors being comparable with
plume width.   On 11 December 1980 several traverses of the plume
were made at roughly the same height at a constant distance downwind of the source in strong westerly flow.   Aircraft winds showed
fluctuations, superimposed on a synoptic-scale trend, sufficient
to account for a considerable part of the observed errors.   On
28-19 January 1981 the same portion of plume was sampled at about
100km and again at just over 600km downwind: the error grew only
slowly over this interval compared with the initial increase.
Thus the general picture that emerges is that the major part of
the error occurs in the early stages of advection, and may be
associated with mesoscale accelerations and circulations set up by
nearby topography or by the land/sea boundary.

     More direct evidence of a mesoscale flow pattern is shown in
Fig.2, which compares the surface isobars at 1200 GMT with the
streamlines of the airflow measured by the aircraft at a height of

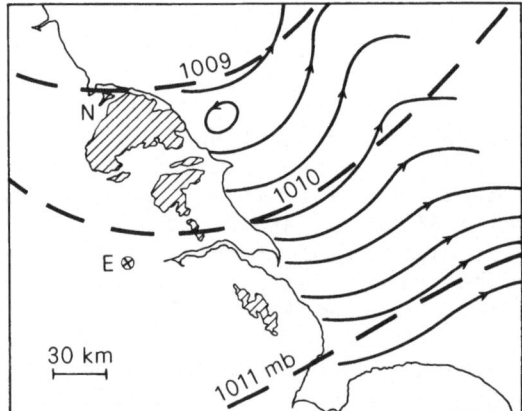

Fig. 2. Surface isobars at 1200 GMT and streamlines of
        airflow at 150m above sea level on 18 June 1980.
              ⬚   Ground above 100m
              N   North Yorkshire Moors
              E   Eggborough

150m just off the east coast of England. The most striking effect
is a large eddy downwind of the North Yorkshire Moors, a phenomenon
observed on at least two other flights.

Errors of the trajectory speeds

     For most of the experiments there was no indication of the
time at which the sampled gases were emitted from Eggborough, so a
strict comparison of predicted and actual speeds is not possible.
However, a limited check was made by comparing the observed aircraft winds with the predicted mean transport winds derived from
the six-hour trajectory elements: the results are shown in Fig.3.
The predicted wind speeds were generally about eight per cent less
than the observed, possibly a consequence of smoothing in the coarse-
mesh forecast model. On one occasion a "puff" of a second tracer
was surveyed in detail and a mean transport wind obtained from the
interval between the start of emission and the first appearance of
the tracer in the sampling area: the predicted and observed winds
agreed in this instance to well within 1 m/s.

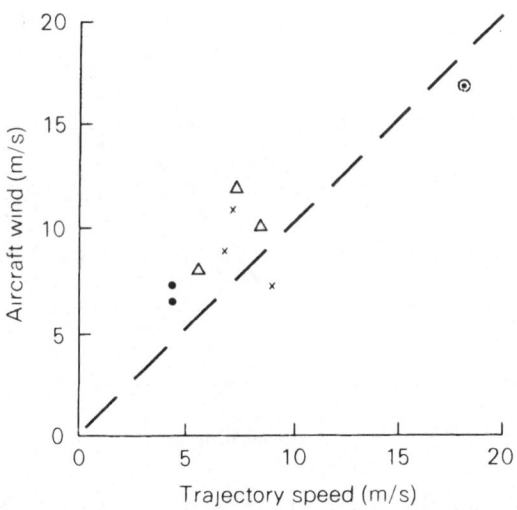

Fig. 3. Comparison of aircraft and trajectory winds.

Symbols as Fig. 1.

HORIZONTAL DISPERSION

The tracer gas enabled the plume to be uniquely identified, so plume profiles could be quite accurately determined, to distances of several hundreds of kilometres: peak concentrations of tracer even at the furthest ranges flown were still more than an order of magnitude above background levels. Assuming the profile to follow a Gaussian distribution (in many instances this seemed a not unreasonable approximation), the cross-wind dispersion coefficient, σ_y , was derived. This is shown in Fig.4 as a function of time of travel (the latter derived from the trajectories). Also shown is the envelope of points derived from studies of the plume from the Mount Isa smelter, Queensland, Australia (see Gifford, 1981).

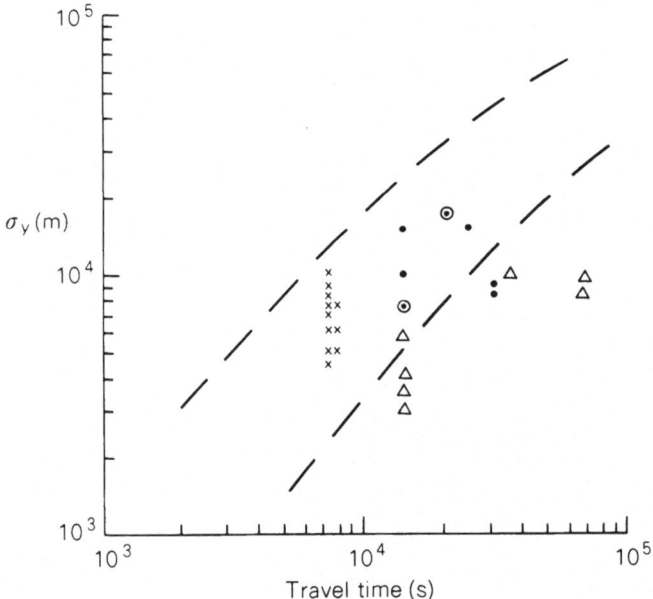

Fig. 4. Comparison of dispersion of Mount Isa plume
with that of the plume from Eggborough.
σ_y as a function of travel time, T.
/ /   Envelope of Mt Isa points.
Eggborough plume, symbols as Fig. 1.

Most of the North Sea values lie within this envelope, except
for those observed on the second day of the two-day experiment.
Although the widths of the two plumes are comparable after travelling for 2-4 hours over land, there is a suggestion that subsequently, as the Eggborough plume moves out over the sea, it grows
less rapidly than the Mount Isa plume over land.

The aircraft data are also being used to evaluate turbulent
intensities and fluxes, with the overall aim of deriving
relationships between the growth of the plume and observed
turbulence parameters and, eventually, between these quantities
and external, synoptic-scale variables. Fig.5 shows the results
of an attempt to relate observed plume widths and turbulence
intensities through the statistical theory of turbulence
(Taylor, 1921). For a Lagrangian autocorrelation function of
the form

$$R\ (\xi) = e^{-\xi/\tau_L}$$

(where ξ is the time interval and τ_L is the Lagrangian
time scale) the angular width of the plume, σ_Θ after travel time
T, should be related to the crosswind turbulent intensity, σ_v/u
by the equation

$$\frac{\sigma_\Theta}{\sigma_v/u} = \sqrt{2}\ \frac{\tau_L}{T}\ (e^{-T/\tau_L} - 1 + \frac{T}{\tau_L})^{1/2}$$

The large scatter, particularly for short travel times, may
reflect the fact that $\sigma_{v/u}$ is measured over the sea and so
does not really describe the boundary layer over land where the
plume originates.

Turbulence spectra are also being examined, and relationships
will be sought between the growth of the plume itself and the
high-frequency components of the turbulence, and between the
meandering of the plume and the low-frequency part of the
spectrum.

ACKNOWLEDGEMENT

The author wishes to thank the members of the Meteorological
Research Flight, Farnborough, and the staff of the Central Electricity Research Laboratories, Leatherhead, for carrying out the
experiment, and the latter for helpful discussions.

REFERENCES

1.   Burridge, D. M. and Gadd, A. J., 1975, The Meteorological
        Office ten-level numerical weather prediction model.
        Scient.Pap.Met.Off.London, No.34.

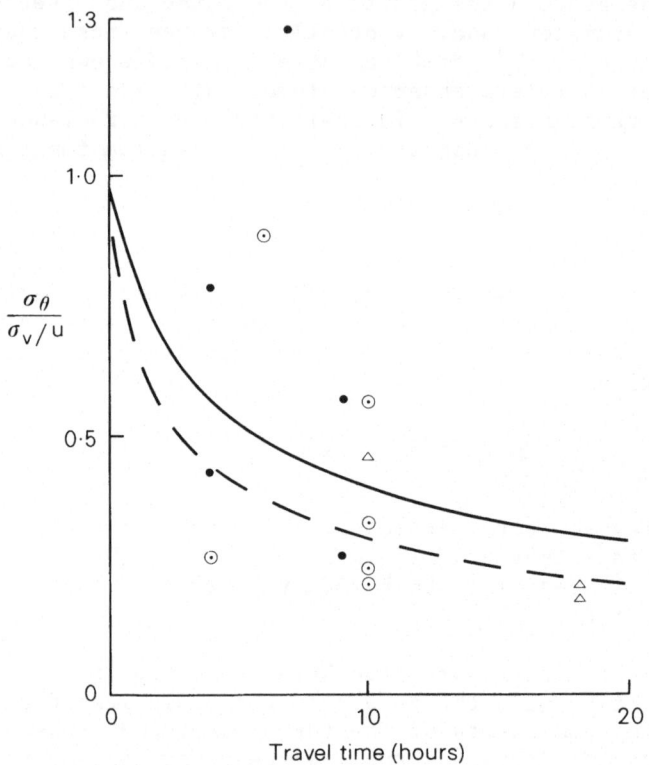

Fig. 5. Ratio of plume angular width, σ_θ , to crosswind
turbulence intensity (σ_v/u) as a function
of travel time. Symbols as Fig. 1.
Theoretical values

$$\underline{\hspace{2cm}} \quad \tau_L = 1 \text{ hour}$$
$$\text{-------} \quad \tau_L = 1/2 \text{ hour}$$

2. Crabtree, J. and Marsh, A. R. W., 1981, Instrumentation of
        the Hercules aircraft of the Meteorological Research
        Flight. Turbulence and Diffusion Note No.127.
        Unpublished. A copy is available in the Meteorological
        Office Library, Bracknell.

3. Findlater, J., Harrower, T. N. S., Howkins, G. A. and
        Wright, H. L., 1966. Surface and 900mb wind relation-
        ships. Scient.Pap.Met.Off., London, No.23.

4. Gifford, F. A., 1981, Horizontal Diffusion in the atmosphere:
        a Lagrangian-dynamical theory. Atmospheric
        Environment, 16: 505.

5. Nicholls, S., 1978, Measurements of turbulence by an
        instrumented aircraft in a convective atmospheric
        boundary layer over the sea. Q.J.R.Met.Soc., 104: 653.

6. Smith, F. B., 1975, Brief notes on the specification of the
        boundary layer in terms of external readily assessed
        parameters. Turbulence and Diffusion Note No.63.
        Unpublished. A copy is available in the Meteorological
        Office Library, Bracknell.

7. Taylor, G. I., 1921, Diffusion by continuous movements.
        Proc.London Math. Soc., 20: 196.

DISCUSSION

T. WARNER (USA)   1. Was it possible to
        obtain any observations of the vertical structure
        of the plume ?
                                         2. Would the trajectory
        predictions have been improved if model output had
        been used more frequently than every 6 hours ?

J. CRABTREE   1. It has not yet been
        possible to obtain observations of the detailed
        vertical structure of the plume, and this is now
        one of our prime objectives.
                                         2. Trajectory predictions
        may well be improved a little if the model output
        could have been available more frequently than
        every six hours, but the linear interpolation
        between grid-points would be equally important.
        It appears that both were probably less important
        than the mesoscale (sub-grid scale) variability
        that the forecast model could not handle.

M. CHAN   How were the SF_6 data
        collected, by grab-sampling or a continuous
        analyser ?

J. CRABTREE   The SF$_6$ concentration was
measured continuously, the system having a time
response of 1 to 2 minutes. Grab samples were
used as a back-up.

M.K. MISRA   It seems to me that a single
trajectory in the atmosphere would be a random
path of an air parcel. Therefore, comparison of
statistics of an ensemble of trajectories with
model results would be more meaningful than a one
to one comparison.

J. CRABTREE   I agree that it would have
been valuable to compare a large number of tra-
jectories with observations. However, the
project was confined to a relatively small number
of flights, and it was possible to carry out only
the limited comparison described.

LAGRANGIAN MODELLING OF DISPERSION IN THE PLANETARY BOUNDARY LAYER OF PARTICULATE RELEASED BY A LINE SOURCE

Pietro Melli

IBM Scientific Center
Via del Giorgione, 129 - 00147 ROMA
Italy

INTRODUCTION

Formulating and solving the diffusion-deposition-resuspension problem of particulate released in the planetary boundary layer is of primary importance in assessing their environmental impact. Solid particles are originated, as it is well-known, both by industrial processes (releases from industrial stacks) and by natural processes (e.g. formation of aerosols droplets by chemical reactions in the atmosphere). The physical mechanisms governing the evolution of particles in the atmosphere have been largely investigated in recent years both in laboratory studies and on the field. For instance, as far as the mechanism of dry deposition is concerned, a detailed review of available results has been compiled by Sehmel (Sehmel G.A., 1980). Modelling of dispersion of solid particles has been generally achieved by suitable modifications of the models available to treat dispersion of gaseous pollutants. Most of these models are therefore based on the Gaussian formulation and are the most used in real world applications. More complex modelling based on the eddy-diffusivity approximation has been recently developed to study the evolution of particles in the plumes of coal-fired power plants (see e.g. Hobbs et al., 1979). This complex model includes, besides the usual mechanisms of advection and diffusion, significant phenomena as particle coagulation, and gas-to-particle conversion both in homogeneous and in heterogeneous phase.

The approximations and the inadequancies inherent in the eddy-diffusivity formulation have been discussed by several authors (see e.g. Lamb et al., 1979), who have suggested to overcome these drawbacks of the k-theory approaches by developing Monte-Carlo algorithms simulating the dispersion process by means of particles whose velocity processes reproduce the statistics of the considered turbulent flow (Reid, 1978; Lamb et al. 1979). In the present paper the connection of the above algorithms with the theory of stochastic differential equations is discussed and their application to the case of settling particles is investigated.

ATMOSPHERIC DISPERSION WITH GRAVITATIONAL SETTLING

As in the case of hot effluent gases from a stack the dispersion process of an airborne material consists of two distinct phases. In the first one emitted particles behaviour is greatly influenced by the turbulent motion produced by hot buoyant stack gases. Moreover significant mechanism such as coagulation, hindered motion because of high particle density, chemical reactions, water vapour condensation on particles etc. affect the evolution of the solid phase in this stage. At a certain point a separation starts between the 'particle plume' and the 'gas plume' which is obviously complete only after a while because of the different particle sizes and densities present in the released material.

After this phase is completed the airborne material starts to undergo dispersion and settling under the influence respectively of atmospheric turbulence and action of gravity. This paper neglects the first stage above outlined and concentrates on modelling the dispersion and settling process.

DIFFUSION MODEL OF PARTICULATE TRANSPORT

In the following we study the dispersion and settling of a dust plume emitted by a linear source situated at height h (effective height determined by the processes outlined in the previous paragraph) and we assume that the considered processes can be modeled by the following set of stochastic differential equations :

$$dx = u_p dt \tag{1a}$$

$$dz = w_p dt \tag{1b}$$

$$dw_p = -A (w_p - w_a) - \tilde{g} \ dt \tag{1c}$$

$$dw_a = -(w_a/T) \ dt + f (x,z,t) \ dt + \sigma(x,z,t) \ d\beta_t \tag{1d}$$

$$\text{with} \quad \tilde{g} = g (1 - \rho_s/\rho_a) \tag{2a}$$

$$\text{and} \quad A = 18\mu/(D_p^2 \ \rho_s) \tag{2b}$$

where :

x, z are horizontal and vertical coordinates (m)

u_p, w_p are horizontal and vertical components of particle
                 velocity (m/s)

w_a is air velocity fluctuation (m/s)

T is correlation time, which may be a function of
                 spatial coordinates (s)

f is a suitable function to be defined in the fol-
                 lowing (m/s^2)

σ is a suitable function depending on local air
                 velocity fluctuations (m/s$^{3/2}$)

$d\beta_t$ is a Wiener process which can be represented for-
                 mally as the derivative of a Brownian motion

D_p is particle diameter (μm)

g is gravity acceleration (m/s^2)

ρ_a is air density (kg/m^3)

ρ_s is particle density (kg/m^3)

μ is air viscosity (kg/m.s)

Eq.(1c) expresses the balance between particle acceleration, friction and gravity acting on particle during its motion. It is
well-known that friction cannot in general be expressed as a linear
function of the difference between particle and air velocities.
In fact, this is true only for low values of the above difference
(let's say for particle Reynolds numbers lower than 1), while for
values of Re>1000 a quadratic dependence must be used. For intermediate values the exponent must lie in the range between 1 and 2
(see e.g. Seinfeld, 1975). For a careful quantitative analysis of
airborne material dispersion in a real case the above considerations
should be accounted for, but since we are here interested more to
investigate the principles of modelling dispersion using stochastic
differential equations rather than making realistic simulations,
we neglect all the above complications and assume that friction can
be assumed to depend always linearly on velocity differences
between particle and air. The considerations drawn from Eqs.(1-2)
are, however correct for small particles ($D_p \leqslant 0.1$ μm). These
particles, in fact, possess velocities, during most part of their
motion, such that their Reynolds number is lower than 1 and only
in negligible fractions of time they move with Re>1.

Eq. (1d) represents air velocity fluctuatios as a first order Markov process with exponential correlation (because of the presence of the term $- (1/T)$ w) driven by a deterministic acceleration f (whose presence will be justified in the following) and a stochastic acceleration $\sigma d\beta_t$. The absence of a stochastic term in eq. (1a) and the absence of a supplementary stochastic differential equation for velocity component u indicates that we are neglecting horizontal velocity fluctuations.

Study of system (1) for the case of non-settling particles

Although system (1) should be integrated as it is, to give rigorous results, we can nevertheless make some reasonable assumptions to introduce useful simplifications. Parameter A is usually large (e.g. for $D_p = 0.1$ μm, $A \cong 5.10^7$ s^{-1}), therefore eq. (1c) can be simplified to give :

$$w_p = w_a - w_f \tag{3}$$

where $w_f = \tilde{g}/A$ is usually called the "final settling velocity" and represents the velocity of a particle settling in uniform motion in a still atmosphere. Finally we note that if limit for $g \to 0$ (or $A \to \infty$) is taken we get the following set of stochastic differential equations :

$$dx = u \ dt \tag{4a}$$

$$dz = w_p dt \tag{4b}$$

$$dw = - (w_p/T) \ dt + f(x,z,t) \ dt + \sigma(x,z,t) \ d\beta_t \tag{4c}$$

that can be taken as a model for non-settling particles moving in a non-homogeneous turbulent field. Note that we write w_p for velocity considering that air and pollutant particles have the same velocity.

Apparently eq. (4c) is the continuous counterpart of the finite difference schemes used by several authors who have developed Lagrangian models to simulate atmospheric turbulent dispersion (Hanna, 1980; Janicke, 1981; Lamb, 1979; Reid, 1979).
If, in fact, we apply Euler's scheme and integrate eq. (4c), we get:

$$w \ (t + \Delta t) = (1 - \Delta t/T) \ w_p \ (t) + f \Delta t + \sigma \beta (t + \Delta t) \tag{5}$$

Eq. (4c) represents therefore the simplest attempt to generalise, to the case of non-homogeneous turbulence, the equation :

$$dv = - \gamma v \ dt + d\beta_t \tag{6}$$

which originates the well-known Ornstein-Uhlenbeck process (Jazwinski, 1970) usually called a colored noise in contrast with process β_t usually termed a white noise. Process v, which will be used in the following, is a Gaussian process exponentially correlated in time (correlation time $T = 1/\gamma$), and with variance given by :

$$\sigma_v^2 = (2/\gamma)(1 - e^{-\gamma t}) \tag{7}$$

Let's now go back to system (4) and study its properties in the particularly simple case when u = const. and f and σ are function of z only (and eventually of t). This is by no means a restriction to our analysis as eq. (4a) is purely deterministic and the fact that u is not a function of z does not change the statistical properties of the processes generated by eq. (4b-4c). Moreover horizontal homogeneity of turbulence in contrast with vertical non-homogeneity is a rather common assumption in atmospheric dispersion studies. We thus restrict ourselves to study the one-dimensional dispersion problem :

$$dz = w_p dt \tag{8a}$$

$$dw = \left[-(w_p/T) + f(z,t) \right] dt + \sigma(z,t) \, d\beta_t \tag{8b}$$

with all parameters depending on spatial coordinate and time.

From the theory of stochastic differential equations we know that both the probability density $p(z,w_p,t)$ and the conditioned probability density $p(z,w_p,t|z_0,w_{p0},t_0)$ are solutions of the Fokker-Planck (or forward Kolmogorov) equation (Jazwinski, 1970) :

$$T \left(\frac{\partial p}{\partial t} + w_p \frac{\partial p}{\partial z} \right) = p + (w_p - fT) \frac{\partial P}{\partial w_p} + \frac{1}{2} \sigma^2 T \frac{\partial^2 P}{\partial w_p^2} \tag{9}$$

with certain initial conditions and appropriate boundary conditions that will be discussed in the following. It must be recalled that the presence of particularly complex boundary conditions causes sometimes troubles in rigorous derivation of eq.(9) from (8) (Ludwig, 1975).

Before proceeding in our analysis we want to stress that from the point of view of the solution of our problem (i.e. computation of ensemble average pollutant concentration) use of system (8) or eq.(9) is totally indifferent. We want, in fact, to compute dust concentration which is given by :

$$<c(z,t)> = \int_o^t \int_o^1 p(z,t|z_s,\tau) S(z_s,\tau) \, dz \, d\tau \tag{10}$$

where $<\cdot>$ stands for ensemble average and $S(z,\tau)$ is a function specifying pollutant emissions. The density function can obviously be
obtained either from (8) or from solution of (9). In the former
case we must use a suitable integration scheme assuring stability
and convergence and compute enough trajectories to achieve stability
of the distribution momenta. The greatest advantage in using (8)
is the fact that mass conservation is rigorously respected.
Eq. (9), on the contrary, has similar stability problems with the
additional disadvantage that in multidimensional problems, the natural dimensionality of the problem is doubled as eq. (9) requires
the presence of velocities as independent variables.
Fokker-Planck equations are however very useful if one wants to
study some properties of the distribution density, as was done,
for instance, by Janicke (Janicke, 1981), who used eq. (9) to check
if the conditions under which system (8) produces an asymptotic
solution $p(z)$ = const., which is obviously required by physical
considerations. If we repeat his analysis, by using eq.(9) itself,
we conclude quite easily that, because of the dependence of σ and
f on z, the only constant possible solution satisfying the above
requirements can only be obtained if σ is constant and f=0.
Janicke's analysis was, on the contrary, based on the system of
equations describing the evolution of the distribution momenta.
This system, easily derived from eq.(9) by multiplication by w_p
and integration with respect to w_p, reads :

$$T\left(\frac{\partial q_n}{\partial t} + \frac{\partial q_{n+1}}{\partial z}\right) = fTnq_{n-1} - nq_n + \frac{n(n-1)}{2} T\sigma^2 q_{n-2} \qquad (11)$$

$$\text{where : } q_n = \int_{-\infty}^{\infty} w_p^n p \, dw_p$$

It is apparent, from an examination of (11), that no choice of the
three functions T,f, and σ can allow us to satisfy the above system.
The misleading conclusions drawn by Janicke depend on his assumption,
made at level n=3, that the correlation time T is short. If such
condition is rigorously applied to the original system (8) we are
led to study the equation :

$$dz = f(z,t)Tdt + \sigma(z,t)Td\beta_t \qquad (12)$$

Process leading from (8) to (12) is usually known in stochastic
differential equations analysis as Smoluchowski's approximation and
it is very useful in studying the asymptotic properties of some problems (Schuss, 1980). It has the obvious physical meaning that
instead of representing atmospheric dispersion by the equations of
the harmonic nonlinear oscillator (8) we are now neglecting temporal
correlation in the velocity w_p, or, which is exactly the same from
a computational point of view, that we are now studying the w_p
process on a time scale larger than the correlation time. Before
studying eq.(12) and showing its connection with diffusion approxi-

mations let's consider if it is possible to get an alternative model to the one represented by system (8).

An alternative to system (8)

An alternative model to represent dispersion in a non-homogeneous turbulent field, i.e. an alternative to eq. (8) could be given by the following system :

$$dz = \sigma(z) \ v \ dt \tag{13a}$$

$$dv = - \ \gamma v \ dt + d\beta_t \tag{13b}$$

i.e. instead of representing particle velocity as a first order Markov process we are attempting to represent displacements as a stochastic non-Markovian process (Jazwinski, 1970), driven by a coloured noise v, having the properties described for process generated by eq. (6). It must be noted, however, that even if z is not Markovian, the vector process $[z,\bar{v}]$ is Markovian and for it the Fokker-Planck equation can obviously be derived :

$$\frac{\partial p}{\partial t} + \frac{\partial}{\partial z} \ (\sigma \ v \ p) = \frac{\partial}{\partial v} \ (\gamma \ v \ p) + \frac{1}{2} \ \frac{\partial^2 p}{\partial v^2} \tag{14}$$

Inspection of eq. (14) immediately shows that solution p(z)=const. is possible only if σ is not a function of z. This proves that also eqs.(13) studied as a possible extension of the Ornstein-Uhlenbeck process to the case of non-homogeneous turbulence lead to the same conclusions drawn in the previous paragraph. It is interesting to point out that the available theoretical results for system (13) have been derived for the limit situation $\gamma \to \infty$. In this case it has been shown (Jazwinski, 1970) that system (13) can be represented by the following stochastic differential equation :

$$dz = \frac{1}{2} \ \frac{d\sigma^2}{dz} \ dt + \sigma(z) \ d\beta_t \tag{15}$$

Eq.(15) is very similar to eq.(12), which shows that the two approaches followed in getting system (8) and (13), although different at a first look, imbed essentially the same concept. Before studying the solution provided by eq.(15) and eq.(12) we remark that (12) differs from (15) and that a certain functional relationship exists between the coefficient of the stochastic and the deterministic part. To point out the consequences of this circumstance we get the Fokker-Planck equation for eq.(15) :

$$\frac{\partial p}{\partial t} = \frac{1}{2} \ \frac{\partial}{\partial z} \ (\ \sigma^2 \ \frac{\partial p}{\partial z} \ ) \tag{16}$$

It turns out that probability density p is in this case the solution
of a diffusion equation, allowing an asymptotic solution p(z)=const.
as required.

THE DIFFUSION EQUATION APPROXIMATION

The analysis developed above shows that both systems (8) and
(13) produce solutions which are physically unacceptable. This
means that while the Ornstein-Uhlenbeck process is a good mathematical description of a physical process such as the Brownian
motion, its extention to the case ot atmospheric turbulence by a
set of two stochastic differential equations is totally unsatisfactory. The above analysis has also shown that description of the
dispersion process using eq.(12) or eq.(15) gives correct results
although, of course, we had to assume (rather unrealistically for
atmospheric turbulence) that velocity is a temporally non-correlated process. As no better representation is, to the author's
knowledge, at present available, we can use this approximation and
model dispersion of settling particles in a two-dimensional turbulent field by the system :

$$dx = u \, dt \tag{17a}$$

$$dz = (w_f + \frac{1}{2} \frac{d\sigma^2}{dz}) \, dt + \sigma d\beta_t \tag{17b}$$

while for non-settling particles we must simply put $w_f=0$. The
Fokker-Planck equation for system (17) is given by :

$$\frac{\partial p}{\partial t} + \frac{\partial}{\partial x} (u \, p) + w_f \frac{\partial p}{\partial z} = \frac{1}{2} \frac{\partial}{\partial z} (\sigma^2 \frac{\partial p}{\partial z}) \tag{18}$$

We find therefore that p is the solution of an advection-diffusion
equation quite similar to the one commonly derived from the continuity equation for a passive scalar in a turbulent field. This
means that the eddy-diffusivity approximation is the simplest
Lagrangian model that can be derived to describe atmospheric
dispersion. As a final consideration we note that both (17) and
(18) express the k-theory in conservative form and with parameters
more general than those commonly used.

The problem of boundary conditions

In the above discussion the problem of boundary conditions
has been deliberately neglected in order not to add a supplementary
complexity to the analysis carried out. Let's briefly consider
it here. As far as eq.(18) is concerned it is well-known that
boundary conditions ranging from total reflection to total absorption can be imposed by specifying the values of p and/or of its
first derivative. If we want to compute p by using system (17),

then we need to impose some conditions on the trajectories once
they reach the boundaries of our computation domain. If we want
to get total reflection, then we simply need to extend f and σ
fields beyond boundaries by imposing $f(-z)=-f(z)$ and $\sigma(-z)= \sigma(z)$,
then proceed in the computation and at the end reflect the obtained
trajectories. Total absorption is obviously obtained by removing
from the computation every particle hitting the absorbing boundary,
while any combination of the two events can be obtained by specifying the probability of each of them. This probability can
obviously be expressed as a function of the "parameters" of the
model (namely wind speed deterministic field and stochastic terms
coefficients) thus allowing to model the deposition-resuspension
phenomenon, on the basis of experimental data.

CONCLUSION

     Starting from the basic equations describing particle motion
under the action of both deterministic and stochastic forces it
has been shown that Lagrangian models proposed to study atmospheric
diffusion are finite-difference approximations of a set of stochastic differential equations. By using results given by the theory
of these equations it has been shown that, when the parameters
of the equations depend on spatial coordinates, solutions obtained
are in contrast with physical considerations. It is finally shown
that if, instead of using for the velocity process the nonlinear
harmonic oscillator equation, we use for the displacements a first
order Markov process, the eddy-diffusivity approach is obtained
in its most general form. This is equivalent to assuming either
that velocity process is non-correlated in time, or that dispersion
is being investigated on a time scale larger than the correlation
time.

ACKNOWLEDGEMENTS

     The author wants to express his appreciation to Dr. R. Benzi
of the IBM Scientific Center of Rome for his helpful discussions.

REFERENCES

Eltgroth M.W. and Hobbs, P.V., 1979 : Evolution of particles in
     the plumes of coal-fired power plants - II. A Numerical model
     and comparisons with field experiments, Atmos. Environ., 13,
     953-975.
Hanna, Steven R., 1981 : Effects of release height on σ_y and σ_z in
     daytime condition. Air Pollution Modelling and its Application
     vol. I., ed. C. De Wispelaere. Plenum Press, New York.
Janicke, Lutz, 1981 : Particle simulation of inhomogeneous turbu-
     lent diffusion. Preprints of the 12th NATO/CCMS Techn. Meeting,
     Palo Alto.

Jazwinski, A.H., 1980 : Stochastic processes and filtering theory
    Academic Press.
Lamb, R.G., Hogo, H., and Reid, L.E., 1979 : A Lagrangian approach
    to modeling air pollutant dispersion, EPA Report,
    EPA-600/4-79-023.
Ludwig, Donald, 1975 : Persistence of dynamical systems under ran-
    dom perturbations, SIAM Review, 17,4 605-540.
Reid, J.D., 1979 : Markov chain simulations of vertical dispersion
    in the neutral surface layer for surface and elevated releases,
    Boundary-Layer Met., 16, 3-22.
Schuss, Zeev, 1980 : Theory and applications of stochastic diffe-
    rential equations. John Wiley and Sons, New York.
Sehmel G.A., 1980 : Particle and gas dry deposition : a review,
    Atmos. Environ. 14, 983-1011.
Seinfeld J.H., 1975 : Air Pollution - Physical and Chemical
    Fundamentals McGraw-Hill.

DISCUSSION

P. ZANNETTI   Where is the cross correlation
            term $\overline{u'w'}$ in your equations ?

P. MELLI   It's the term q_n.

A MONTE CARLO-MODEL FOR THE SIMULATION OF LONG-RANGE TRANSPORT OF AIR POLLUTANTS

J. Lehmhaus, E. Roeckner, I. Bernhardt and J. Pankrath[1]

Meteorologisches Institut der Universität Hamburg
and [1]Umweltbundesamt Berlin, Fed. Rep. Germany

INTRODUCTION

Most trajectory models developed for simulating long-range transport of e.g. sulfur components are based on the assumption that a vertical air column is advected horizontally without being distorted by horizontal or vertical wind shears. Within a column advected with some representative layer averaged wind speed, the change of concentration c with transport time τ is given by

$$\frac{\partial c}{\partial \tau} = -w\frac{\partial c}{\partial z} + \frac{\partial}{\partial x}(K_x\frac{\partial c}{\partial x}) + \frac{\partial}{\partial y}(K_y\frac{\partial c}{\partial y}) + \frac{\partial}{\partial z}(K_z\frac{\partial c}{\partial z}) + Q-S \qquad (1)$$

where x, y, z are the space coordinates, K_x, K_y, K_z the eddy diffusion coefficients, Q, S the sources and sinks, respectively, and w the mean velocity. Generally, (1) is further simplified by neglecting vertical advection and assuming a complete mixing within each column up to some specified height which is either constant (Eliassen, 1978) or may increase slightly with time (Bhumralkar et al., 1981). The sink at the surface depends simply on the relation v_d/h, where v_d is a constant deposition velocity and h the height of the column. In the OECD model (Eliassen, 1978) horizontal diffusion is neglected while Bhumralkar et al. (1981) assume an increase of the horizontal area of the column with time depending on the local deformation of the large-scale velocity field.

In statistical models equation (1) is solved by assuming some v_d and K_z which are either representative for the whole integration area and simulation time (e.g., Bolin and Persson, 1975) or prescribed in a diurnal cycle (Shannon, 1981). In these models the horizontal dispersion produced by the variability of the large scale horizontal wind field is described by a normal distribution whose parameters are calculated from a large number of particles starting in 3-hourly intervals.

Most of these simplifying assumptions are due to economic constraints in long-term simulations and/or the insufficient density of meteorological routine observations. It is suspected, however, that some of the above restrictions, e.g. the neglect of wind shear effects may lead to serious errors, especially in episode studies. It seems therefore desirable to develop a model which is basically free of the above simplifying assumptions and which may in principle take into account a large number of spatial and temporal inhomogeneities.

The transport model which is based on a simple Monte Carlo technique is presented in section 2. In section 3 we describe our method of producing the meteorological input by a forecast model and a diagnostic boundary layer model. Some preliminary test results which demonstrate the feasibility of the method are discussed in section 4.

THE TRANSPORT MODEL

A Monte Carlo method is used for solving the advection-diffusion equation

$$\frac{\partial c}{\partial t}+u\frac{\partial c}{\partial x}+v\frac{\partial c}{\partial y}+w\frac{\partial c}{\partial z} = \frac{\partial}{\partial x}(K_x\frac{\partial c}{\partial x})+ \frac{\partial}{\partial y}(K_y\frac{\partial c}{\partial y})+ \frac{\partial}{\partial z}(K_z\frac{\partial c}{\partial z}) \qquad (2)$$

For simulating the process described by (2), a large number of particles is released at some point. Their trajectories are calculated independently according to the meteorological conditions found at the position of each particle. Superimposed upon the mean velocities are random fluctuations describing the effect of turbulence. In the limiting case of an infinite number of particles, infinitesimal timestep and vanishing spatial variations of the eddy diffusion coefficients, the solution coincides with the analytical solution of (2).

Horizontal transport

The horizontal transport is calculated from the components of the mean wind u (x, y, z, t) and v (x, y, z, t) which have to be specified from observations or from a meteorological model simulation.

The trajectory equations

$$x_i(t+\Delta t) = x_i(t)+(u_i(x_i,y_i,z_i,t)+\alpha_i).\Delta t$$

$$y_i(t+\Delta t) = y_i(t)+(v_i(x_i,y_i,z_i,t)+\alpha_i).\Delta t$$

(3)

are integrated for every particle (index i). The α_i are random numbers with a mean equal to zero and variances δ_u , δ_v which represent the turbulent fluctuations. This eddy diffusion process is simulated by calculating the variances tentatively from

$$\delta_u(x,y,z,t) = a.u(x,y,z,t)$$

$$\delta_v(x,y,z,t) = a.v(x,y,z,t) \qquad (4)$$

where a is an empirical constant

In Figure 1 we illustrate the method in 24-hours simulation of sulfur components with a continuous source (puff of 2400 particles every 3 hours) over England. The wind data are taken from a meteorological forecast model. The design of the experiment will be given later. At this time we would only like to point to some differences that might occur if applying the Monte Carlo method instead of the more conventional method discussed in the introduction : In the conventional method the particles within one puff are displaced with one velocity found in the centre of the puff. In our method each particle is displaced with the actual velocity found at the respective position.

In Fig. 1 we compare the vertically integrated concentration of SO_x after 24 h for the conventional method a) and the Monte Carlo method b). In these experiments a = 0.1 was chosen for the variances in (4), corresponding to a mean eddy diffusion coefficient of $K_h \sim 10^4 m^2/s$. Additionally, in Figs. 1c, d we show the respective results for a = 0.5, corresponding to $K_h \sim 2.10^5 m^2/s$. The distribution in Fig. 1a) shows clearly the individual puffs while the Monte Carlo method in Fig. 1b) gives a much smoother distribution due to additional dispersion caused by inhomogeneities of the wind field. The increase of the turbulent fluctuations (Fig. 1c, d) causes an increase of the area affected by the puffs, especially in the case of the conventional method c). The horizontal extent obtained with our method is much smaller which may be understood from convergences in the wind field. Thus, the Monte Carlo method appears to be more sensitive to inhomogeneities of the wind field than the conventional method.

Vertical Transport

The vertical transport (and also the transformation and sink terms) are calculated in the centre of the horizontal trajectory, i.e. at

$$x_{iM} = 0.5 (x_i (t + \Delta t) + x_i (t))$$

$$y_{iM} = 0.5 (y_i (t + \Delta t) + y_i (t)) \qquad (5)$$

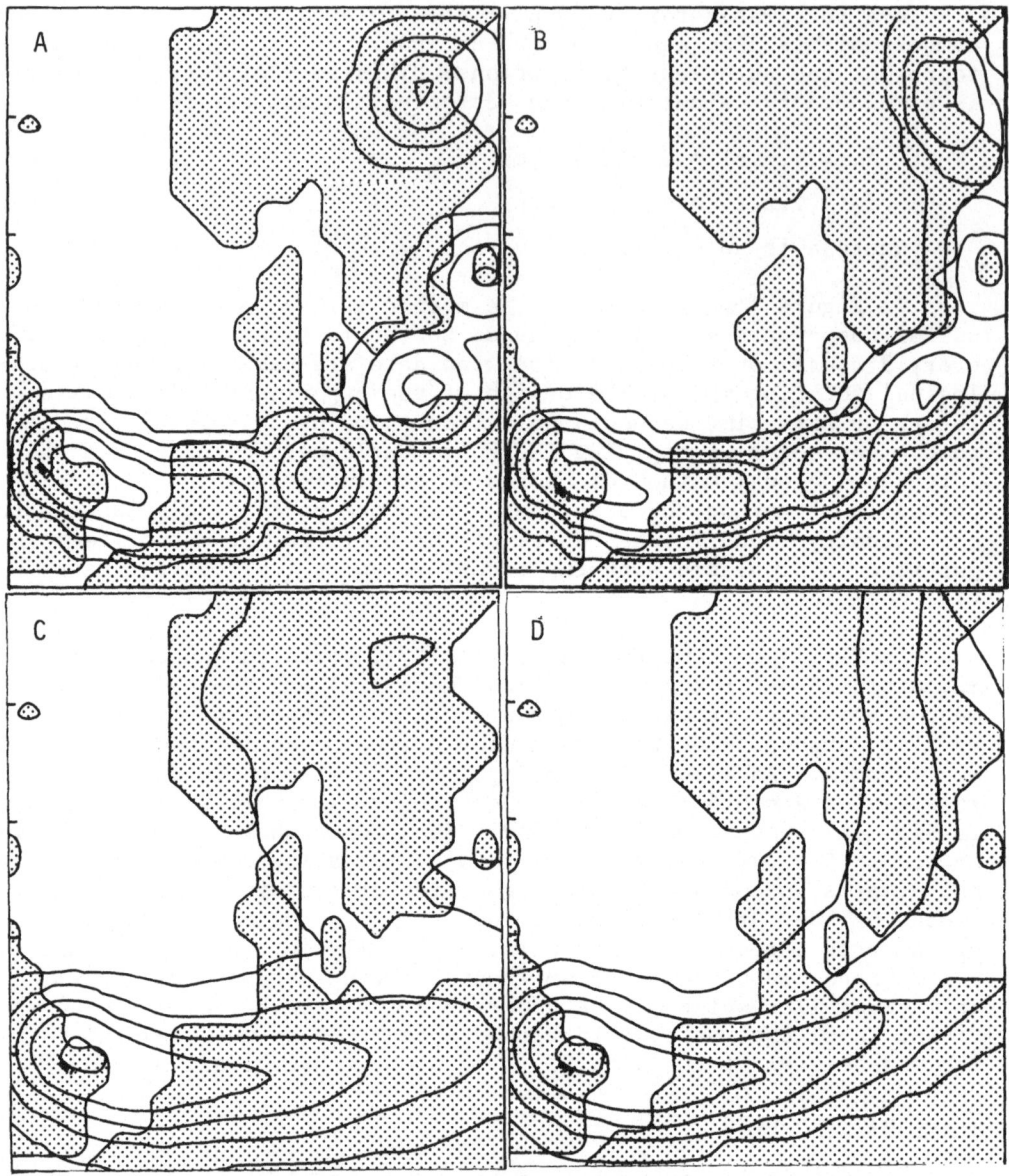

Fig. 1 Vertically integrated concentration of SO_x after 24 hours. Puffs of 2400 particles/volume were emitted from a source over England. Wind data from a meteorological model. a) and c) : all particles are displaced with the same velocity within one puff. b) and d) : all particles are displaced with the actual velocity within the puff. a) and b) : $K_H \sim 10^4 m^2/s$. c) and d) : $K_H \sim 2.10^5 m^2/s$. Contour spacing : $1,15,50,100,200,...$ particles per column (area : ~ 50 km x 50 km).

The vertical transport is carried out in two steps.

First the particles are displaced with the mean vertical velocities according to the trajectory equation

$$z_i (t+\Delta t) = z_i (t)+w(x_{iM},y_{iM},z_i,t).\Delta t \tag{6}$$

Particles which are found below z = 0 after the integration, are reflected.

In a second step the turbulent diffusion is simulated in the following way.

For calculating the displacement of a particle in the vertical direction one must know the probability distribution for the new position after one time step. This distribution is given by the solution of the one dimensional diffusion equation

$$\frac{\partial c}{\partial t} = \frac{\partial}{\partial z}(K_z \frac{\partial c}{\partial z}) \tag{7}$$

Using an eddy diffusion coefficient K_z independent of height and assuming initial and boundary conditions according to

$$c(z,o) = \delta(z-h) \tag{8}$$

$$(K_z \frac{\partial c}{\partial z} - v_d \cdot c) \Big|_{z=o} = 0 \tag{9}$$

where v_d is the deposition velocity, h the source height and δ the Kronecker symbol, the solution of (7) is given by

$$c(z,t) = \frac{1}{\sqrt{4\Pi Kt}} \{ e^{\frac{-(z-h)^2}{4Kt}} + e^{\frac{-(z+h)^2}{4Kt}} \} - \frac{v_d}{K} e^{\frac{v_d}{K}(z+h+v_dt)}$$

$$\cdot \text{ erfc } (\frac{z+h+2v_dt}{\sqrt{4Kt}}) \tag{10}$$

The washout effect is simply included by assuming an exponential decay in c so that the diffusion equation (7) takes the form

$$\frac{\partial c}{\partial t} - \lambda c = \frac{\partial}{\partial z} (K_z \frac{\partial c}{\partial z}) \tag{11}$$

with a constant decay coefficient λ.

For solving (7) or (11) with the Monte Carlo method the vertical coordinate is devided into discrete layers of equal depth Δz centered at z_i (i = i'th layer). Equation (1/0) is used to evaluate the probabilities of the particles to reach the different layers according to

$$P_i = 0.5 \ (c \ (z_i + \frac{\Delta z}{2}, \Delta t) + c(z_i - \frac{\Delta z}{2}, \Delta t)) . \Delta z \qquad (12)$$

The probability ε for a particle to be absorbed is given by

$$\varepsilon = 1 - \int_0^\infty c(z,t)dz = 1 - \Sigma_i P_i \qquad (13)$$

From an equal distribution over the interval $0,1$ we choose a random number to determine the new position or absorption according to the P_i's and ε.

The method was tested by comparing the flux of particles at z = 0 with the analytical solution of (7) together with the boundary condition (9). In Fig. 2 we show the integral over the surface flux obtained analytically together with the number of particles absorbed at the surface calculated from 10 identical Monte Carlo realizations (mean and standard deviation), normalized with the source intensity. The mean Monte Carlo solution is close to the analytical solution though only 100 particles have been used in each realization (i.e. 1000 particles totally).

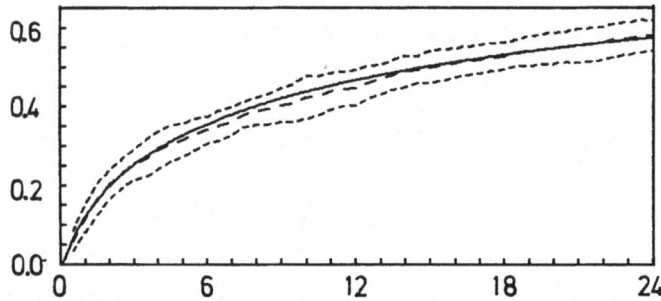

Fig. 2. Material accumulated at the surface, normalized with source intensity, as a function of time in hours. Parameters : K = 5m²/s, h = 100 m, v_d= 0.01m/s, Δt= 1800s.
——— analytical solution
— — — average of 10 Monte Carlo simulations
------ standard deviation

Transformation rate

The transformation $SO_2 \rightarrow SO_4$ is calculated from a constant rate of k_T = 0.065 %/30 min. After each timestep, which is 30 min. normally, a random number δ_i is chosen for each particle. If $\delta_i > k_T$, the particle still represents a mass of SO_2, otherwise the particle is transformed to SO_4.

Emission

We assume that all particles are equipartitioned at the moment of the emission in a volume which has a typical size of 1° x 1° horizontally and extends between 30 m and 200 m vertically.

Dry deposition

The parameterization of dry deposition is based on the resistance concept which relates the deposition velocity to a resistance of the surface r_S and of the lowest meters of the atmosphere r_A according to

$$v_d = (r_S + r_A)^{-1} \tag{14}$$

The surface resistance is assumed to be constant except for a modulation by prescribed diurnal and seasonal cycles. Over sea the surface resistance vanishes. For the atmospheric part we use the relation

$$r_A = u/u_x^2 \cdot \eta \tag{15}$$

where u is the mean velocity in the surface layer and u_x the friction velocity. The factor η represents the effect of stability derived from a series of experiments with a boundary layer model (Dunst, 1982) according to

$$\eta = (1.35 - Ri)^{-1} \tag{16}$$

where Ri is the Richardson number.

The way of using the deposition velocity in the Monte Carlo model has already been discussed in section 2.2.

Wet deposition

According to Chamberlain (1960) the decay constant λ in (11) can be related to the precipitation amount J (mm/h).

At present we use

$$\lambda = 10^{-5} \cdot \sqrt{J} \qquad \text{for } SO_2$$

$$\lambda = 10^{-4} \cdot \sqrt{J} \qquad \text{for } SO_4$$

However, in the experiments described below this process is neglected.

METEOROLOGICAL INPUT DATA

At the present stage of model development we use input data from a large scale meteorological forecast model. The main advantage of model data with respect to observational data are (i) dynamical consistence of the three-dimensional velocity field, (ii) high spatial and temporal resolution, and (iii) inclusion of subgrid-scale processes like turbulence, convection, radiation and cloud processes. The main disadvantage for practical applications, e.g. comparison of simulated and observed sulfur deposition, is that weather forecast models produce errors which may be substantial after a few days.

The model used for creating the data set for the transport model is a regional version of the Hamburg University genereal circulation model (Roeckner, 1979). The model covers an area of $\sim$ 4500 km x 4500 km centred at Ireland approximately with 8 layers of constant mass ($\sim$ 125 mbar). Actually, a normalized pressure coordinate $\delta = (p-p_\Gamma)/(p_s-p_\Gamma)$ where p is the pressure and p_Γ, p_s are the pressures at top and bottom of the model respectively. The horizontal resolution is $\Delta\lambda \sim 2.8°$, $\Delta\phi \sim 1.4°$ (λ, ϕ = longitude, latitude). A realistic orography is used. The lateral boundary conditions are fixed in time in the experiment described below. Turbulent surface fluxes of momentum, heat and water vapour are parameterized from a generalized resistance law. Convective fluxes of heat and water vapour are obtained from a simple cloud model. Radiational fluxes are neglected in the version used here.

The initial conditions are derived from analyses of the German Weather Service at 18 November 1973, 12 GMT by applying a dynamical initialization procedure for calculating the wind field. Surface temperatures over sea and land are interpolated from routine obser-vations. The synoptic situation is characterized by a strong cyclone deepening further and travelling eastwards (Fig. 3). The relatively strong winds over western Europe are turning from southwesterly to northwesterly direction. This particular initial field was chosen for convenience because it was already initialized for a different project. Possibly, this synoptic situation characterized by strong horizontal advection may not be very suitable for testing a tran-sport model. The result of the sensitivity experiments presented in section 4 should also be viewed in this perspective.

    All variables that might be relevant for the simulation of
sulfur transport have been stored at 1/2-hourly intervals in the
region 11°W – 24°E ; 42° – 69°N at the lower 6 levels of the model
up to a height of ∿ 5 km (3 wind components, temperature, moisture,
pressure, mixing height, cloud coverage, precipitation, friction
velocity etc.). This data set has been extended by calculating the
wind structure in the boundary layer below 500 m at 5 additional

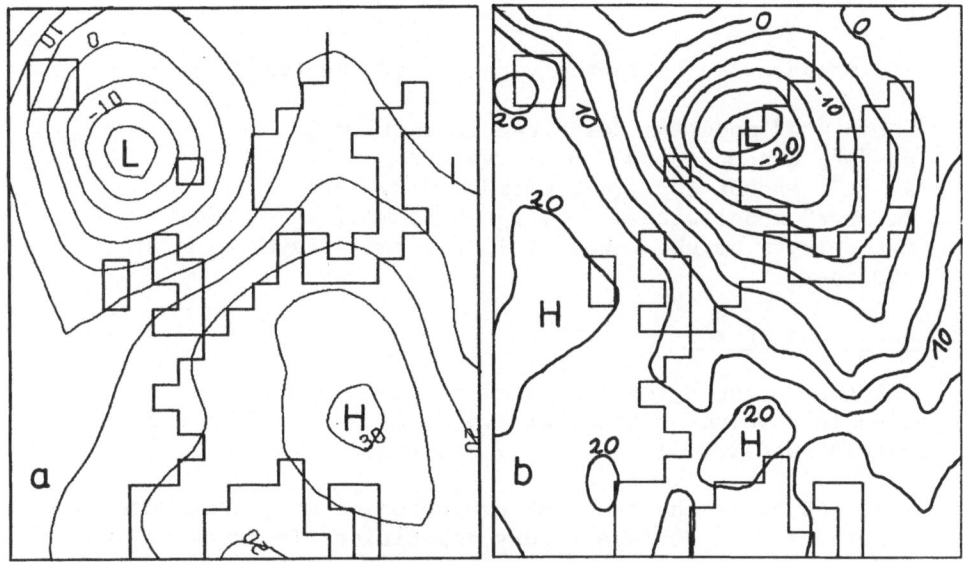

Fig. 3. Surface pressure (deviations from 1000 mbar)
        a) t=0 (18 November 1973, 12 FMT)
        b) t = 24 h (model simulation)

levels (30, 100, 200, 300, 400 m) from a 1-dimensional diagnostic
boundary layer model. The boundary values at 500 m and the surface
(wind, geostrophic wind and temperature) are prescribed by the forecast model every 30 min. The wind profile depends crucially on the
eddy diffusion formulation. Here we follow Dunst (1982)who developed
a concept for the eddy diffusion coefficient depending on the
Richardson number and height mainly.

    More satisfactory would be a complete prognostic simulation of
the boundary layer structure. The diagnostic model can only give some
first guess of the wind structure developing under the large-scale
conditions simulated by the forecast model.

To summarize, a comparatively complete data set has been produced during a 24-hours simulation for application in the Monte Carlo model discussed in section 2. According to these data the transport model covers a large part of Europe and extends up to $\sim$ 5 km height with a 10-layer resolution. The lowest 1 km is represented by 6 layers. A time step of 30 min is used.

EXPERIMENTS

We compare the following versions of the transport model

A : "complete" model as described in section 2
B : as "A", except for $w = 0$ everywhere
C : as "A", except for $v_d = 0.01$ m/s everywhere
D : "simple" model with $w = 0$, $v_d = 0.01$ m/s, $K_z = 5 m^2/s$, mixed layer height equal to 1500 m and 1-layer wind at 950 mbar.

Instantaneous source

At the beginning of each simulation 5000 particles are released in a volume centred at 1.5°W, 54°N and 115 m height (between 30 and 200 m).

Fig. 4a shows the temporal evolution of area integrated SO_2 accumulated at the surface by dry deposition. The results of exp. C and D coincide nearly, therefore only exp. C is shown. Most striking is the reduction of deposition in ex. C. In exp. A and B the deposition velocity is calculated from (14) and (15) showing a linear relationship between v_d and u at neutral conditions. The particular synoptic situation used in these experiments is characterized by strong winds (cf. section 3) which lead to deposition velocities up to 0.03 m/s explaining the differences between A and C.

The influence of the mean vertical velocity on the total SO_2 deposition is comparatively moderate and becomes substantial only during the last phase of the experiment when the puff enters a region of strong upward wind within a front (Figs. 3b and 4b). It should be noted further that the forecast model needs about 6 hours for producing realistic vertical wind velocities.

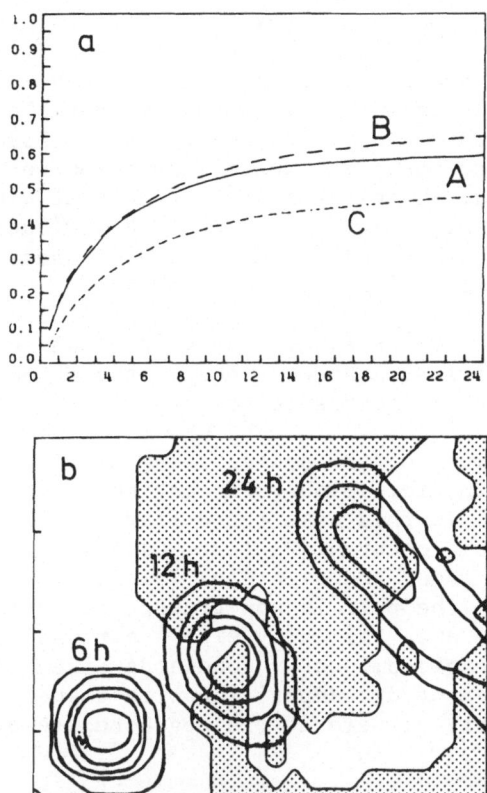

Fig. 4. a) SO_2 deposited at the surface, normalized by source intensity, as a function of time for the experiments A, B, C (see section 4).

b) Vertically integrated concentration for exp. A after 6, 12 and 24 h (puff emitted over England). Contour spacing as in Fig. 1.

Continuous source

      Continuous sources are simulated by releasing 400 particles
at each time step which is 30 min normally. We use the same source
point over Englang as in section 4.1. Fig. 5 shows the spatial
distribution of the SO_2 dry deposition after 24 hours for the
experiments A-D. The differences between A and B are rather weak
because the area of strong vertical winds in the northeasterly
part of the area is reached only by a few particles. The main
differences between A and C occur in the area close to source,
especially over the North Sea where deposition velocities up to
0.03 m/s are found. The neglect of the vertical wind shear (Fig.
5d) leads to a displacement of the whole distribution which is
obvious especially when the 820-mbar wind is used (Fig. 6) instead of the 950-mbar wind.

      The SO_2 budget for the whole area is shown in Fig. 7a. As
in the case of the instantaneous source, the effect of the variable
deposition velocity is most evident.

      The displacement of the source area may lead to different
results as shown in Fig. 7b where a continuous source at 7°E,
51.5° N is chosen. In this case the differences between a variable
v_d (exp. A, B) and a constant v_d (exp. C) are much smaller and
partly opposed to those found in Fig. 7a. This may be understood
from the fact that the puff is now transported mainly over land,
in a high pressure area initially with light winds. Moreover, the
vertical winds are predominantly downwards directed in this area
causing an increase of dry deposition with respect to the case
where the mean vertical velocity is neglected (exp. C).

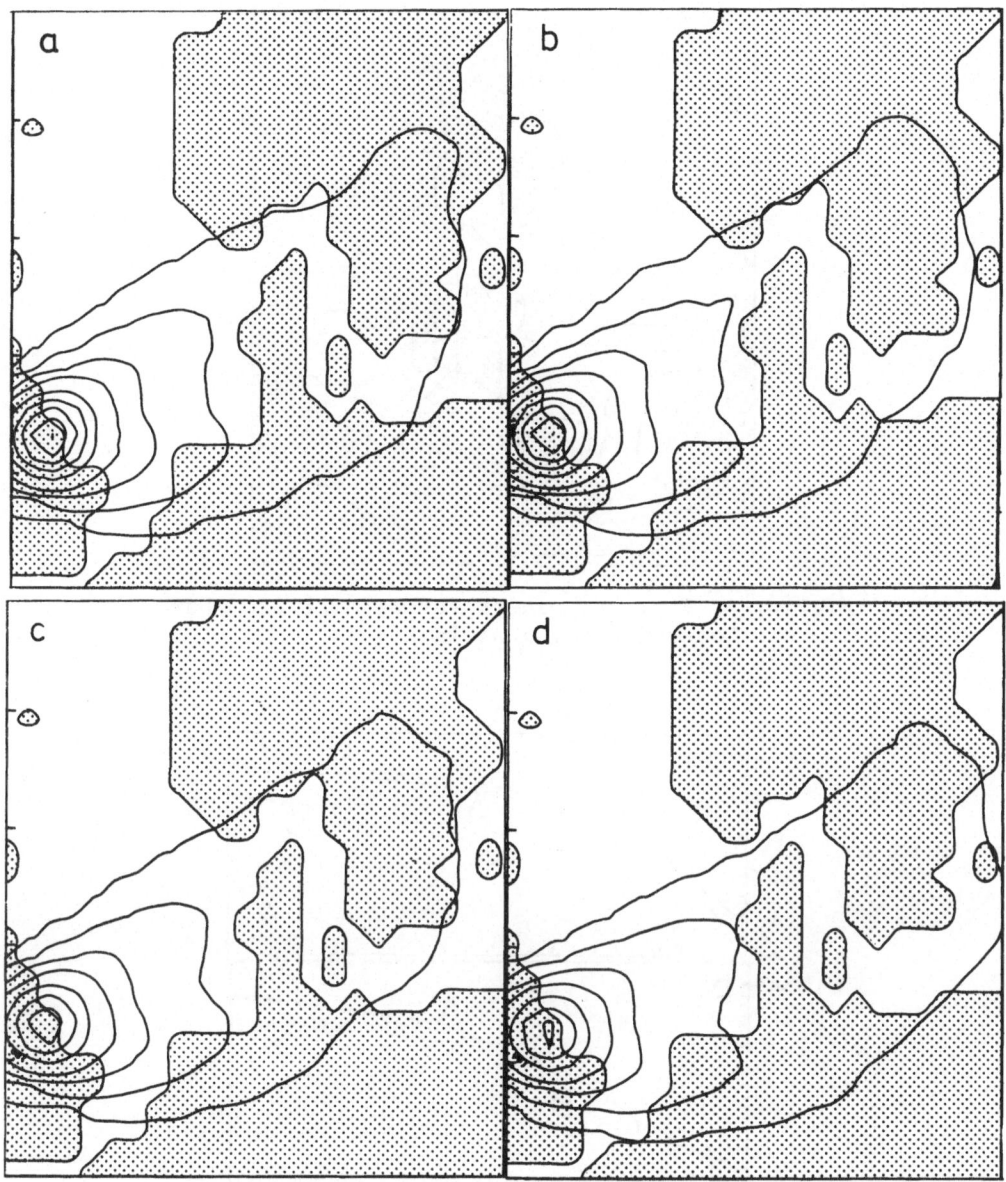

Fig. 5. Deposition patterns for a continuous source over England
        after 24 h for experiments A, B, C, D (see section 4).
        Contour spacing as in Fig. 1.

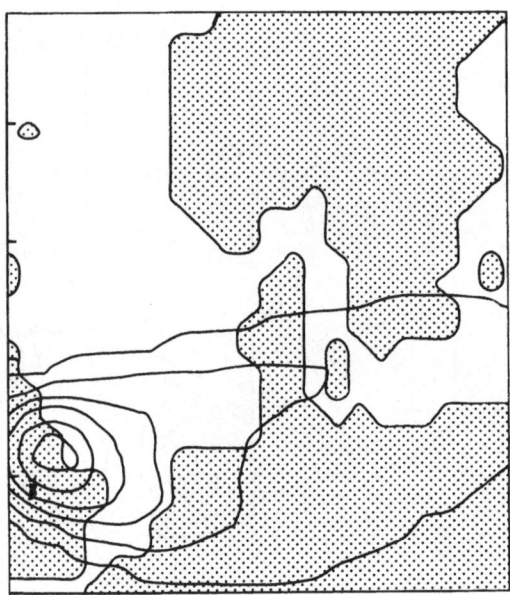

Fig. 6. As Fig. 5d, except for wind at 820 mbar.
Contour spacing as in Fig 1.

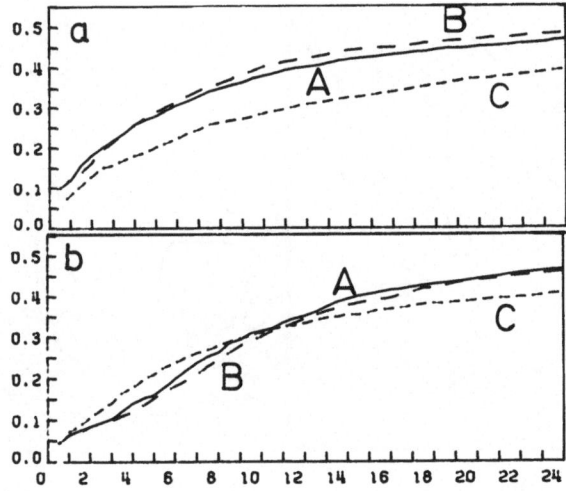

Fig. 7. SO_2 deposited at the surface, normalized by the amount
of particles released until time t as a function of t
for experiments A, B, C (see section 4.) Continuous
source over a) England   b) W. Germany.

CONCLUDING REMARKS

A preliminary version of a 3-dimensional transport model has been presented for simulating long-range transport of sulfur components. The model is based on a simple Monte Carlo technique which allows to simulate a variety of atmospheric processes that may influence sulfur dispersion without restrictive assumptions.

For a first test performed to demonstrate the feasibility of the method, the meteorological input was produced by a 24-hours simulation with a regional forecast model. The wind structure in the boundary layer was estimated by a diagnostic 1-dimensional boundary layer model.

The calculations performed with different versions of the Monte Carlo model indicated the importance of wind inhomogeneities, mean vertical velocity and deposition velocity depending on the structure of the boundary layer for the simulation of sulfur transport.

ACKNOWLEDGEMENTS

This study was sponsored by Umweltbundesamt Berlin (West), Fed. Rep. Germany.

REFERENCES

Bhumralkar, C.M., Mancuso, R.L., Wolf, D.E., Johnson, W.B., and Pankrath, J., 1981, Regional air pollution model for calculating short-term (daily) patterns and transfrontier exchanges of airborne sulfur in Europe, Tellus, 33:142.

Bolin, B., and Persson, C. 1975, Regional dispersion and deposition of atmospheric pollutants with particular application to sulfur pollution over Western Europe, Tellus, 27:281.

Chamberlain, A.C., 1960, Aspects of the deposition of radioactive and other gases and particles, Int. J. Air Pollut., 3:63 ;

Dunst, M., 1982, On the vertical structure of the eddy diffusion coefficient, Atmos. Environ., in press.

Eliassen, A., 1978, The OECD study of long range transport of air pollutants : Long range transport modeling, Atmos. Environ., 12:479.

Roeckner, E., 1979, A hemispheric model for short range numerical weather prediction and general circulation studies. Beitr. Phys. Atmosph., 52:262.

Shannon, J.D., 1981, A model of regional long-term average sulfur atmospheric pollution, surface removal, and net horizontal flux, Atmos. Environ., 15:689.

DISCUSSION

P. ZANNETTI

           You should change Eq (4)
which does not seem correct to me.

E. ROECKNER

          You are right. We have used
this relation only tentatively. In the future we
shall use some eddy diffusion concept to relate
the δ_u, δ_v to properties of the mean velocity
field.

PRELIMINARY ANALYSIS OF LAGRANGIAN MONTE-CARLO

METHODS FOR NONHOMOGENEOUS TURBULENCE*

E. Runca and M. Posch

IIASA, A-2361 Laxenburg, Austria

G. Bonino

Laboratorio di Cosmogeofisca del CNR
Corso Fiume 4, Torino, Italy
Istituto di Fisica, Generale
della Universita, Torino, Italy

INTRODUCTION

This paper reports on preliminary numerical and analytical results obtained as a part of a study underway for the development of a Lagrangian point source model. The analysis, for the sake of simplicity and computer time saving has been so far restricted to the two-dimensional case of dispersion from an infinite crosswind line source between two solid boundaries (ground and inversion layer) as illustrated in Fig. 1.

Drawing from the work of Lamb et al. 1979, it can be shown with reference to Fig. 1 that for an inert pollutant the following equation holds:

$$[C] \equiv \frac{1}{\Delta x \Delta z} \int_{z_p - \frac{\Delta z}{2}}^{z_p + \frac{\Delta z}{2}} \int_{x_p - \frac{\Delta x}{2}}^{x_p + \frac{\Delta x}{2}} <C(x,z,t)>dxdz = \frac{Q<T(x_p,z_p,t|x_s,z_s,0)>}{\Delta x \Delta z}$$

(1)

---

*Research partly supported by Commission of the European Communities, Contract ENV-598-I(S)

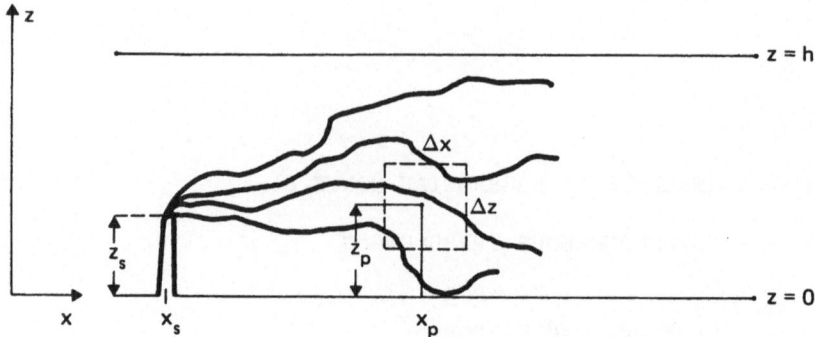

Fig. 1.   Geometry of the model: h = height of the inversion layer;
          z_s = source height; Δx and Δz widths of the rectangular
          cell centered on the receptor point p.

where:

Q   : emission rate [kg/ms]

$<T(\cdot|\cdot)>$: average time spent by a pollutant particle in the
                 cell $[x_p \pm \frac{\Delta x}{2}; z_p \pm \frac{\Delta z}{z}]$ in the time interval $[0,t]$.

$[C] \equiv \frac{1}{\Delta x \Delta z} \iint <C>$ dxdz: ensemble average concentration averaged
                 over the cell $[x_p \pm \frac{\Delta x}{2}; z_p \pm \frac{\Delta z}{2}]$.

To put equation (1) into practice a suitable algorithm must
be developed to compute the average time spent by a pollutant
particle in a given cell. This implies the simulation of an ensem-
ble of air pollutants trajectories from which then $<T(\cdot|\cdot)>$ can
be estimated. The simplest way to proceed is to use the following
recursive equations:

$$x_{n+1} = x_n + u_{n+1} \cdot \Delta t$$
$$z_{n+1} = z_n + w_{n+1} \cdot \Delta t$$

(2)

where (x_n, z_n) is the particle position at time equal to $n \cdot \Delta t$,
and (u_{n+1}, w_{n+1}) is the particle velocity from $n\Delta t$ to $(n+1)\Delta t$.

To apply (2), the particle velocity must be specified at each time step. In this analysis we have assumed a) the horizontal component equal to the mean wind speed, b) the vertical component to be due only to atmospheric turbulence. Thus we have

$$u \equiv <u>$$

$$w \equiv w'$$

The stochastic nature of w is the principal difficulty in applying (2). Generally it is assumed that w can be described in terms of a Markov process. Using suffix e and ℓ to indicate Eulerian and Lagrangian quantities respectively, w is then given by

$$w^{\ell}(\Delta t, z_s) = w^e(z_s) \qquad\qquad n = 1$$
$$w^{\ell}((n+1)\Delta t, z_s) = \alpha w^{\ell}(n\Delta t, z_s) + \rho((n+1)\Delta t) \quad n > 1 \qquad (3)$$

That is, at the first time step w is selected at random from the Eulerian distribution of w^e at source height. In the next time step w is a function of the previous value plus a quantity ρ which has to be chosen at random from a given distribution. The parameter α is positive, less than one and is given by

$$\alpha = 1 - \frac{\Delta t}{T_L}$$

where T_L is the Lagrangian time scale.

Algorithm (3) has been originally proposed by Smith (1968), and has been applied by many authors (see, e.g., Thompson (1971), Jonas and Bartlet (1972), Reid (1979), Lamb et el. (1979), Wilson et al. (1981a)). In the form reported above it has been discussed by Runca et al. (1981). The reader could refer to this paper for more details on the application of (3). Briefly (3) has proved to be suitable to describe diffusion in homogeneous turbulence by taking the probability density of ρ Gaussian with

$$<\rho> = 0; \quad <\rho^2> = (1-\alpha^2) <w^{e^2}> \qquad\qquad (4)$$

Algorithm (3) has also proved to be applicable to those cases for which

$$<w^{e^2}> = \text{constant}$$

$$T_L = T_L(z)$$

See, e.g., Reid (1978) and Wilson et al. (1981a). Extension of algorithm (3) to the inhomogeneous case in which $<w^{e^2}>$ is a function

of z is still a matter of research. Janicke (1981), Runca et al.
(1981), Wilson et al. (1981b) through different ways reached the
conclusion that the moments of ρ given by (4) must be modified.
While Runca et al. (1981) and Wilson et al. (1981b) modified only
the first moment of ρ, Janicke (1981) suggested also a modification
of $\langle\rho^2\rangle$.

Following Janicke and using for the standard deviation of the
vertical Eulerian velocity fluctuations the more familiar symbol
σ_w instead of $\sqrt{\langle we^2\rangle}$, (4) becomes:

$$\langle\rho\rangle = \Delta t \ \frac{d\sigma_w^2}{dz}$$

$$\langle\rho^2\rangle = (1-\alpha^2) \ [\sigma_w^2 - \frac{1}{4} T_L \frac{d}{dz} T_L \frac{d}{dz} \sigma_w^4]$$

(5)

In the following we will report some numerical results provided
by (1), (2) and (3), and we will show the influence of using the
moments of ρ modified as from (5). Then we will discuss the simulation of dispersion in a convective boundary layer under the assumption that atmospheric motion is such that the pollutant particles
behave like bouncing balls between the ground and the top of the
inversion layer.

NUMERICAL RESULTS

The results presented in this section were obtained assuming
σ_w and u given by the power laws:

$$\sigma_w = \sigma_{w_R} (\frac{z}{z_R})^\gamma, \ u = u_R(\frac{z}{z_R})^\delta$$

and T_L = constant.

Taking z_R = 10m, σ_{w_R} = 0.41 m/s, u_R = 2 m/s, γ = 0.3, δ = 0.2 and
T_L = 30s, 2000 trajectories were generated by making use of (2) and
(3) with ρ having Gaussian distribution and moments given by (4).
Such an ensemble of trajectories was then used to compute [C] over
a regular rectangular mesh having horizontal and vertical resolution
of 125m and 50m respectively.

Isolines of [C] are given in Fig. 2 for a simulation time of
1800s. The inversion layer height was taken equal to 1000m and
the source was located at z_S = 250m. Reported values of [C], x and
z are scaled to $\frac{Q}{u_R h}$, $\frac{u_R h}{\sigma w_R}$ and h, respectively. The figure shows, as
expected, accumulation of particles at the ground.

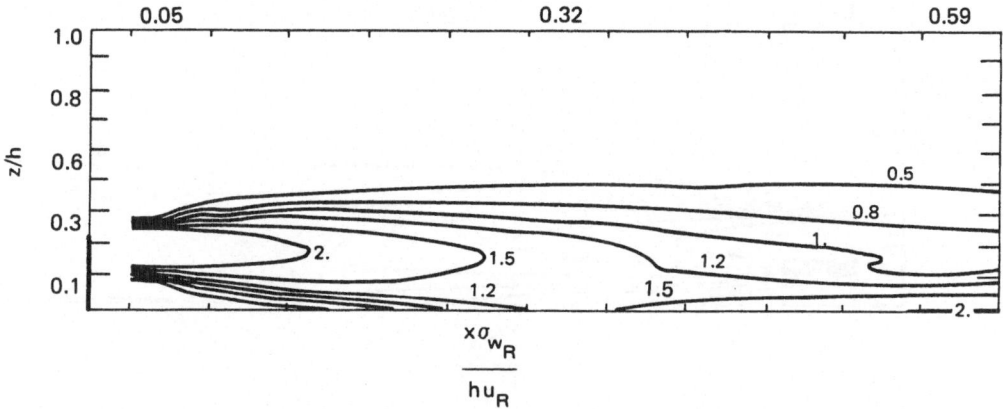

Fig. 2. Isolines of [C] after a simulation time of 1800s for the
conditions specified in the text. Moments of ρ given by
(4).

Repetition of this numerical experiment by taking the moments
of ρ given by (5) yields results of Fig. 3. The accumulation on the
ground does not occur. The simulated distribution of the concentration appears to be correct.

To quantify this statement we observe that after a travel time
larger than the Lagrangian time scale the ensemble of trajectories
must produce a distribution of [C] which is a close approximation
to the solution of the diffusion equation:

$$\frac{\partial C}{\partial t} + u \frac{\partial C}{\partial x} = \frac{\partial}{\partial z}(K \frac{\partial C}{\partial z}) \tag{6}$$

with

$$u = u_R (\frac{z}{z_R})^\delta \text{ and } K = \sigma_w^2 T_L = \sigma_{w_R}^2 (\frac{z}{z_R})^{2\gamma} T_L.$$

Ground concentration values obtained by solving (6) numerically
(see Sardei and Runca (1975)) are compared in Fig. 4 with the corresponding values from the simulations of Figs. 2 and 3. Results from
trajectories generated with moments of ρ given by (5) approximate
closely the solution to (6). It appears therefore that the modification of the moments of <ρ> can be effective to generate "realistic"
ensemble of trajectories. However, further analysis is necessary
in order to assess the validity of this approach.

Due to the work of Willis and Deardorff (1978) and Lamb (1978)
considerable attention has been recently given to Lagrangian modelling of air pollutants dispersion in a convective boundary layer.

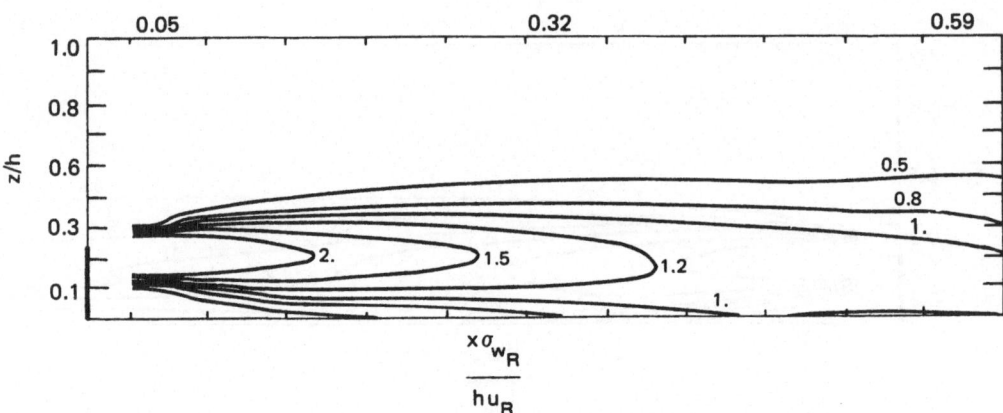

$$\frac{x\sigma_{w_R}}{hu_R}$$

Fig. 3.   Isolines of [C] as in Fig. 2, but for moments of ρ
          given by (5).

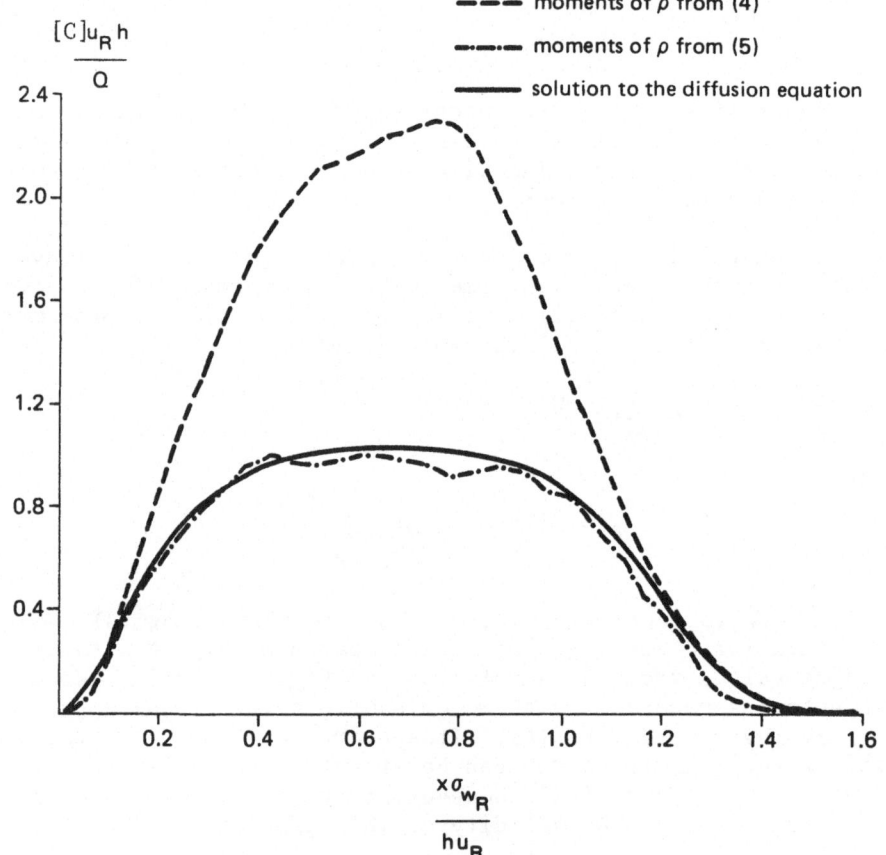

$$\frac{x\sigma_{w_R}}{hu_R}$$

Fig. 4. Comparison of the ground level concentration corresponding
        to Figs. 2 and 3 with the solution to the diffusion equa-
        tion.

So far we have not verified, for this specific situation, a procedure similar to the one discussed above. Our work has been centered on the assumption that the atmospheric motion in a convective boundary layer is such that pollutant particles behave like bouncing balls between the boundaries. This has led to the analytical results discussed below.

ANALYTICAL RESULTS

It has been shown by Lamb (1978) that the Lagrangian approach can provide an adequate basis to model the behaviour of plume dispersion in a convective boundary layer. He utilized the velocity flow field from the numerical turbulence model of Deardorff (1974) to build an ensemble of air pollutant trajectories and obtained results in substantial agreement with the experimental simulations performed by Willis and Deardorff (1978).

Drawing on these studies Weil and Furth (1981) concluded that the plume dispersion could be simulated by assuming that a pollutant particle in a convective boundary layer behaves like a bouncing ball between the ground and the top of the inversion layer.

In practice according to Weil and Furth the ensemble of trajectories is generated as follows: a) the particle's initial velocity is chosen at random from a given distribution at source height and b) the particle's velocity is changed only at the boundaries where the vertical component is reverted and multiplied by a positive factor: α_o at the ground and α_h at the top of the inversion layer.

With the above assumptions, Weil and Furth reproduced Lamb's results (1978) by taking $\alpha_o = 3/2$, $\alpha_h = 2/3$ and the function represented by the dashed line of Fig. 5 as the probability density of the vertical velocity at source height. (The reader should refer to their paper for the arguments which led to this choice.)

As shown in a paper in preparation the assumption on the basis of Weil and Furth's model allow the analytical derivation of the probability density

$$F(w, z, t | w_o, z_s, 0) \tag{7}$$

of finding the particle, released at z_s with velocity w_o, at z with velocity w after a travel time t. In terms of (7) the ensemble average concentration and the Lagrangian autocorrelation can be computed respectively from the following expressions:

$$<C(x, z)> = \frac{Q}{u} \int\limits_{-\infty}^{+\infty}\!\!\!\int g(w_o) F(w, z, \frac{x}{u} | w_o, z_s, 0) \, dw_o \, dw \tag{8}$$

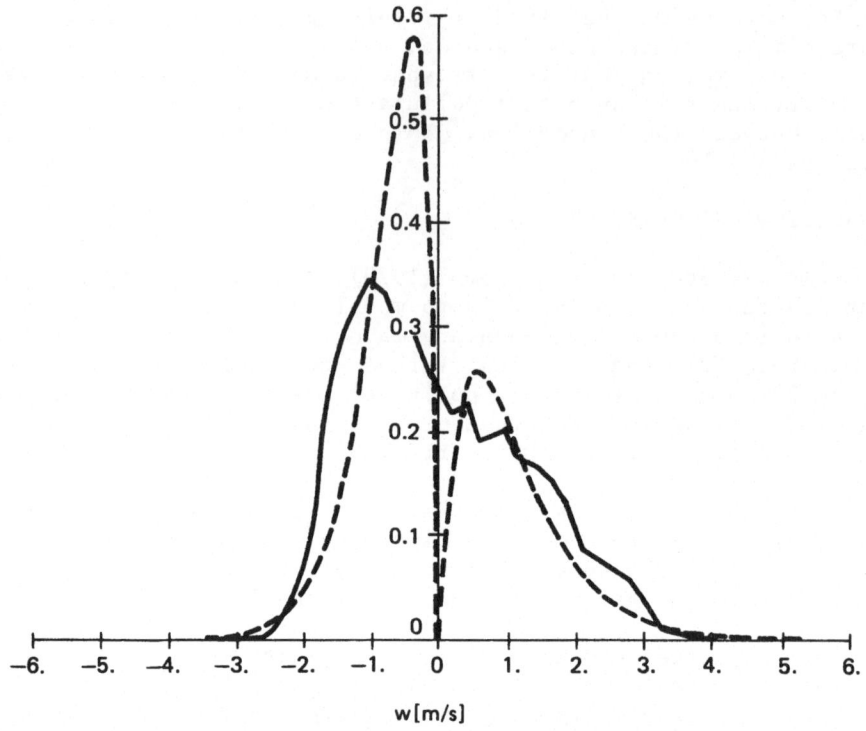

Fig. 5. Probability density functions for w at z_s. Dashed line:
function adopted by Weil and Furth; solid line: function
computed from Deardorff's data.

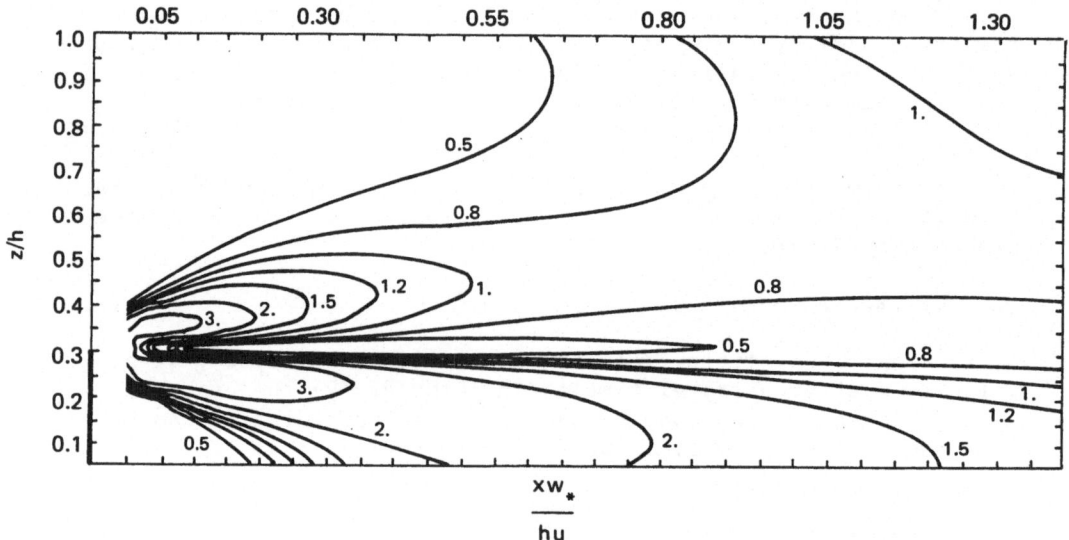

Fig. 6. Isolines of <C> for $\alpha_0 = \frac{3}{2}$, $\alpha_h = \frac{2}{3}$ and $g(w_0)$ given by
the dashed line of Fig.5.

and

$$\frac{\overline{\langle w(t)w(o)\rangle}}{\langle w(o)^2\rangle} = \frac{1}{\langle w(o)^2\rangle} \iiint\limits_{-\infty}^{+\infty} ww_o g(w_o)F(w,z,t|w_o,z_s,0)\,dw_o\,dw\,dz \quad (9)$$

In (8) and (9), $g(w_o)$ is the probability density function of the vertical velocity at z_s; equation (8) has been derived making use of the fact that, in the convective situation simulated by

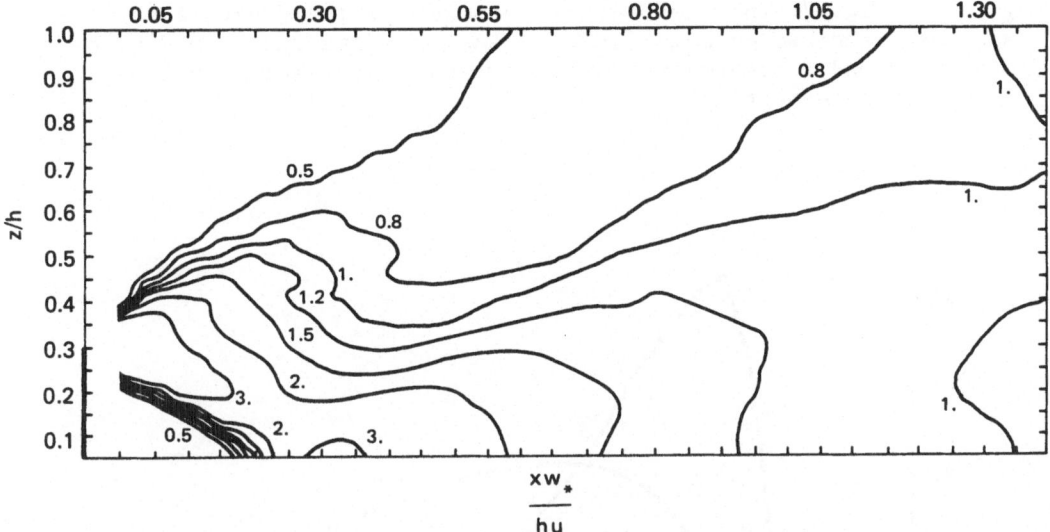

Fig. 7. Isolines of $\langle C \rangle$ for $\alpha_o = \alpha_h = 1$ and $g(w_o)$ given by the solid line of Fig. 5.

Deardorff (1974) and to which both Lamb's and Weil and Furth's papers refer, the horizontal wind can be considered uniform. Repeating in practice the work done by Weil and Furth, (8) has been applied to this convective situation, characterized by u = 2.9 m/s, w* (the convective velocity scale) = 1.9 m/s and h = 1150 m, in order to describe dispersion from a source at z_s = 0.26 h. The results achieved by adopting the choice of $g(w_o)$, α_o and α_h made by Weil and Furth are displayed in Fig. 6. The reported values of $\langle C \rangle$, x and h are scaled to $\frac{Q}{uh}$, $\frac{uh}{w*}$ and h, respectively.

The unrealistic distribution of <C> shown in Fig. 6 is due to the chosen $g(w_o)$ (dashed line of Fig. 5). However, some of the features of this distribution such as concentration at ground level and plume centerline height are in perfect agreement with the corresponding results of Lamb (1978) (see Weil and Furth's paper (1981)). This observation leads to a basic question connected with the generation of an ensemble of trajectories, that is: which statistics of the ensemble have to be considered in order to conclude that the ensemble is a satisfactory representation of reality?

To gain some insight into this question, (8) has been applied with $\alpha_o = \alpha_h = 1$ (case of ground and inversion layer perfect reflectors of the diffusing particles), and $g(w_o)$ (solid line of Fig. 5) computed from the velocity flow field simulated by Deardorff (1974). With this choice of $g(w_o)$, α_o and α_h, (8) produced the isolines given in Fig. 7. The qualitative behaviour of the plume is the same as described by Lamb (1979). However, comparison in Fig. 8 of the ground concentration values with Lamb's results shows a strong disagreement.

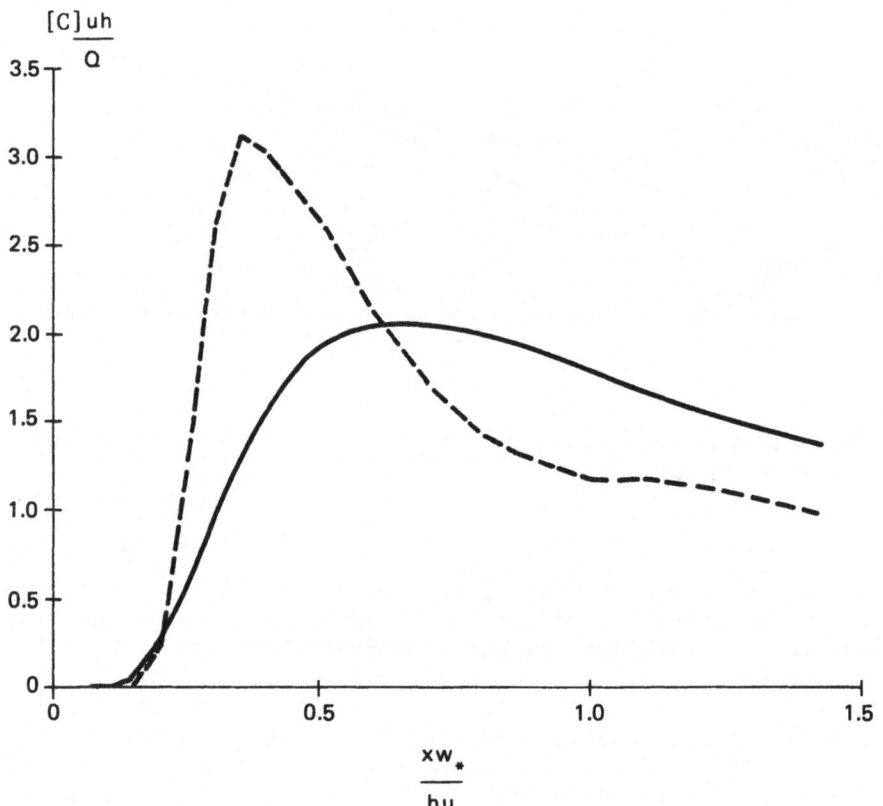

Fig. 8. Comparison of the ground level concentration corresponding to Fig. 7 (dashed line) with Lamb's results (solid line).

The analysis was also extended to the Lagrangian autocorrelation function given by (9). Comparison of the autocorrelation* derived from Deardorff's data (solid line) with the autocorrelations computed by (9) for both Weil and Furth case (dashed line) and the case to which Fig. 7 refers (dash dotted line) is made in Fig. 9. The choice of $g(w_0)$, α_0 and α_h made by Weil and Furth provides a better approximation. However, taking $\alpha_0 = 0.5$ and $\alpha_h = 1$ in the case of Fig. 7, the corresponding autocorrelation becomes (see Fig. 10) very close to the autocorrelation derived from Deardorff's data. Unfortunately computation of $<C>$ with this choice of α_0 and α_h showed results which departed from Lamb's results more than for $\alpha_0 = \alpha_h = 1$.

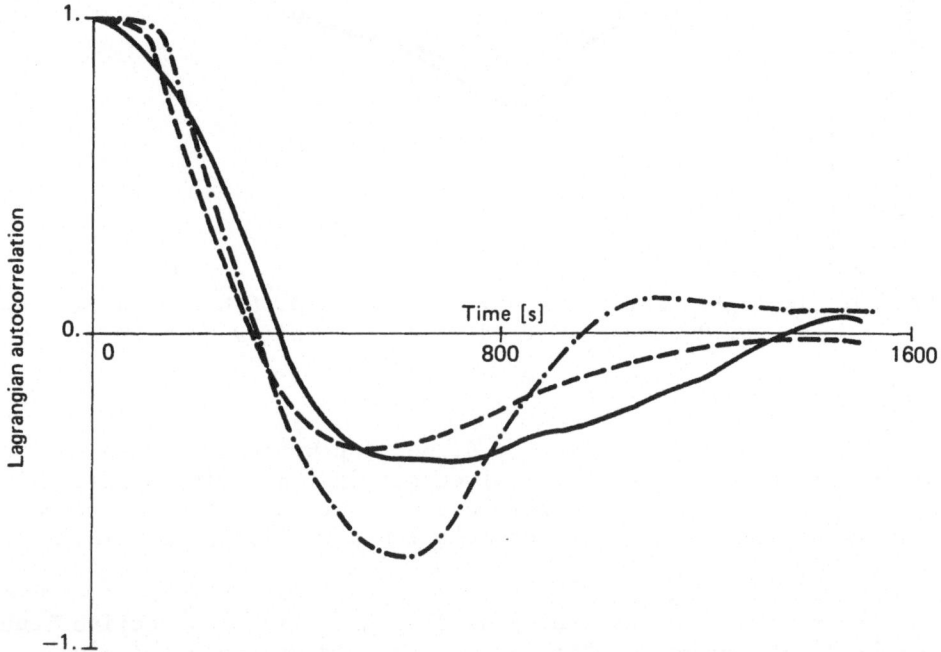

Fig. 9. Lagrangian autocorrelation at $z_s = 0.26$ h. Solid line
         has been derived from Deardorff's data; dashed line
         refers to Weil and Furth's choice of $g(w_0)$, α_0 and α_h;
         dashed-dotted line to $\alpha_0 = \alpha_h = 1$ and $g(w_0)$ computed
         from Deardorff's data (solid line of Fig. 5).

---

*This autocorrelation was derived from the same ensemble of trajectories used by Lamb (1978) in his paper.

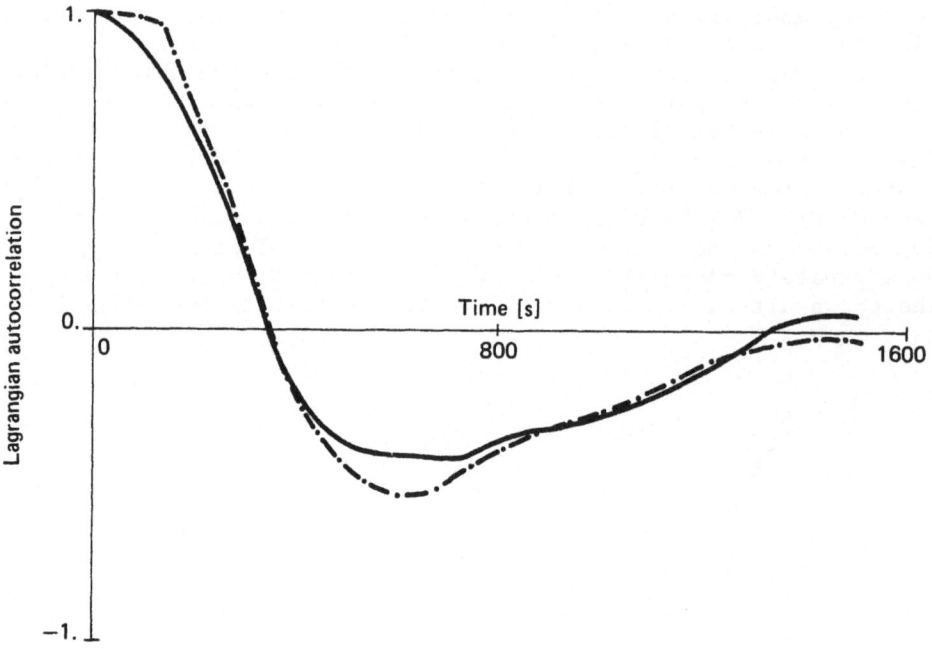

Fig. 10. Lagrangian autocorrelation for z_s = 0.26 h. Solid line as
in Fig. 9, dashed-dotted line as in Fig.9 but for α_o = 0.5.

CONCLUSION

Simulation of air pollutant dispersion by generation of an
ensemble of trajectories is still an art in its early stages.
Progress has been made in order to avoid, at least in some cases,
unrealistic results such as accumulation of pollutant particles on
the ground.

The improved understanding of dispersion in a convective boundary layer has suggested that in this situation the dispersion
process could be described by assuming that the pollutant particles
behave like bouncing balls between the ground and the top of the
inversion layer. However, this assumption provides only a partial
image of the "real" process, since simulation results proved satisfactory for some quantities, but incorrect for others.

It can be generally concluded that additional research must be
done to develop a suitable algorithm for the generation of air pollutant particles trajectories. Mainly, it should be ensured that
the statistics of the simulated ensemble of trajectories are consistent with the "real" process.

ACKNOWLEDGEMENTS

The authors are indebted to Dr. Robert G. Lamb for helpful
discussions and the data utilized in this study.

REFERENCES

Deardorff, J.W., 1974, Three-dimensional numerical study of the
    height and mean structure of a heated planetary boundary
    layer. Boundary-Layer Met. 7: 81-106.
Janicke, L., 1981, Particle simulation of inhomogeneous turbulent
    diffusion. Preprints volume of NATO/CCMS 12-th Int. Tech.
    Meeting on Air Pollution Modelling and its Application,
    SRI, Menlo Park, California.
Jonas, P.R., and Bartlett, J.T., 1972, The numerical simulation of
    particle motion in a homogeneous field of turbulence. J.
    Comput. Phys. 9, 290-302.
Lamb, R.G., 1978, A numerical simulation of dispersion from an
    elevated point source in the convective planetary boundary
    layer. Atmospheric Environment, 12: 1297-1304.
Lamb, R.G., Hugo, H., and Reid, L.E., 1979, A lagrangian approach
    to modeling air pollutant dispersion. EPA Report, EPA-600/4-
    79-023.
Reid, J.D., 1979, Markov chain simulations of vertical dispersion
    in the neutral surface layer for surface and elevated relea-
    ses, Boundary-Layer Met. 16: 3-22.
Runca, E., Bonino, G., and Posch, M., 1981, Lagrangian modelling
    of air pollutants dispersion from a point source. Preprints
    volume of NATO/CCMS 12-th Int. Tech. Meeting on Air Pollution
    Modelling and its Application. SRI, Menlo Park, California.
Sardei, F., and Runca, E., 1975, An efficient numerical scheme for
    solving time dependent problems of air pollution advection
    and diffusion. Proceedings of the IBM Seminar on Air
    Pollution Modeling held in Venice, Italy, November 1975.
    IBM Italy Scientific Center, Report No. 48.
Smith, F. B., 1968, Conditioned particle motion in a homogeneous
    turbulent field. Atmospheric Environment, 2, 491-508.
Thompson, R., 1971, Numeric calculation of turbulent diffusion.
    Quart. J. Roy, Meteorol. Soc. 97, 93-98.
Weil, J.C., and Furth, W.F., 1981, A simplified numerical model
    of dispersion from elevated sources in the convective
    boundary layer. Proceedings of the Fifth AMS Symposium
    on Turbulence, Diffusion and Air Pollution, Atlanta.
Willis, G.E., and Deardorff, J.W., 1978, A laboratory study of
    dispersion from an elevated source within a modeled convec-
    tive planetary boundary layer, Atmospheric Environment, 12:
    1305-1311.

Wilson, J.D., Thurtell, G.W., and Kidd, G.E., 1981a, Numerical
    simulation of particle trajectories in inhomogeneous tur-
    bulence, I: Systems with Constant Turbulent Velocity Scale,
    Boundary Layer Met. 21, 295-313.
Wilson, J.D., Thurtell, G.W., and Kidd, G.E., 1981b, Numerical
    simulation of particle trajectories in inhomogeneous
    turbulence, II: Systems with variable turbulent velocity
    scale. Boundary Layer Met. 21, 423-441.

DISCUSSION

A. VENKATRAM   You suggest that the model of Weil and Furth does
                  not reproduce Deardorff's simulations. This appears
                  to contradict the results in the original paper by
                  Weil and Furth. Please comment.

E. RUNCA   The model of Weil and Furth reproduces in a very
                  satisfactory way Deardorff's results as to ground
                  concentration and plume centerline height. However
                  the display of the isopleths of the crosswind
                  integrated concentration is very unrealistic,
                  showing a minimum at source height.

PPI-THEORY FOR PARTICLE DISPERSION

Ib Troen, Søren Larsen and Torben Mikkelsen

Physics Department. Risø National Laboratory
DK-4000 Roskilde, Denmark

INTRODUCTION

For many problems with turbulent dispersion the Lagrangian approach as introduced by Taylor (1921) is the most appropriate. Extension of this theory to atmospheric dispersion problems is, however, complicated by the inhomogeneity and instationarity of atmospheric turbulence. Here we first describe a method based on the Preferred Path Integration (PPI) theory relating the probability distribution of particle displacement to the probability for a particle to move along the most probable or preferred path. The theory is based on the assumption that the particle velocity fluctuations are governed by a first order autoregressive process equivalent to the Langevin model of dispersion (Lin and Ried (1962), Novikov (1963), Smith (1968), Hanna (1978) and Gifford (1982)). The PPI-model yields conditional probability distributions in the case of stationary conditions. The Langevin model has been proposed also for instationary conditions in Monte-Carlo simulations of particle dispersion (Durbin (1980), Wilson et al. (1981)), and here are presented some analytical results based on this type of model for the case of dispersion in decaying turbulence.

PREFERRED PATH INTEGRATION

The assumption that the velocity of a single particle follows a first order autoregressive process can be written

$$v_{n+1} = \rho v_n + \sigma\sqrt{1-\rho^2}\ G_n \tag{1}$$

where v_n is the turbulent velocity at time step n, $\rho = \exp(-\Delta t/T)$ is the autocorrelation corresponding to the

time lag Δt, σ^2 is the velocity variance and G_n is a gaussian
white noise process with unit variance and zero mean. We
are interested in the probability that a particle released at
$t = 0$ at the source point at $y = 0$ will follow a given trajectory
$y(t)$ to the receptor point at $y(t_1) = y_1$. We imagine the path
subdivided into N steps for which eq. (1) applies for each step.
The probability that the particle follows the path $y(n\Delta t)$,
$n = 1,N$ is equivalent to the probability that it experiences a
certain sequence of velocities v_n

$$\Pr(y(n\Delta t),\ n = 1,N) = \Pr(v_n,\ n = 1,N) \ . \tag{2}$$

This probability is given by the product of the probabilities
of each step in the sequence which by virtue of eq. (1) can be
written

$$\Pr(y(n\Delta t),n=1,N) = \prod_{n=1}^{N} \frac{1}{\sqrt{2\pi}}\ \exp\left(-\frac{(v_{n+1}-\rho v_n)^2}{2\sigma^2(1-\rho^2)}\right) \ , \tag{3}$$

or by taking the logarithm and assuming $\Delta t \ll T$ (and therefore
$\rho \simeq 1 - \Delta t/T$):

$$\ln\Pr = -\sum_{n=1}^{N} \frac{(\Delta v_n + v_n\,\Delta t/T)^2}{2\sigma^2\ 2\Delta t/T} - \frac{1}{2}\sum_{n=1}^{N}\ln(2\pi) \ , \tag{4}$$

where $\Delta v_n = v_{n+1} - v_n$. Taking the limit as $\Delta t \to 0$ (and $N \to \infty$ with
$N\Delta t = t_1$) and defining the scaled quantities $d\eta = dy/(\sigma T)$ and
$d\tau = dt/T$ we get from the first term in eq. (4).

$$\ln\Pr(y(t)) = -\frac{1}{4}\int_0^{\tau_1}(\ddot{\eta}+\dot{\eta})^2\ d\tau \ . \tag{5}$$

Here $\tau_1 = t_1/T$, and the dots means differentiation with
respect to τ. The second term in eq. (4) only depends on the
number of subdivisions to the path, and eq. (5) defines a probability density for the path $y(t)$. The probability that a particle,
released at $t=0$ at the source point $y=0$, hits the receptor
at $y=y_1$ at time t_1 can formally be written as the integral of
probabilities over all possible paths P :

$$\Pr(y_1,t_1|0,0) = \int_P\ \exp\left(-\frac{1}{4}\int_0^{\tau_1}(\ddot{\eta}+\dot{\eta})^2\ d\tau\right)\ dP \ . \tag{6}$$

We have here neglected a normalization factor by the neglect of the second term in eq. (4), but this normalization is recovered below. It is impossible directly to perform the integration in the infinite dimensional space of all paths, and in order to evaluate the integral we proceed along the lines originally developed in quantum mechanics by Feynmann and Hibbs (1965). We write a path as the sum of a particular path $E(\tau)$ to be specified below, and a deviation $\varepsilon(\tau)$ i.e:

$$\eta(\tau) = E(\tau) + \varepsilon(\tau) \quad . \tag{7}$$

Expanding the integrand in the exponent in eq. (6) and integrating by parts to remove derivatives of ε gives

$$\int_0^{\tau_1} (\ddot{\eta}+\dot{\eta})^2 d\tau = \int_0^{\tau_1} (\ddot{E}+\dot{E})^2 d\tau + \int_0^{\tau_1} (\ddot{\varepsilon}+\dot{\varepsilon})^2 d\tau$$

$$+ 2 \left\{ \int_0^{\tau} \varepsilon(\dddot{E} - \ddot{E}) d\tau + \left[\dot{\varepsilon}(\ddot{E}+\dot{E})-\varepsilon(\ddot{E}-\dot{E})\right]_0^{\tau_1} \right\}. \tag{8}$$

Thus, if we define $E(\tau)$ as the solution to the differential equation $\dddot{E} - \ddot{E} = 0$ with the boundary conditions $E(0) = 0$ (source point), $\dot{E}(o) = \dot{\eta}_o$ (start velocity), $E(\tau_1) = \eta_1$ (receptor), and $(\ddot{E}+E)(\tau_1) = 0$, then the last term in eq. (8) vanishes for all paths which connects the source and receptor points, and which has the prescribed start velocity $\dot{\eta}_o$. From eq.(6) we get

$$Pr(\eta_1,\tau_1|0,0,\dot{\eta}_o) = \exp(-\frac{1}{4} \int_0^{\tau_1} (\ddot{E} + \dot{E})^2 d\tau) \cdot \int_\varepsilon \exp (= \int_0^{\tau_1} (\ddot{\varepsilon}+\dot{\varepsilon})^2 d\tau) d\varepsilon \tag{9}$$

where the integral over $\varepsilon(\tau)$ must be taken over all $\varepsilon(\tau)$ with $\varepsilon(o) = \dot{\varepsilon}(o) = \varepsilon(\tau_1) = 0$. This integral is therefore a function of only one parameter namely the travel time τ_1, and we therefore finally write

$$Pr(\eta_1,\tau_1|0,0,\dot{\eta}_o) = F(\tau_1) \cdot \exp(-\frac{1}{4} \int_0^{\tau_1} (\ddot{E} + \dot{E})^2 d\tau). \tag{10}$$

The function $F(\tau_1)$ recovers the normalization, and is determined by the requirement that a particle must reach some point η_1:

$$\int_{-\infty}^{\infty} Pr(\eta_1,\tau_1|0,0,\dot{\eta}_o) d\eta_1 = 1 \tag{11}$$

The differential equation determining $E(\tau)$ is the Euler differential equation which appears in the variational problem of finding the extrema for the integral in eq. (5), and what we have shown therefore is that the probability for a particle released with a given start velocity, to reach a certain point in a given time is – apart from a normalization factor – given as the probability that it follows the most probable path. The solution for $E(\tau)$ is easily found as

$$E(\tau) = a\exp(\tau) + b\exp(-\tau) + c\tau + d \ , \tag{12}$$

where the coefficients are determined by the boundary conditions listed above. The integral in eq. (10) can therefore be written

$$\int_0^{\tau_1} (\ddot{E} + \dot{E})^2 d\tau = 2a^2(\exp(2\tau_1)-1) + c^2\tau_1 + 4ac(\exp(\tau_1)-1). \tag{13}$$

Solving for a and c from the boundary conditions then gives

$$-\frac{1}{4} \int_0^{\tau_1} (\ddot{E} + \dot{E})^2 d\tau = -\frac{1}{2} (\eta_1 + \dot{\eta}_0(\exp(-\tau_1)-1))^2 \ \sigma_1^{-2} \tag{14}$$

with

$$\sigma_1^2 = \left[\exp(-2\tau_1) - 4\exp(-\tau_1) + 3 - 2\tau_1\right] \quad . \tag{15}$$

Inserting into eq. (10) and integrating over the initial velocity $\dot{\eta}_0$ assuming this to be Gaussian and given as $Pr(\dot{\eta}_0, 0) = \exp(-\dot{\eta}_0^2/2)/\sqrt{2\pi}$ we get the result

$$Pr(\eta_1, \tau_1 | 0, 0) = \frac{1}{\sqrt{2\pi} \ \sigma_y}\exp(-\frac{1}{2} \frac{\eta_1^2}{\sigma_y^2}) \tag{16}$$

with

$$\sigma_y^2 = 2(\tau_1 + \exp(-\tau_1) - 1) \ .$$

This well known result could of course be obtained a lot more easy by directly inserting the exponential autocorrelation function into Taylors expression for the plume dispersion in homogeneous turbulence. However eqs. (10) and (14) give the conditional probability for a particle which has a given initial speed to reach the receptor, and by changing the boundary conditions for $E(\tau)$ we can obtain solutions for other conditional probabilities. This will not be pursued here, where our main

intention has been to present an alternative method to the
treatment of dispersion governed by the first order autoregressive process for the velocities.

NONSTATIONARY CASE, DECAYING TURBULENCE

     In a number of studies (e.g. Wilson et al. (1981), Durbin,
(1980) and Hanna (1982)) the first order process has been used for
the simulation of particle dispersion also under inhomogeneous
conditions. In this case eq. (1) is generalized by assuming R
and σ to be functions either explicitly of the time or as
functions of the position of the particle as can be done in
numerical Monte-Carlo simulations. Here we present some results
using an analytical approach to the problem of dispersion in
decaying turbulence. We assume as above that eq. (1) applies
for the motion of a particle and that we can specify the time
scale T entering into R and the velocity variance as function
of travel time. Squaring on both sides of eq. (1), taking the
ensemble average, and taking the limit as $\Delta t \to 0$, as above,
leads to an equation for the time development of the particle
velocity variance S(t) :

$$\frac{dS}{d\tau} = 2(\sigma^2 - S) \tag{17}$$

where as above $Td\tau = dt$, and σ refers to eq. (1), but now the
integral scale T is a function of time as also σ and S. Under
stationary conditions $\sigma^2 = S$, and this is the defining equation
for σ^2 as the velocity variance of the turbulence. When extending the process to inhomogeneous situations we see from eq. (17)
that this no longer holds. In decaying turbulence, as, for
example, would be experienced by a particle released at a coast
between a heated land and a colder sea and transported by the
mean wind over the sea where the turbulence decays as the surface
heatflux vanishes, the decay process starts at the high frequency
part of the spectrum, and successively fluctuations at lower
frequencies decay. This means that the Lagrangian time scale
T(t) times the velocity variance stays constant viz. K =
S(t)T(t) (see e.g. Lumley and Panofsky (1964)). If we envision
the decay as a step change between two turbulence levels, and
enter this into the process by prescribing step changes in
σ^2 and T at t = 0 with $\sigma^2 T = K_1$ = constant, the response through
the process eq. (1) will be that K(t) = S(t)T(t) =
$K_1 + K_1(\sigma^2_1/\sigma^2_2 - 1)\exp(-2t/T_2)$ for t > 0 which follows from
solving eq. (17) for this case where σ^2 jumps from σ^2_1 to σ^2_2.
We see therefore that associating $\sigma^2(t)$ with the physical
velocity variance leads in this case to an overestimation of the

dispersion, and similarly to an underestimation if we instead considered a situation with $\sigma^2_2 > \sigma^2_1$. In order to calculate the dispersion from a point source we need the correlation of particle velocities at two times t_1 and t_2: $R(t_1,t_2) = <v(t_1)v(t_2)>$, where $<>$ denotes ensemble averaging. This correlation can be obtained from the process equation (1) by multiplying on each side with v_n and taking the ensemble average and noting that by definition we have $<v_n\, G_n> = 0$. After taking the limit as $\Delta t \to 0$ we obtain

$$\frac{dR(t_1,t_2)}{dt_2} = -\frac{R(t_1,t_2)}{T(t_2)} \;,\quad t_1<t_2 \tag{18}$$

and by integration

$$R(t_1,t_2) = S(t_1)\exp(-(\tau_2-\tau_1)) \;, \tag{19}$$

where we have used the identity $R(t_1,t_1) = S(t_1)$ and inserted the scaled time τ defined by

$$\tau = \int_o^t \frac{dt}{T(t)} \;. \tag{20}$$

Following Taylors classical development of the dispersion from a point source we can write

$$\sigma_y^2(t) = 2 \int_o^t \int_o^{t'} R(t',t'')dt''dt' \tag{21}$$

where σ_y is the standard deviation of displacements for particles released at the source point at $t = 0$. Inserting eq. (19) and using eq. (20) we can rewrite to obtain

$$\sigma_y^2(t) = 2 \int_o^\tau \int_o^{\tau'} S(\tau'')T(\tau'')\exp(\tau''-\tau')d\tau''T(\tau')d\tau' \;. \tag{22}$$

This result is valid for all variations of the velocity variance S and the Lagrangian time scale T provided we can assume the autoregressive process eq. (1) to be valid. Specializing now

to the case of decaying turbulence we see that eq. (22) becomes
greatly simplified because of the relation ST = K = constant
and we can rewrite by integrating to obtain:

$$\sigma_y^2(t) = 2K(t - \int_0^\tau T(\tau')\exp(-\tau')d\tau'). \tag{23}$$

In decaying turbulence we have from scale arguments that the
turbulent energy e decays as (Tennekes and Lumley (1972))

$$\frac{de}{dt} \sim - \frac{e^{3/2}}{\ell} , \tag{24}$$

where ℓ is a characteristic length. This implies the following
equation for S

$$\frac{dS}{dt} = - \frac{a}{\ell} S^{3/2} . \tag{25}$$

The solution for $T(\tau)$ obtained by solving eq. (25) and using
TS = K and eq. (20) can be written

$$T(\tau) = T(o)(1-\gamma\tau)^{-2} \tag{26}$$

with

$$\gamma = \frac{a}{2\ell} T(o)\sqrt{S(o)} \tag{26a}$$

and

$$\tau = \frac{t/T(o)}{1+ \gamma t/T(o)} \tag{26b}$$

which gives from eq. (23)

$$\sigma_y^2 = 2KT(o)\left[\frac{t}{T(o)} - \frac{1}{\gamma}\left(\frac{\exp(-\gamma)}{1-\gamma\tau} -1 + \frac{\exp\left(-\frac{1}{\gamma}\right)}{\gamma} Ei\left(\frac{1}{\gamma}\right)\right.\right.$$
$$\left.\left.-Ei\left(\frac{1}{\gamma} - \tau\right)\right)\right] \tag{27}$$

Ei(x) is the Exponential Integral function defined as

$$Ei(x) = \int_{-\infty}^{x} \frac{\exp(t)}{t} \, dt \quad . \tag{28}$$

The dispersion calculated using a conventional K-model with the diffusivity defined as $S(o)T(o) = K$ is given by the first term in eq. (27) viz. $\sigma_y = \sqrt{2Kt}$ as usual. In the case of nonhomogeneous or nonstationary dispersion calculations a diffusivity value based on scale argument pertaining to local quantities is usually adopted. In the case considered here the only choice is $K(t) = b\ell\sqrt{S}$. Using as above ℓ constant and the form of $S(t)$ obtained from eq. (25) the solution using the diffusion equation for the spread is easily found as

$$\sigma_y^2 = 2\gamma^{-1} \, KT(o) \, \ln(1+\gamma\frac{t}{T(o)}) \tag{29}$$

where we have introduced $K = K(o) = S(o)T(o) = b\ell\sqrt{S(o)}$ in order that the result from eq. (29) in the limit $a \to 0$ (no decay) is identical to the result $\sigma_y^2 = 2Kt$. The time development of the plume spread σ_y from the Langevin model eq. (27), and the K-model formulation eq. (29) are shown on fig. 1 for different values of the decay parameter γ. Instead of the assumption ℓ constant we can use $T \sim \ell/\sqrt{S}$ (Tennekes and Lumley (1972)). Then the equation for S becomes

$$\frac{dS}{d\tau} = - cS \tag{30}$$

which gives

$$T(\tau) = T(o)\exp(c\tau) \tag{31}$$

and

$$\tau = \frac{1}{c} \ln(c \, t/T(o) + 1) \quad , \tag{32}$$

$$\sigma_y^2 = \begin{cases} 2KT(o)(t/T(o) - \frac{1}{c-1}\{(c \, t/T(o)+1)^{1-1/c} -1\}) & \text{for } c \neq 1 \\ \\ 2KT(o)(t/T(o) - \ln(t/T(o) + 1)) & \text{for } c = 1. \end{cases} \tag{33}$$

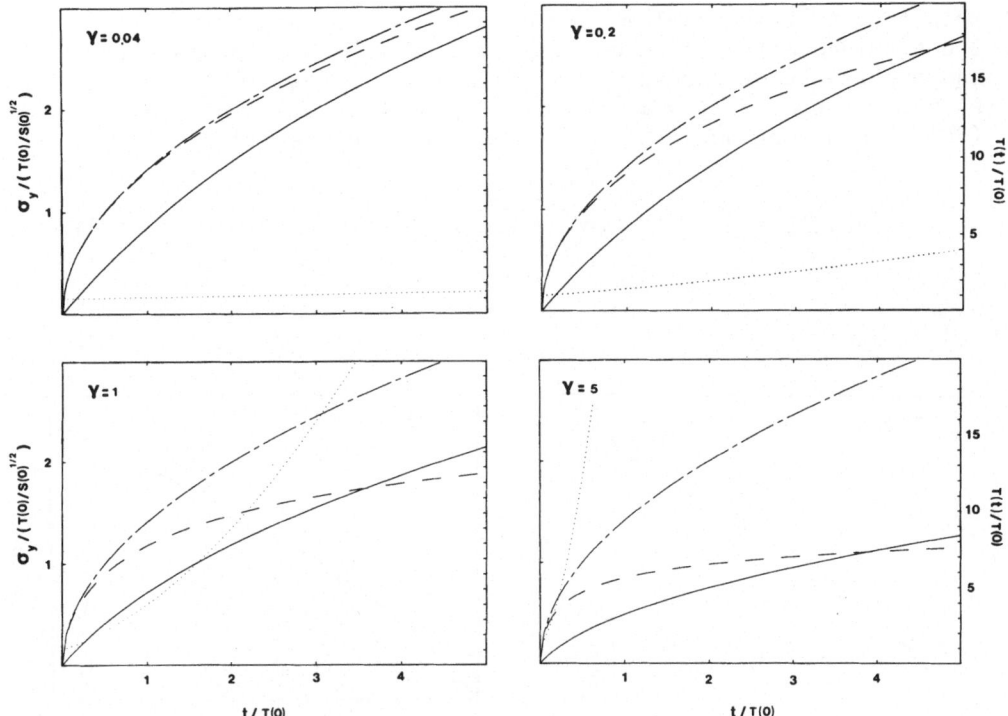

Fig. 1. The standard deviation of particle displacements
(plume width) σ_y as function of travel time for
different values of turbulent decay rate γ. (———):
Langevin model eq. (27), (- - -) K-model with
$K \sim \ell\sqrt{S(t)}$ eq. (29), (——•——•): K-model with
constant $K = S \cdot T$. The dotted curve (••••) shows
the Lagrangian time scale $T(t)$ as function of
travel time. The latter curve refers to the
scale on the right and the other curves to the
left scale.

The result obtained using a K-model formulation with $K \sim \ell\sqrt{S}$ and $T \sim \ell/\sqrt{S}$ becomes simply $K = TS = $ constant. In boundary layer models (as for example Yamada and Mellor (1975)) the length scale above the surface layer comes out as essentially constant when the turbulent energy can be assumed nearly homogeneous irrespective of the time development of this quantity. For the case considered here application of a boundary layer model of this type to this problem we would have $\ell = $ constant and with S determined either from eq. (25), which is model consistent; and the resultant spread σ_y^2 is again given by eq. (29). As another possibility we can use S from the solution of eq. (30) in conjunction with ℓ constant with the result:

$$\sigma_y^2 = 2KT(o) \frac{2}{c} (\sqrt{c\ t/T(o) + 1} - 1) \quad . \tag{34}$$

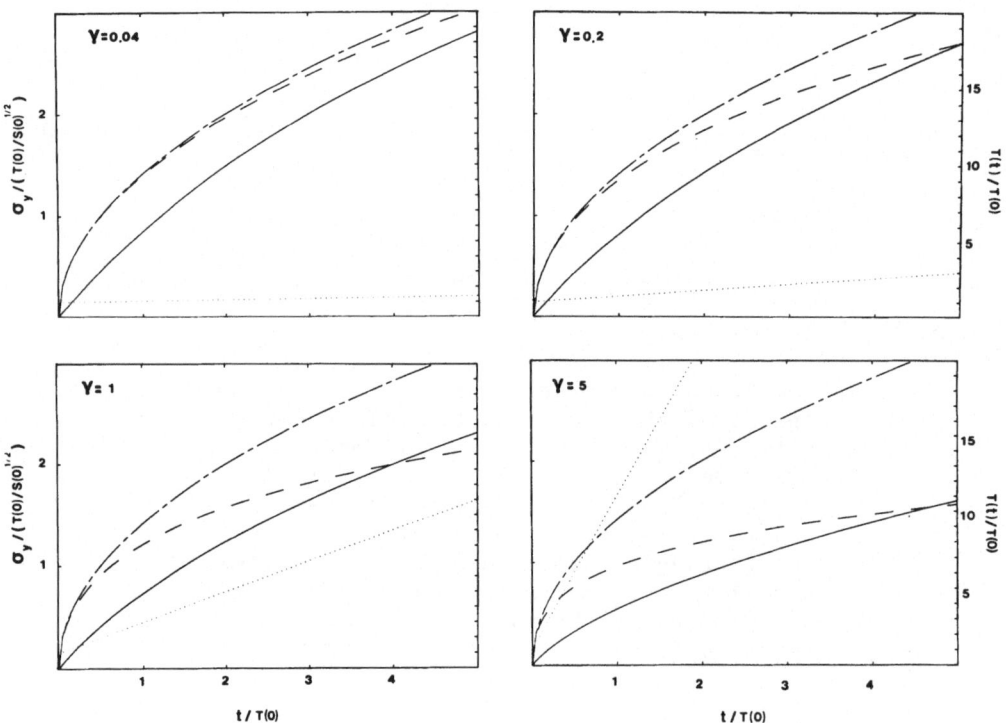

Fig. 2. As fig. 1 but based on the Langevin model results from eq. (33): (———),(- - -): K-model with $K \sim \ell\sqrt{S}$ from eq. (34).

Requiring that the initial diffusivity K(o) equals S(o)T(o) in this model gives $\ell = d\sqrt{S(o)T(o)}$ and using for consistency the same value in the definition of γ (eq. 26a) gives $2\gamma = c$. Fig. 2 shows the plume width σ_y^2 as function of time based on these models with $2\gamma = c$.

CONCLUSIONS

The preferred path theory shows, given possible boundary conditions on the initial and/or final velocity, that under homogeneous conditions the Langevin model of particle dispersion is equivalent to the condition, that the probability for a particle to traverse between two points is proportional to the probability for the particle to move along the most probable path.

The extension of the Langevin model to nonstationary conditions by the prescription of a time dependent timescale and velocity variance leads to a problem, particularly when using a simulation technique, because of the different way these parameters enter into the process. The timescale can be changed instantaneously, but entering the variation of velocity variance through the noise process leads to an exponential approach of the particle velocity variances to the prescribed value. In the case of decaying turbulence the combination of the turbulence energy equation and the Langevin model leads to simple analytical expressions for the time development of plume spread which can be compared with results from using K-models with time dependent diffusivities. The results show the usual overestimation of the spread near the source by the K-models. The effect of the increasing correlation time, as the high frequency fluctuations decay, gives a dispersion in the Langevin model, which at larger travel times may exceed that from the K-models.

REFERENCES

Durbin P A (1980): A random flight model of in homogeneous turbulent dispersion. Physics of Fluids, Vo. 23, No. 11, 2151-2153.
Feynmann R P, Hibbs A R (1965): Quantum mechanics and path integrals. (McGraw-Hill, New York).
Gifford F A (1982): Horizontal diffusion in the atmosphere: A Lagrangian dynamical theory. Atmospheric Environment, Vol. 16, No. 3, 505-512.
Hanna S R (1979): Some statistics of Lagrangian and Eulerian wind fluctuations. J. Appl. Meteor., Vol. 18, 518-525.
Hanna S R (1980): Applications in air Pollution Modelling, in : Atmospheric Turbulence and air Pollution Modelling. F T M Nieuwstadt and H van Dop eds. Riedel, Dordrecht Holland. 277-310

Kristensen K, Jensen N O and Petersen E.L. (1981): Lateral dis-
    persion of pollutants in a very stable atmosphere - The
    effect of meandering. Atmospheric Environment, Vo. 15, No.
    5, 834-844.
Lin C C, Ried W H (1963): Turbulent flow. In Handbuch der Physik,
    VIII/2, 438-543, Springer, Berlin.
Lumley J L, Panofsky H A (1964): The structure of atmospheric
    turbulence, John Wiley & Sons, 231 pp.
Novikow E A (1963): Random force method in turbulence Theory,
    Soviet Phys. JETP, vol. 17, 1449-1454 (translation).
Pasquill F (1974): Atmospheric Diffusion, 2nd ed., John Wiley
    & Sons, New York, xi + 429 pp.
Smith F B (1968): Conditioned particle motion in a homogeneous
    turbulent field. Atmospheric Environment, vol. 2, 491-508.
Taylor G I (1921): Diffusion by contineous movements. Proc.
    London Math. Soc., vol. 20, p. 196-202.
Tennekes H, Lumley J L (1972): A first cource in turbulence, The
    MIT Press, Cambridge, 300 pp.
Wilson J D, Thurtell G W and Kidd G E (1981): Numerical Simula-
    tion of particle trajectories in homogeneous turbulence. I
    Systems with constant turbulent velocity scale, Boundary
    Layer Meteorology, vol. 21, No. 3, 295-314.
Yamada T, Mellor G (1975): A simulation of the Wangara atmos-
    pheric boundary layer data. Journal of the Atmospheric
    Sciences, vol. 32, No. 12, 2309-2329.

DISCUSSION

F.B. SMITH   Thank you for a very
                        interesting paper. I have just one query.
                        Equation (16) gives the concentration distribu-
                        tion downwind from a point source, and it is
                        Gaussian in form. My integral-equation theory
                        indicates that the distribution should be the
                        sum of two sub-distributions: the first takes
                        the form of the turbulent distribution $Pr(\dot{n}_0,0)\eta$
                        and the second in Gaussian (and dominates at short
                        ranges). Even if $Pr(\dot{n}_0,0)$ is Gaussian, the sum
                        of 2 Gaussian distributions is in general not
                        Gaussian. Would you like to comment on this ?

I. TROEN   In this treatment the
                        physical assumption in the autoregressive process
                        is that the noise process is Gaussian; without
                        further assumptions this treatment by the PPI-
                        method shows that the distribution is a multi-
                        variate Gaussian at all distances. If the
                        initial velocity is specified as non-Gaussian,

then of course there will be a region (in the near field) where there will be non-Gaussian distributions. In my opinion this, however, is inconsistant with the model.

2: MODELING COOLING TOWER AND POWER PLANT PLUMES

Chairman: L. Santomauro

Rapporteur: J. Kretzschmar

AN IMPROVED ANALYTICAL MODEL FOR VAPOUR

PLUME VISIBLE OUTLINE PREDICTIONS

G.A. Davidson, W. Jager, and A.B. Strong

Dept. of Mechanical Engineering
University of Waterloo
Waterloo, Ontario, Canada N2L 3G1

ABSTRACT

The derivation of one-dimensional conservation equations for a buoyant wet plume in the atmosphere is reviewed, and a small correction to the conservation of energy statement is suggested. Subsequently, an analytical solution for plume trajectory and radius is obtained under the Boussinesq, bent-over plume, and constant windspeed, lapse rate, and humidity gradient restrictions. Unlike previously published solutions, however, latent heat effects are explicitly retained in this solution for all lapse conditions, and the significance of these effects on plume predictions can be examined quantitatively. Consideration is also given to the formulation of source conditions consistent with the approximations introduced into the governing equations. It is shown that a consistent formulation of source fluxes leads to a significant change in plume outline predictions. In particular, when the analytical model presented in this paper is applied to cooling tower data from Paradise steam plant, the predicted visible plume length is increased by up to a factor of three, and brought into reasonable agreement with measurements.

INTRODUCTION

Although integral models for plume rise and spread in the atmosphere have been quite successful when applied to dry, buoyant plumes, they have been less satisfactory when applied to moisture laden plumes undergoing condensation and evaporation processes. Generally it appears that if predicted and observed plume centre-line trajectories are matched, the visible plume length is consist-

ently underpredicted by these models unless some ad hoc modifications are introduced. To match both trajectory and visible length,
Slawson (1978), for example, reduced the entrainment constant, which
increased both the visible length and the plume rise; it was then
necessary to introduce a trajectory modification to bring observed
and predicted trajectories back into agreement. Hanna (1976) over-
came the problem by postulating different entrainment rates for the
momentum, heat, and vapour constituents of the plume. Carhart et
al. (1980) discuss other modifications introduced by different
authors in order to improve model performance.

The aim of this paper is to present a new integral model for
wet plume predictions in the atmosphere. The model may be regarded
as an extension of the work by Slawson (1978) in that, while the
same basic approach is used, (i) analytical expressions for plume
variables are developed retaining latent heat effects; (ii) source
flux corrections are derived to account for the Boussinesq and bent-
over plume assumptions; and (iii) a small correction is made to
the conservation of energy equation. It is shown that, with these
improvements, predictions of an integral model for wet plume behaviour in the atmosphere are in good agreement with observations
of short cooling tower plumes.

GOVERNING EQUATIONS

Integral equations governing wet plume rise and spread in the
atmosphere can be developed by applying conservation of total mass,
water mass, vertical momentum, horizontal momentum, and energy to a
cylindrical plume element, in which an integrated or average value
of each plume variable is assigned at each cross-section. The
element is assumed to interact with its surroundings through turbulent mixing, which is modelled as an entrainment process with
characteristic velocity v_e, and through buoyancy and drag forces.
In its visible or condensed phase, the plume is assumed to contain
water droplets and to remain saturated with water vapour. The
hydrostatic law, the velocity definition ($V^2 = V_x^2 + w^2$), and a relationship between saturated specific humidity and temperature,
such as that of Richards (1971),

$$q_{sat} = \frac{0.62 \ e_{sat}}{P - e_{sat}} \tag{1}$$

with $e_{sat} = 101325. \ \exp(13.3185\tau - 0.1967\tau^2 - 0.6445\tau^3 - 0.1299\tau^4)$

and $\tau = 1 - 373.15/T$

complete the equation set. (All variables are defined in the
Nomenclature). With the substitution of hypotheses for entrainment

velocity and drag force; the specification of ambient wind, temperature, and humidity profiles; and the specification of source conditions, these equations can be solved numerically for the eight plume
variables ρ_p, R, V, V_x, w, σ, q_p, T as functions of travel time t
from the source. The plume centreline position may also be calculated simultaneously by including the kinematic relationships

$$\frac{dx}{dt} = V_x \tag{2}$$

$$\frac{dz}{dt} = w \tag{3}$$

Rather than proceeding with a numerical solution, however, an
analytical solution can be derived following the procedure used for
dry plumes (Slawson and Csanady, 1971; Briggs, 1975). By introducing
the Boussinesq, bent-over plume, and negligible drag force assump-
tions along with the entrainment hypothesis $v_e = \beta|w|$, the governing
equations reduce to a set of linear equations in terms of flux
variable groups:

$$\frac{dR^3}{dx} = 3\beta \frac{|M|}{U^2} \tag{4}$$

$$\frac{d\Delta Q}{dx} + \frac{d\Sigma}{dx} = - G \frac{M}{U} \tag{5}$$

$$\frac{dM}{dx} = \frac{F}{U} - g \frac{\Sigma}{U} \tag{6}$$

$$\frac{dF}{dx} - K \frac{d\Sigma}{dx} = - N^2 \frac{M}{U} \tag{7}$$

where constant values for U, N^2, and G must be specified. A remaining difficulty is to incorporate a $q_{sat}(T)$ correlation into this
simplified equation set. In order to linearize Eq. 1 in terms of
flux variables, it is first assumed that a power law substitution
can be made

$$q_{sat} = aT^b \tag{8}$$

As is illustrated in Fig. 1, the accuracy of this assumption can be
verified through a least squares curve fit procedure. The introduction of flux variables into Eq. 8 and a binomial expansion for
$T_p - T_a \ll T_a$ then yield

$$\Delta Q = \frac{c_1}{R} R^3 + c_2 F \tag{9}$$

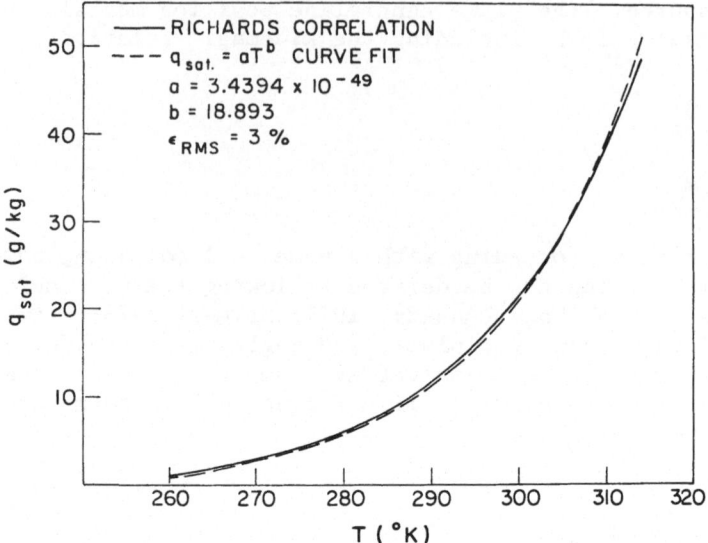

Fig. 1 Richards' (1971) correlation for $q_{sat}(T)$ and the least
squares power law fit.

where the constant $\bar{R}$ factor has been introduced to enable lineariza-
tion. Comparison of analytical model predictions with those of a
numerical solution suggests that a good choice of $\bar{R}$ is the radius
of the plume near its final rise point (Jager, 1982). The visible,
saturated section of the plume is described by Eqs. 4 to 9, while
the dry plume equations are recovered when the plume becomes unsaturated ($\Sigma \to 0$) and invisible.

It should be noted that Eq. 7, the conservation of energy statement, is slightly different than that used by previous authors
(e.g. Csanady, 1971; Wigley and Slawson, 1975). According to our
derivation, the factor K should be

$$K = \frac{h_L}{c_p T_{ao}} - 2.36 \tag{10}$$

whereas previously only $h_L/c_p T_{ao}$ was used. The additional 2.36
factor, which reduces K by about 25%, arises through algebraic

rearrangement of terms involving the specific heats of dry air, water vapour, and liquid water.

AN IMPROVED ANALYTICAL MODEL FOR WET PLUMES

Formulas predicting wet plume behaviour in the atmosphere can be obtained by seeking analytical solutions of Eqs. 4–7 and 9, which can be combined to yield the single equation

$$\frac{d^4 F}{dx^4} + \lambda^2 \frac{d^2 F}{dx^2} = 0 \tag{11}$$

for $\lambda^2 > 0$, which occurs under essentially all atmospheric conditions of practical interest, the solution of Eq. 11 is

$$F(x) = F_o + (gA - U\lambda^2)[\frac{M_o}{\lambda} \sin \lambda x + \frac{1}{U\lambda^2} (F_o - g\Sigma_o)(1 - \cos \lambda x)] \tag{12}$$

Substitution of F(x) back into the governing equations leads to the remaining solutions

$$M(x) = M_o \cos \lambda x + \frac{1}{U\lambda} (F_o - g\Sigma_o) \sin \lambda x \tag{13}$$

$$\Sigma(x) = \frac{A}{\lambda} M(x) \tag{14}$$

$$z(x) = \{ \frac{3}{U\beta^2 \lambda} [F(x) - F_o] + (\frac{R_o}{\beta})^3 \}^{1/3} - \frac{R_o}{\beta} \tag{15}$$

$$R(x) = R_o + \beta z(x) \tag{16}$$

Unlike previous solutions, Eqs. 12–16 explicitly retain latent heat effects. In the unsaturated or no liquid water limit, it is easy to show that these solutions reduce to those which have been previously published for dry plume predictions (e.g. Briggs, 1975; Davidson and Slawson, 1982).

The source flux parameters F_o, M_o, Σ_o, and R_o in Eqs. 12–16 refer to source conditions for the Boussinesq, bent-over plume, as opposed to conditions for the actual source. The specification of these source fluxes consistent with the approximations introduced into the governing equations has been shown to have a significant effect on the performance of integral plume models (Davidson and Slawson, 1982; Davidson, 1982). Matching total mass and water mass fluxes for the actual and model source, as illustrated in Fig. 2, yields the relationships

$$R_o = R_s \sqrt{\frac{T_{ao}}{T_s} \frac{w_s}{U} \frac{1-q_s}{1-q_o} \frac{1+0.62q_o}{1+0.62q_s}} \qquad (17)$$

$$\sigma_o = \frac{q_s - q_o}{1 - q_s} \qquad (18)$$

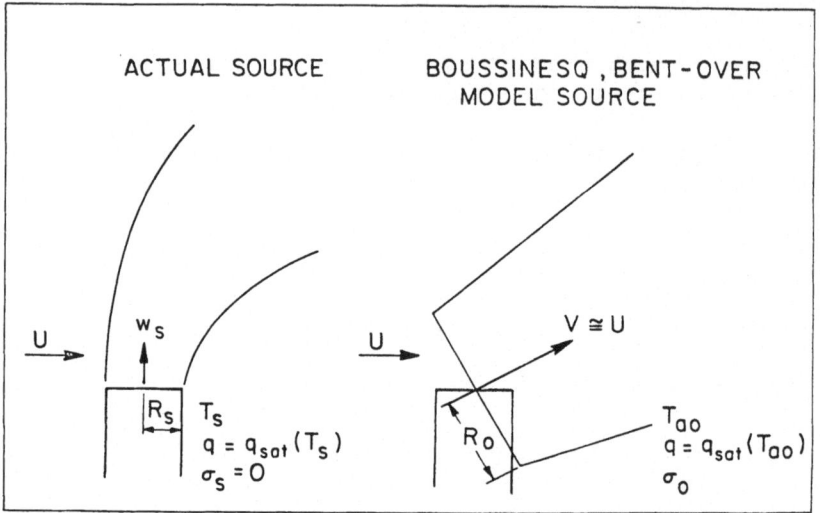

Fig. 2 The actual and model sources.

where $q_s = q_{sat}(T_s)$ and $q_o = q_{sat}(T_{ao})$. Eq. 18 implies that an initial condensation of the saturated plume has occurred. As discussed by Hanna (1972), the corresponding release of latent heat would add energy to the plume, reduce its rate of temperature decrease, and add to the source buoyancy flux. However, such an increase in buoyancy would be inconsistent with the analysis presented previously, since the Boussinesq assumption has already assumed that the plume temperature falls to the air temperature immediately when the plume enters the atmosphere. Thus no latent heat correction of the source buoyancy flux is applied here.

By substitution of Eqs. 17 and 18 into the source flux definitions, the appropriate source flux parameters consistent with the simplifications applied to develop Eqs. 12-16 therefore become

$$M_o = R_s^2 \, w_s \, T_{ao}/T_s \tag{19}$$

$$F_o = R_s^2 \, w_s g \, (T_s - T_{ao})/T_s \tag{20}$$

$$\Sigma_o = R_s^2 \, w_s \, \sigma_o \, T_{ao}/T_s. \tag{21}$$

and the R_o expression of Eq. 17, which to good accuracy can be reduced to

$$R_o = R_s \sqrt{\frac{T_{ao}}{T_s} \frac{w_s}{U}} \tag{22}$$

COMPARISON WITH OBSERVATIONS

Some preliminary tests of the model described above have been performed by comparing its predictions with the cooling tower measurements made at Paradise steam plant by Slawson (1978). In these comparisons, the visible outline was calculated by replacing the top-hat $\sigma(x)$ profile by a Gaussian profile with the same mean and with standard deviation $R(x)/2.15$. Assuming that the plume becomes invisible when its liquid water content falls to some small specified value σ_E, yields a visible radius

$$R_V(x) = 0.659 \, R(x) \ln \sqrt{\frac{2.559\sigma(x)}{\sigma_E}} \tag{24}$$

In the downwind direction, the plume becomes invisible when $\sigma(x)$ falls to $\sigma_E/2.559$ and $R_V = 0$. Predictions were made using an entrainment constant $\beta = 0.6$, in good agreement with the recommendations of Briggs (1975) for dry plumes.

In Fig. 3, trajectory and radius predictions are shown for the Feb. 2, 0715-0813 data set. The upper dashed curve shows the trajectory $z(x)$ and a typical radius R as predicted by a dry plume model without latent heat effects or F_o and M_o corrections. Correction of F_o and M_o reduces the rate of rise slightly to the dotted central curve, while application of both F_o and M_o corrections and latent heat effects (Eq. 15) leads to the bottom solid curve. Predicted trajectories all become parallel after the end of the visible plume, indicated by vertical bars in Fig. 3, since water vapour then becomes a passive contaminant of a dry plume. From the figure, it is evident that trajectory predictions are in good agreement with observations whether or not the improvements suggested in this paper are applied. The approach of Wigley and Slawson (1975) and Slawson (1978) which neglects the effects of latent heat and source flux corrections on plume dynamics therefore appears to be adequate for trajectory predictions of cooling tower plumes.

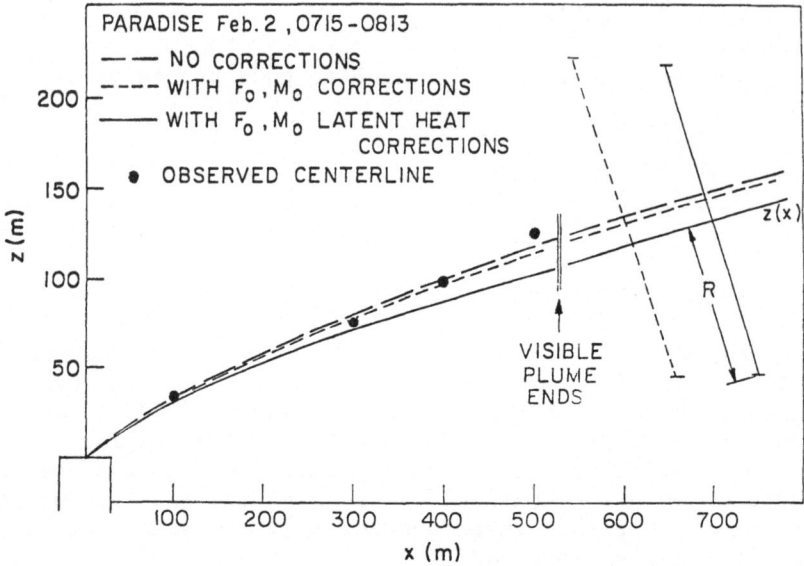

Fig. 3   Predictions and observations of plume centerline trajectory at Paradise steam plant, Feb. 2, 0715-0813.

In Figs. 4 and 5, visible outline predictions for the same data set are compared with observations. Comparison of these figures shows that, while latent heat and source flux corrections may alter the trajectory prediction only slightly, the visible outline prediction is strongly affected. As Hanna (1976) has pointed out, the use of a dry plume trajectory and $\sigma_s = 0$ leads to the underestimated outlines of Fig. 4. Application of the improved model, Eqs. 12-16 with Eqs. 20-24, however, leads to the much larger outline predictions of Fig. 5, which are in better agreement with observations. From Fig. 5 it is also evident that the replacement of a top-hat $\sigma(x)$ profile (which yields the dotted outline in Figs. 4 and 5) with an equivalent Gaussian profile (which yields the solid σ isopleths in Figs. 4 and 5) leads to better agreement with measurements. Based on the Paradise data sets summarized in Table 1, the $\sigma_E = 7.5 \times 10^{-4}$ isopleth was chosen as the visibility cut-off. This edge criterion may seem somewhat high relative to cloud and fog measurements which show σ values in the 5×10^{-5} to 5×10^{-3} range (e.g. Ludlam, 1980). However, because of the Boussinesq assumption, the model will tend to overpredict σ_0, and, because of the round cross-section assumption, the model will tend to overpredict the vertical extent of the plume. A high σ_E assumption would compensate for such model simplifications. It is therefore perhaps more realistic to view σ_E as a model constant as opposed to a physical visibility criterion.

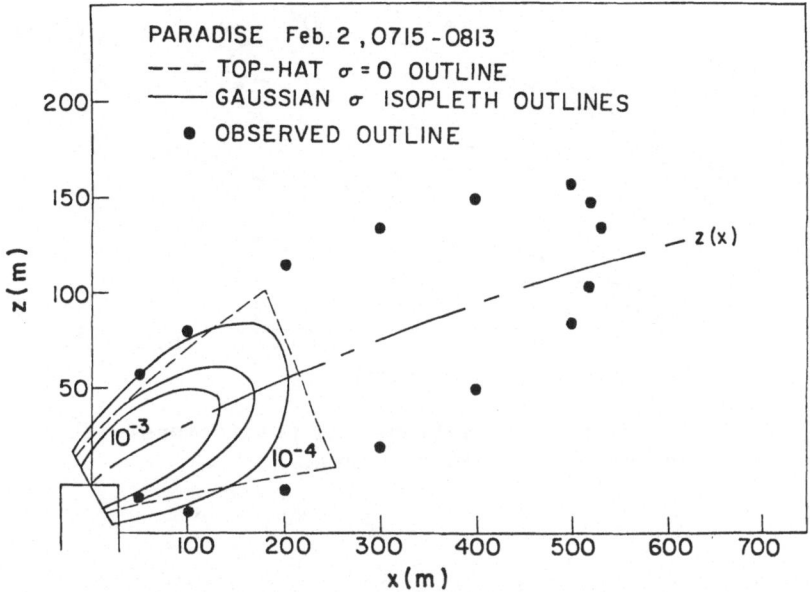

Fig. 4 Visible outline observations and predictions without latent
        heat and source flux corrections.

In Table 1, the performance of the improved model is summarized for other Paradise data sets. Attention has been confined to short visible plumes, which disappear before the maximum rise point is reached, in order that the initial phase analysis presented here remains valid, and no atmospheric diffusion phase is required. For the six cases considered, the improved model with $\beta = 0.6$, $\sigma_E = 7.5 \times 10^{-4}$ predicts both trajectory and visible length reasonably well. Without latent heat and source flux corrections, trajectories can still be predicted to about the same accuracy by allowing for the scatter in input data to the model and/or by changing the entrainment constant

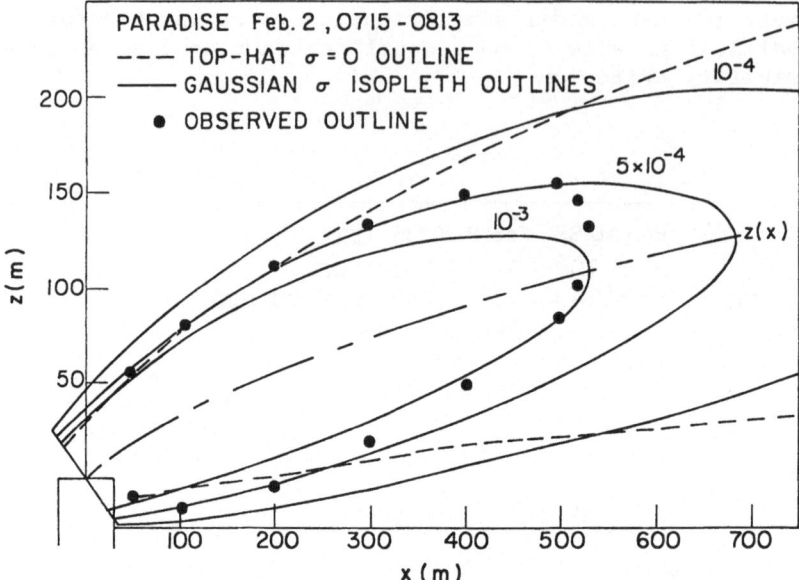

Fig. 5 Visible outline observations and predictions with latent
        heat and source flux corrections.

slightly. However, visible plume lengths are consistently underestimated. Thus, it can be concluded the simple integral model, Eqs. 12-
16, with the same entrainment constant $\beta = 0.6$ recommended for dry
buoyant plumes, can successfully predict both trajectory and visible
length of short condensing plumes in the atmosphere as long as the
source flux corrections, Eqs. 19-22, are applied.

ACKNOWLEDGEMENTS

    The authors would like to thank Dr. P.R. Slawson for many useful
discussions on the cooling tower problem, and Mrs. Colleen Janer for
typing the manuscript. The work was supported by grants from the
Natural Sciences and Engineering Research Council of Canada.

Table 1. Plume trajectory and visible length predictions and observations for Paradise steam plant data.

Case	Trajectory Prediction RMS Percent Error	Observed Visible Length (m)	Predicted Visible Length (m)	
			with corrections	without corrections
Jan 31 0800	13.9	400	425	110
Feb. 2 0715	8.9	530	580	150
Feb 4 0700	9.4	480	515	220
Feb 5 0715	13.0	520	500	146
Feb 11 1226	18.1	225	150	70
Feb 12 1408	12.0	200	270	63

REFERENCES

Briggs, G.A. (1975). Plume rise predictions. AMS Workshop on Meteor-
        ology and Environmental Assessment, Boston, 29 Sept. - 3 Oct.
        1975.
Carhart, R.A., Policastro, A.J., Wastag, M., Haake, K. (1980). An
        improved model for natural and mechanical draft cooling tower
        plume rise. Proc. 1980 IAHR Cooling Tower Workshop. San
Csanady, G.T. (1971). Bent-over vapor plumes. J. Appl. Meteor. 10,
        36-42.
Davidson, G.A. (1982). Source flux corrections in analytical vapour
        plume models. J. Appl. Meteor., to be published
Davidson, G.A. Slawson, P.R. (1981). Effective source flux parameters
        for use in analytical plume rise models. Atmos. Environ.
        16, 223-227.
Hanna, S.R. (1972). Rise and condensation of large cooling tower
        plumes. J. Appl. Meteor. 11, 793-799.
Hanna, S.R. (1976). Predicted and observed cooling tower plume rise
        and visible length at the John E. Amos Power Plant. Atmos.
        Environ. 10, 1043-1052.
Jager, W. (1982). An analytical wet plume model with latent heat and
        source flux corrections. M. A. SC. Thesis, University of Waterloo
Ludlam, F.H. (1980). "Clouds and Storms". Pennsylvania State Univer-
        sity Press, p. 185.
Richards, J.M. (1971). A simple expression for the saturation vapour
        pressure of water in the range -50 to 140°C. J. Phys. D:
        Appl. Phys., L15-L18.
Slawson, P.R. (1978). Observations and predictions of natural draft
        cooling tower plumes at Paradise Steam Plant. Atmos. Environ.
        12, 1713-1724.
Slawson, P.R., Csanady, G.T. (1971). The effect of atmospheric condi-
        tions on plume rise. J. Fluid Mech. 47, 33-49.
Wigley, T.M.L., Slawson, P.R. (1972). A comparison of wet and dry
        bent-over plumes. J. Appl. Meteor. 11, 335-340.
Wigley, T.M.L., Slawson, P.R. (1975). The effect of atmospheric condi-
        tions on the length of visible cooling tower plumes. Atmos.
        Environ. 9, 437-445.

NOMENCLATURE

A model constant $= (\dfrac{N^2}{U} c_2 - \dfrac{G}{U} - \dfrac{3c_1\beta}{\bar{R}\,U^2}) (1 + Kgc_2)^{-1}$

a, b constants in the correlation $q_{sat} = aT^b$

c_p specific heat of dry air at constant pressure

c_1 model constant $= U(a\,T_{ao}^b - q_{ao})$

c_2 model constant $= abT_{ao}^b/g$

F	buoyancy flux parameter	$= UR^2 g(T_p - T_a)/T_a$
G	atmospheric humidity gradient	
g	gravity	
h_L	latent heat of vaporization	
K	model constant	$= h_L/c_p T_{ao} - 2.36$
M	momentum flux parameter	$= UR^2 w$
N	Brunt–Vaisala frequency	
q	specific humidity	
R	plume radius	
T	temperature	
U	windspeed	
V	total plume velocity	
V_x	plume x-velocity	
v_e	entrainment velocity	
w	plume vertical velocity	
x	downwind distance from source	
z	plume centerline elevation above source	
β	entrainment constant	
ΔQ	humidity flux parameter	$= UR^2 (q_p - q_a)$
λ^2	solution constant	$= N^2/U^2 - g(K-1)A/U$
ρ	density	
Σ	liquid water flux parameter	$= UR^2 \sigma$
σ	liquid water concentration (kg water/kg gas)	

Subscripts

a	atmospheric property
E	edge of visible plume
o	property at the source
p	plume property
sat	saturated vapour property
s	actual source property
V	visible plume property

DISCUSSION

G. RESELE   You only looked at short plumes.
Did you ever look at low plumes. Since your entrain-
ment is proportional to the vertical velocity, you
will get into troubles !

G.A. DAVIDSON   The assumption $v_e = \beta|\omega|$
required to obtain analytical solutions is only
valid in the initial phase of plume behaviour where
self-generated turbulence as opposed to ambient
atmospheric turbulence controls plume spread.
In comparing model predictions with observations,
attention was therefore restricted to short plumes
which evaporated completely well within this initial
phase. In this way, the additional uncertainties
ties of adding an atmospheric diffusion phase to the
model were avoided for the purposes of this paper.

F. FANAKI   There is difficulty in choosing
the suitable value for the entrainment coefficient β;
β varies from 0.1. to 0.7. Would you like to comment
on what value of β one should use ?

G.A. DAVIDSON   One aim of this work was to
show that $v_e = \beta|w|$ with $\beta \cong 0.6$ works well in analy-
tical models for both dry and wet, bent-over plumes.
In my opinion, the variability of β which you dis-
cribe is due, at least in part, to inconsistencies
between the assumptions built into the model and the
source flux definitions.

SIMULATION OF ATMOSPHERIC EFFECTS OF INDUSTRIAL HEAT RELEASES :

THE "ARTIFICIAL CONVECTION" PROGRAM

A. Saab (1) - T. Rasoamanana (2) - B. Benech (3)

(1) Electricité de France, D.E.R. 78400 Chatou
(2) Université Paul Sabatier 31062 Toulouse
(3) Université Clermont II 65300 Lannemezan

INTRODUCTION

The work reported here was initiated in 1977 both by the Electricité de France and the Observatoire du Puy de Dôme in order to provide a better theoretical understanding of waste heat effects in the atmosphere such as those caused by dry cooling towers. The approach followed was two-fold :

1 - To develop and to carry out an experimental project on the environmental impact of dumping waste dry heat from an oil-fired 1000 MW source of heat named the Meteotron (Benech, 1976). Approximately 20 hours of operation by the Meteotron have given us physical data on the dynamical, thermal and microphysical processes caused by these artificial heat releases and allow us to elucidate these processes under a wide variety of meteorological conditions.

2 - To verify and to test the capabilities of a three-dimensional numerical model to simulate the atmospheric response to waste heat already observed. This model which uses variable eddy exchange coefficients to simulate subgrid-scale turbulence and a cloud physic parameterization yields the thermodynamic and dynamic fields as well as bulk microphysical properties.

Through this experimental and numerical approach, our final goal is to build a numerical model which combines useful predictive capability and operating characteristics in order to assess the environmental impact for power plant cooling towers involving much larger waste heat.

THE EXPERIMENTAL PROJECT

In terms of heat source and flux density, the Meteotron is comparable to a large power plant cooling tower. The modifications it produces in clouds are similar to those found in the vicinity of large coal-fired electric power plants, oil refinery complexes and forest slash fires (Radke, 1980 ; Koenig, 1979 ; Auer, 1976).

During this program, the Meteotron consisted of 105 fuel-oil burners arrayed over an area of 15 000 m² and producing a heat flux density of 66 000 $W.m^{-2}$. During each 30 minute experiment, various experimental techniques were used to obtain descriptions of the heat source by temperature sensors, of the ambient atmosphere conditions by radiosoundings and radar wind measurements, and of the ascending hot plume and the induced artificial cloud by instrumented aircraft (thermodynamic and microphysical parameters) as well as lidar, meteorological radar and infrared thermography.

Forty experiments were carried out under a wide variety of background meteorological conditions : June 19 to July 8, 1978 ; October 3 to 20, 1978 ; and May 29 to June 23, 1979 (Benech et Al., 1980). These experiments may be distributed in four classes relating to the progressive development of effects caused by a waste-heat perturbation :

Type 1 : Dry plume produced in a dry atmosphere (humidity less than 60 %) characterized by a neutral or stable vertical thermal structure. Only the temperature and the wind fields are modified over the heat source (16 experiments).

Type 2 : Isolated artificial cloud produced within an extended convective mixing layer with important humidity (more than 80 %) and the presence of few natural cumuli clouds. The microphysical characteristics of the artificial cloud (condensation level, cloud depth, liquid water content and droplet size) are similar to those found in the natural clouds (10 experiments).

Type 3 : Interaction of artificial cloud with natural convective clouds which are numerous. The humidity is more than 90 % and the wind velocity is weak. The microphysical characteristics of resultant cloud are modified in comparison with the natural clouds ones : cloud depth and liquid water contents are slightly larger (4 experiments).

Type 4 : Interaction of artificial cloud with the strato-cumulus
layer (humidity $\simeq$ 100 %). Some particular effects are
observed on the modification of the strato-cumulus layer
such as a cloud free ring around the plume or a coupling
effect between dynamics and microphysics (10 experiments).

These experimental results are used to validate the numerical
model described here after through the following procedure : the
first validation is based on comparison between observed and predicted orders of magnitude for various parameters. The second step
will consider the physical representation of the phenomenon evolution and complex coupling effects likely to occur in the presence
of specific atmospheric conditions.

THE 3D CLOUD MODEL

Conservation Equations and Model of Turbulence

The model is based on the equations of motion with the
Boussinesq approximation, the conservation of mass, energy and
water substance. The microphysical processes included in the
conservation equations for cloud droplets and rain water contents,
i.e. evaporation-condensation, autoconversion and accretion, are
parameterized following Kessler (1969).

Using constant eddy coefficients, the values of which are
valid for the description of the subcloud region, was a first step
toward understanding the dynamical and microphysical development
of hot plumes and induced convective clouds (Saab et Al., 1980).

Presently, turbulence parameterization of subgrid motions is
included using a sophisticated formulation of the deformation eddy
viscosity. This formulation investigated by Tag et Al. (1979) with
a 2D cloud model and extended to both the deformation and buoyancy
fields can be represented in a 3D model as :

$$
\nu_T = (\Delta)^{2/3} \left\{ k_d^2 \left[\left(\frac{\partial u}{\partial y} + \frac{\partial v}{\partial x} \right)^2 + \left(\frac{\partial w}{\partial x} + \frac{\partial u}{\partial z} \right)^2 \right. \right.
$$

$$
+ \left(\frac{\partial w}{\partial y} + \frac{\partial v}{\partial z} \right)^2 + 2 \left(\left(\frac{\partial u}{\partial x} \right)^2 + \left(\frac{\partial v}{\partial y} \right)^2 + \left(\frac{\partial w}{\partial z} \right)^2 \right) \right]^{1/2}
$$

$$
+ k_b^2 \left| \frac{g}{\theta} \frac{\partial \theta}{\partial z} \right|^{1/2} \left. \right\}
$$

where :

$\Delta = (\Delta x \, \Delta y \, \Delta z)^{1/3}$, length scale,

$\theta = T_O \left(\dfrac{1000}{P_O}\right)^{2/7}$, potential temperature where T_O and P_O are the
initial temperature and pressure,

$k_d = 0.22$, dimensionless constant,

$k_b = 0.2-0.7$, dimensionless constant, if $\dfrac{\partial \theta}{\partial z} < 0$,

$k_b = 0$, if $\dfrac{\partial \theta}{\partial z} > 0$.

  Sensitivity tests on the influence of buoyancy term on the eddy viscosity coefficients provided us different values for the dimensionless constant k_b depending on thermal stability conditions and the coupling effects between dynamics and microphysics. Thus the value of 0.7 for k_b suggested by Tag seems too large for strong thermal instability, and when the artificial cloud interacts with the natural clouds or the strato-cumulus layer. The best comparisons with observations were producing k_b values of 0.4 for type 3 experiments and 0.20 for type 4 experiments. A value of 0.22 for k_d slightly larger than 0.2 value used by Tag was adequate for most experiments involving neutral conditions (Saab et Al., 1982).

Solution Procedure and Boundary Conditions

  The numerical "SMAC" method used in this work has been developed at Los Alamos and applied in France by Gaillard (1978) ; it is explicit in time and implicit in pressure. The spatial discretisation in momentum equations is a weighted mean of CDM2 (Centered momentum flux) and UDV (Upwind Velocity flux). The pressure/continuity treatment is based on a pressure correction iteration involving a mutual adjustment of velocities and pressures, in sequence :

a/ A provisional velocity field is derived from explicit momentum equations.

b/ Velocity divergence is evaluated at each grid point and used to calculate a pressure change ΔP and an associated ΔV for obtaining continuity balance.

c/ This treatment is repeatedly until local balance prevails everywhere.

The initial and boundary conditions are specified as follows :

a/ The perturbation consists of continual injection of heat and moisture at specified grid points. The 1000 MW heat source is simulated as an 10 000 m² (100 × 100) aera, with a 20 °C temperature excess with respect to the ambient air, a 4 ms^{-1} mean vertical velocity and a 30% mean relative humidity.

b/ Ambient atmospheric values at the upwind boundary.

c/ No slip conditions on the ground.

d/ Open boundary conditions at the downwind boundaries.

The model needs about 20 minutes of CPU time to reach quasi-dynamical equilibrium with 48 × 12 × 48 grid points on a CRAY 1 computer. Considering the spatial extent needed to simulate this type of perturbation (2500 meters along each direction), a grid resolution (Δx, Δy, Δz) of 40 m to 60 m was selected. The time step Δt is adjusted in order to ensure stability of the numerical scheme (i.e. Δt = 2 to 3 seconds for most of the runs).

MODEL RESULTS AND DISCUSSION

Comparison with Experiments

Fourteen experiments corresponding to different meteorological conditions and plume types have been used to validate the model according to the following criteria : cloud top, cloud base, maximum water liquid content and vertical velocities in the plume.

Table 1 summarizes for seven experiments producing dry plumes or isolated artificial clouds (plume types 1 and 2) the comparison between observed and predicted top plume height and condensation occurrence. The observed plume tops are estimated using three remote sensing systems (lidar, radar plus chaff and photogrammetry). The simulated condensation agrees well with cloud experimental data except when small clouds (fractocumulus) are generated. Whenever a large part of the plume is condensed, the agreement is observed between computed and observed condensation levels and water liquid contents.

Table 1. Comparison between observed and predicted top plume heights and condensation occurrence (plume types 1 and 2).

Meteotron experiments	Condensation		Top plume height (m)	
	Observed	Predicted condensation	Observed	Predicted
June 22, 78	Fractocumulus	No	1400	1500
July 3, 78	Dry plume	No	$\simeq$ 900	820
July 5, 78	Fractocumulus	No	1100	1150
July 8, 78	Cumulus	Yes	$\simeq$ 850	920
October 5, 78	Cumulus	Yes	1200	1020
October 7, 78	Dry plume	No	1100	1000
June 18, 79	Fractocumulus	No	1150	1200

Considering the meteorological conditions when artificial cumulus clouds interact with natural clouds or stratocumulus layers, Table 2 compares the observed and predicted microphysical characteristics for seven experiments (plume types 3 and 4). In general, the computed and observed values are in agreement as well as their trend, keeping in mind that the observed values are punctual (aircraft measurements) whereas the predicted ones are averaged in space. Predictions of liquid water content and cloud depth are slightly overestimated (20 per cent) in most cases. Figure 1 summarizes the correlations obtained between computed and observed values for cloud depth the maximum vertical velocities and the liquid water contents at different heights within the condensation layer.

An example on the physical representation of phenomenon evolution and complex coupling effects occuring in the presence of interaction between the artificial cloud and the natural stratiform cloud layer is shown on Figures 2a, 2b, 2c. It represents the spectacular effect observed on June 21, 1979 by the formation of a cloud free ring around the plume. The plume penetration into the cloud layer can be shown on Figures 3a and 3b ; it is characterized by an upward motion upwind the artificial cloud compensated by a downdraft motion downwind.

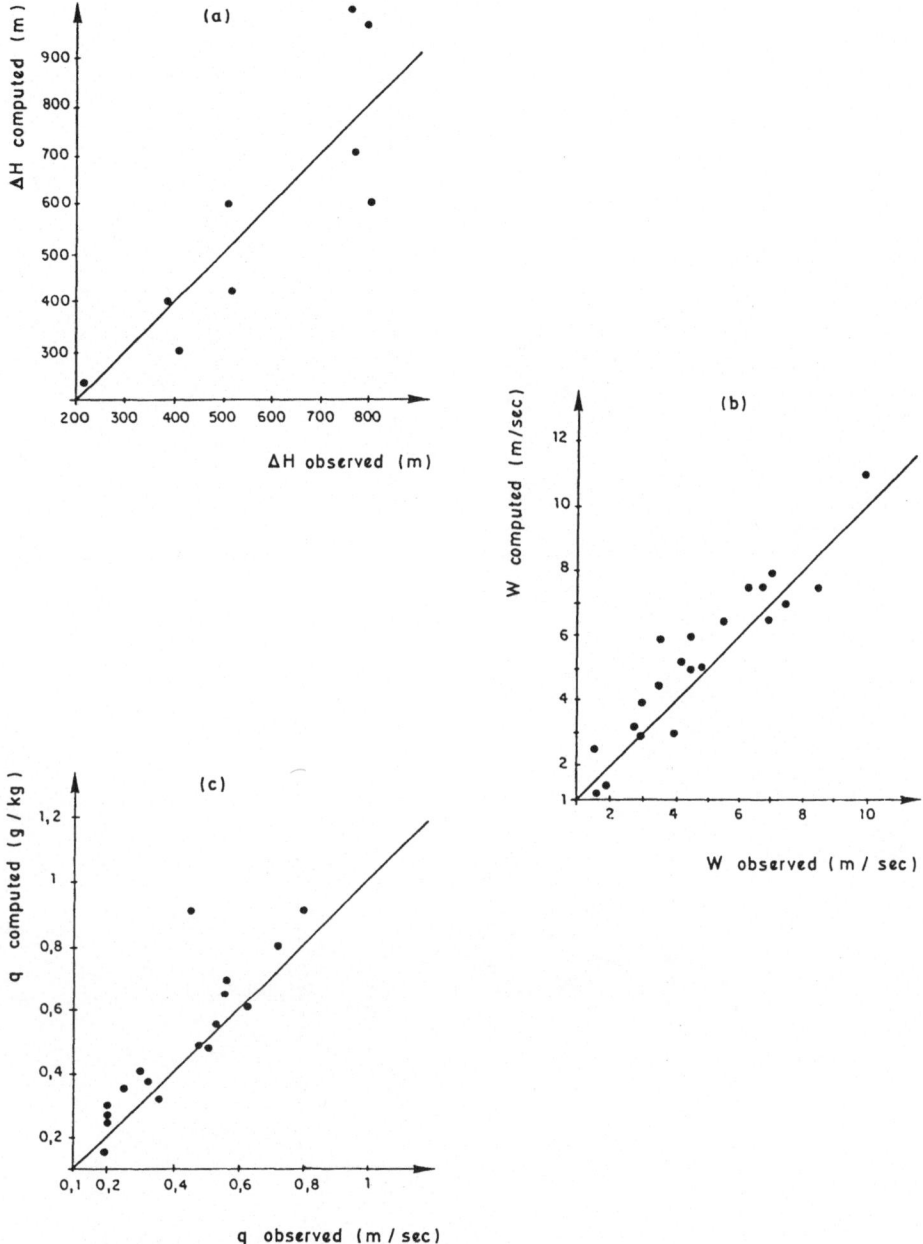

Fig. 1. Comparison between computed and observed values of dynamical and microphysical characteristics within the artificial clouds :
(a) : Cloud depth (ΔH, m),
(b) : Maximum vertical velocity (W, m/s),
(c) : Liquid water content (q, g/kg).

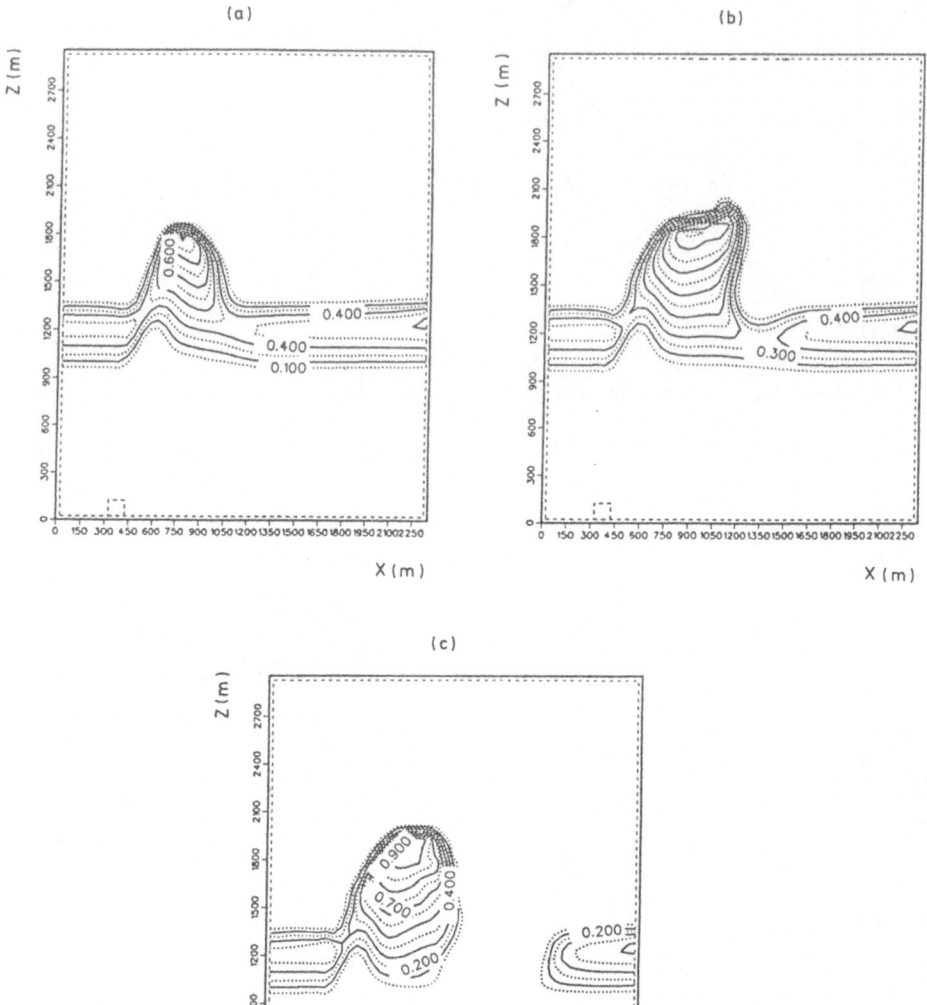

Fig. 2   Time evolution of contour values of cloud water content
         (g/kg) in the longitudinal cross section over the
         heat source center.

         June 21, 1979

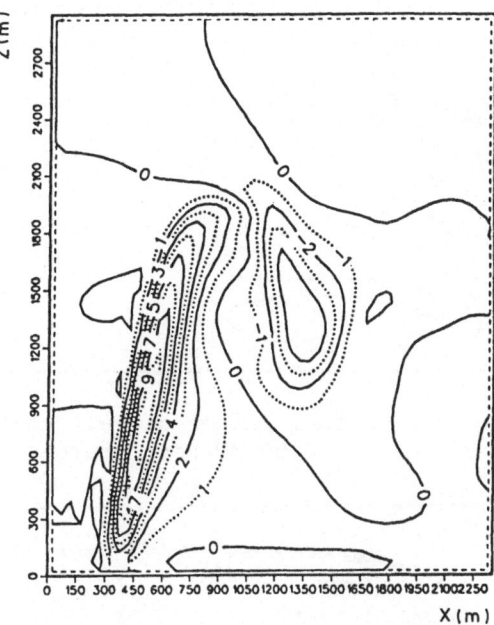

Fig. 3a   Isovalues of vertical velocity (m/s) in longitudinal
          cross section over the heat source center – June 21, 1979.

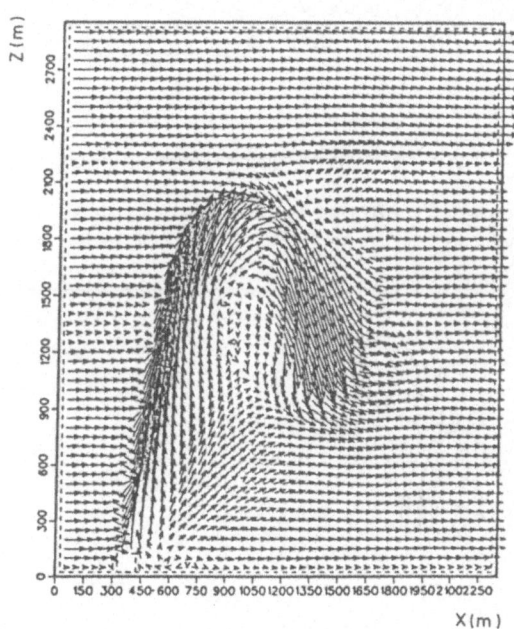

Fig. 3b   Wind field in longitudinal cross section over the heat
          source center – June 21,1979.

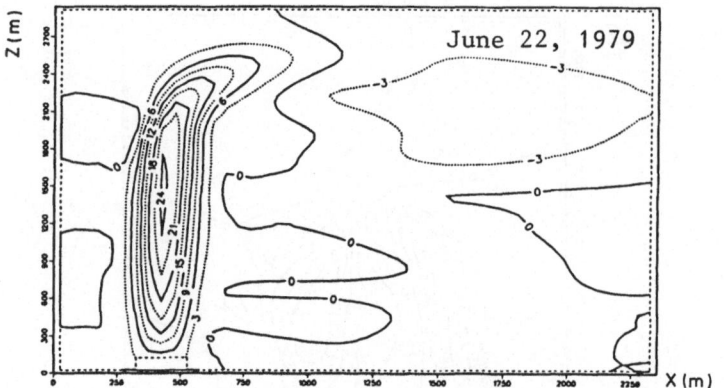

Fig. 4a Expectation of contour values of vertical velocity (m/s)
over a hypothetical 8000 MW heat source.

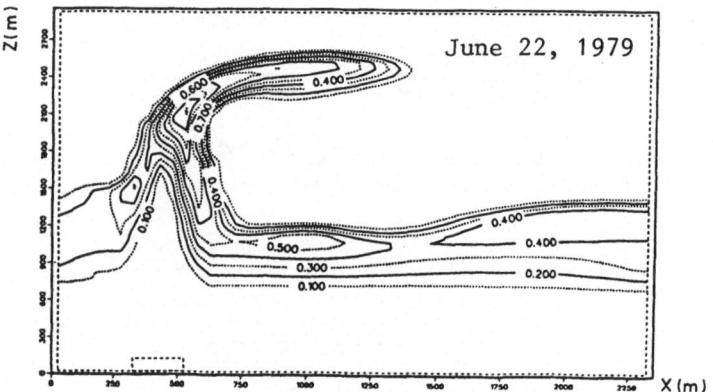

Fig. 4b Expectation of contour values of liquid water content
(g/kg) over a hypothetical 8000 MW heat source

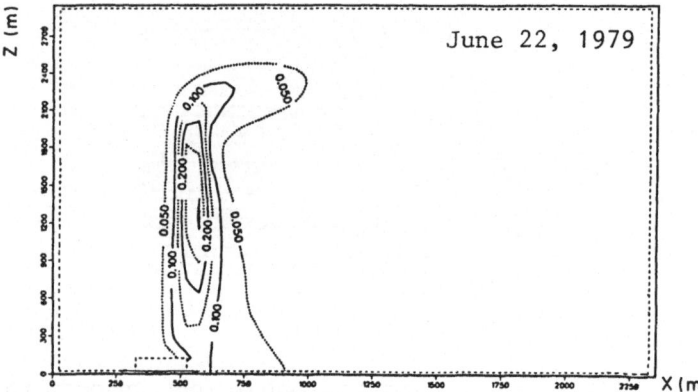

Fig. 4c Expectation of contour values of precipitating water
contents (g/kg) over à hypothetical 8000 MW heat source.

Table 2. Comparison between observed and predicted microphysical characteristics of artificial clouds
interacting with natural clouds or statrocumulus
layers (plume types 3 and 4).

Meteotron experiments	Data	Cloud top (m)	Cloud base (m)	Maximum liquid water content (g/kg)
June 28, 78	Observed	1400	990	0.52
	Predicted	1450	1250	0.56
June 2, 79 a.m.	Observed	750	360	0.20
	Predicted	960	560	0.26
June 2, 79 p.m.	Observed	1340	820	0.32
	Predicted	1320	900	0.37
June 7, 79	Observed	1770	1000	0.46
	Predicted	1550	850	0.90
June 14, 79	Observed	1630	820	0.56
	Predicted	1520	920	0.68
June 21, 79	Observed	1580	800	0.72
	Predicted	1900	900	0.80
June 22, 79	Observed	>1285	520	0.66 (rain ?)
	Predicted	1900	600	1.23 (rain ?)

Numerical Tests on Large Waste Heat Releases

Future application of this model will concern the assessment
of the environmental impact for power plant cooling towers involving much larger waste heat. To illustrate the ability of the
model to simulate the atmospheric response to large heat releases
in amounts of 8000 MW encountered in present power parks, some
tests are performed during critical meteorological situations
such as the probable rainfall case observed on June 22, 1979.
On Figure 4, one can show the major modifications in the dynamics
and the microphysics :

a/ Strong convective perturbation provoking large increase in
   vertical velocities (up to 25 m/s) and whirling motions,
   Figure 4a,

b/ Deep cloud development (top level at 2400 m) with a slight
   modification in the microphysical characteristics in comparison with a 1000 MW source : a maximum value of 1.1 g/kg for
   liquid water content producing light precipitation,
   Figures 4b and 4c.

SUMMARY

    The general goal of the "Artificial Convection" program was
to study the atmospheric effects of large heat rejection from dry
cooling towers using both experimental and numerical approach.
On one side, experiments performed on the Meteotron (a 1000 MW
sensible heat source) under a wide variety of meteorological
situations provided us quantitative information on the dynamical
and microphysical processes caused by artificial heat releases.
On the other side, a three-dimensional numerical model has been
developed and validated through comparisons with experiments.
Thus the model produced in most cases a simulation that agreed
very well with the real cloud in size and shape and reasonably
well in temperature and velocity (updraft) fields, and liquid
water contents.

REFERENCES

Auer, A., Observation of an industrial cumulus, J. Appl. Meteor.,
    15, 406-413, 1976.
Benech, B., Experimental study of an artifical convective plume
    initiated from the ground, J. Appl. Meteor., 15, 127-137,
    1976.
Benech, B., Dessens, J., Charpentier, C., Sauvageot, H., Druilhet,
    A., Ribon, M., Dinh, P., Mery, P., Thermodynamic and micro-
    physical impact of a 1000 MW heat released source into the
    atmospheric environment, Proc. Third WMO Scientific Conference
    on Weather Modification, 1, 111-118, 1980.
Gaillard, P., Modelisation numérique des panaches d'aéroréfrigé-
    rants, Ph. D. Thesis, Lyon, 1978.
Kessler, E., On the distribution and continuity of water substance
    in atmospheric circulation, Meteor. Monogr.,10, n° 32, 1969.
Koenig, L.R., Anomalous cloudiness and precipitation caused by
    industrial heat rejection, Rand Corporation, Report R-2465,
    DOE, 1979.
Raske, L.F., Benech, B., Dessens, J., Eltgroth, M., Henrion, X.,
    Hobbs, P., Ribon, M., Modifications of cloud microphysics by
    a 1000 MW source of heat and aerosol (the Meteotron project),
    Proc. Third WMO Scientific Conference on Weather Modification,
    1980.
Saab, A.E., Caneill, J.Y., Sarthou, P., Rasoamanana, T., Benech,
    B., A three dimensional numerical model of convective clouds,
    J. Rech. Atmos., 14, 423-432, 1980.
Tag, P., Murray, F., Koenig, L.R., A comparison of several forms
    of eddy viscosity parameterization in a 2D cloud model, J.
    Appl. Meteor., 18, 1429-1441, 1979.

DISCUSSION

J. KNOX   Could you please discuss the
June 22 case in which observed and predicted precipitation was shown as in Table 2.2.
Did remote sensing (e.g. radar) show the time of formation of precipitation comparable to the formation
time in the model ?

E. SAAB   First, the (?) means a probable rainfall case since precipitation data on
ground are partly questionable. But it was quite
sure from remote sensing and aircraft measurements
that precipitation occurred in altitude (as we observed rain drops).
In the model, we observe up to 10 or 15 min. of computing time an oscillation in the microphysical characteristics and we reach a quasi equilibrium state
for dynamics and water liquid content after 20 to
30 minutes. Comparison with the real time of observation is speculative.

F. FANAKI   In Canada with similar conditions like the one represented, the plume initiated
gravity waves; did the same thing happen in your
case at the condensation level ?

E. SAAB   Yes, it did. Many times, we
observed excitement of gravity waves, but we lack
quantitative information.

K.E. GRØNSKEI   Based on your model, could
you indicate a lower bound on the heat release from
an industrial complex when the extra atmospheric effects become negligible ?

E. SAAB   The atmospheric effects are
not linear in function of the amount of heat releases.
They depend on the vertical atmospheric structure.
Thus, in dry atmosphere, a 8000 MW source does not
induce an artificial cloud; whereas in wet atmosphere (H > 90 %), 250 MW could initiate an artificial cloud.

THE COOLING TOWER MODEL SMOKA AND ITS APPLICATION TO A LARGE SET

OF DATA

Bruno Rudolf

Deutscher Wetterdienst
6050 Offenbach a.M.
Federal Republic of Germany

ABSTRACT

SMOKA, an entrainment model for moist heat sources, contains a set of seven differential equations : horizontal and vertical motion, enthalpy, vapour, cloud water, rain water, any admixture. It respects buoyancy, conversion of sensible and latent heat, mixing of plume air and environmental air as well as mutual mixing of up to ten separate sources. The entrainment rate depends on wind shearing and RICHARDSON number and is based on modern boundary layer theories. The model is calibrated by cooling tower plume observations. The results of the verification are shown.

The dependence on the marginal conditions (waste heat, wind speed, atmospheric stability, temperature, humidity) is demonstrated.

The model is applied to radiosonde data of five stations (about 3650 data sets of each one). This large number allows statistical declarations on cooling tower plumes respecting regional and seasonal particularities.

SMOKA can also be used for other heat sources. Calculations for the purpose of calibrating and testing the model applied to hot smoke are in preparation.

DESCRIPTION OF THE MODEL

SMOKA contains ordinary differential equations for the parameters horizontal movement, vertical movement, enthalpy, total water content, liquid- water content and concentrations of pollutants (see synopsis of the equations).

Synopsis of Equations

(1) $\qquad \dfrac{dv_h}{ds} = -\mu \cdot v_{hD} - c_1 \cdot \left[\dfrac{v_{hD}}{v}\right]$

(2) $\qquad \dfrac{dw}{ds} = -\mu \cdot w_D + \dfrac{b}{v} - c_2 \cdot \left[\dfrac{w_D}{v}\right]$

(2a) $\qquad b = g\left[\dfrac{T_D}{T_U} + 0.61 \cdot q_D - l\right]$

(3) $\qquad \dfrac{dh}{ds} = -\mu \cdot h_D$

(3a) $\qquad h = c_p \cdot T + g \cdot z + L \cdot q$

(4) $\qquad \dfrac{d(q+l)}{ds} = -\mu (q+l)_D - \dfrac{P}{v}$

(5) $\qquad \dfrac{dl}{ds} = -\mu \cdot l_D + \dfrac{Gen}{v} - \dfrac{Evap}{v} - \dfrac{P}{v}$

(5a) $\qquad Gen = 0 \qquad bei\ q < q_s$

$\qquad\qquad\quad = v\,\dfrac{d(q+l)}{ds}\ bei\ q \geq q_s$

(5b) $\qquad q_s = \dfrac{0.622 \cdot e_s}{p - 0.378 \cdot e_s}$

$\qquad\quad e_s = 6.1078 \cdot 10^{\frac{7.5(T-273.16)}{T-35.86}}$

EVAP, P nach KESSLER

(6) $\qquad \mu = -\dfrac{1}{M}\dfrac{dM}{ds}$

$\qquad\qquad = \dfrac{2}{R}\dfrac{w}{v}\left[\dfrac{m_1^2\,w_D^2 + m_2^2\,v_{hD}^2}{w_D^2 + v_{hD}^2}\right]^{\frac{1}{2}} + m_3 \cdot \dfrac{2}{R} \cdot \dfrac{i_U}{v}$

(6a) $\qquad i_U = \left[\dfrac{1}{\alpha} \cdot \dfrac{z \cdot K}{1+\dfrac{z \cdot K}{\lambda\infty}} \cdot \dfrac{\partial v_{hU}}{\partial z}\right]^2 \cdot f(Ri)$

$\qquad f(Ri) = \dfrac{1}{2}\left[1-(\gamma_1 + \gamma_2)Ri + \left[(1-(\gamma_1+\gamma_2)Ri)^2 + 4\gamma_2 Ri\right]^{\frac{1}{2}}\right]$

$\qquad Ri = \dfrac{g}{\theta_U}\dfrac{\partial\theta_U}{\partial z} \cdot \left[\dfrac{\partial v_{hU}}{\partial z}\right]^{-2}$

(6b) $\qquad M = \pi \cdot R^2 \cdot \rho \cdot v$

(7) $\qquad dx = \dfrac{v_x}{v}ds, \quad dy = \dfrac{v_y}{v}ds, \quad dz = \dfrac{w}{v}ds$

(8) $\qquad v_h = (v_x^2 + v_y^2)^{\frac{1}{2}}, \quad v = (v_x^2 + v_y^2 + w^2)^{\frac{1}{2}}$

symbols

b	buoyancy
c_1, c_2	drag coefficients
c_p	specific heat at constant pressure
e	vapor pressure
g	gravity
h	enthalpy
i	turbulence intensity
L	evaporation heat
l	specific liquid water content
M	mass flux
$m_1 - m_6$	mixing coefficients
P	precipitation rate
p	pressure
q	specific humidity
R	characteristic plume radius
Ri	RICHARDSON number
s	coordinate along the plume axis
T	temperature
v	wind velocity along the plume axis
w	vertical component of v
x, y, z	cartesian coordinates
$\alpha, \gamma_1, \gamma_2$	boundary layer coefficients
$\varkappa$	KARMAN constant
λ_∞	mixing length in free atmosphere
μ	mixing rate
ρ	density
θ	potential temperature

indices

D	difference between plume and environment
h	horizontal
s	saturation
U	environment

The model equations describe the changes of these parameters along the plume axis beginning with the air conditions at the cooling tower exit (figure 1).

The model equations are derived [1,2] by integrating the atmospheric basic equations in the cross section normal to the plume axis respecting the following assumptions :

- All processes are stationary.
- Plume and environment are in a quasi-hydrostatic balance.
- The wind direction within the plume is identical with the wind direction in the environment in the corresponding height.
- The distribution of all parameters in the plume normal to the plume axis is determined by empirical functions on the basis of the values computed for the plume axis.

Because of these assumptions, the small-scale or short-term variations of the structure within the plume cannot be computed by SMOKA, e.g. transverse circulation or the dissolution of the plume into different cells. Like other stationary models, SMOKA should only be used when the marginal conditions remain more or less unchanged during a period of about an hour.

The mixing rate is defined as relative increase of the mass flux along the plume axis. It depends empirically on the wind shear between plume and ambient air as well as on the turbulence of the ambient air. The intensity of the turbulence is parameterized as function of the RICHARDSON number [3]; the used value is averaged vertically over the dispersion layer. The characteristic plume radius can be derived from the definition of the mass flux.

The evaporation and precipitation rates are calculated on the basis of empiric statements according to Kessler [4].

The wake effect occuring at the lee of a building during strong wind is considered empirically by superposing a vertical deformation over the horizontal homogeneous environmental flow. The vertical deformation depends on the extension of the building and the wind velocity at the top of the building.

The equations are solved in finite differences with a step of 1 m.

The procedure described allows for the calculation and representation of the plume of one cooling tower. The plumes of several cooling towers are calculated alternately step by step so that interactions between neighbouring plumes (e.g. mutual entrainment) may be considered (figure 2).

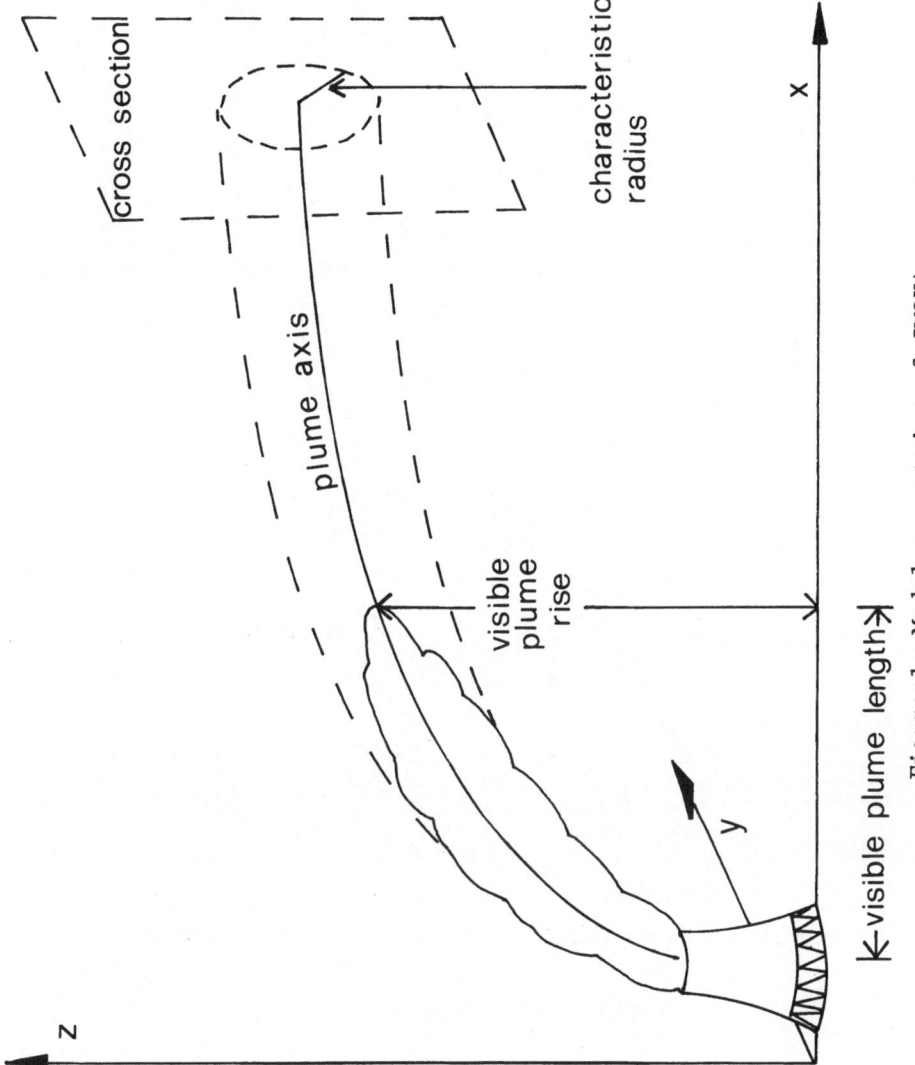

Figure 1: Model geometrics of SMOKA.

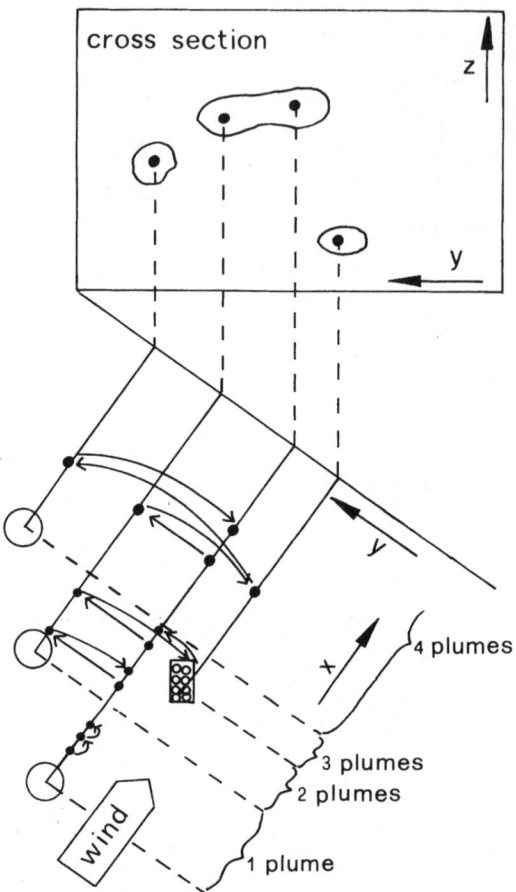

Figure 2: Computation of several plumes by SMOKA.

⭕ ▓▓ cooling towers

●   point of the axis

Model results are for example :
- the outline of the visible cooling tower plume caused by the liquid water droplets,
- the spatial distribution of the increase of the specific water content,
- the areal distribution of the precipitation reaching the ground,
- the shadow of the visible plume depending on the altitude of the sun,
- the spatial distribution of a pollutant emitted by a cooling tower or by a stack.

Apart from the simulation of cooling tower plumes the model may also be used for other heat or humidity sources. It may be applied to the computation of the hot stack plume rise respecting vertical profiles of temperature, humidity and wind as well as the emission of latent heat.

CALIBRATION AND VERIFICATION OF THE MODEL

The equations of the model contain a number of empiric parameters, i.e. the entrainment and drag coefficients, the flow deformation in the lee of a building etc. Their order is known from laboratory tests.

A calibration of the model, i.e. a more exact determination of the parameters, was made in field tests.
For this purpose the released heat, water vapour and droplets as well as the meteorological marginal conditions and the cooling tower effects are measured. These data sets should be exhaustive so the differences between the model results and measurements may be judged. For a comparison between measured and computed cooling tower effects, the outline of the visible plume is an especially good criterion as it is not only depending strongly on the marginal conditions but may also be easily photographed.

Six well-founded cases were used for the calibration. Five data sets originate from measurements carried out at Neurath in 1973 (three natural draft cooling towers, height 100 m, total heat release approximately 1250 MW). A sixth case is based on data obtained at Lünen in 1972 (a natural draft cooling tower, height 109 m, release 500 MW). The model coefficients are fitted to yield model results corresponding well to the observations (figures 3 and 4).

Two of these examples (Neurath, 28 September 1973, 15.00 Central European Time (CET)and 16. December 1972, 11.30 CET) show how in case of a strong wind speed the plumes are pressed down by the wake effect.

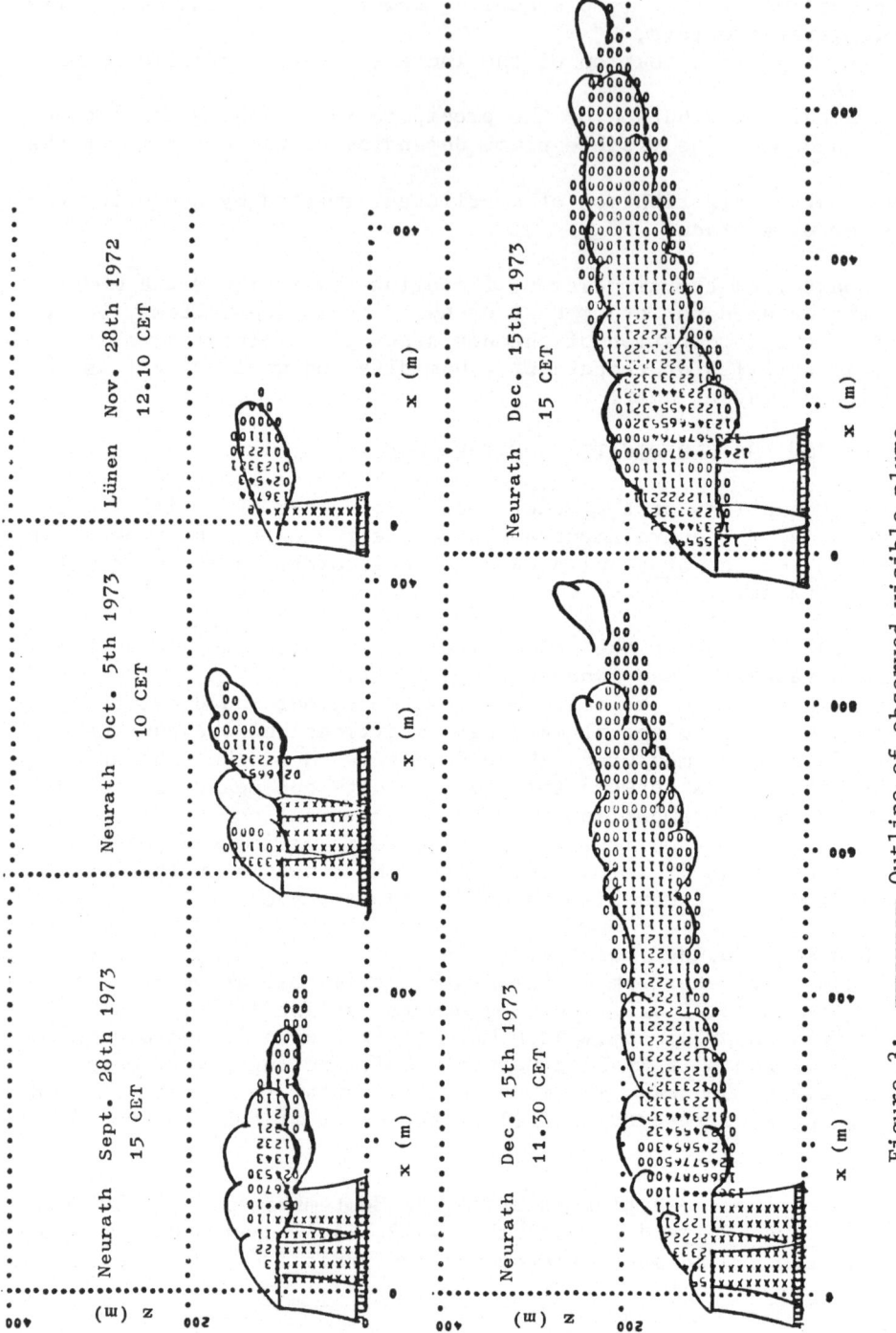

Figure 3:   —— Outline of observed visible plume
             numbers: predicted visible plume after calibration of SMOKA

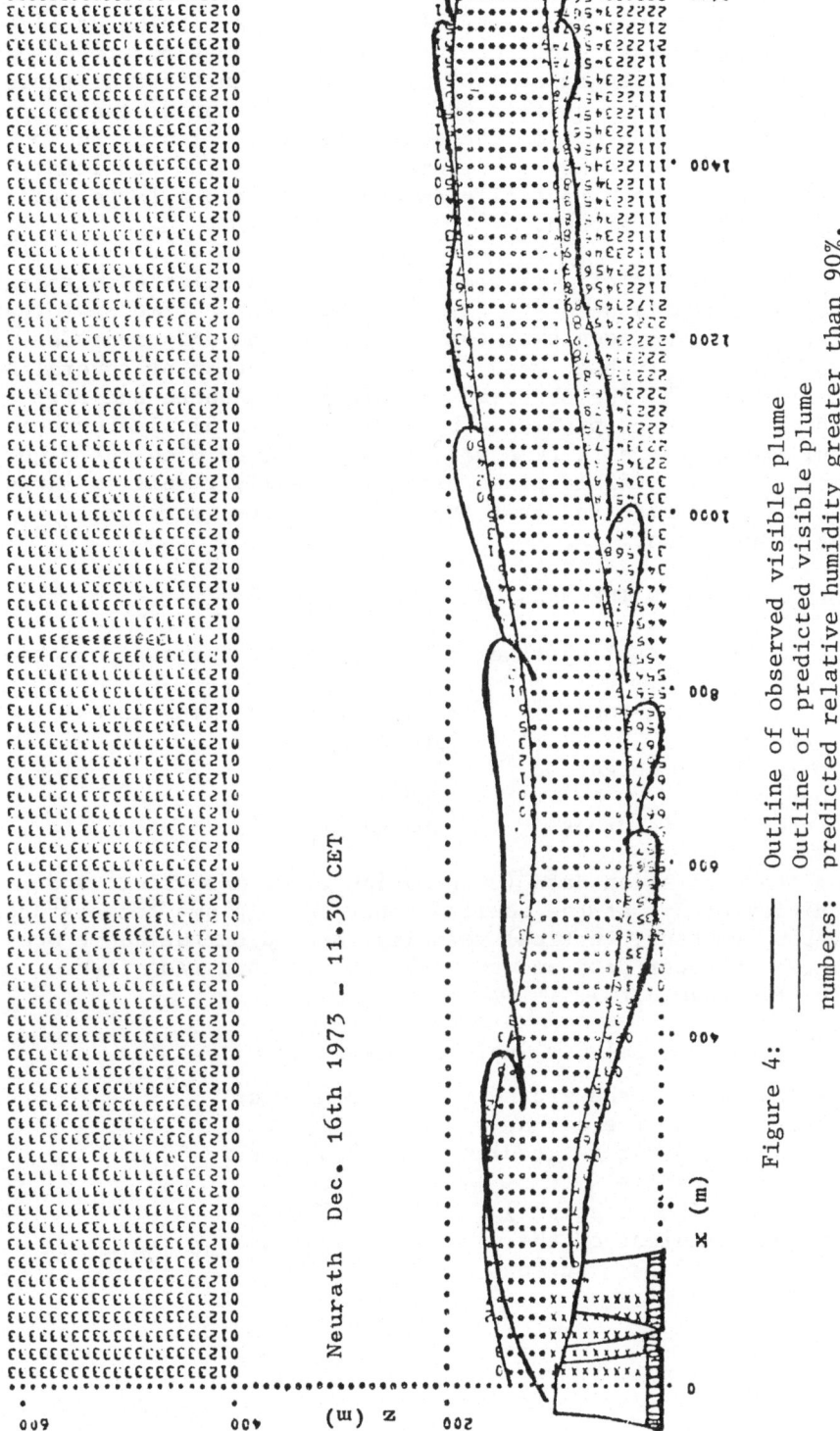

Figure 4:

Neurath Dec. 16th 1973 - 11.30 CET

	Outline of observed visible plume
	Outline of predicted visible plume
numbers:	predicted relative humidity greater than 90%.

The case which occured on 16 December 1973, 11.30 CET (figure 4) clearly shows that the simplifications of the model have to be considered in the interpretation of the results. The plumes observed reach in a compact form a length of 1500 m to 2000 m, in farer distances the plume dissects into smaller puffs the larger the distance is. Between 500 m and 1000 m from the cooling towers parts of the observed plume touch the ground. The plume computed by the stationary model, however, shows a spatially smoothed form and a large relative humidity near the ground.

A greater number of measured data sets is available containing some not sufficiently accurate data (differences in time or larger spatial distances between the measurements of the different data). The data have already been used for the verification of several cooling tower models [5]. The observed visible plume length and rise are compared with the ones predicted by SMOKA (figure 5). Inaccuracies of measurements and short time variations combined with the high sensitivity of the visible plume extension easily result in larger uncertainties; considering this the model results agree fairly well with the observations.

MODEL SENSITIVITY ON THE MARGINAL CONDITIONS

The following cooling tower data were given for the computation :

type of cooling tower	natural draft cooling tower
waste heat	2500 MW
cooling tower height	160 m
inner radius	42 m
ejected drizzle drops	4400 g/s

The other emission data (evaporation loss, mass flux) and the condition of air at the cooling tower exit (plume temperature, water vapour content, vertical velocity) depend on the above mentioned cooling tower data and the ambient air condition (temperature, relative humidity).

The meteorological marginal conditions are :

air pressure at the ground	1010	mbar
temperature at the ground	10	°C
vertical temperature gradient	−0.6	K per 100 m
relative humidity	80	%
wind speed (200 m height)	3	m/s
vertical gradient of the wind speed	0.25	m/s per 100 m

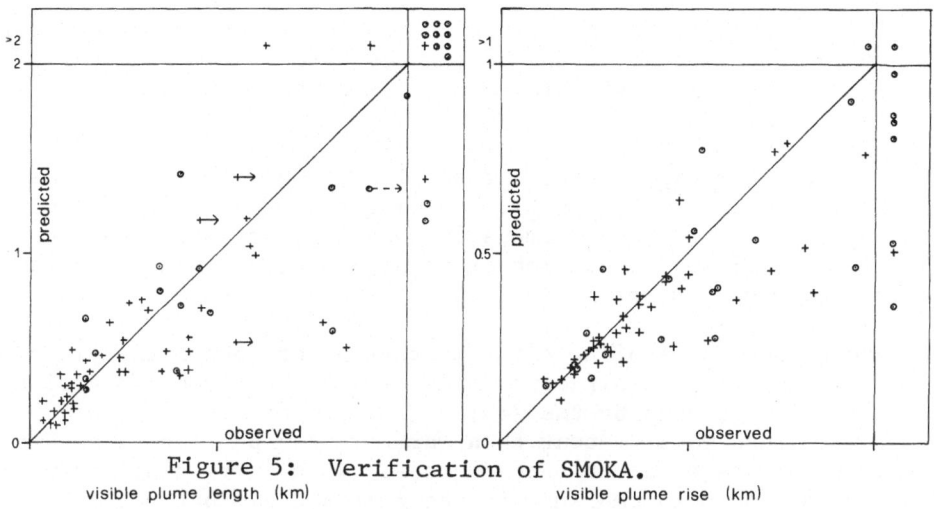

Figure 5:   Verification of SMOKA.

visible plume length (km)   visible plume rise (km)

Figure 5: Verification of SMOKA

⊙ 3 cooling tower cases (Amos, Neurath)

+ 1 cooling tower cases (Chalk Point, Paradise, Lünen, Philippsburg)

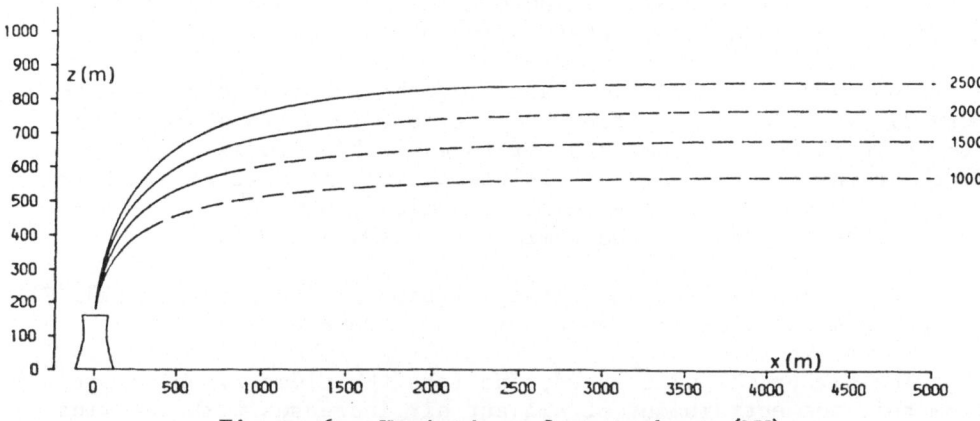

Figure 6:   Variation of waste heat (MW)

Figure 7: Variation of ambient air temperature (^{O}C)

The following paragraphs show the influence of parameter variation on the model results. In the figures 6 to 14 the plume axis is
solid, if the plume is visible, and it is dashed, if the plume is
not visible.

The less the waste heat, the smaller the plume rise and the
visible plume length (figure 6). Reducing the waste heat without
changing the cooling tower dimensions has a great impact on the
exit air conditions : reduction of temperature, vertical velocity
and water vapour content.

The higher the ambient air temperature the longer the visible
plume (figure 7). Exit air conditions as well as plume dispersion
are heavily influenced by the ambient air temperature. As cold
air absorbs less water vapour than warm air the portion of sensible
heat on total waste heat amount decreases with increasing temperature. In case of cold ambient air the released air has not only a
greater emission velocity but also a larger buoyancy; consequently
the plume ascends higher, and, because of the additional condensation, the visible plume is longer. The entrainment of cold ambient air causes a slower evaporation of the visible plume.

The larger the relative humidity, the longer the plume (figure 8). The increased latent heat release in moist ambient air
causes an increased plume rise of long visible plumes compared
to the plume rise of shorter visible plumes. At a given air temperature it depends on the relative humidity how much water vapour can be absorbed by the humid air to get saturated.

For these examples an isothermal temperature gradient (10 °C)
was assumed, in order to avoid any temperature effect.

The higher the wind speed, the less the plume rises. Apart
from this the entrainment of ambient air increases with the wind
speed. For this reason long, high rising plumes are computed for
low wind speeds (figure 9). In case of strong wind a depression
is caused in the air current in the lee of the cooling tower. The
plume axis is pushed down (wake effect).

The more the temperature decreases with the height in the ambient air, the higher the plume rises and the longer the visible
plume gets (figure 10).

The latent heat released in very long, high-rising visible
plumes initiates an additional vertical pulse as may be seen in the
cause of the plume axis with a vertical temperature gradient of
0.8 K/100 m.

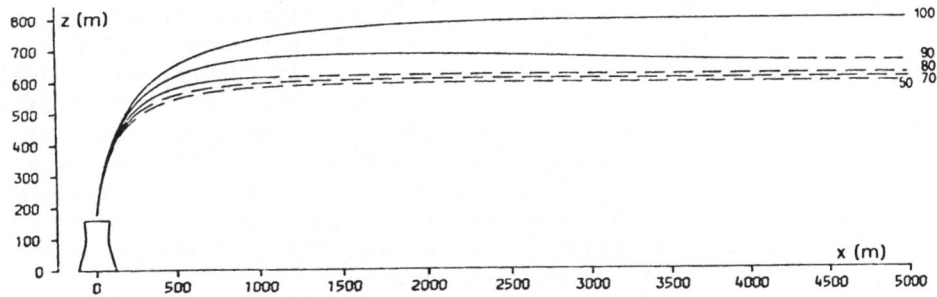

Figure 8: Variation of ambient relative humidity (%).

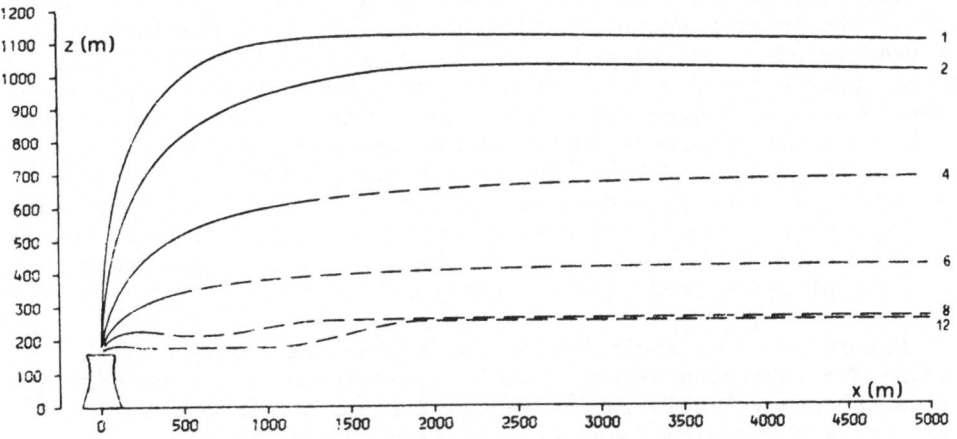

Figure 9: Variation of wind speed (m/s).

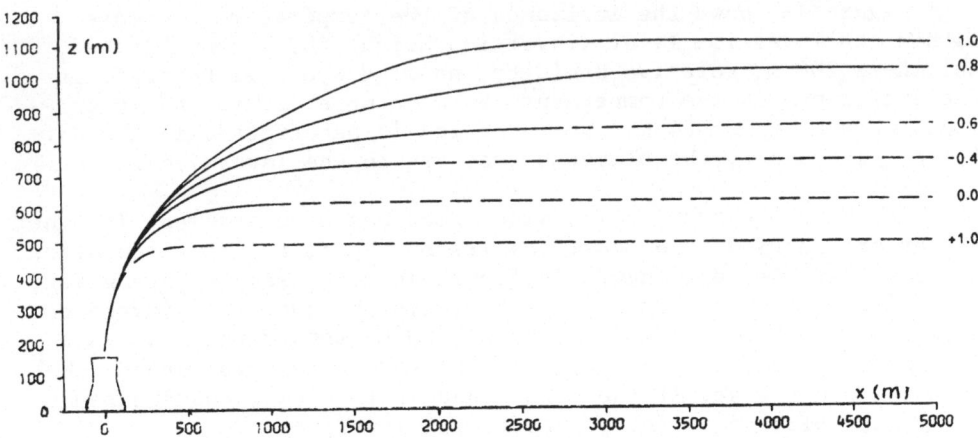

Figure 10: Variation of vertical temperature gradient (K/100 m).

The linear vertical gradients of the parameters considered reflect the actual atmospheric conditions in a very simplified form.
In most cases several layers of different conditions are superposed.
Of special interest are cases of inversions preventing to a large
degree the vertical exchange of air. Cooling tower plumes are able
to penetrate inversion layers; this is determined by the inversion
height above ground, the thickness of the inversion layer, and the
temperature increase within the inversion. Apart from this the meteorological conditions below the inversion layer as well as the
cooling tower emissions are of significance.

Figure 11 shows the behaviour of the plume axis for different
inversion heights. The inversion thickness is 200 m, the temperature increase is 4 K. The relative humidity is 70 % on the ground,
90 % at the lower bound of the inversion, 60 %, above the inversion,
the wind speed on the ground is 1 m/s, at the lower bound 2 m/s,
at the upper bound 2.5 m/s, and above the inversion 0.5 m/s. With
these conditions in the 250 m case the low inversion is penetrated
by the high rising plume. In the 400 m case the plume penetrates
the inversion, but does not have enough buoyancy to reach a greater height. In the dry air above the inversion the droplets of the
plume evaporate rapidly; therefore the plume cools and sinks to
the upper bound of the inversion. In the cases with higher inversions the plumes spread in the inversion layers.

Figure 12 shows the behaviour of a plume at various thick-
nesses. The inversion height is 400 m above ground, the temperature
within the inversion is increasing by 2 K per 100 m height. Relative humidity and wind speed are the same as before. The inversion of a thickness of 100 m is penetrated by the plume. When the
thickness is 300 m or more, the plumes enter the inversion layer
approximately 200 m, and spread.

Figure 13 shows the influence of the temperature increase
within the inversion layer (inversion height 400 m above ground,
thickness 200 m, relative humidity and wind speed as before). If
the difference in the temperature between upper bound and lower
bound is smaller than 4 K, the inversion is penetrated; if the difference is larger, the plumes spread within the inversion layer.

In the above cases a low wind speed has been assumed. If other
values are assumed, the model results of the diffusion calculation
are considerably influenced. In figure 14 some examples are given
of different wind speeds below and within an inversion (inversion
height 400 m above ground, thickness 100 m, temperature increase
2 K, relative humidity as before). The larger the wind speed, the
less the plume rise. If the wind speed at the lower bound extends
5 m/s, the plume does not penetrate the inversion.

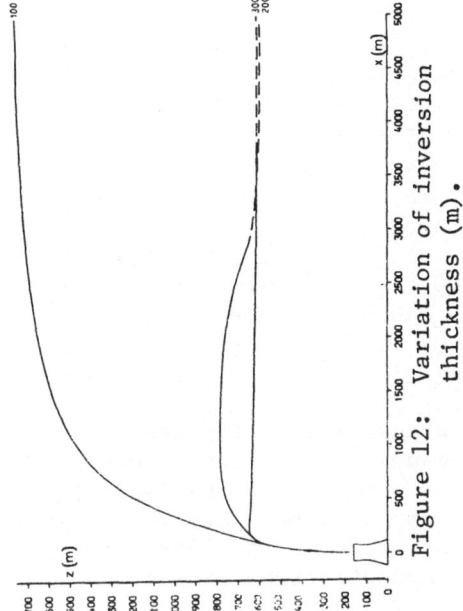

Figure 12: Variation of inversion thickness (m).

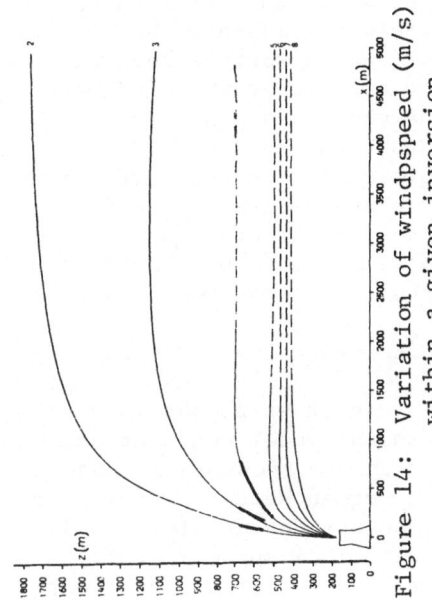

Figure 14: Variation of windspeed (m/s) within a given inversion.

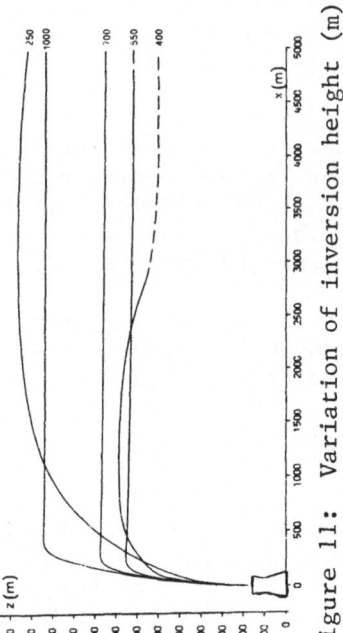

Figure 11: Variation of inversion height (m).

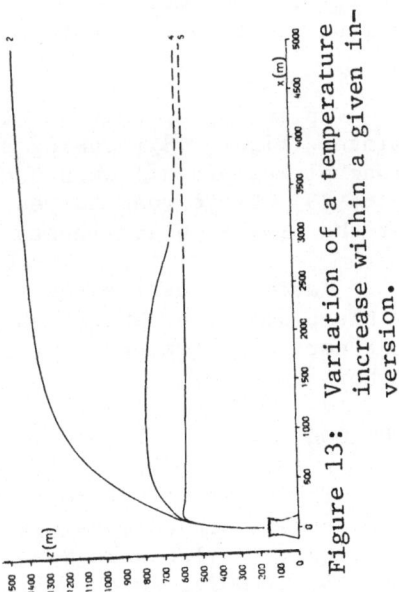

Figure 13: Variation of a temperature increase within a given inversion.

The visible plume length is mostly influenced by the meteorological parameters wind speed, saturation deficit and vertical temperature gradient. The saturation deficit depends on the temperature and describes the water vapour capacity of the air. For this the air temperature influences the plume length indirectly. For the complex dependency shown in figure 15 the following marginal conditions have been chosen : air temperature on the ground 10 °C, vertically constant, relative humidity resulting from air temperature and saturation deficit at the ground, vertically constant, wind speed increasing linearly to the height with the gradient corresponding the ground value per 1000 m, waste heat 2000 MW; the other cooling tower data correspond to those mentioned above.

MODEL RESULTS FOR LARGER DATA SETS

In order to obtain statistical statements on the plume extensions model computations have been made for large data sets. For this purpose radiosonde data of five aerological stations of the German Weather Service (see figure 16) have been used. Having 2 ascents per day (00 and 12 GMT) in the period March 1959 - February 1964 approximately 3650 model computations have been made for each one of the stations.

Figure 17 shows the frequency distribution of the visible plume lenght of the five stations. Plumes longer than 4000 m have been computed for around 30 % of all observations. Due to the higher humidity the frequency of the shortest plume class (up to a length of 250 m) is less at the seashore sited stations, Schleswig and Emden (about 15 % of all cases); it is higher at the inland stations at Stuttgart and Munich (about 30 %).

Splitting up the data the long plumes are most frequent in winter (figure 18); taking account of the low and medium clouds long plumes prevail when the sky is very cloudy or overcast (figure 19). Large year to year changes in the frequency distribution of the visible plume length have to be expected (figure 20).

Table 1 contains the frequency distribution of computed visible plume rise. Due to the frequently higher wind speeds at the northern stations more low plumes are computed than for the other stations.

OUTLOOK

The treatment of cooling tower effects will be continued. The actual computations deal with aerological data from Berlin; vertical profiles were measured there four times a day (03, 09, 15, 21 GMT), so the daily course of the meteorological conditions and their influence on cooling tower effects - especially

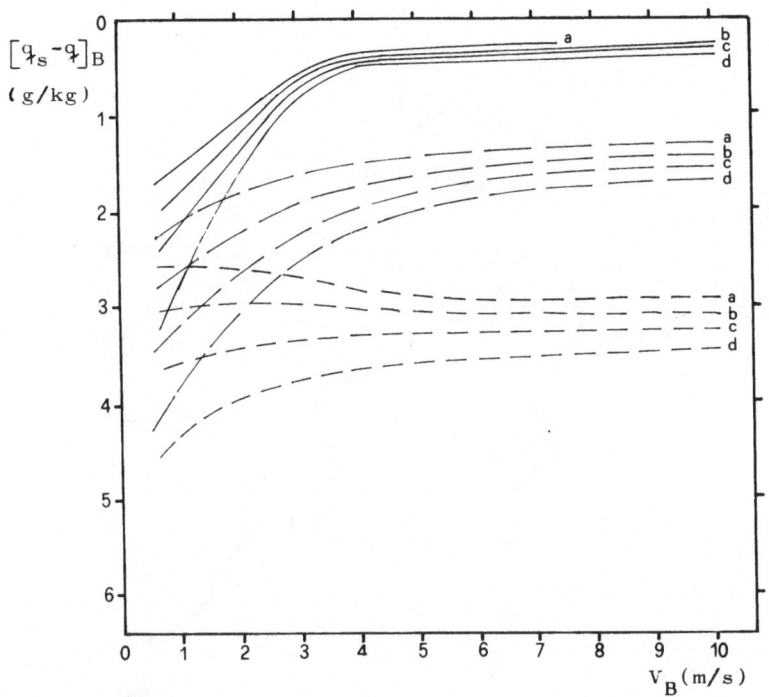

Figure 15.

Visible plume length ——————————————— 2000 m
                      — —— —— —— —— —— 500 m
                      — — — — — — — — — 300 m

depending on saturation deficit $q_s - q$ and wind speed at
the ground V_B for different vertical temperature gradients

a) dT/dz = 0.4 K/100 m   c) dT/dz = -0.4 K/100 m
b) dT/dz = 0.0 K/100 m   d) dT/dz = -0.8 K/100 m

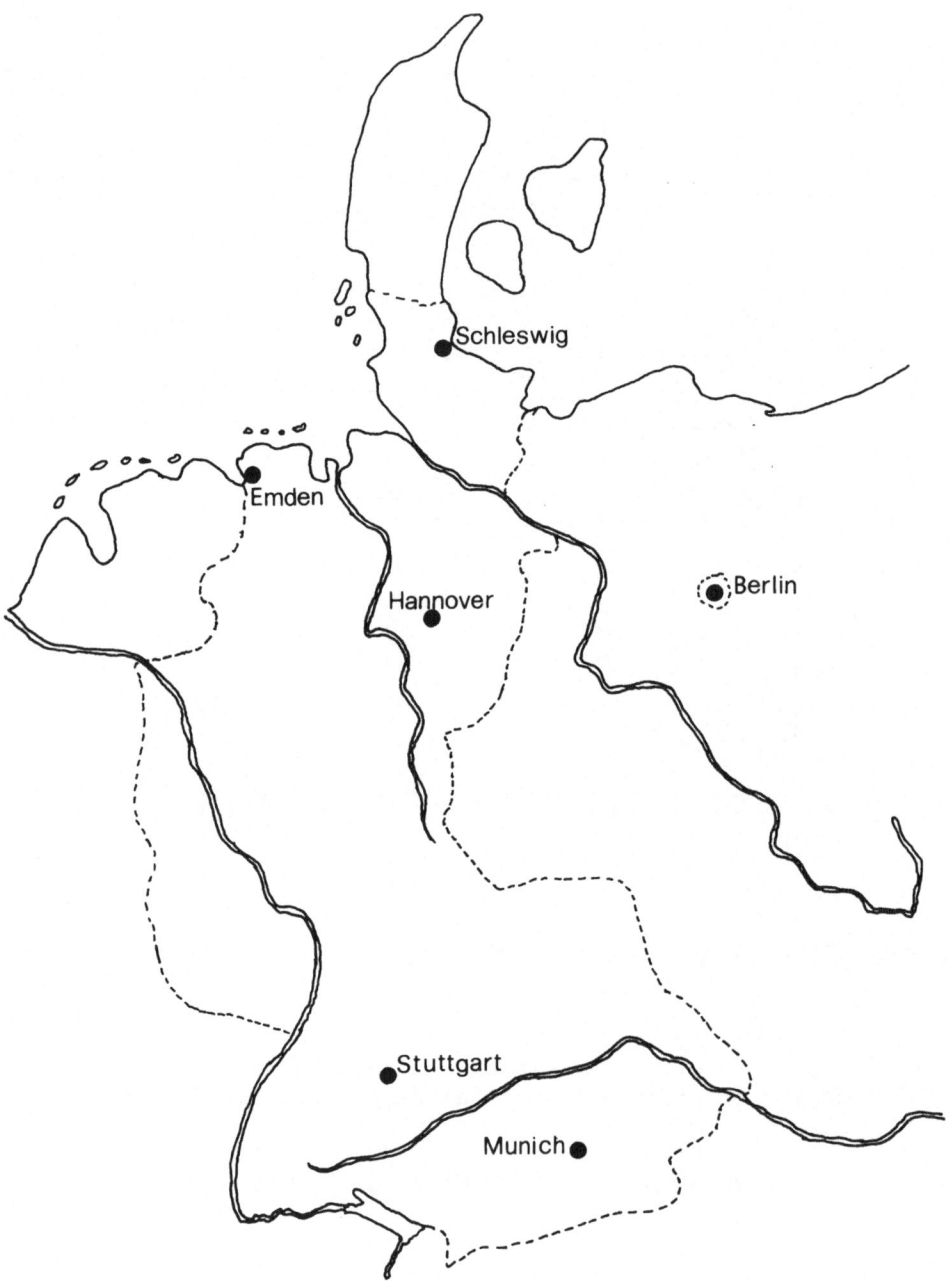

Figure 16.   Sites of the aerological stations

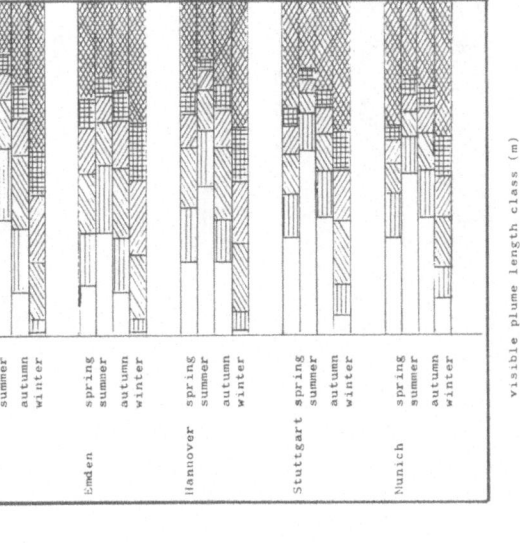

Figure 18:

Frequency (%) of calculated visible plume length depending on season natural draft cooling tower (waste heat 2500 MW) data from Schleswig, Emden, Hannover, Stuttgart, Munich period 1959 (III) – 1964 (II)

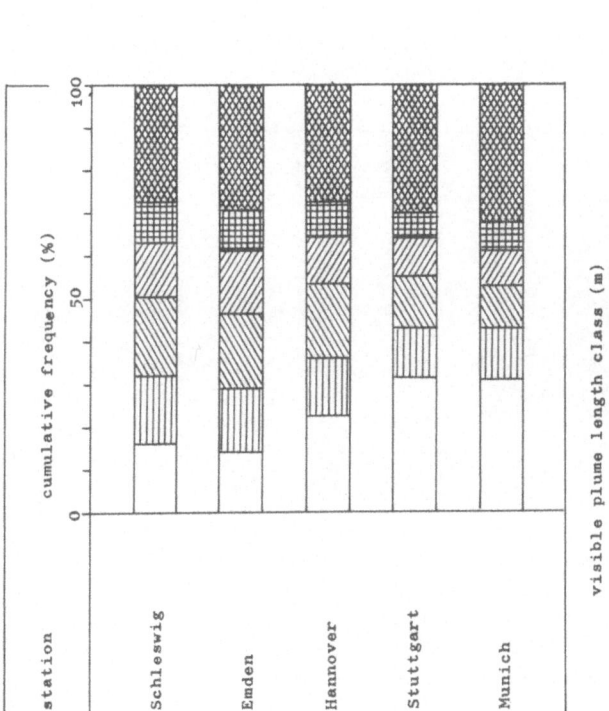

Figure 17:

Frequency (%) of calculated visible plume length natural draft cooling tower (waste heat 2500 MW) data from Schleswig, Emden, Hannover, Stuttgart, Munich period 1959 (III) – 1964 (II)

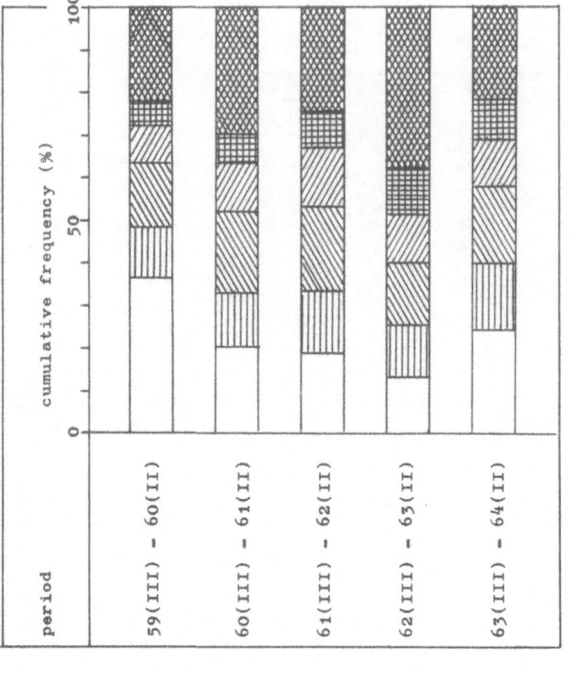

Figure 20:

Frequency (%) of calculated visible plume length
in different one year periods
natural draft cooling tower (waste heat 2500 MW)
data from Hannover
period 1959 (III) – 1964 (II)

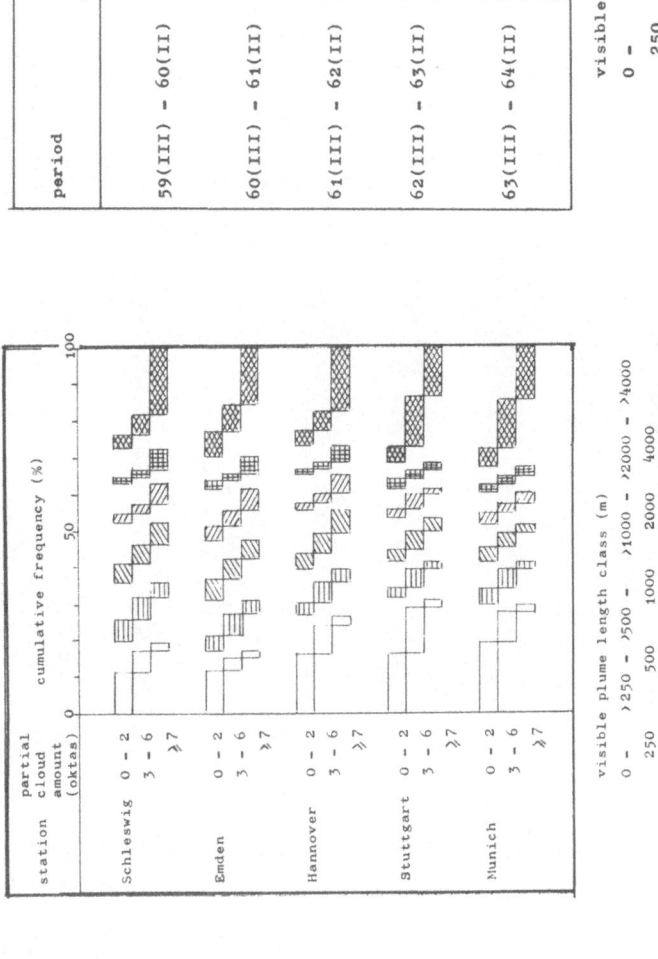

Figure 19:

Frequency (%) of calculated visible plume length
depending on cloud amount
Natural draft cooling tower (waste heat 2500 MW)
data from Schleswig, Emden, Stuttgart, Hannover,
Munich
period 1959 (III) – 1964 (II), 12 GMT

Table 1 : Relative frequency (%) of the computed visible
plume rise above ground

natural draft cooling tower
(waste heat 2500 MW,
height of tower 160 m)

data from Schleswig, Emden, Hannover, Stuttgart
and Munich,
period 1959 (III) - 1964 (II)

Plume rise (m)	Schleswig	Emden	Hannover	Stuttgart	Munich
> 150 - 200	10.4	8.1	9.4	3.6	7.7
> 200 - 250	21.8	17.7	20.6	11.9	13.8
> 250 - 300	15.1	14.3	13.9	13.1	10.1
> 300 - 400	17.4	17.1	15.5	16.6	13.5
> 400 - 500	8.4	10.1	8.2	10.2	10.1
> 500 - 700	9.6	10.6	10.2	12.7	13.8
> 700 - 1000	7.1	8.7	7.9	12.6	9.9
> 1000	10.3	13.3	14.3	19.3	21.1
> 150 - 300	47.3	40.1	43.9	28.6	31.6
> 300 - 500	25.8	27.2	23.7	26.8	23.6
> 500	27.0	32.6	32.4	54.6	44.8

the sunshine reduction due to the visible plume - can be examined.
Different types of cooling towers shall be taken into account, actually SMOKA has been tested based on data measured at a group of
12 low ventilator draft cooling towers. Finally computations for dry
and wet hot stack plumes are in progress respecting the vertical
profiles of temperature, humidity and wind as well as the released
latent heat.

ACKNOWLEDGMENT

The assistance of Klaus Hoffmann by preparing the needed meteorological data and carrying out the computer runs is gratefully
acknowledged.

REFERENCES

1. Rudolf, B., Ein numerisches Modell zur Berechnung der Wolkenbil-
   dung über einem Kühlturm, Diplomarbeit, Met. Inst. d. Uni. Bonn
   (1974)

2. Fraedrich, K., Nitsche, H., Rudolf, B., Thommes, W. and Wergen,
   W., Konvektion und Diffusion über einem Kühlturm - Entwurf
   eines Modells, MBA 25 (ISSN 0006-7156), Met. Inst. d. Uni.
   Bonn (1978).

3. Mellor, G.L. and Yamada, T., A. Hierarchy of Turbulence Closure
   Models for Planetary Boundary Layers, Journ. Atmos. Sci. 31
   (1974).

4. Kessler, E., On the Distribution and Continuity of Water Substance in Atmospheric Circulations, Met. Monogr. 10, Am. Met.
   Soc. (1969).

5. Policastro, A.J., Carhart, R.A., Ziemer, S.E. and Haake, K.,
   Evaluation of Mathematical Models for Characterizing Plume
   Behavior from Cooling Towers, Argonne National Laboratory,
   NUREG/CR-1581 Vol. 1 (1980).

DISCUSSION

G. RESELE   You mentioned a shadowing
            time of 75 min/day in 1 to 2 km distance. This
            number seems very high. How is shadowing defined
            and for what situations are the 75 min valid.

B. RUDOLF   The shadowing time (or the
            loss of sunshine duration, what I said) is defined
            by the time, an observer standing at the considered
            point cannot see the sun, because it is hidden by
            the plume or by the cooling tower building.
            The mentioned 75 minutes per day are valid for win-
            ter, and are averaged for all days in which the maxi-
            mum observed cloud amount did not exceed 6 oktas.
            This value does not contain the first and the last
            17 minutes of a day (the time the sun is lower than
            2° above the astronomic horizon).
            The used data are from Berlin, March 1951 to Februa-
            ry 1954, the considered released heat is 2500 MW.

DYNAMIC INTERACTION OF

COOLING TOWER AND STACK PLUMES

K. Nestor and H. Verenkotte

Nuclear Research Center
Karlsruhe
Fed. Rep. of Germany

INTRODUCTION

In most cases modern thermal power plants are equipped with cooling towers for heat removal. With distances of several 100 m between the cooling tower and the stack, interactions may occur between the plumes of both sources. A distinction can be made between two main possibilities of influencing:

(1) The chemical reactions within the stack plume are modified by the humidity excess and by the presence of droplets in the cooling tower plume.

(2) The rise and the dispersion of the stack plume are influenced by the cross circulations and turbulence in the cooling tower plume.

The second aspect which may be called dynamic interaction will be treated in the following sections.

THE COOLING TOWER PLUME MODEL

The three-dimensional model WALKÜRE (Nester, 1976, 1979) is used to calculate the cooling tower plume. The calculations rely on:

- the Navier-Stokes equations of motion
- the continuity equation
- the first theorem of thermodynamics
- the law of conservation for water vapor and droplets
- the equation for the turbulent energy k
- the equation for the dissipation ε of the turbulent kinetic energy.

This gives a set of seven partial differential equations.

The following essential assumptions are made to solve these equations:

- Steady-state conditions.
- The turbulent exchange in the direction of transport is neglected in favor of the horizontal advection.
- Changes in density are taken into account via the Boussinesq approximation; otherwise the medium is considered to be incompressible.

To calculate the cooling tower plume these equations are converted into difference equations and solved numerically in steps of Δx in the direction of transport.

The model WALKÜRE calculates the fields of velocity, temperature, specific humidity, droplet content and dispersion coefficients of induced turbulence on planes normal to the transport direction of the plume. The dispersion coefficients in the ambient atmosphere must be known for the calculation. The latter coefficients are based on values used in Gaussian diffusion models. The induced and ambient diffusion coefficients add up to give the effective diffusion coefficient.

THE MERGING MODEL FOR STACK PLUMES WITHOUT BUOYANCY

The dispersion of a non-reactive pollutant gas without buoyancy in the atmosphere can be described by the following diffusion equation:

$$u \frac{\partial c}{\partial x} + v \frac{\partial c}{\partial y} + w \frac{\partial c}{\partial z} = \frac{\partial}{\partial y} \left(K_y \frac{\partial c}{\partial y} \right) + \frac{\partial}{\partial z} \left(K_z \frac{\partial c}{\partial z} \right) , \qquad (1)$$

where u, v, w are the time averaged velocities in the x, y, z directions. K_y, K_z are the turbulent dispersion coefficients in the lateral and vertical directions, respectively. For the computation of superposition $u = u(z)$ represents the horizontal boundary layer flow whilst v, w are velocities induced by a cooling tower plume. Bevore solving (1), the v-w field must be determined by an appropriate cooling tower plume model, in our case WALKÜRE.

To solve the diffusion equation numerically a discretization is performed for (1). The diffusion equation then gives a system of linear equations solved in steps of Δx in the transport direction.

RESULTS OBTAINED WITH THE MERGING MODEL FOR A STACK PLUME WITHOUT BUOYANCY

Dispersion calculations were made with stack plume without

buoyancy under neutral stability conditions. The stack height
selected was 200 m. With a cooling tower height of 130 m and a waste
heat output of 1000 MW no mixing of the two plumes takes place in
the immediate vicinity of the cooling tower orifice. Consequently,
mixing is not disturbed by the flow in the wake of the cooling tower.

The following approach was chosen for the diffusion coefficients:

$$K_y\ (x) = \frac{u}{2}\ \frac{d\sigma_y^2(x)}{dx}, \quad K_z\ (x) = \frac{u}{2}\ \frac{d\sigma_z^2(x)}{dx} \quad . \qquad (2)$$

The dispersion parameters $\sigma_y(x)$ and $\sigma_z(x)$ were taken from the
evaluations of the Karlsruhe dispersion experiments applicable to
180 m source height (Kiefer et al., 1981).

The influence of the lateral spacing between cooling tower and
stack on the concentration at ground-level was investigated. It was
assumed that the cooling tower is located 200 m downstream of the
stack. The configurations are illustrated in Fig. 1. The result
for the maximum ground-level concentration as a function of the
distance from the source is shown in Fig. 2.

In the case characterized by the spacing 0 m the strongest
decrease in exhaust gas concentration at ground-level is calculated.
Here the stack plume is sucked into the cooling tower plume and
rises with the latter. This reduction compared with the undisturbed
case is by more than a factor 5.

The cases characterized by lateral displacement of the stack
first show a marked increase in concentration near the source. Due
to the displaced position of the cooling tower the zones of high
pollutant contents attain the downwind region of the cooling tower
plume. This effect appears in the distribution of concentration and
the velocity field at 1100 m distance (Figs. 3 and 4). At greater
distance and with the spacing becoming smaller the process of
dilution through suction into the cooling tower plume dominates,
implying a reduction of the maximum concentration at ground-level.
With a large spacing the stack plume is hardly any longer sucked
into the cooling tower plume so that the latter effect disappears.
The pollution at ground-level is maximum for a lateral spacing of
400 m. There is an increase about the factor 1.6 compared to the
undisturbed case.

COMPARISON WITH WIND TUNNEL TESTS

The results cannot be validated with the help of observations
made in nature since such investigations are not available. This
means that, at present, only a comparison with measurements made
in a wind tunnel is possible.

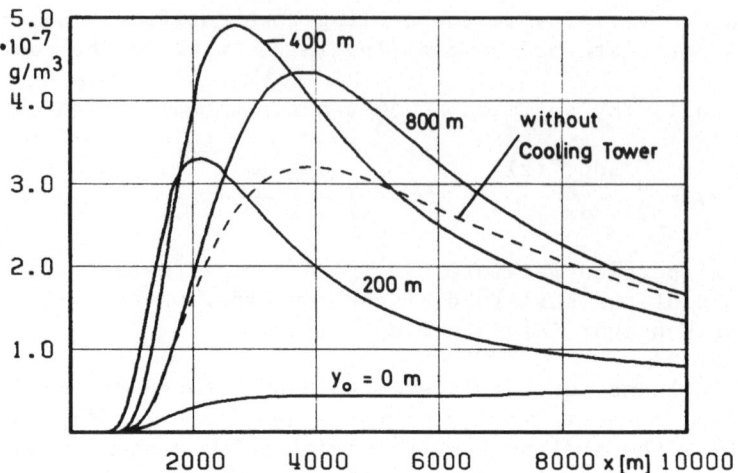

Fig. 1.   Variation of the spacing between stack
and cooling tower

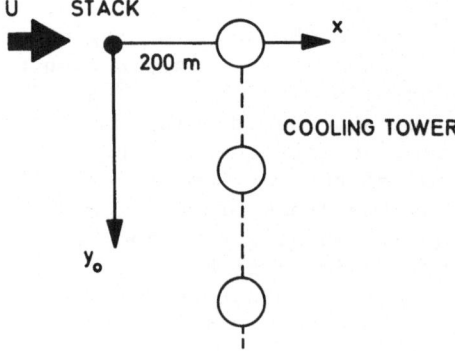

Fig. 2.   Maximum ground-level concentration as a function of the
lateral spacing between stack and cooling tower

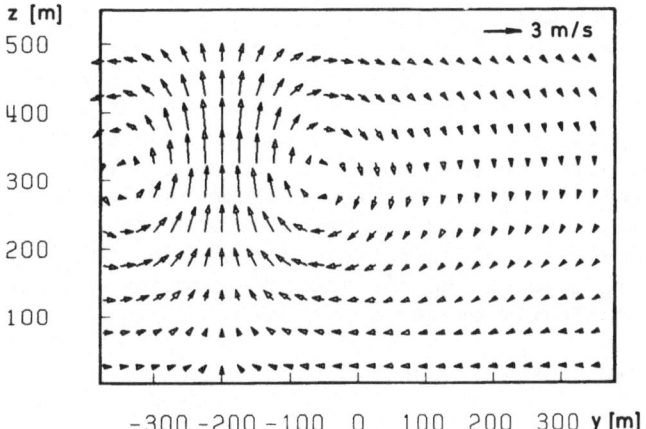

Fig. 3. Velocity field in a cross section at 1100 m distance from the source

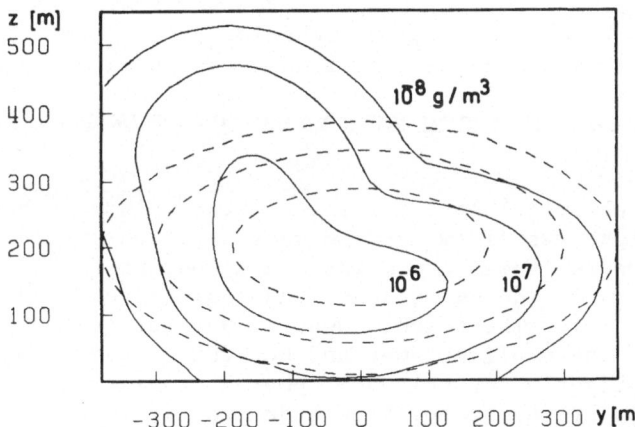

Fig. 4. Cross section of the concentration field at 1100 m distance from the source, spacing stack-cooling tower 200 m, (----) without influence of the cooling tower plume

For this purpose, special wind tunnel tests were performed at
the Institut für Wasserbau III of the Karlsruhe University. The
200 m high stack was located in a lateral position with respect to
the cooling tower at 200 m distance. The direction between stack
and cooling tower was normal to the direction of flow. The cooling
tower with a waste heat output of approx. 300 MW (only sensible
heat simulated by helium) was 140 m high. The stack plume was
measured in three cross sections with and without the influence
of the cooling tower plume. Fig. 5 shows as an example the
concentration distribution at 2000 m distance. The influence of
the circulation in the cooling tower plume is clearly visible.
The maximum of concentration occurs nearer to the ground and the
maximum ground-level concentration increases by the factor 2. At
the upper left edge of the distribution suction into the cooling
tower plume is clearly evident.

THE STACK PLUME MODEL WITH BUOYANCY

If the buoyancy of the stack plume has to be taken into account,
the diffusion equation (1) is no longer sufficient. Now the equations for momentum, energy and humidity must be introduced additionally into the model. This is done by incorporation of the
diffusion equation into the WALKÜRE model. Since this model is
already capable of taking into account the superpositions of
several plumes, the stack plume was treated as an additional
"cooling tower plume". But this is only possible beyond several
deca meters of dispersion. Only at that distance the conditions
in the stack plume (spread, temperature, vertical velocity) allow
to employ the WALKÜRE model. The behavior of the stack plume during
rise up to this distance is described by an integral model.

RESULTS OBTAINED WITH THE MERGING MODEL FOR STACK PLUMES WITH BUOYANCY

For the calculations involving a stack plume with buoyancy
the stack height was 170 m and the waste heat output 160 MW. The
130 m high cooling tower had a waste heat output of 850 MW. The
configurations considered of stack and cooling tower corresponded
to that in Fig. 1. The stack, however, was located immediately
downstream of the cooling tower and not 200 m away from it.
Neutral and slightly instable atmospheric conditions were studied.
σ_y was reduced by the factor 2 in order to eliminate some of the
long-wave movements of the plume and the shear effects. These
effects do not play a role in the rise and merging of plumes but
they are included in the σ_y parameters derived from ground-level
measurements. If the stack is situated downstream of the cooling
tower without lateral displacement the ground-level concentration
decreases clearly under conditions of neutral stratification as
compared with the case without cooling tower (see Fig. 6). Both

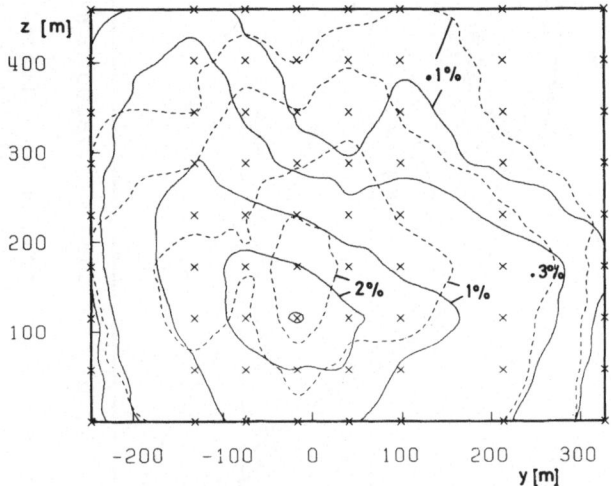

Fig. 5. Cross section of tracer-concentration measured in a wind
tunnel (——) with and (----) without a cooling tower
plume. Location of cooling tower at y = -200 m,
distance from source = 2000 m (full scale).
x = location of probes

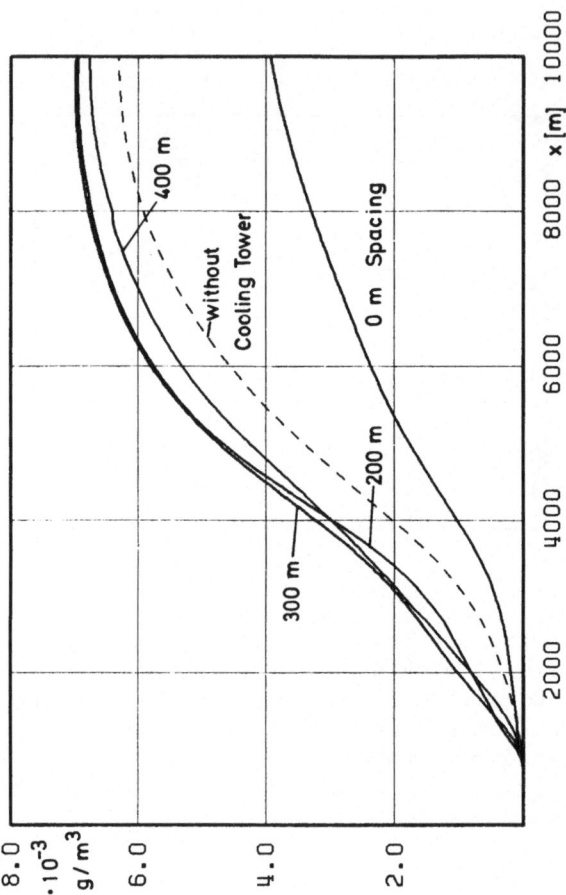

Fig. 6.   Distribution of ground-level concentration as a function of the spacing
          between stack and cooling tower under neutral diffusion conditions.

plumes combine into one plume which rises higher than the stack plume alone. This reduction is much greater than the increase in ground-level concentration, if the cooling tower is displaced laterally. The maximum of ground-level concentration rises only by about 10 %. The pattern of ground-level concentration with growing distance from the stack does not greatly differ from that without cooling tower.

This case shows that the dynamic superposition under neutral diffusion conditions does not lead to a noticeable increase in ground-level concentration. This statement should be valid also under conditions of stable stratification.

However the effects are greater under slightly instable atmospheric conditions. Fig. 7 shows the result of a computation made for a lateral spacing of 300 m. Compared with the case without cooling tower the maximum of ground-level concentration increases by 50 % at the maximum and is shifted again towards the source. At 1 km distance an interaction can already be detected in the concentration distribution of both plumes. Full mixing has already taken place after 2 km. This is apparent from Fig. 8. The circulation consists of only one double vortex. The concentration distribution on this plane (see Fig. 9) shows clearly a rotation of the vertical axis. This is due to the inflow of the lower part of the stack plume into the plume of the cooling tower. Moreover, caused by the downward flow at the edge of the cooling tower plume the concentration distribution gets displaced towards the ground. This leads to an increase in concentration near the ground.

CONCLUDING REMARKS

The results obtained give a first impression of the effect of dynamic interaction of stack and cooling tower plumes on the ground-level concentration of the stack effluents. In case of buoyant exhaust gases the ground-level concentration distribution is not so greatly influenced as in case without buoyancy. Besides the meteorological conditions, the effect of dynamic interaction of both plumes on the ground-level concentration in both cases depends highly on the configuration of stack and cooling tower with respect to the wind direction.

Considering the results, the spacing between the cooling tower and the stack should be selected to be as small as possible since then the ground-level concentration can be reduced compared with the case of non-interaction. But a decrease in the distance of stack and cooling tower leads to an increase of the chemical interactions so that relevant statements can be made only after these effects have been taken into account.

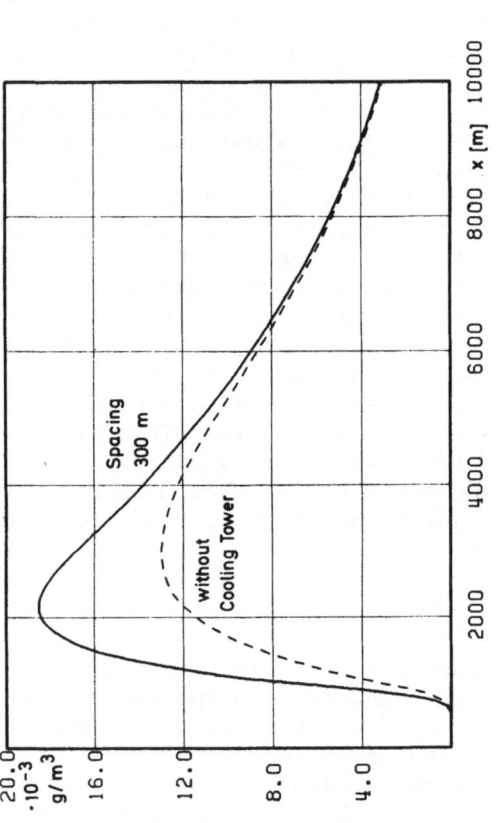

Fig. 7. Ground-level concentration under slightly instable diffusion conditions.

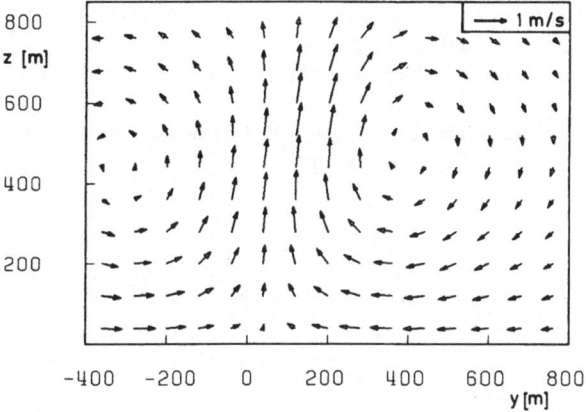

Fig. 8.   Velocity field in a cross section at 2000 m distance from the source

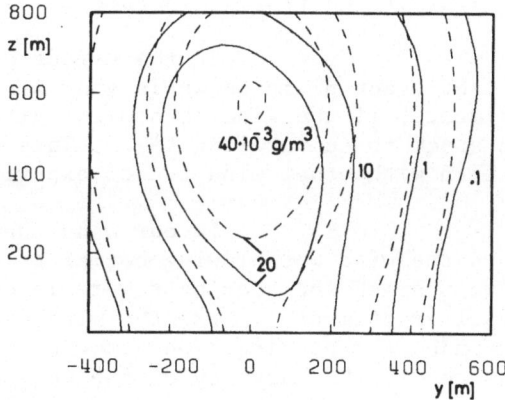

Fig. 9.   Concentration distribution in a cross section at 2000 m distance from the source

REFERENCES

Kiefer, H., Koelzer, W., König, L. A. (Edit.), 1981: Jahresbericht
    1980 der Hauptabteilung Sicherheit. Report of the Nuclear
    Research Center Karlsruhe, Fed. Rep. Germ., KFK 3113.

Nester, K. 1976: WALKÜRE - Simulation der Wirbelstruktur der Ab-
    luftströmung aus Kühltürmen mit einem Rechenprogramm. Rep.
    Nuc. Res. Center Karlsruhe, Fed. Rep. Germ., KFK 2249.

Nester, K., 1979: Simulation von Kühlturmfahnen. B+D Ing. Ges. mbH,
    Aldenhoven, Fed. Rep. Germany.

ACKNOWLEDGEMENTS

This work was supported by the UMWELTBUNDESAMT under
contract No. 104 07 353.

DISCUSSION

F. FANAKI   Was your study restricted only
              to low wind speed ? What happens at U > 5m. sec $^{-1}$

K. NESTER   We have performed our study
              with wind speeds between 5 m/s and 10 m/s. These
              are wind speeds which are usual under neutral
              and slightly unstable atmospheric conditions.

G. RESELE   Is the effect of rising non
              buoyant stack plume shown in your wind tunnel ex-
              periments, if the wind is blowing directly from
              the stack to the cooling tower. This is in contra-
              diction with other wind tunnel experiments.

K. NESTER   In our wind tunnel experi-
              ments the wind was blowing normal to the direction
              between stack and cooling tower. In other wind
              tunnel experiments, where the wind direction cor-
              responded to the direction between stack and coo-
              ling tower, the buoyancy of the cooling tower was
              not simulated. This may be the reason for the dis-
              agreement between our calculations and the results
              of these experiments.

A. SAAB   How do you specify the ini-
              tial conditions for stack emissions ?

K. NESTER   To specify the initial conditions of the stack plume in the cooling tower
model WALKÜRE it was necessary to calculate the stack
plume up to a distance of some decameters from the
source with an integral model. At this distance
the conditions (spread, vertical speed and temperature) are reduced to values comparable with those
in a cooling tower plume. These values are then
used as initial conditions in the WALKÜRE model to
simulate the stack plume together with the cooling
tower plume.

P. SLAWSON   In your future work, do you
intend to include the effects of the particulates
from the stack plume ?

K. NESTER   In our future work we wish to
take the chemical reactions into account. This
means that a lot of further equations will have to
be incorporated into the model. One of these equations will relate to the aerosol concentration in
the stack plume.

NUMERICAL MODELING OF COOLING TOWER PLUMES :

COMPARISON WITH EXPERIMENTS

A. Hodin, J.Y. Caneill, F. Iffenecker and P. Biscay

Direction des Etudes et Recherches

Electricité de France

SUMMARY

In order to calibrate numerical models of cooling tower plumes, EDF has developed a wide range experiment near the BUGEY nuclear power plant, including aircraft and remote sensing measurements in the plumes over short periods as well as routine measurements over long periods (photographs of the plumes, micro-meteorological networks).

With all the data collected, two types of models have been validated :

1/ The first type of model, used to build statistics of visible plume characteristics (length, reduction of insolation and solar radiation), was compared to routine measurements and show a fairly good agreement with observations.

2/ The second type (a 3D model and a spectral model of microphysics) designed for studying with more details the dynamical and microphysical processes in the plume was used to simulate particular meteorological conditions observed during the campaigns.

I - INTRODUCTION

When EDF plans to construct any nuclear power plant, an impact study of the equipment on the environment is required. Particularly when atmospheric cooling systems have to be installed, a study of the effect of the resulting plumes on the atmosphere has to be carried out.

The most convenient tool to reach this objective consists in using numerical simulations of the plume behaviour by appropriate models.

If we have to evaluate modifications of the local climate, it is necessary to use fairly simple numerical models which can simulate long series of meteorological conditions. These models give integrated values or statistical distributions, in term of frequencies of the characteristics of the plumes, or of their effects (reduction of sunshine duration for example).

If we have to study critical situations, well defined from a meteorological point of view, we can use more sophisticated models, taking into account more physical phenomena ; for example, this type of model must be able to simulate artificial rain coming from the plumes.

In this paper, we present three models currently used and their comparison to experimental data. Experimental data were obtained near the nuclear power plant of BUGEY (two units of 900 MWe, two natural draft cooling towers per unit) : they include both routine data and data collected from short experimental campaigns.

II - EXPERIMENTAL NETWORK (A. HODIN, 1982)

The following experiments have been undertaken from 1978 to 1980 in order to get :

- a one-year photographic record of the plumes, from November 1979 to October 1980, three times a day, coupled with meteorological radio-soundings,

- solar radiation measurements at the ground surface (one reference station and two target stations), between 1978 and 1980, completed by the measurements of a mobile station,

- short duration campaigns (aircraft experiments) in order to study the risks of precipitations due to the plumes.

III - CLIMATIC SIMULATION OF THE PLUMES

An integral model (1D) has been developped to assess the effects of the plumes over a long period : in a first step, the geometry of the plume is calculated, knowing the meteorological conditions (given from routine data) and the output characteristics of the towers. In a second step, the monthly and yearly reduction of insolation are evaluated.

III.1 - <u>Plume Model</u>

The water vapor saturated plume is assumed to follow a three-stage development scheme :

. In a first phase, the effect of the rejected air mass properties is prevailing and mixing with the ambient air is governed by the plume turbulent characteristics. The only effect of the wind is to bend the plume trajectory. This phase ends up when the plume vertical velocity approximately equals the wind velocity.

. The second phase is a transitional one : the plume is influenced both by its own generated turbulence and by atmospheric turbulence. This phase ends up when the vertical plume velocity is of the same order of magnitude than the vertical wind fluctuations.

These two phases are considered as jet phenomena and their basic equations are those of an incompressible fluid motion, with the simplified BOUSSINESQ theory.

. The third phase is considered as governed essentially by the atmospheric turbulent properties. During this phase, the model is based on the diffusion of the plume enthalpy, which includes the effects of water both in gaseous and in liquid state. As WESSELS and WISSE (1971), we shall assume that the enthalpy diffusion law is gaussian.

Microphysical processes inside plume are parameterized by assuming that the main phenomenon governing droplet formation is condensation. This assumption can be used when the residence time of the drop is not too long (half an hour to one hour, which corresponds to a plume of several kilometers long).

The actual model allows us to simulate a large number of plumes because of the short CPU time required (1 to 3 s for one case with a IBM 370 Computer). It takes into account the emissions from several sources and the interactions between the resulting plumes (HODIN and KAUFHOLD, 1979).

III.2 - <u>Results</u>

The input data come from radiosoundings carried out at SATOLAS, 20 km from the BUGEY site, between November 1979 and October 1980. The numerical results have been compared to photographic record of the plumes. Inappropriate cases, resulting from bad weather conditions (mist or mixing with natural cloud deck), have not been taken into account : 612 observations have been analysed among the 1000 photographs available.

Comparison of the plume geometry

The plumes are classified according to their length and their culmination height :

- Length : 3 classes :

  . short   ($\leqslant$ 1 km),
  . medium (1 to 5 km),
  . long   ($>$ 5 km).

- Culmination height : 3 classes :

  . low   ($\leqslant$ 0.5 km),
  . medium (0.5 to 1.5 km),
  . high   ($>$ 1.5 km).

Monthly distribution of both calculated and observed length classes are given in figure 1. The agreement is fairly good, mainly for plumes shorter than 1 km, which represents 67 % of the cases. We can observe, however, that the model tends to slightly overestimate the length of the plumes.

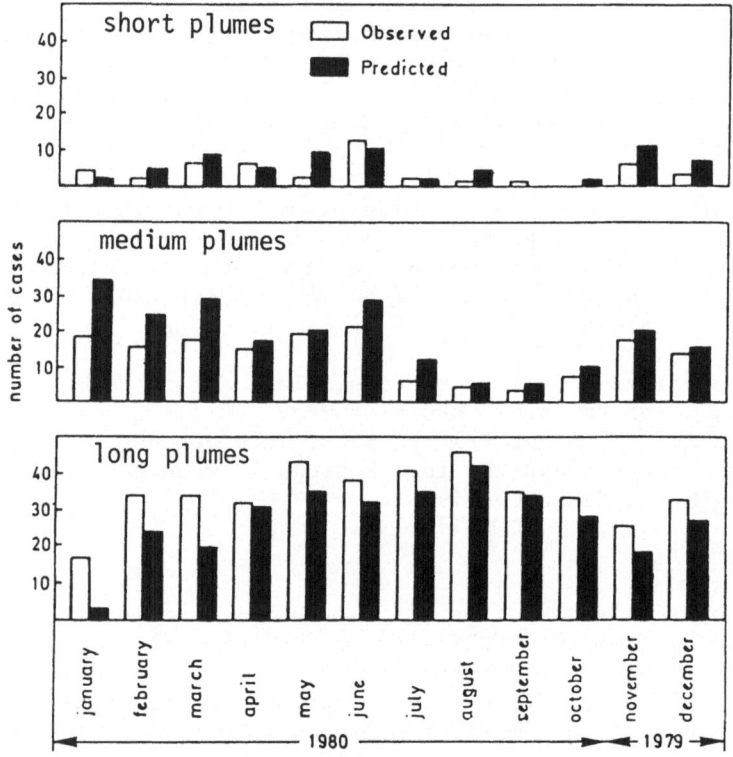

Fig. 1. Monthly distribution of measured and predicted plume
          lengths: a/ short plume, b/ medium plume,
                    c/ long plume.

Yearly distribution of the plume lengths and culmination heights have been plotted in figure 2. Over a one year period, the prediction of the plume dimensions is rather good : for example, 10 % of long plumes are predicted by the model and 7 % are observed.

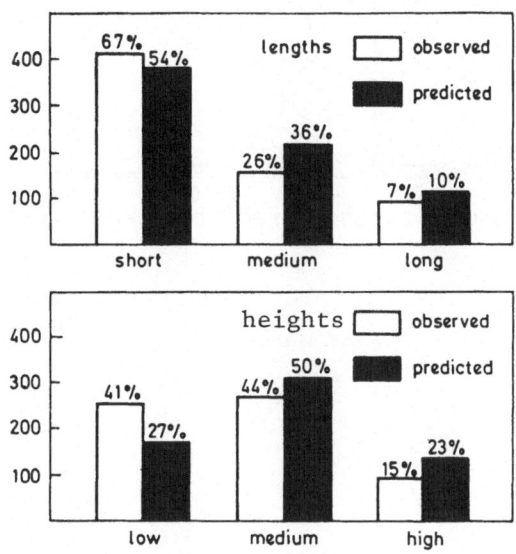

Fig. 2.   Comparison of measured and predicted plume lengths and heights over a period of one year.

The statistical distribution of the deviations between calculated and experimental values is shown in figure 3. We can see, for instance, that 79 % of the length deviations are lower than 2 km, and 85 % of the height deviations lower than 1 km.

Comparison of the reduction of insolation

The main application of the integral model consists in evaluating the reduction of insolation due to the plumes over a long period (generaly on a monthly and on a yearly average). For this purpose, we use routine data given by the French meteorological network, and cooling tower input data, calculated with a thermodynamical model (BOURILLOT, 1981).

The results are presented under the form of isoreduction curves around the power plant.

This method has been applied to the BUGEY site, for the period between November 1979 and October 1980, for which three daily radiosoundings where usually available.

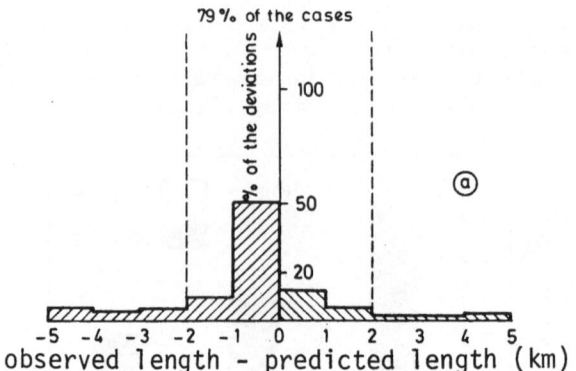

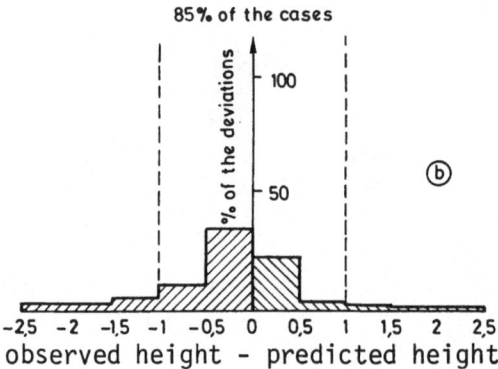

Fig. 3.   Deviation distribution between observed and
           predicted values :

           a/ lengths,

           b/ heights.

     The plume dimensions have been estimated :

1/ From the photographic records.

2/ From the integral model.

     The yearly results of the two simulations are given in
figures 4 and 5. We must note that these simulations are rather
approximate for three main reasons : first of all, no more than
three daily radiosoundings and records of the plumes were available. So it was necessary to interpolate the plume dimensions
between two successive observations. Secondly, some observations
being unavailable (because of sensor failures for instance), real
effects are underestimated. Finally, we have assumed that the

plume width taken into account in the shadowing calculations was
identical to its thickness. Nevertheless, we can observe that the
yearly attenuations calculated from the model are in good agreement with the yearly ones estimated from the photographic record
of the plumes.

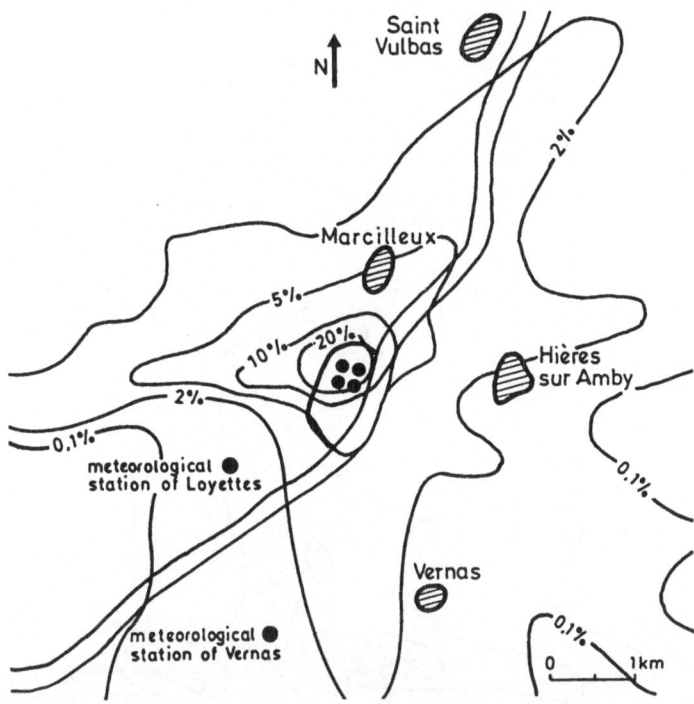

Fig. 4.   Yearly isoreduction of isolation curves deduced from
          photographic records of the plumes (November 1979 to
          October 1980).

On a monthly scale, deviations between the model and observations are more important (see table 1 relating to the two locations LOYETTES and VERNAS), but the orders of magnitude are generally comparable.

As previously mentioned, the results given on figures 4 and 5 underestimate the real effects. This one have been estimated from the micrometeorological network in operation for three years. For instance, at 3 km south of the power plant, the attenuation is about 2 to 4 % of the yearly insolation (i.e. 40 to 80 hours).

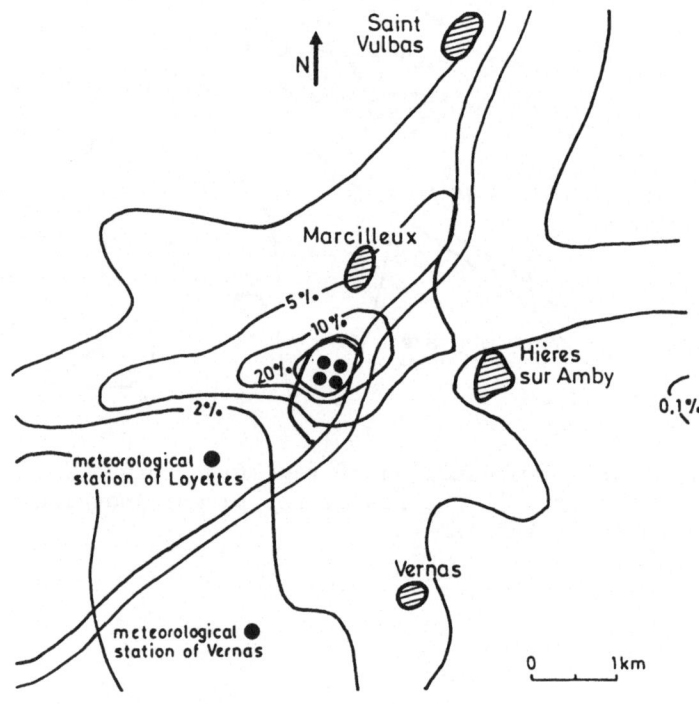

Fig. 5. Yearly isoreduction of insolation curves deduced from numerical simulation of the plumes.
(November 1979 to October 1980).

Table 1. Monthly reductions of insolation (hours) at the meteorological stations of LOYETTES and VERNAS :

a/ from the numerical model,

b/ from the photographic record of the plumes.

Period	Loyettes		Vernas	
	a/ Predicted	b/ Observed	a/ Predicted	b/ Observed
November	0.5	0.2	0.5	0.2
December	0.1	0.1	0.1	0.1
January	1.0	0.4	0.2	0.3
February	0.5	0.5	0.5	0.4
March	2.0	1.0	4.5	0.5
April	1.0	2.0	1.5	2.0
May	0.1	0.1	0.1	0.1
June	3.5	3.5	0.2	0.5
July	0.5	0.4	0.5	0.5
August	1.0	0.1	1.0	0.1
September	0.5	0.4	0.2	0.5
October	1.0	0.4	0.2	2.0
Year	11.7	9.1	9.5	7.2

IV - MICROPHYSICAL MODEL

The aim of this model (box model) is to describe the time evolution of a warm cloud (positive temperature) fed by a cooling tower plume (PASTRE, 1978).

Space homogeneity of physical parameters in the cloud (temperature, liquid water content, supersaturation) is assumed. The phenomena taken into account are the following : condensation, coalescence, aerodynamic breakdown, collision breakdown, activation of condensation nuclei and precipitations. The range of droplet diameters is 1.5 to 2000 µm and droplets are distributed into 80 classes. Advection (entrainment of the cloud by the wind) is also considered.

The model, in its present version (HODIN, PASTRE, SAUSSET, 1982) predicts the following cloud characteristics : temperature, liquid water content, droplet spectrum and precipitations at the ground surface. Coalescence and breakdown are considered from a stochastic point of view and the integrals are solved by the trapeze method. The other processes, considered from a deterministic point of view, are solved numerically by a Runge Kutta method of 4th order. One hour simulation requires about 1 CPU minute on a IBM 370 Computer.

Model application

We have made a comparison of the model results with experimental data obtained from an aircraft campaign on March 3, 1980 (HODIN, 1980) near the BUGEY site.

The weather is anticyclonic, with a north flux in altitude. The plume reaches a stratocumulus deck located at a height between 1000 and 1400 meters.

Several traverses (11) are done in the natural cloud, in the plume and in the mixing zone. The system cloud-plume has been shared into 5 distinct areas (figure 6).

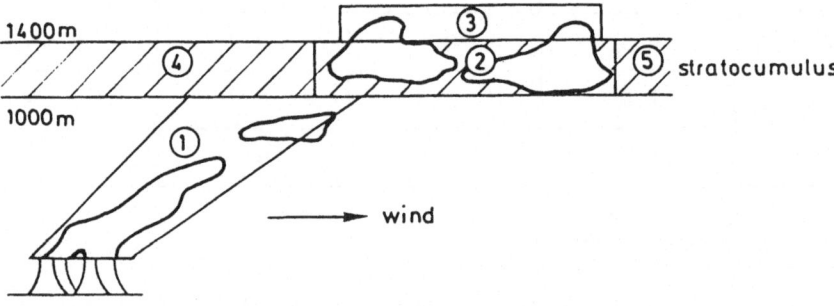

Fig. 6   Observed plume on March 3, 1980
at 11.30 A.M.

The first area gives us the plume characteristics below the cloud deck. The comparison model-experiment is done in area ②. The first traverse measurements relative to this area have been considered as initial values.

Areas ④ and ⑤ are associated to the natural cloud and an overshoot phenomenon can be observed in area ③.

Mean values of the plume characteristics (4 traverses) are given in table 2. In the last column of this table we have indicated the initial plume characteristics taken into account in the numerical simulation.

Table 2. Plume characteristics.

Characteristics	Measurement				Model
	Traverse A	Traverse B	Traverse C	Traverse D	
Hour	11 h 07	11 h 13	11 h 25	12 h 04	
Duration of the plume traverse (s)	36	25	38	8	
Pressure (mb)	880	881	881	879	880
Temperature (°C)	-3.2	-3.1	-2.9	-2.4	-2.5
Droplets concentration (cm^{-3})	223	340	295	293	293
Liquid-water content (g.m-3)	0.04	0.06	0.04	0.05	0.05

We observe a rather good stability of the plume characteristics during all the experiment (about one hour).

The similar characteristics in the cloud are given in table 3. Initial cloud characteristics taken into account in the model are in the last column.

Two numerical simulations of one hour have been carried out. In the first one, initial supersaturation σ_0 was supposed to be zero, and in the second one, supersaturation was -0.3 % (slightly undersaturated cloud).

Table 3. Cloud characteristics.

Characteristics	Measurement				Model
	Traverse A'	Traverse B'	Traverse C'	Traverse D'	
Hour	11 h 03	11 h 19	11 h 31	11 h 57	
Duration of the plume traverse (s)	47	34	34	54	
Pressure (mb)	865	863	864	863	865
Temperature (°C)	-4,0	-3.8	-3.8	-4.0	-4.0
Droplets concentration (cm^{-3})	440	324	589	623	440
Liquid water content ($g.m^{-3}$)	0.12	0.08	0.15	0.17	0.12

Figures 7 and 8 show numerical results compared to observations. We note that the time evolution of liquid water content into the cloud fed by the plume is fairly well predicted (figure 7). However, the model tends to accumulate droplets near a given size, whereas the observed spectrum enlarges with time (figure 8).

Finally, we can mention that neither observations nor numerical simulations show a sensible growth of the cloud droplets size.

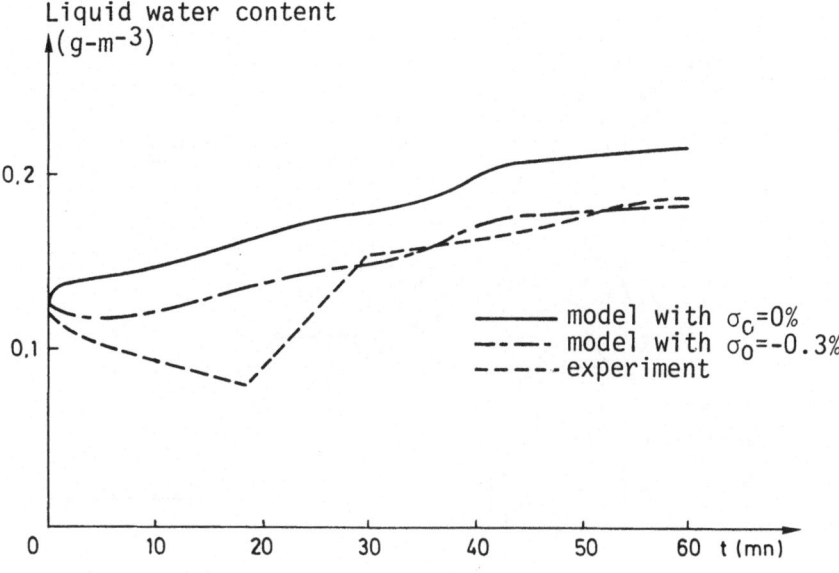

Fig. 7. Time evolution of liquid water content in
the cloud at BUGEY on March 3, 1980.

V - 3D PLUME MODEL

V.1 - Description of the Model

The atmospheric flow over a cooling tower may be considered
as a perturbation of the ambient atmosphere by a continuous
injection of heat and moisture. The equations of the model are
the equations of motion with the Boussinesq approximation, the
conservation of mass, energy and water substance. The microphysical processes included in the conservation equations for cloud
droplets and rain water content are evaporation, condensation,
autoconversion and accretion (SAAB, CANEILL, 1980), taken into
account with the aid of KESSLER's formulation (1969). Subgrid
scale turbulence is modelled with SMAGORINSKI's formulation
extended to both the deformation and buoyancy fields, as investigated by TAG and Al. (1979).

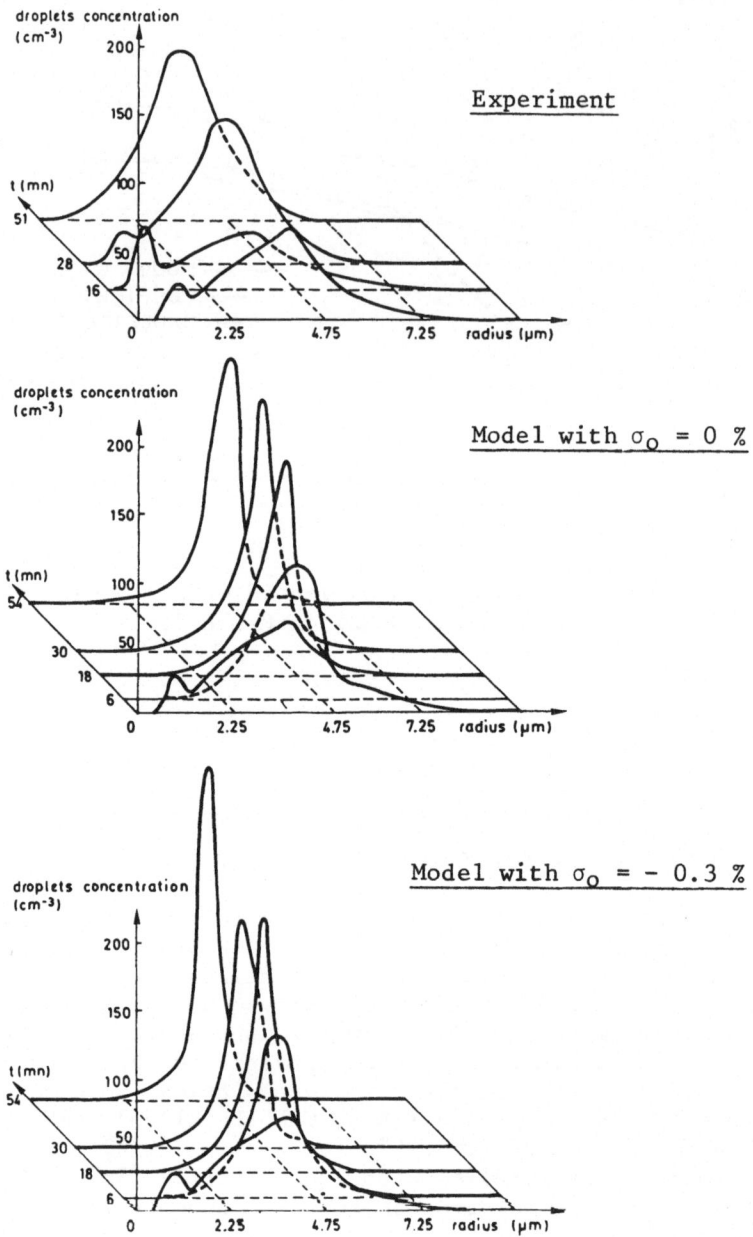

Fig. 8. Time evolution of measured and predicted cloud droplets spectra at BUGEY on March 3, 1980.

V.2 - Solution Procedure and Boundary Conditions

The numerical "SMAC" method used in this work, and presented in a more detailed way by SAAB (1982) has been developed at LOS ALAMOS and applied in FRANCE by GAILLARD (1978). It is a finite difference method, explicit in time and implicit in pressure. The spatial discretization is performed through a weighted mean of centered and upwind differencing. The pressure continuity treatment is carried out in a pressure correction iteration involving a mutual adjustment of velocities and pressure in order to make the velocity field compatible with the continuity equation.

Initial and boundary conditions are specified as follows :

a/ The perturbation consists of continuous injection of heat and moisture at the cooling tower exit.

b/ Ambient atmospheric values at the upwind boundary.

c/ No slip conditions on the ground.

d/ Open boundary conditions at the downwind boundaries.

The model needs about 20 minutes of CPU time on a CRAY-1 computer to reach quasi dynamical equilibrium with $40 \times 12 \times 48$ grid points. Considering the spatial extent needed to simulate this type of perturbation (2500 meters along each direction) a grid resolution $(\Delta x, \Delta y, \Delta z)$ of 50 m was selected. The time step is adjusted in order to ensure stability of the numerical scheme (i.e. $\Delta t = 2$ seconds).

V.3 - Results

At present, the model undergoes a validation on the BUGEY observations previously mentioned, which provided us a basis for comparison between model results and experimental observations. As example, model results on March 4, 1980, 11 h 30 a.m. are presented.

Meteorological conditions

The meteorological situation was characterized by an anti-cyclone centered on the North of FRANCE. The weather was clear and sunny. The atmospheric structure was characterized by the existence of an unstable layer, topped with an inversion of temperature (0.6 °C/100 m), associated to a wind shear (figure 9).

Emission conditions
=========

      The 4 × 900 MW source is simulated as a 15 000 m² (150 × 100)
aera, with a 20.5 °C temperature excess with respect to the
ambient air, a 3.9 m/s mean vertical velocity and a 0.8 g/kg
cloud water content. In fact, the BUGEY's case includes four
3700 m² cooling towers. But they are rather close and for this
meteorological situation the plume was almost unique as soon as
the towers exit . So, this case, exhibiting a plume with a rather
strong vertical development justifies the use of an equivalent
source conserving the power and the area of the real source.

      In order to investigate more general cases, a multiple source
version of this code is on stage of development.

Results
======

      Figures 10 and 11 (velocity field and isovalues of vertical
velocities) show, downwind the plume, a zone of downdraft motion
the intensity of which is in good agreement with measurements
(figure 12). But, upward vertical velocities within the plume are
overestimated, compared to the experimental ones (figure 12).

      The isovalues of cloud water content predicted by the model
are plotted in figure 13. They reproduce fairly well the plume
observations (plume top, plume basis, length, shape). The vertical distribution of the calculated liquid water content
(0,79 g/kg) is in good agreement with the experiment (0,76 g/kg)
(figure 14).

VI - CONCLUSION

      In order to study the impact of cooling tower plumes, EDF
has developed three operational numerical models . :

- An integral model for a statistical evaluation of plume characteristics and their cumulative effect (reduction of insolation).

- A spectral microphysical model, to study the interaction processes between a natural cloud and the plume. These processes
  may produce large droplets eventually precipitant.

- A 3D plume model, involving both dynamics, microphysics and
  their coupling, to investigate the problems of plumes development, especially in convective situations (cumuli formation).

      The experimental routine data collected in the site of the
BUGEY nuclear power plant from 1978 to 1980 allowed us to compare
the results of our models to observations. A rather good agreement has been shown.

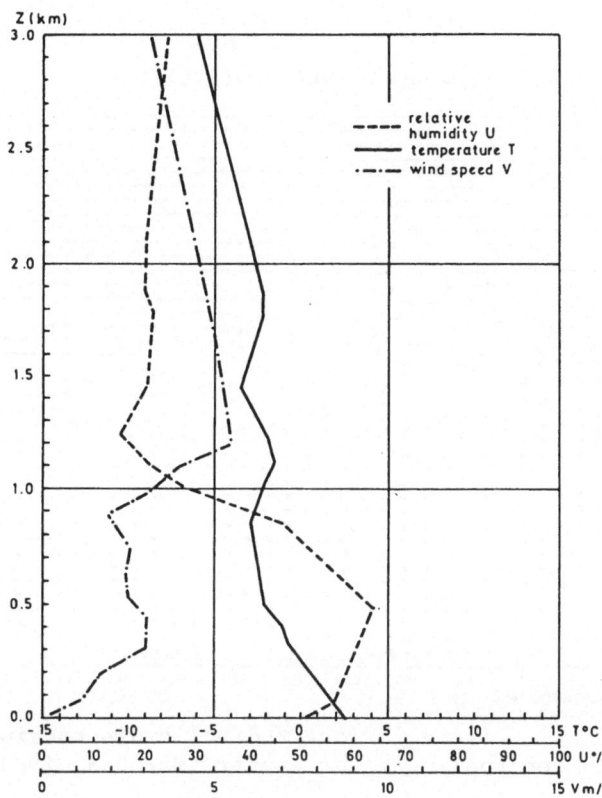

BUGEY. 4.03.1982. (195) 11h 30

Fig. 9. Meteorological conditions at
BUGEY on March 4, 1980.

CHAMP DE VECTEURS VENT

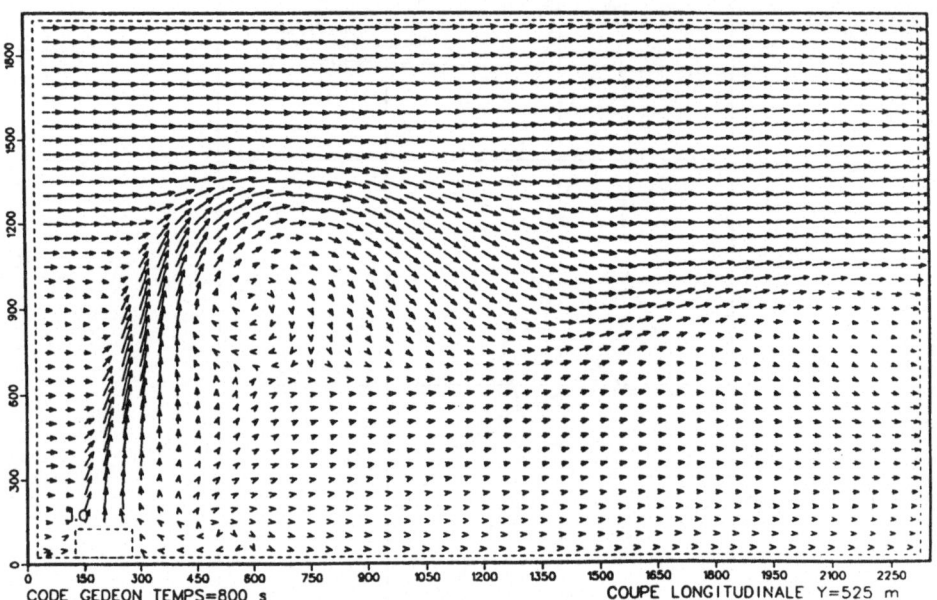

Fig. 10. Wind field in longitudinal cross section over
the cooling tower center (March 3, 1980).

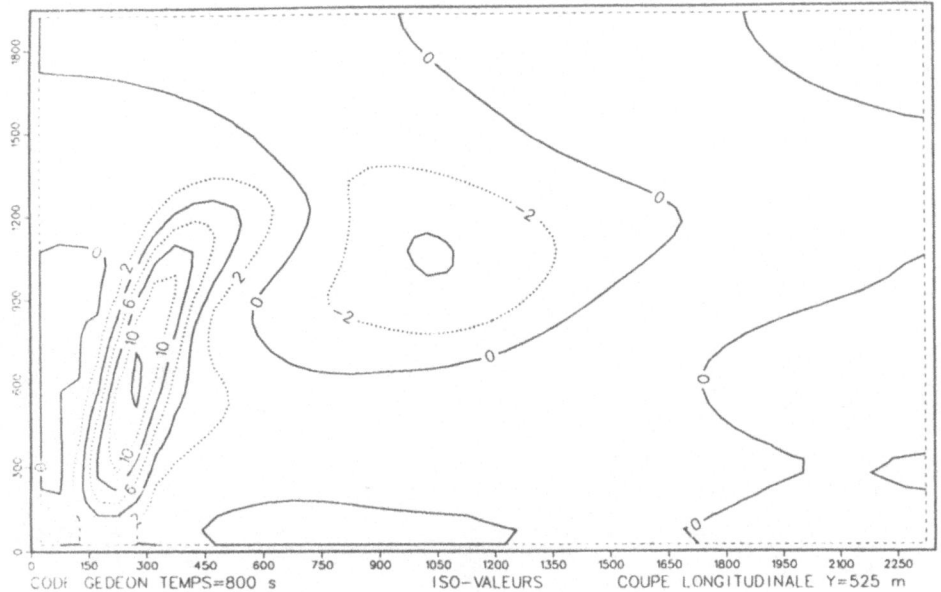

Fig. 11.   Isovalues of vertical velocity (m/s) in
           longitudinal cross section over the
           cooling tower (March 4, 1980).

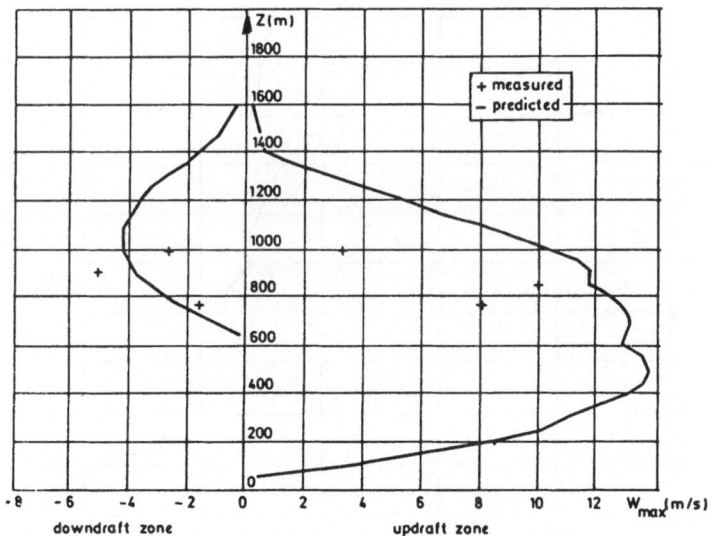

Fig. 12.   March 4, 1980 experiment :
           - Vertical profile of the vertical
             velocity inside the plume.
           - W_{max} : Maximum value.

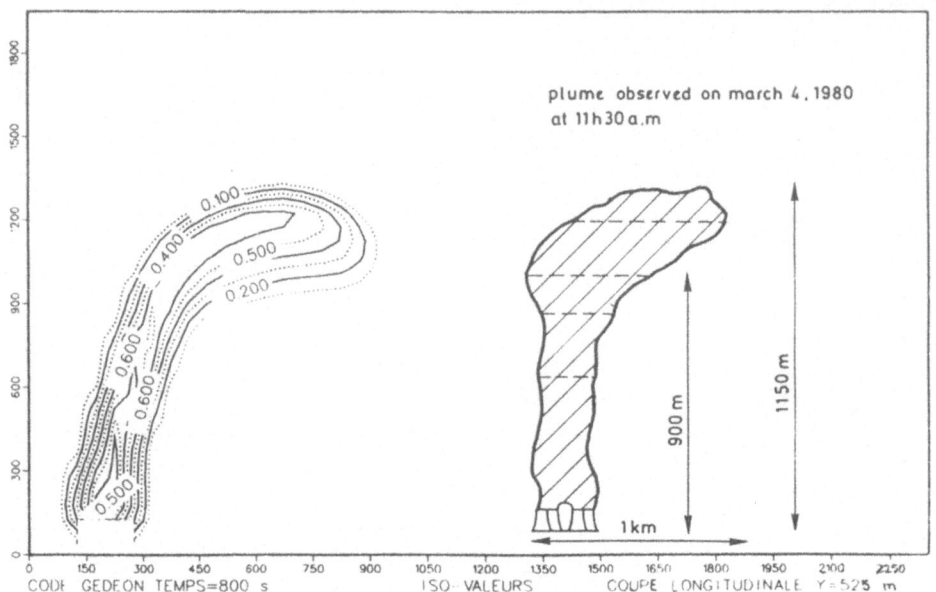

Fig 13.   Isovalues of cloud water content (g/kg) in the
          longitudinal cross section over the cooling
          tower.

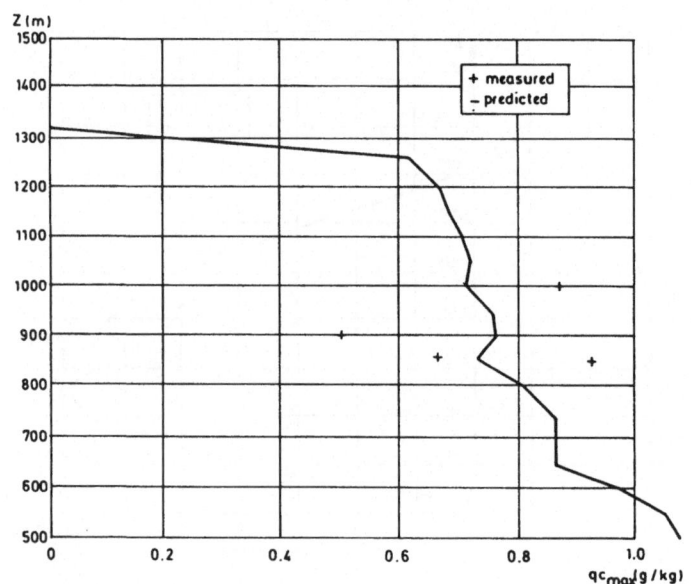

Fig. 14.   March 4, 1980. Vertical profile
           of liquid water content.

. Morphological characteristics of plumes and their effects on sunshine duration evaluated by the integral model agree fairly well with experimental data at least on a yearly scale.

. Time evolution of the liquid water content of a stratocumulus is correctly predicted by the spectral model of microphysics, when the cloud is fed by the plume (March 3, 1980).

Finally, when the plume has a rather strong vertical development, the 3D model correctly predicts the dynamical structure of the plume and its mean liquid water content.

REFERENCES

BOURILLOT C.
Modèles numériques pour le calcul des performances des réfrigérants atmosphériques humides.
Rapport interne EDF HT/31-81 n° 6, 1981.

GAILLARD P.
Modélisation numérique des panaches d'aéroréfrigérants.
Thèse Docteur-Ingénieur, Université de LYON, 1978.

HODIN A. - KAUFHOLD R.
A fast numerical model to predict the statistical behaviour of multiple source cooling tower plumes up to long range extension.
EDF internal proceeds, 1979.

HODIN A.
Etude expérimentale des panaches des réfrigérants atmosphériques du BUGEY 4-5.
Premier bilan des mesures physiques effectuées entre le 25 février et le 13 mars 1980.
Rapport interne EDF HE/32-80.33, 1980.

HODIN A.
The influence of a power station on local climate : a study of the BUGEY cooling tower plumes.
IAHR Cooling Tower Workshop BUDAPEST, 12-15 october 1982.

HODIN A. - PASTRE J. - SAUSSET M.
Interaction entre un panache humide et un nuage : étude de sensibilité d'un modèle numérique et comparaison avec des données expérimentales.
Rapport interne EDF, en préparation.

PASTRE J.
Description microphysique d'un nuage chaud. Application au problème de l'interaction entre un panache humide et un nuage.
Thèse de 3ème cycle soutenue à l'Université de CLERMONT-FERRAND en 1978.

SAAB A. - CANEILL J.Y.
Modélisation numérique des nuages convectifs induits par des rejets de chaleur sensible dans l'atmosphère (Application aux expériences réalisées sur le site du Météotron en 1978).
Rapport EDF - HE/32.80.29.

SAAB A. - RASOAMANANA T. - BENECH B.
Simulation of atmospheric effects of industrial heat releases :
The artificial convection program.
13th International technical meeting on air pollution modeling and its application, 1982.

SMAGORINSKY J.
General circulation experiments with the primitive equations :
I. The basic experiment. Mon. Wea. Rev. 91, 99-164, 1963.

TAG P. - MURRAY F. - KOENIG L.R.
A comparison of several forms of eddy viscosity parametrization in a 2D cloud model.
J. Appl. Meteor., 18, pp 1429-1441, 1979.

WESSEL H.R. - WISSE J.A.
A method for calculating the size of cooling tower plumes.
Atmospheric Environment, Vol 5, pp 743-750, 1971.

DISCUSSION

T. WARNER   In table 1 you represent monthly reductions in insolation in terms of "hours".
How is this calculated ?

A. HODIN   The model takes into account the radiosounding data obtained every six hours.
So. if one point on the ground is shadowed by the plume, which is estimated : from the photo or from the model, from the wind direction and from the position of the sun, we suppose that this shadow remains six hours, three hours before and three hours after the time of the radiosounding. The same procedure is carried out for every radiosounding and so we can estimate, in hours, on a monthly basis for instance, the reduction in insolation induced by the plumes.

J. KNOX   Do your measured data or model
           calculations give evidence of artificial (or plume
           related) precipitation reaching the ground ?

A. HODIN   Precipitations were measured
           by two means : with a micrometeorological network
           in the surroundings of the power plant and with an
           aircraft. Neither the statistical analysis of the
           network data nor the aircraft droplet measurements
           gave evidence of detectable artificial precipitations
           reaching the ground.
           Numerical simulations of precipitation processes gave
           the same results. Of course, these conclusions are ap-
           plied to a 2 x 900 MWe power plant in given climatic
           conditions, near the Lyon area.

M. PIRINGER   Can you explain why your model
           underpredicts the frequency of short and low plumes and
           overestimates all the others ?

A. HODIN   Because the plumes are shared
           in length and height classes. The total number of cases
           corresponding, for instance, to the 3 length classes
           corresponds to 100 %. Obviously, if the model under-
           predicts the frequency of short plumes, it must over-
           estimate at least one of the two other classes fre-
           quencies.

3: MODELING THE DISPERSION OF HEAVY GASES

Chairmen: H. E. Turner
H. van Dop

Rapporteurs: H. Apsimon
P. J. H. Builtjes

THE MODELING OF DISPERSION OF HEAVY GASES

Joseph B. Knox

Lawrence Livermore National Laboratory
P. O. Box 808
Livermore, CA 94550

EXTENDED ABSTRACT

The motivation for the modeling of heavy gas dispersion lies in the need to estimate the impacts on the surroundings in the event of a leak of heavy gas during normal operations or a much larger release in the event of a major accident. In this presentation, we will consider the phenomenology and/or the evolution of a heavy gas spill from the early times of release to when it is diluted below a prescribed level of toxicity or lower flamability limit (LFL). To the extent the processes contributing to this phenomenology are known, they are described. From this discussion, salient processes emerge whose quantification leads to the structuring of models for heavy gas dispersion. There have been numerous efforts to model the dispersion of such heavy gases, in one to three space dimensions [Van Ulden (1974), Kaiser and Walker (1978), Zeman (1980), Colebrander (1980), Eidsvik (1980), Blackmore et al. (1982), Havens (1982), Taft et al. (1982), Hertel and Teuscher (1982), Ermak et al. (1981), and Chan et al (1982)]. We will describe some typical or representative models in each dimensionality and for various modes of solution (e.g. finite difference and finite element methods).

The principal heavy gases of interest in the U.S. include NH_4, H_2S, liquefied natural gas (LNG) (which after release and phase change becomes a heavy gas), Cl_2, and various chemicals related to missile propellents. Although our discussion and examples may appear to dwell on LNG, enough commonality of physics and solution methodology remains so that the discussion is appropriate to those with a broader interest in heavy gases. In

fact, the models described herein can, with appropriate physical
properties and initial conditions, be applied to a spectrum of
different heavy gases. It is pertinent to discuss the types of
questions that models of differing dimensionality and/or complexity can address. For instance, a hierarchy of models of
various complexity and of differing sophistication in physics and
solution resolution can be envisaged; they each could meet differing assessment needs. A 1-D time dependent model can be
appropriate for describing a radially symmetric cloud evolution in
the absence of ambient winds, provided vertically integrated concentrations were satisfactory. Whereas a 2-D hydrodynamic model,
for the same conditions, could provide spatial resolution in the
solution consistent with the model zoning and the nature of solution methodology. A simple slab time dependent model (which is
mathematically 1-D but pseudo 3-D) gives information on the rate
of advance of a heavy gas cloud due to its gravity flow outwards
from a point of release and information regarding the depth-
averaged concentration in the cloud. Finally, a heavy gas release
into an atmospheric boundary layer with speed and directional wind
shear in the presence of complex terrain would require a 3D time
dependent model to describe the distribution of gas concentration
in space and time.

Our knowledge of the processes and the data base for model
validation and testing, Ermak (1982) and Woodward (1982), must
come from well conceived experimental programs whose experiments
span a spectrum of atmospheric conditions, sizes of release, perhaps mode of release, and type of underlying surface. The experimental results from the British efforts at Porton Downs, Picknett
(1981), at Maplin Sands, Puttock and Blackmore (1982), and by the
U.S. in the DOE LGF Test Program, Cederwall (1982), are most
important for conceptualizing models. For atmospheric conditions
involving low atmospheric wind speeds and stable thermal stratification (cases with weak atmospheric turbulence), the experimental evidence shows that the initial heavy-gas cloud spreads out
radially with the speed of advance of the heavy gas proportional
to $\sqrt{gh \, \Delta\rho / \rho}$, and that the pressure-volume solenoids concentrated at the forward edge of the heavy gas cloud create vorticity. This vorticity generation results in a vortex just behind
the leading edge of the heavy gas. An alternative way to view
this vorticity generation is that if $\nabla\rho$ is not parallel to $\vec{g}$,
then vorticity is produced. The characteristic speed of the
vortex circulation can exceed ambient wind speeds. At least in
some of the Porton Downs experiments, which were conducted under
calm atmospheric conditions, the highest concentration of tracer
gas appears to be semi-confined in the "spinning" edge vortex. In
such cases, the ambient wind and levels of turbulence appear to
have little to do with the early cloud evolution. Whereas, if the
heavy gas release is small, and the atmosphere unstable with

modest wind speeds, the evolution of the gas cloud is dominated by
the ambient wind and its turbulent processes. The former gravity
flow dominated regime resembles a dissipative structure, Prigogine
(1980); the latter regime dominated by ambient transport-
turbulence processes is the familiar point or volume-source air
pollution problem. In general, our experimental findings at this
time are insufficient to define the boundary in "event-space"
between these two regimes. By "event-space" we mean a space
defined by type and size of release, atmospheric wind speed and
stability, and surface type.

To date the largest volume LNG spill experiment conducted by
the DOE-LGF program is about 40 m^3 of liquid (or about 10,000
m^3 of heavy gas), with only one experiment (in the DOE-LGF field
program) clearly conducted in the gravity-flow dominated regime.
Crucial tests at higher release volumes and within this regime are
lacking, and 40 m^3 is indeed a small spill volume in comparison
to those of 25,000 m^3 or more envisaged in some accident scenarios. As the size of the spill increases, the gravity-flow
domain extends into event space in a manner that is poorly known
today. The duration of the gravity-flow domain is clearly extended in time with increasing spill size, so that sloping terrain
and heat transfer from the underlying surface become more
important. Clearly, well coupled experimental and theoretical
programs are required to bridge the gap between 40 m^3 and the
largest sizes for which practical assessments ($\sim$ 25,000 m^3)
may be required. Models with the potential of addressing this
assessment gap are just beginning to emerge, e.g. 3-D time dependent simulations of the evolution of an LNG spill in the presence
of terrain have been performed by Chan et al. (1982) and is to be
presented at the meeting in this session. To our knowledge, this
is a pioneering FEM simulation for there is clear evidence of the
vortex formation, some evidence in the solution of vortex trapped
high concentrations, and gas cloud deflection by the sloping terrain. Certain deficiencies of the model including the treatment
of surface heat transfer, a turbulence prescription suitable to
the complex thermal stratification in the cloud, and the entrainment mechanism at the cloud top have been identified, and are
being improved, Chan et al. (1982).

Decision makers are faced with the assessment of the consequences of postulated large LNG spills at coastal sites and in the
presence perhaps of complex or sloping terrain; these assessment
needs unfortunately lie beyond the range of experience of the
available data or the confident use of advanced models at this
moment. Further, it is not at all clear that assessment of safety
performed with standard air pollution models, (the Gaussian puff
or plume models, for instance, that essentially ignore or mimic in
some questionable manner the gravity flow regime) are sufficiently

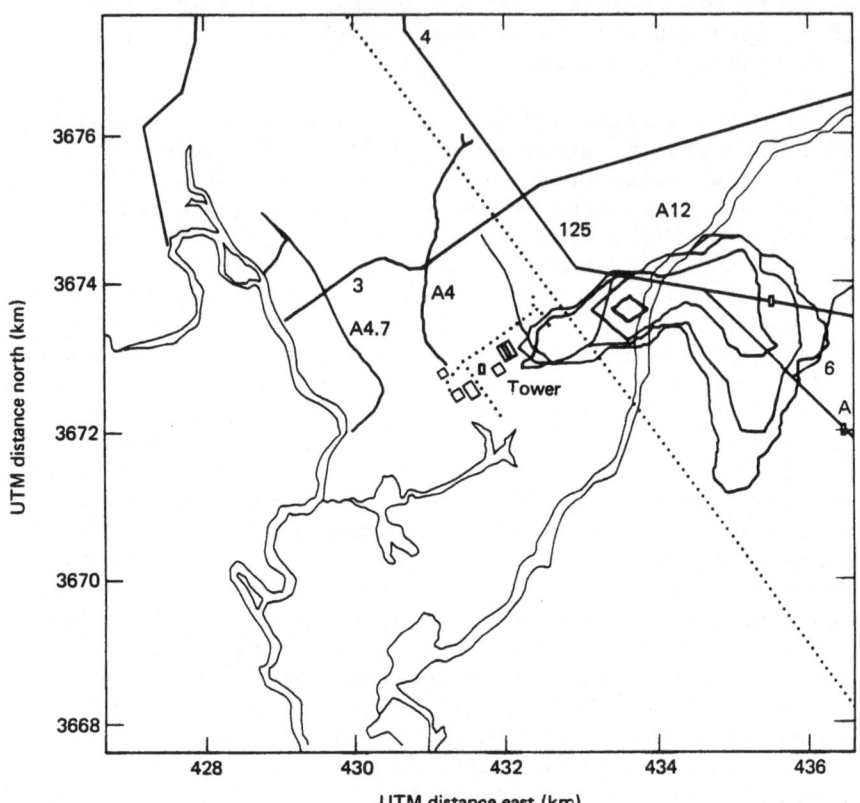

Fig. 1.   Depiction of isopleth pattern for instantaneous air concentration of hydrogen sulfide (H_2S) at two meters
above ground level, as calculated for three hours after
start of simulated continuous leak (initial time 1600
GMT, 21 May 1980; valid time 1900 GMT, 21 May 1980).
Isopleths are 1000, 100, 10, and 1 ppm, respectively.

valid.   There is a pressing need technically to develop validated
models of sufficient fidelity to simulate the gravity flow regime,
the transition phase, and the final phase of dilution by atmospheric transport-diffusion processes.   The predicted capability
of such models should be tested against experimental data in
regard to its ability to estimate the 3D distribution of heavy gas
to a concentration level considerably below the lower combustible
limit, and for test spills of modest size (e.g. 100 to perhaps
1000 m^3).

It would be our goal that emerging scaling laws combined with
validated and tested numerical simulation model for the gravity
flow portion of event space contribute to a sound basis for facility planning for LNG within a few years, as well as a confident
prediction capability for expected smaller operational spills in a
variety of meteorological scenarios.   In the meantime, we must
contend with existing models but should use them cautiously until
a fuller validated set of tools are available.

Promising avenues of research for both experimenters and
modelers include at least the following, particularly for larger
volume spills, (a) heat transfer effects for spills over soil
surfaces including soil heat budgets coupled to the 3-D fluid
flow, (b) an improved turbulence model driven in part by predictive soil surface temperature, local Richardson number or proxy,
and perhaps Reynolds number, (c) an improved treatment of entrainment of air into the gas cloud paying particular attention to the
region just behind the edge vortex, and (d) inclusion, also of
atmospheric water vapor (or rain) in the energy budget.   In addition, the synergistic interaction and analysis involving simple
models (e.g. slab) and multidimensional models should be pursued.
The edge vortex phenomena and entrainment process explicitly calculated in the multi-dimensional model is parameterized in the
simpler slab model; hence, the benefits of shared results between
the models has considerable potential.

Detailed diagnostic studies of the edge vortex phenomena,
while perhaps difficult, might be very strategic in that the decay
of the vortex is promoted by surface friction, heat transfer from
the underlying surface, and entrainment of ambient air.   The latter two processes destroy the pressure volume solenoids that generate the vorticity; the detailed measurements of the edge vortex
contains the effect of the aggregate of these processes.   Assuming
that the heat transfer process and the frictional effects are
quantifiable, the history of the edge vortex circulation would
then contain some quantitative information on the entrainment
process.

An example of a recent use of existing models for the

assessment use in a hazardous condition follows. In 1981, a 100
ton H_2S storage facility developed a hazardous condition in
which there was a potential for the release of the pressurized
hydrogen sulfide at the SRP (Savannah River Plant) and at the head
of a sloping valley. The Department of Energy's Atmospheric
Release Advisory Center, located at LLNL and operated by my division, was called to assist. Regional scale predictions were made
of the transport-diffusion of the H_2S by the atmosphere including the terrain modeling of the gravity flow out to a range of
several kilometers. The isopleths produced by a particle-in-cell
transport-diffusion model (ADPIC) are shown in the attached figure, Knox and Orphan (1980). The heavy gas was crudely simulated
by particles given a small negative bouyancy and by increasing the
stability of the atmosphere in regard to selection of diffusion
parameters. The safe handling of the H_2S hazardous condition by
plant staff lead to the non-acquisition of any confirming data.
However, the assessment calculation performed early in the three
day period of remedial action illustrates the character of the
existing state-of-the-art in dealing with significant heavy gas
dispersion problems.

All of the matters dealt with in the executive summary are to
be covered in more detail in the full presentation.

REFERENCES

Blackmore, D. R., Herman, M. N., Woodward, J. L., 1982, Heavy Gas
    Dispersion Models, J. Hazardous Materials, 6:1/2.

Cederwall, R. T., Ermak, D. L., Goldwire, H. C. Jr., Koopman, R.
    P., McClure, J. W., McRae, T. G., Morgan, D. L., Rodean, H.
    C., Shinn, J. H., 1982, Analysis of Burro Series 40-m^3
    LNG Spill Experiments, J. of Hazardous Materials, 6:1/2.

Chan, S. T., Rodean, H. C., Ermak, D. L., 1982, Numerical Simula-
    tions of Atmospheric Releases of Heavy Gases Over Variable
    Terrain, NATO/CCMS 13th International Technical Meeting on
    Air Pollution Modeling and Its Applications, Ile d'Embiez,
    France, September 14-17, 1982.

Colenbrander, G. W., 1980, A Mathematical Model for the Transient
    Behavior of Dense Vapour Clouds, The 3rd International
    Symposium on Loss Prevention and Safety Promotion in the
    Process Industries, Basel, Switzerland, September 15-19,
    1980.

Eidsvik, K. J., 1980, A Model for Heavy Gas Dispersion in the
    Atmosphere, Atm. Envir., 14, pp. 769-777.

Ermak, D. L., Chan, S. T., Morgan, D. L., and Morris, L. K., 1982,
    A Comparison of Dense Gas Dispersion Model Simulations with
    Burro Series LNG Spill Test Results, J. of Hazardous
    Materials, 6:1/2.

Havens, J. A., 1982, A Description and Computational Assessment of
    the SIGMET LNG Vapor Dispersion Model, J. of Hazardous
    Materials, 6:1/2.

Hertel, J. and Teuscher, L., 1982, Advances in Heavier-Than-Air
    Vapor Cloud Dispersion Modeling, AGA Gas Transmission
    Conference, Chicago, Ill., May 17-19, 1982.

Kaiser, G. D. and Walker, B. C., 1978, Releases of Anhydrous
     Ammonia from Pressurized Containers — the Importance of
     Denser than Air Mixtures, Atm. Envir., 12, pp. 2289-2300.

Knox, J. B. and Orphan, R. C., 1980, Program Report for FY-1980:
     Atmospheric and Geophysical Sciences Division, LLNL Report,
     UCRL-51444-80.

Picknett, R. G., 1981, Dispersion of Dense Gas Puffs Released in
     the Atmosphere at Ground Level, Atm. Envir., 15, pp.
     509-525.

Prigogine, I., 1980, From Being to Becoming, Time and Complexity
     in the Physical Sciences, San Francisco, Freeman.

Puttock, J. S. and Blackmore, D. R., 1982, Field Experiments on
     Dense Gas Dispersion, J. of Hazardous Materials, 6:1/2.

Taft, J. R., Ryne, M. S., and Weston, D. A., 1982, MARIAH: A
     Dispersion Model for Evaluating Realistic Heavy Gas Spill
     Scenarios, AGA Gas Transmission Conference, Chicago,
     Illinois, May 17-19, 1982.

Van Ulden, A. P., 1974, On the Spreading of a Heavy Gas Released
     Near the Ground, First International Loss Symposium, The
     Hague, Netherlands.

Woodward, J. L., Havens, J. A., McBride, W. C., and Taft, J. R.,
     1982, A Comparison with Experimental Data on Several Models
     for Dispersion of Heavy Vapor Clouds, J. of Hazardous
     Materials, 6:1/2.

Zeman, O., 1982, The Dynamics and Modeling of Heavier-Than-Air
     Cold Gas Releases, Atm. Envir., 16:4, pp. 741-751.

This work was performed under the auspices of the U. S. Department
of Energy by the Lawrence Livermore National Laboratory under
contract No. W-7405-Eng-48.

DISCUSSION

J. PUTTOCK   In your separation of the
flow into three components, following Townsend,
what is the mean flow for this time-dependent cloud ?
Do you refer to the time-averaged atmospheric flow
prior to the start of the experiment, or the ensemble-averaged behaviour of the cloud?

J. KNOX   In preparing this review paper on heavy gas dispersion, I have not adressed
the formalism of obtaining.$\bar{w}$, $\tilde{w}$, or w'.
The thrust has been rather that as the dispersion
model physics is improved and captures more of the
"organized portion of turbulence-$\tilde{w}$ ", the simulation may well improve in details of pollutant distribution while at same time requiring less of the
transport to be represented by fine-scale (random)
turbulence.

E. RUNCA   Can you comment on the application of dispersion models of heavy gases ?

J. KNOX   Heavy gas dispersion models
may play the following roles :
(1) planning-assessments for the evaluation of the
    low probability but high risk accidential re-
    leases of large volume (e.g. 25.000 m3 of LNG).
(2) for management of emergencies involving reme-
    dial actions and their timing for meteorologi-
    cal periods favorable for reducing or minimi-
    zing risk to surrounding environs and popula-
    tion.
(3) real-time assessments of small release occurring
    in normal operation and their risk management
    (even on site).

A. VENKATRAM   For dispersion of toxic gases
the variance of the concentration is at least as
important as the mean concentration. Do you know
of any models that provide estimates of the variance.

J. KNOX   As 2D and 3D heavy-gas dispersion models are further developed and improved
that is models with good physics and high resolution, that time-series information at model modes
could well give information on both the mean and
variance of concentration.

DISCLAIMER

This document was prepared as an account of work sponsored by an agency of the United States Government. Neither the United States Government nor the University of California nor any of their employees, makes any warranty, express or implied, or assumes any legal liability or responsibility for the accuracy, completeness, or usefulness of any information, apparatus, product, or process disclosed, or represents that its use would not infringe privately owned rights. Reference herein to any specific commercial products, process, or service by trade name, trademark, manufacturer, or otherwise, does not necessarily constitute or imply its endorsement, recommendation, or favoring by the Unites States Government or the University of California. The views and opinions of authors expressed herein do not necessarily state or reflect those of the United States Government thereof, and shall not be used for advertising or product endorsement purposes.

NUMERICAL SIMULATIONS OF ATMOSPHERIC RELEASES

OF HEAVY GASES OVER VARIABLE TERRAIN

Stevens T. Chan, Howard C. Rodean, and Donald L. Ermak

Lawrence Livermore National Laboratory
University of California
Livermore, California 94550, U.S.A.

INTRODUCTION

The Lawrence Livermore National Laboratory (LLNL), under the
sponsorship of the U.S. Department of Energy, is conducting safety
research related to the possible consequences of liquefied natural
gas (LNG) spills. Under this program, LLNL and the Naval Weapons
Center (NWC) jointly conducted the Burro and Coyote series of LNG
spill experiments at China Lake, California during 1980 and 1981
respectively (Koopman et al., 1982; Koopman, 1982). LLNL is
concurrently developing models for use in predicting the vapor
dispersion from LNG spills (Ermak et al., 1982).

In this paper, we compare two sets of numerical simulations of
LNG vapor dispersion with the results of the Burro 8 and 9 experiments. We used our FEM3 model for both sets; one with the variable
terrain at the NWC test site and the other with flat terrain.

BURRO LNG SPILL TESTS

The Burro series consisted of nine field experiments at NWC
(Koopman et al., 1982). The first experiment, Burro 1, was conducted with liquid nitrogen; the others, Burros 2 - 9, were conducted
with LNG. The LNG spill volume ranged from 24 to 39 m^3, the spill
rate from 11.3 to 18.4 m^3/min, the wind speed from 1.8 to 9.1 m/s,
and the atmospheric stability from unstable to slightly stable.
An extensive array of instrumentation was deployed both upwind and
downwind of the spill pond (Fig. 1). Wind speed and direction, gas
concentration, temperature, humidity, and heat flux from the ground

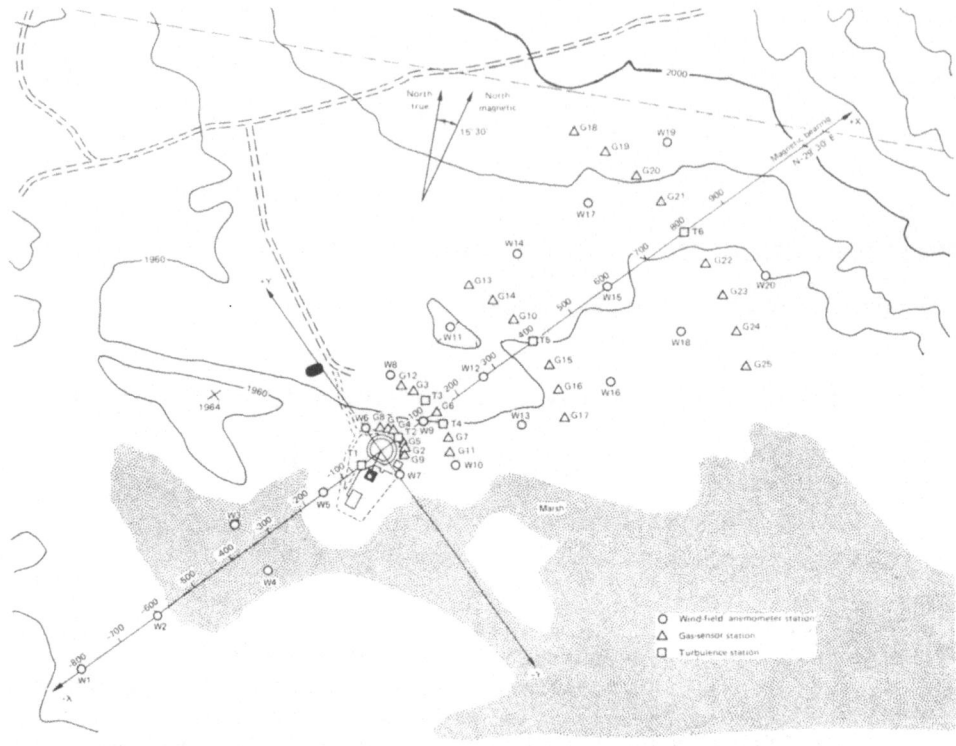

Fig. 1.   Instrumentation array for the 1980 LNG dispersion
          experiments at the NWC, China Lake, California.

were measured at different distances from the spill point and at
different elevations relative to ground level.

     We emphasized Burros 3, 7, 8, and 9 in our earlier studies
because each represented a different category of experimental
conditions, the gas clouds of Burros 3, 8, and 9 were fairly well
aligned with respect to the instrumentation array, and the array of
operational gas sensors was most complete for Burros 7 - 9.   We
selected Burros 8 and 9 for use in this study because of the above
and, in particular, because these two experiments had the highest
LNG spill rates, very different atmospheric conditions, and
distinctly different gas cloud dispersion phenomena.   The test
conditions are summarized in Table 1.

Table 1. Spill and Meteorological Parameters for Burros 8 and 9.

Parameter	Burro 8	Burro 9
Spill Volume (m^3)	28.4	24.2
Spill Rate (m^3/min)	16.0	18.4
Wind Speed at 2 m (m/s)	1.8 ± 0.3	5.7 ± 0.7
Wind Direction (deg)	235 ± 6	232 ± 4
Temperature at 2 m (°C)	33.1	35.4
$T_* = dT/d \ln z$ (°C)	+ 0.145	− 0.100
Friction Velocity, Diabatically Adjusted (m/s)	0.074	0.252
Sensible Heat Flux, Negative Upward (W/m^2)	+2.2	−10.0
Turbulent Prandtl Number	0.623	1.05
Momentum Diffusivity at 2 m, Diabatically Adjusted (m^2/s)	0.037	0.212
Richardson Number at 2 m	+ 0.121	− 0.014
Monin-Obukhov Length Scale (m)	+ 16.5	− 140
Relative Humidity (%)	4.7	11.7

NUMERICAL MODEL

    In recent years, there have been numerous efforts to model the
dispersion of heavy gas releases and some representative models have
been reviewed by Knox (1982). It is generally recognized that any
numerical model capable of treating the most general spill scenarios
associated with heavy gas releases are bound to be complex and compu-
tationally expensive. Therefore, appropriate simplifications to the
complex model are generally sought. On the opposite extreme, an
overly simplified model may be so limited that its range of applicability would be severely restricted.

    This report describes a three-dimensional conservation equation
model which involves certain simplifying assumptions mainly to
expedite the numerical procedures, but not to seriously compromise
the true physics. The model is being developed for simulating the
vapor dispersion associated with LNG spills onto water and/or land,
including the treatment of variable terrain (of course, with appropriate physical properties, and initial and boundary conditions, the
model applies equally well to other heavy gas releases). The
present model is based on solving the set of three-dimensional,
time-dependent, conservation equations of mass, momentum, energy and
species. Spatial discretization is performed via a modified
Galerkin finite element method, and time integration is carried out
via the forward Euler method (pressure is computed implicitly,
however).

The following sections briefly summarize and describe several salient features of the model, including the governing equations, related numerical procedures, initial and boundary conditions, and the submodels currently used for turbulence and ground heat transfer. An earlier version of the model, together with related numerical procedures, was described in Chan et al. (1982a). More details on the numerical aspects of the present model can be found in Chan et al. (1981), Gresho et al. (1982), and Chan and Gresho (1982b).

Governing Equations

In the present model, the spread and dispersion of heavy gases is predicted by solving the three-dimensional conservation equations governing incompressible flows. However, in order to accomodate large density changes in both space and time, and yet to preclude sound waves, we have generalized the "anelastic" approximation (Ogura and Phillips, 1962) and obtained a set of equations which are slightly different from those usually derived for incompressible flows (the standard Boussinesq approximation is considered inadequate for our purposes, see for example, Daley and Pracht, 1968 and Lee et al., 1981). These equations, written for the mean (time-averaged) quantities in a turbulent flow field, are:

$$\frac{\partial(\rho\underline{u})}{\partial t} + \rho\underline{u} \cdot \nabla\underline{u} = -\nabla p + \nabla \cdot (\rho\underline{\underline{K}}^m \cdot \nabla\underline{u}) + (\rho - \rho_h)\ \underline{g} \tag{1a}$$

$$\nabla \cdot (\rho\underline{u}) = 0 \tag{1b}$$

$$\frac{\partial\theta}{\partial t} + \underline{u} \cdot \nabla\theta = \frac{1}{\rho c_p} \nabla \cdot (\rho c_p\ \underline{\underline{K}}^\theta \cdot \nabla\theta) + \frac{c_{p_N} - c_{p_A}}{c_p} (\underline{\underline{K}}^\omega \cdot \nabla\omega) \cdot \nabla\theta \tag{2}$$

$$\frac{\partial\omega}{\partial t} + \underline{u} \cdot \nabla\omega = \frac{1}{\rho} \nabla \cdot (\rho\underline{\underline{K}}^\omega \cdot \nabla\omega) \tag{3}$$

and

$$\rho = \frac{PM}{RT} = \frac{P}{RT\ (\frac{\omega}{M_N} + \frac{1 - \omega}{M_A})} \tag{4}$$

where $\underline{u}$ = (u, v, w) is the velocity, ρ is the density of the mixture, p is the pressure deviation from an adiabatic atmosphere at rest, with corresponding density defined as ρ_h, $\underline{g}$ is the acceleration due to gravity, θ is the potential temperature deviation from an adiabatic atmosphere, ω is the mass fraction of NG vapor, and $\underline{K}^m$, $\underline{K}^\theta$, and $\underline{K}^\omega$ are the eddy diffusion tensors (which are parameterized using K-theory) for the momentum, energy, and NG vapor, respectively, and C_{P_N}, C_{P_A}, and $C_p = \omega\, C_{P_N} + (1 - \omega)\, C_{P_A}$ are the specific heats for NG vapor, air, and the mixture, respectively. In the equation of state, P is the absolute pressure, R is the universal gas constant, T is the absolute temperature ($T/(\theta + \theta_0) = (P/P_0)^{R/MC_P}$), and M_N, M_A are the molecular weights of NG and air, respectively. The above set of equations, together with appropriate initial and boundary conditions, are solved to yield velocity, pressure, temperature, mass fraction of NG vapor, and density of the mixture as functions of time and space. It is worth pointing out that the diffusion terms in Eqs. (2) and (3) are written somewhat differently from those in our earlier reports, Chan et al. (1982a) and Lee et al. (1981). The present form is more appropriate for problems involving large variations in density and heat capacity and has been verified numerically to better conserve energy and species.

Spatial Discretization and Time Integration

Equations (1) through (3) are discretized spatially by the finite element method in conjunction with the Galerkin method of weighted residuals. The primary unknowns, U, V, W, (U = ρu, V = ρv, W = ρw), p, θ, and ω (ρ is computed subsequently by Eq. (4)), are approximated as:

$$U = \sum_{j=1}^{n} \phi_j(\underline{x})\, U_j(t) \tag{5a}$$

with similar expressions for V, W, θ, ω, and

$$p = \sum_{j=1}^{m} \psi_j(\underline{x})\, p_j(t) \tag{5b}$$

where, in the discretized domain, there are n nodes for velocity, temperature, and concentration, and m nodes for pressure. The approximation functions, $\{\phi_i(\underline{x})\}$, are piecewise continuous polynomials which are one degree higher than those for the pressure approximation, $\{\psi_i(\underline{x})\}$. Currently, we are using the 8-node isoparametric "brick" element with trilinear velocity and piecewise

constant pressure. After substituting Eq. (5) into Eqs. (1) through
(3), premultiplying each of the equations by appropriate weighting
functions, and integrating the diffusion and pressure gradient terms
by parts, we obtain a coupled system of nonlinear, first-order
ordinary differential equations. This system of equations, written
in a compact matrix form, is:

$$M\dot{U} + [K + N(U)]u + CP = F \tag{6a}$$

$$C^T U = 0 \tag{6b}$$

$$M_s \dot{\theta} + [K_\theta + N_s(u)] \theta = F_\theta \tag{7}$$

$$M_s \dot{\omega} + [K_\omega + N_s(u)] \omega = F_\omega \tag{8}$$

where now U and u are global vectors of length 3n containing all
nodal values of ρu and $\underline{u}$, respectively, P is a global vector of
length m containing pressure values, M, K, and N (all of size 3n x
3n) are the mass matrix, the diffusion matrix, and the advection
matrix, respectively, C is the 3n x m pressure gradient matrix and
its transpose, C^T, is the m x 3n divergence matrix, F is a global
vector of length 3n incorporating natural boundary conditions
(tractions) and the buoyancy force. The matrices in Eqs. (7) and
(8) for temperature and the concentration of NG vapor are defined
similarly except their "size" is n instead of 3n.

The forward Euler method of time integration, applied to Eq.
(6a) gives:

$$U_{n+1} = U_n + \Delta t M^{-1} [F_n - Ku_n - N(U_n)u_n - CP_n] \tag{9}$$

where U_n is the vector of nodal mass fluxes (u_n are nodal
velocities) at time t_n and Δt is the step-size. Before this
equation can be used to advance the velocity, however, the pressure
at time t_n must be computed. This is done by combining Eq. (6a)
with a time-differentiated version of Eq. (6b) ($C^T \dot{U} = 0$ since
$C^T U = 0$ for all time) to generate the consistent discretized
Poisson equation for the pressure, evaluated at time t_n,

$$(C^T M^{-1} C)P_n = C^T M^{-1} [F_n - Ku_n - N(U_n)u_n] \tag{10}$$

The sequence of steps for advancing the velocity and pressure from
t_n to t_{n+1} is thus (given that U_n is available and that it
satisfies $C^T U_n = 0$):

(1) Form the acceleration vector (sans the pressure gradient)

$$A_n = M^{-1}[F_n - Ku_n - N(U_n)u_n] \tag{11}$$

(2) Solve the linear algebraic system (discrete Poisson equation) for the compatible pressure via

$$(C^T M^{-1} C)\, P_n = C^T A_n \tag{12}$$

(3) Update the mass flux, accounting for the pressure gradient,

$$U_{n+1} = U_n + \Delta t\, (A_n - M^{-1} C P_n) \tag{13}$$

(4) Finally, in an "uncoupled step," update the temperature and concentration, again using the forward Euler method,

$$\theta_{n+1} = \theta_n + \Delta t\, M_s^{-1}[F_{\theta_n} - K_\theta \theta_n - N_s(u_n)\theta_n], \tag{14}$$

$$\omega_{n+1} = \omega_n + \Delta t\, M_s^{-1}\, [F_{\omega_n} - K_\omega \omega_n - N_s(u_n)\omega_n]. \tag{15}$$

Several cost-effective techniques, including one-point quadrature, subcycling, balancing diffusion, and others, which help reduce the computing costs drastically and yet do not seriously compromise the solution accuracy, have been developed and used in the present model. These techniques were discussed in Chan et al. (1981), Gresho et al. (1982) and Chan and Gresho (1982b).

Boundary Conditions and Ground Heat Transfer

Typical boundary conditions used in our numerical simulations are illustrated in Fig. 2. The origin of the Cartesian coordinate system is normally placed at the center of the source area and the mean wind is assumed to be parallel to the x-z plane.

On the upwind surface, which is sufficiently far away from the source area, a wind profile, expressed as a function of height above ground, is specified. The specific form we used is a parabola based on measured data at heights of 1 m, 3 m, and 8 m. Such a profile was selected, rather than using a power law or logarithmic function, because field measurements are available only at the above heights and, for computational efficiency, we want to use a generalized "partial slip" boundary condition to permit better simulation of the ambient wind profile without requiring fine resolution of the surface "logarithmic region". The remaining variables on this surface are specified to correspond to the ambient conditions.

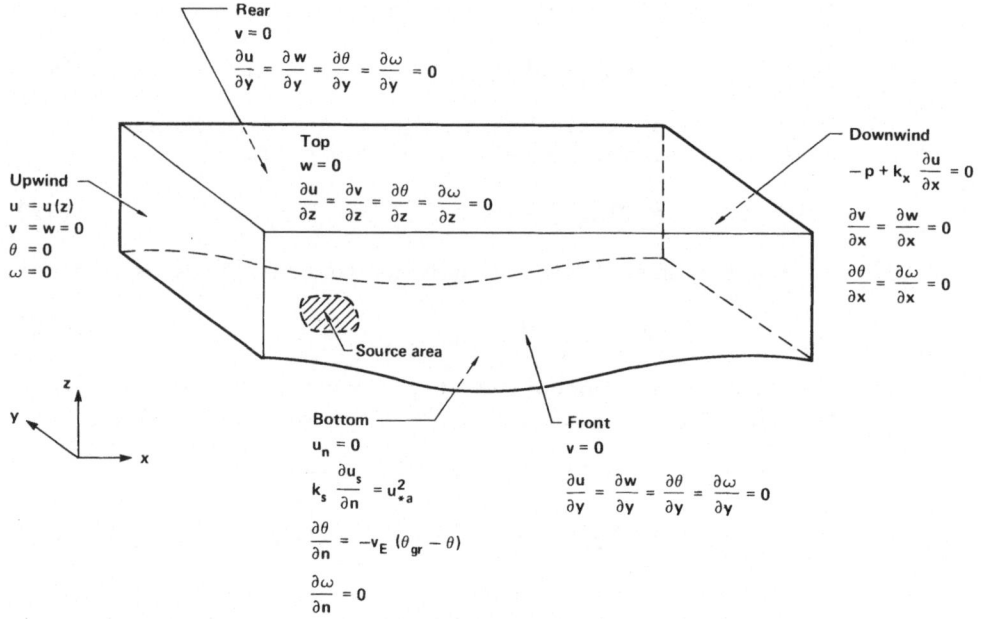

Fig. 2.   Boundary conditions for the governing equations.

On the downwind surface, natural boundary conditions are specified. These include zero traction in the normal direction and zero gradients for the remaining variables. On the top, front, and rear surfaces, which presumably are far enough from the vapor cloud, the respective normal velocity component and normal gradients of the remaining variables are set to be zero.

On the ground surface (either flat or rugged) excluding the source area, generalized "partial slip" boundary conditions are specified for the momentum equations. Specifically, the normal velocity component is set to be zero and shear stresses are specified in the tangential directions. For the case involving variable terrain, this would require the determination of appropriate local normal and tangential directions for the nodal points involved. We basically follow the procedures recommended by Engelman et al. (1982) but choose one of the tangential directions to be parallel to the x-z plane (and of course, be perpendicular to the local normal direction). In this direction, a shear stress of ρu_{*a}^2 (u_{*a} being the friction velocity of the ambient air) is specified. Along the other tangential direction, zero shear stress is currently imposed due to the lack of field data in this direction and because the associated shear stress is presumably much smaller. Outside the source area, the appropriate boundary condition for the concentration is $\partial \omega / \partial n = 0$ so that there is no loss or gain of species at the ground surface.

For the energy equation, a bulk coefficient submodel is used to account for the heat flux from the ground surface, i.e.,

$$K_v \frac{\partial \theta}{\partial n} = -V_E (\Theta_{gr} - \Theta).$$

(16)

In this formula, V_E is an effective energy transfer velocity obtained from field measurements, Θ_{gr} is the ground temperature, and θ is the temperature immediately above the ground (within the computational domain). The effective heat transfer velocity is expressed as either an empirical constant or in the form $V_E = C_f U_r$ where C_f is a surface friction coefficient and U_r is a reference velocity (for example, see Zeman, 1979). In the recent LNG spill tests at China Lake, V_E appeared to be independent of wind speed. The average value of V_E was found to be approximately 0.0125 m/s (Koopman et al., 1982) and this value is used in the calculations presented here.

In the case of flat terrain, due to symmetry of the solution about the vertical center plane, only one-half of the domain needs to be considered. In this case, the center plane becomes the front surface and symmetry boundary conditions are applied.

Characterization of Vapor Source

Due to the lack of a well-validated pool spread and vapor generation model, a relatively simple approach is currently employed to characterize the vapor source. Specifically, a liquid pool of constant area together with constant vapor generation rate is assumed to provide the source (via injection-like boundary conditions) to the present vapor dispersion code over a finite duration of time. The source area is calculated as $A = \dot{V}/W$, where $\dot{V}$ is the spill rate of LNG and W is the liquid regression rate of evaporation. Currently W is taken to be 4.2×10^{-4} m/sec, which is equivalent to an NG vapor injection velocity of $V_I \simeq 0.1$ m/s. The time interval of vapor injection is set equal to the spill duration.

During the period of vapor injection, we specify

$$w = V_I$$

(17a)

and

$$K_v \frac{\partial \omega}{\partial n} = -V_I (1 - \omega).$$

(17b)

At the same time, θ is specified to correspond to the LNG temperature. After the spill is terminated, the above boundary conditions are set to be $\partial\omega/\partial n = \partial\theta/\partial n = 0$.

In some of our previous simulations, a general heat transfer boundary condition in the form of Eq. (16) has been used but was found to be unsatisfactory. Generally, with the energy transfer velocity set equal to V_I, the simulated vapor cloud often appeared to be too light and the predicted downwind distances to the lower flammability limit (LFL) were found to be significantly larger than those measured in the field. It appears that some tuning in the effective energy transfer velocity based on relevant field measurements is required.

Initial Conditions

Prior to the start of a vapor dispersion simulation, initial conditions corresponding to the ambient atmosphere must be provided to the dispersion code. These initial conditions may correspond to either an isothermal or a stratified atmospheric flow, depending on the pre-existing temperature field. Proper initialization of a stratified flow field over variable terrain is rather complex, and is outside the scope of the present discussion. Herein we address only the simpler case--initialization of an isothermal flow field, because the temperature data from the two Burro tests to be simulated indicate that the ambient atmosphere in the region of interest (z < 35 m) can be reasonably approximated as being isothermal.

With an assumed wind field, together with appropriate boundary conditions for the momentum equations as indicated in Fig. 2, we first perform a least-squares mass adjustment to obtain a mass-consistent wind field. Using this as an initial guess, we then integrate in time the momentum equations with continuity constraints until the solution reaches steady state. Such a mass-consistent, steady state wind field and the zero initial fields of temperature and concentration are then used as initial conditions in the vapor dispersion simulation.

Turbulence Submodel

In the present version of the model, the eddy diffusion tensors $\underline{\underline{K}}^m$, $\underline{\underline{K}}^\theta$, and $\underline{\underline{K}}^\omega$ are assumed to be diagonal and further $\underline{\underline{K}}^\theta = \underline{\underline{K}}^\omega$. An ad hoc approach is used for the eddy diffusion coefficients. The vertical diffusion coefficient is essentially an extension of the entrainment velocity concept used by several authors to describe the mixing of air into the cloud in their layer-averaged models of density intrusion flows (i.e., Kato and Phillips, 1969; Tennekes, 1973; Zeman and Tennekes, 1977; and Eidsvik, 1980). This approach is then expressed in the typical formalism used to describe the vertical turbulent diffusion coefficient in the ambient atmosphere near the surface (Businger et al., 1971; Dyer, 1974; and Dyer and Bradley, 1982). The resulting expression for the vertical diffusion coefficient, K_v is:

$$K_v = k \ [(u_*z)^2 + (w_*\ell)^2]^{1/2}/\Phi \qquad (18)$$

where

k = von Karman's constant = 0.4,

u_* = friction velocity = $u_{*a} \ [q(z)/q_a(z)]$, with subscript "a" designating the ambient atmosphere,

q = wind speed,

z = height above ground surface,

w_* = in-cloud "convection velocity" = $\alpha_1[(g/T) \ V_E \ (T_{gr} - T) \ \ell]^{1/3}$,

ℓ = cloud height function = $h \cdot \exp(1 - z/h)$,

h = characteristic cloud height = $\int \omega \cdot z \ dz / \int \omega \ dz$,

g = acceleration of gravity,

T = cloud temperature,

T_{gr} = ambient ground temperature, and

V_E = effective heat transfer velocity from the ground into the cold vapor cloud.

The form of the Monin-Obukhov profile function Φ is taken from Dyer (1974). When the Richardson number R_i, is equal to or greater than zero, Φ is defined as:

$$\Phi = 1 + 5R_i, \qquad\qquad\qquad R_i \geq 0 \quad (19a)$$
for all three (momentum, energy, and species) diffusion coefficients. When the Richardson number is less than zero, Φ is given as:

$$\Phi = \begin{cases} (1 - 16 \ R_i)^{-1/4} & \text{for momentum,} \\ (1 - 16 \ R_i)^{-1/2} & \text{for energy and species.} \end{cases} \qquad R_i < 0 \quad (19b)$$

With this model for Φ, as the Richardson number becomes increasingly negative, the effects of convection are more strongly felt on the turbulent transport of heat and species than on momentum. The Richardson number R_i, is defined as:

$$R_i = u_*^2 \frac{R_{ia}}{(u_*^2 + w_*^2)} + \alpha_2 \frac{(\rho - \rho_a)}{\rho} \cdot \frac{g\ell}{(u_*^2 + w_*^2)} \qquad (20)$$

where R_{ia} and ρ_a are the Richardson number and density of the ambient atmosphere. ($R_{ia} \simeq Z/L$, L being the Monin-Obukhov length scale).

The horizontal diffusion coefficient K_h, is expressed as:

$$K_h = \beta k u_* z/\Phi . \qquad (21)$$

In the above formulae, α_1, α_2, and β represent three empirical (tunable) parameters, which are presently set to be 0.5, 0.05, and 6.5, respectively.

The effects of both density stratification and ground heat transfer into the cold vapor cloud are considered in modifying the pre-existing ambient turbulence field. While these two effects are coupled in the model, the density stratification effect tends to decrease the eddy diffusion coefficient by increasing the Richardson number. The ground heating effect tends to increase the eddy diffusion coefficient via the convective velocity term in the numerator of the expression for vertical diffusion, Eq. (18), and by mitigating the density stratification effect on the Richardson number.

As can be seen, in the absence of the NG vapor cloud, the current submodel recovers the ambient diffusivities, i.e.,

$$K_v = k u_{*a} z/\Phi, \quad \Phi = \Phi (R_{ia}) \qquad (22)$$

and

$$K_h = \beta K_v. \qquad (23)$$

NUMERICAL RESULTS

In this section, we compare the model predictions with the experimental results from two of the field tests, namely, Burros 8 and 9. Attention has been given to the downwind distance to the LFL, the time variation of temperature and concentration at representative locations, and the size and shape of the NG concentration contours within various horizontal and crosswind surfaces. Concentration contours and velocity projections on some

of the surfaces, as predicted by the numerical model, are also given
to illustrate the nature of gravity flow associated with heavy gas
dispersion.

As discussed in Koopman et al. (1982), the field data has been
slightly smoothed by using a 10-second moving average. This time
interval was chosen somewhat arbitrarily; the intent was to use an
averaging time that is long enough to smooth out short wavelength
(much less than cloud width) fluctuations, but short enough to
preserve cloud meander. The time-averaged data was then interpolated in space to generate concentration contour plots.

Consequently, the concentration data can be compared with the
model results in the form of time history plots at specific locations within the cloud and in the form of two-dimensional contour
plots at selected times during cloud dispersal. The advantage of
the time history plots is that there is no space interpolation
uncertainties; however, time history data are generally quite
sensitive to the location within the cloud due to the effects of
cloud meander, turbulence fluctuations of shorter wavelengths, and
vapor source generation fluctuations (all of which are not simulated
by the present model). Since the two tests considered herein had
fairly steady mean winds throughout each experiment, the comparisons
of time history plots are believed to be useful even though there is
some evidence of the above mentioned effects on the time history
signal.

The concentration contours provide a more global view of the
dispersing NG cloud and show the size and shape of the cloud as it
develops with time, the steepness of the gradients, and, in
particular, the extent of the flammable zone. However, there is
uncertainty in the location of the contours due to space
interpolation between sensors. For the tests under consideration,
Koopman et al. (1982) estimated the interpolation uncertainties
within a row of instruments to be generally less than 1 m in the
vertical and only a small fraction of the instrument spacing in the
horizontal. The interpolation uncertainties between rows of instruments, however, were considerably greater since the distance between
rows of sensors is much larger (see Figure 1). For instance, the
uncertainty in the downwind distance to the LFL was estimated to be
approximately -40 m to 20 m.

Using our numerical model, two simulations, one with a flat
terrain and one with a variable terrain designed to simulate the
actual topography at the test site, were performed for each of the
selected tests. Shown in Fig. 3 is the topography in the vicinity
of the spill pond. The extent of the computational domain varies
slightly from case to case. For Burro 8, the case with variable
terrain has the following horizontal dimensions (with the origin of

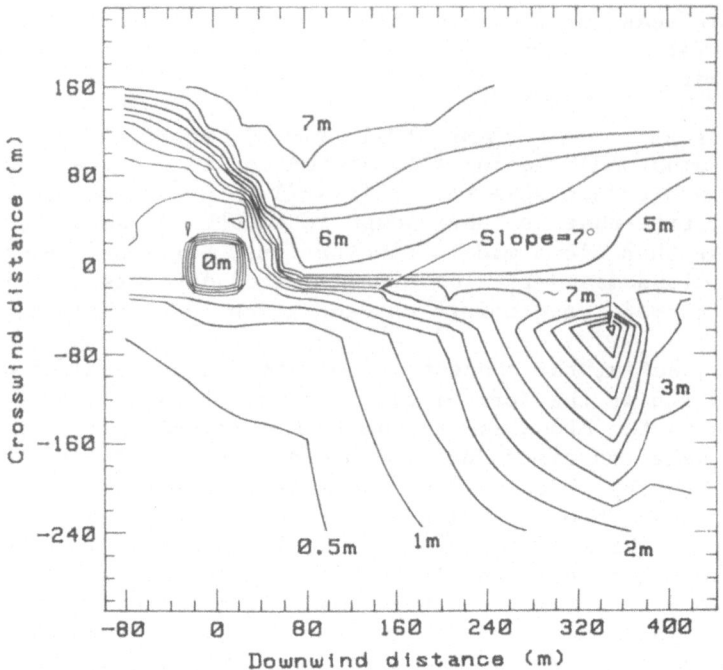

Fig. 3. Topography in the vicinity of the LNG spill facility, China Lake, California.

the coordinates placed at the spill point): x = -100 m to 600 m, y = -280 m to 200 m. The vertical dimension is from the ground to z = 35 m. A graded mesh consisting of 10,350 mesh points (46 x 25 x 9) was used in the computations. Similarly, for Burro 9, the computational domain extends from x = -100 m to 600 m, y = -210 m to 100 m, and the same vertical dimension. A graded mesh of 8,280 points (46 x 20 x 9) was used in this case. For the two simulations with flat terrain, only about half the mesh points were used due to the symmetry of the gas cloud. For Burro 8, a velocity profile in the form u = 1.445 + 0.195 z - 0.00975 z^2 for z < 10 m and u = 2.42 m/s for z $\geq$ 10 m, z being the height above ground, was specified on the upwind surface. Similarly, a velocity profile of

the form u = 4.88 + 0.432 z - 0.0216 z^2 for z < 10 m and u = 7.04 m/s
for z $\geq$ 10 m, was used in the Burro 9 simulations. All cases were
run with the turbulence submodel described earlier except lower
limits (K_v = 0.02 and K_h = 2.0 m^2/s) were used to circumvent
certain numerical difficulties (wiggles) associated with insufficient spatial resolution in some of the flow regions.

Burro 9

The Burro 9 test conditions include a high spill rate, moderate
wind speed, and neutral atmospheric stability. The mean wind was
fairly steady and the gas cloud was approximately central within the
sensor array. However, a series of rapid-phase-transition (RPT)
explosions occurred during the spill test, and as a consequence, the
spill was terminated after only 79 s. The maximum downwind distance
to the LFL, X_{LFL}, was observed to be approximately 325 m at t $\simeq$
80 s (after spill initiation) in the experiment. The calculated
results were $\sim$ 340 m at t $\simeq$ 100 s for the case with flat
terrain, and $\sim$ 325 m at t $\simeq$ 120 s for the case with variable
terrain. The agreement on the values of X_{LFL} is apparently very
good, although some discrepancies exist in the time at which these
distances are reached. The uncertainties associated with the
interpolation of field data and the relatively crude submodel
currently used for characterizing the vapor source are probably the
major contributors of the above discrepancies. The turbulence
mixing and heat transfer submodels currently used might have some
effects also but they are probably less significant, because the
vapor dispersion process of Burro 9 was largely dominated by the
ambient atmospheric boundary layer flow.

Time history plots of the temperature and concentration are
shown in Fig. 4. These data are taken from sensors located at
heights of 1 m and 3 m along the mean wind direction and in the
140-m and 400-m rows. Therefore, these data represent near peak
values of the NG cloud concentration for their respective downwind
distances.

At the 3-m height (Figs. 4a and 4b), both model simulation
results agree quite favorably with the data, especially at the 400-m
row where the agreement is excellent. The variable terrain
concentration result at 140 m (Fig. 4a) agrees very well with the
data for the first 80 s and achieves the same peak value of about
5.5%, but does not decay back to zero as rapidly as the measured
value. For the same case, the flat terrain simulation achieves a
somewhat lower peak concentration value (about 4%) and decays back
to zero after 100 s in a similar manner as the variable terrain
simulation result. The rapid decay of the measured concentration
signal in comparison to the model results may be due to either plume

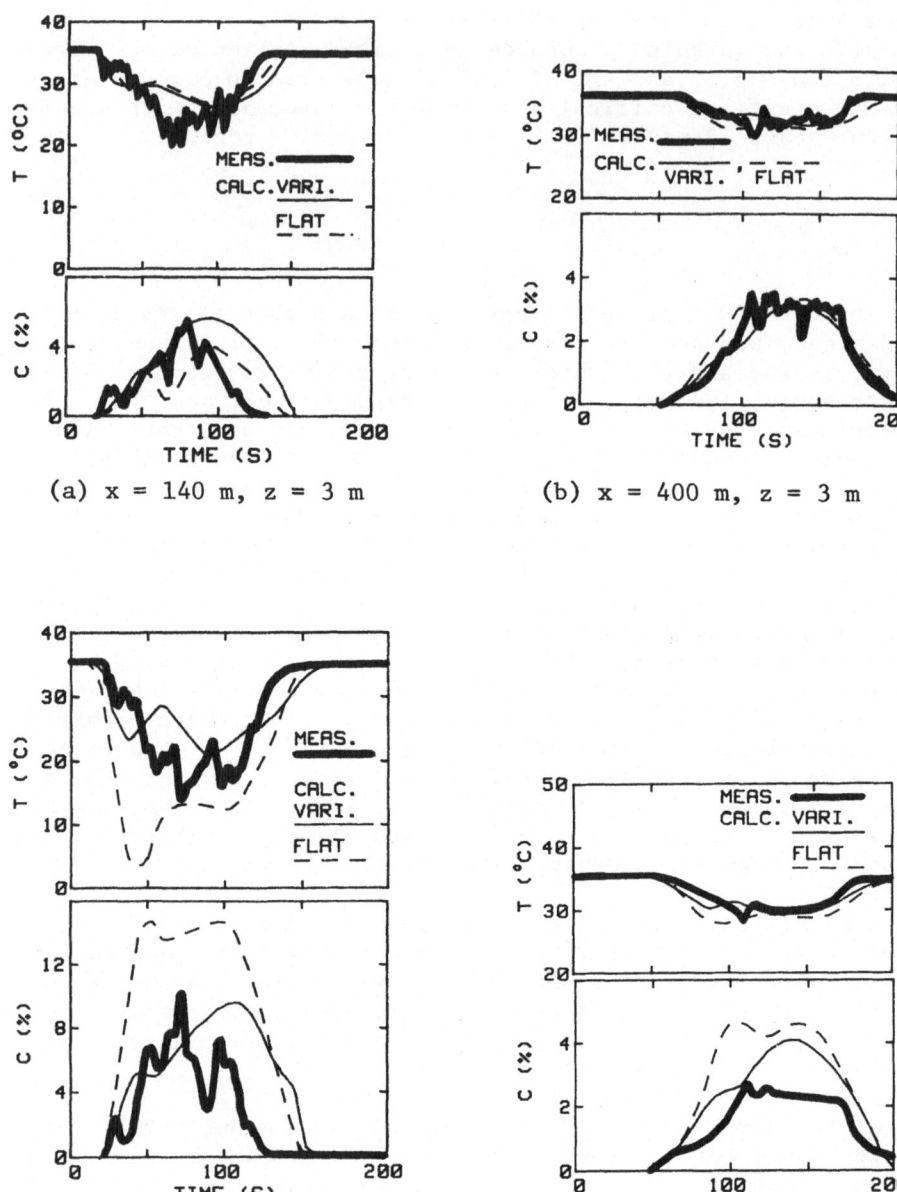

(a) x = 140 m, z = 3 m   (b) x = 400 m, z = 3 m

(c) x = 140 m, z = 1 m   (d) x = 400 m, z = 1 m

Fig. 4.   Burro 9 time history plots of temperature and
          concentration at several representative locations
          on the center plane.

meander, a significantly decreasing NG vapor source rate in the final stages of the spill or some combination of the two effects. Whatever the cause, it also occurs at the 1 m height in the 140-m row (Fig. 4c).

The flat terrain simulation significantly overpredicted the NG concentration at the 1 m height in both cases by 50-100% of the measured value (see Figs. 4c and 4d). The variable terrain simulation did somewhat better especially in the 140-m row (Fig. 4c), where it predicted a peak concentration value of about 10% which is in very good agreement with the measured value. In general, the measured concentration signal has much more structure than the model results as would be expected due to the variations in the evaporation rate and the fluctuations of wind speed and direction during the actual experiment.

Concentration contours of the NG vapor cloud at t = 60 s are compared in Fig. 5 for the "horizontal" surface 1 m above the ground. In the experimental result, the contours near the source (Fig. 5a) are unreliable because RPT explosions occurring in the spill pond threw mud and water on many of the gas sensors in the 57-m arc. Taking the 1% contour to represent the edge of the cloud, both numerical predictions are seen to be in reasonable agreement with the data with respect to cloud shape, length, and width. The differences between the two numerical results (Figs. 5b and 5c) indicate that the effect of the terrain on cloud dispersion in this test is mainly to shorten the progress of the higher concentration levels (especially the 10, 15, and 25% levels) in the downwind direction and to cause a slight lateral shift of the NG cloud with respect to the mean wind direction. Both of these effects observed in the variable terrain simultion are in good agreement with the measured test results.

A comparison of the crosswind contours in the 140-m arc, also at t = 60 s, is shown in Fig. 6. Again, using the 1% contour to indicate the outer edge of the cloud, both model simulations agree fairly well with the measured result with respect to cloud height and width, although the variable terrain simulation is somewhat narrower than the other two plots. The lateral shift of the NG cloud and the maximum concentration of slightly higher than 5% are both well predicted by the variable terrain simulation. In contrast to this, the flat terrain simulation predicted a much higher maximum concentration, namely, greater than 15%.

Burro 8

The Burro 8 test conditions included a high spill rate, very low wind speed, and slightly stable atmosphere. Nearly all of the gas

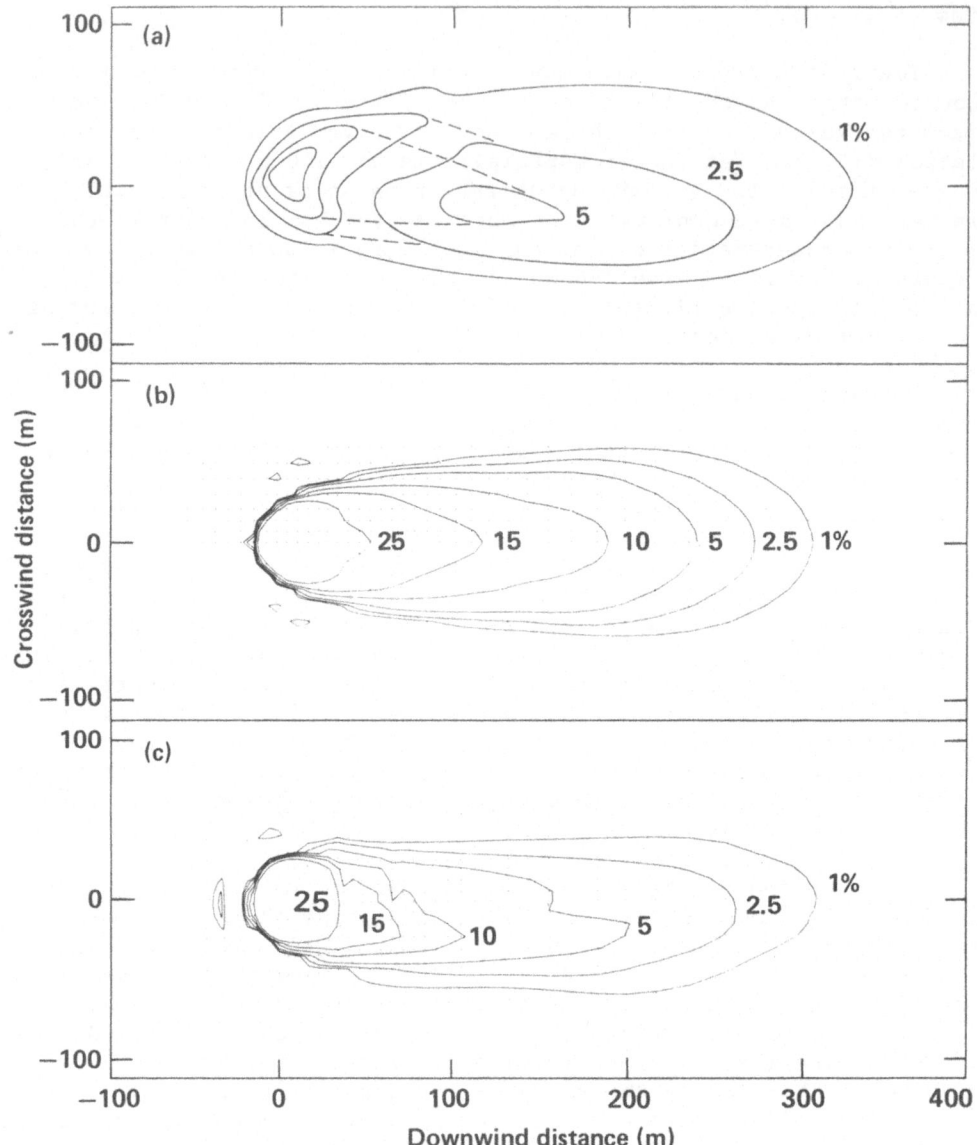

Fig. 5.   Burro 9 contour plots of concentration 1 m above
          ground at t = 60 s: (a) measured; (b) calculated
          (flat terrain); (c) calculated (variable terrain).

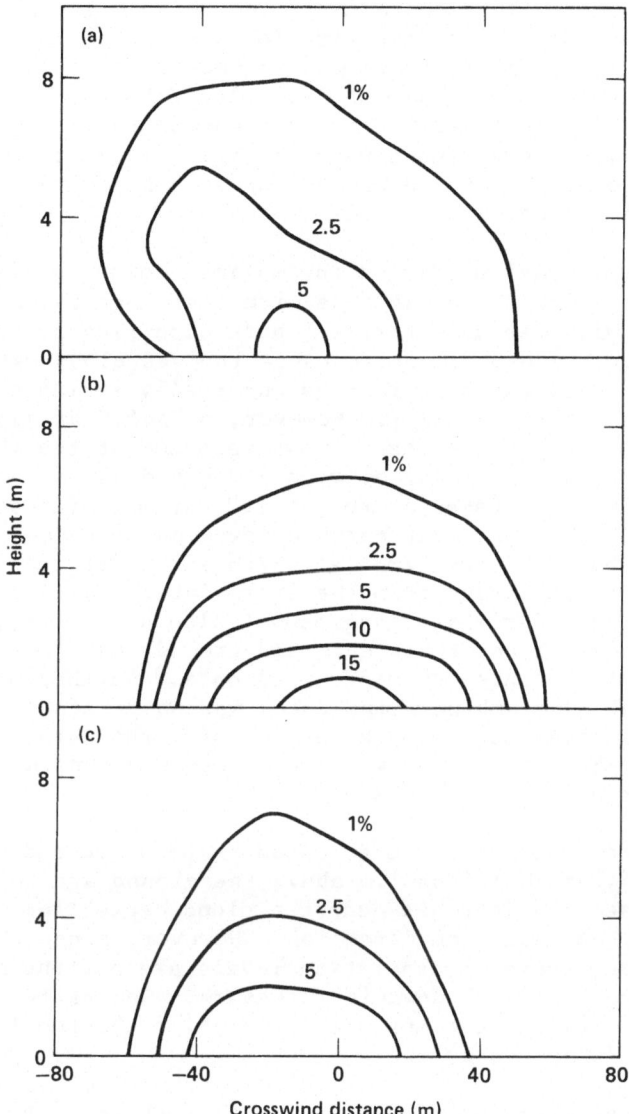

Fig. 6.   Burro 9 crosswind contour plots of concentration
          140 m downwind at t = 60 s: (a) measured; (b) cal-
          culated (flat terrain); (c) calculated (variable
          terrain).

sensors were operational. The cloud was nearly centered within the sensor array but extended beyond both sides. However, mass-balance calculations (Koopman et al., 1982) indicate that most of the gas was accounted for within the array. The Burro 8 gas cloud is especially interesting because the very low wind speed permitted the gravity flow of the cold, dense gas to become so dominant that the flow within the cloud was almost decoupled from the surrounding atmospheric boundary layer. The cloud spread in all directions, including upwind, developed a very distinct bifurcated structure, and lingered over the source region for more than 100 s after the spill was terminated.

The present model predicted the maximum downwind distances to the LFL to be ~ 460 m for the case with flat terrain and ~ 400 m for the case with variable terrain, both occurring approximately at t = 300 s. In the experiment, because the gas cloud extended well beyond the edge of the sensor array, no really reliable data was available for estimating X_{LFL}. However, a "puff" of vapor with gas concentration greater than 5% was recorded at the 400-m row instruments at the 3-m level between 380 s and 440 s. Consequently, a value of 420 m was taken as the "best" estimate (Ermak et al., 1982). Again, the agreement between downwind distances to the LFL is good but the predicted time (at which such distance is reached) is considerably different from the field data. The arguments presented in Burro 9 for this disagreement also apply here, but now, any inadequacies in the turbulence and ground heat submodels are also important for this low wind speed case. Furthermore, during the experiment, the ambient wind speed was observed to be decreasing in a fairly steady fashion by about 30% over the duration of the test. This decrease in mean wind speed was not considered in the present numerical calculations.

Time history plots of the NG cloud temperature and concentration at three downwind locations 1 m above the ground are compared in Fig. 7. These locations are near the cloud centerline (y = 0) as indicated by the mean wind direction. However, since the NG cloud was bifurcated, these concentration levels are not the maximum recorded values for each downwind distance. Generally, higher concentration levels were measured in each of the two lobes on either side of the centerline.

In the concentration time history plots shown in Fig. 7a at a downwind distance of 57 m, the flat terrain simulation agrees best with the data. A peak concentration level of just over 40% was measured and a value of just under 40% was predicted in the flat terrain simulation. The variable terrain simulation did not predict this peak and had a maximum value of only about 20% at this location. For the comparison at 140 m downwind, the situation is just

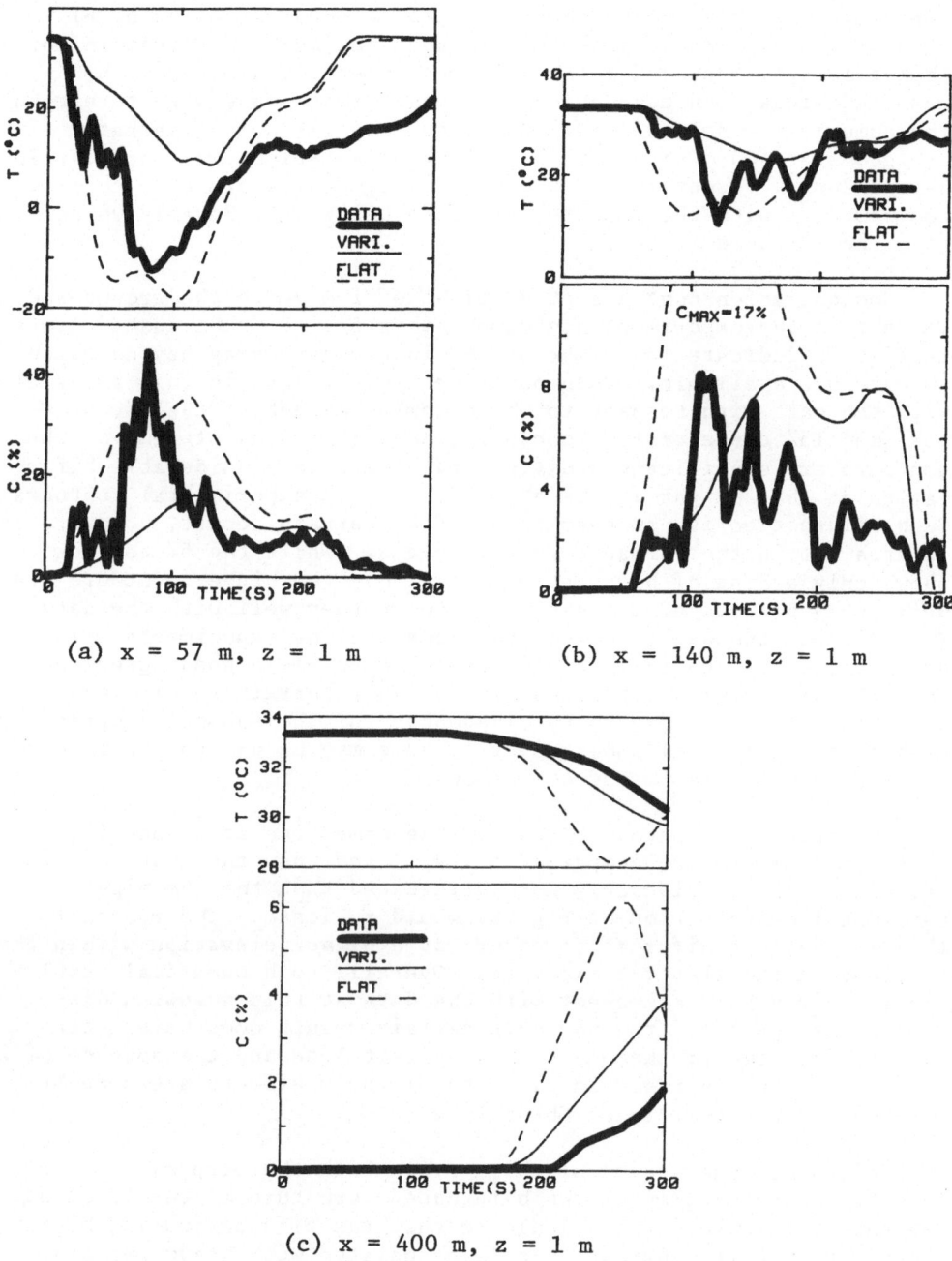

(a) x = 57 m, z = 1 m   (b) x = 140 m, z = 1 m

(c) x = 400 m, z = 1 m

Fig. 7.   Burro 8 time history plots of temperature and concentration at several locations on the center plane.

the reverse. The variable terrain result is in much better agreement with the data, both reaching a peak concentration value of
about 8%. The flat terrain result had a maximum concentration more
than twice the measured value. As was generally the case, the
measured signals in both cases had considerably more structure than
the numerical model calculations. The comparison is somewhat
ambiguous at 400 m since the numerical calculations were not carried
beyond 300 s. However, the two numerical results are seen to rise
too rapidly, with the flat terrain case being considerably worse
than the variable terrain result.

Concentration contours at a height of 1 m above the ground and
180 s into the experiment are compared in Fig. 8. The dashed lines
in Fig. 8a indicate the edges of the instrument array beyond which
no data are available. Both numerical results are in fair agreement
with the data with respect to the downwind extent of the 1%, 5%,
10%, and 15% concentration contours, but the shapes of some of the
contours are significantly different. There is considerable bifurcation at this height in the 5%, 10%, and 15% experimental contours.
Such bifurcation is not seen in the flat terrain result, however, it
is apparent in the variable terrain plots. While the 5% contour
shows only a hint of bifurcation (Fig. 8c), the higher contours are
definitely bifurcated and seem to agree rather well with the data.
In addition, the angle between the lobes of the experimental plot
and the variable terrain solution appears to be in good agreement.
Some of the difference between the variable terrain results and the
data is very likely due to the inadequacy of the submodels currently
used in the numerical code; however, some may be due to the lack of
data between the 140-m and 400-m rows.

The crosswind contour plots for the same time at a downwind
distance of 140 m are compared in Fig. 9 and show the flat terrain
solution to be considerably more bifurcated than the 1 m high,
horizontal contour plot of Fig. 8b would indicate. The reason for
this is that the bifurcation occurs at a higher elevation within the
NG cloud in the flat terrain case. Overall, both numerical results
are in fairly good agreement with the data at this downwind distance. As expected, the variable terrain result does have better
detailed agreement; namely, a higher left lobe and the absence of a
15% contour in the right lobe. Both of these effects are undoubt-
edly due to the terrain at China Lake.

While the interaction of the terrain with the dispersing cloud
appears to have augmented the bifurcated structure of the NG cloud,
the numerical calculations indicate that the bifurcation is chiefly
due to the gravity current usually associated with heavy gas dispersion and the importance of this current relative to the

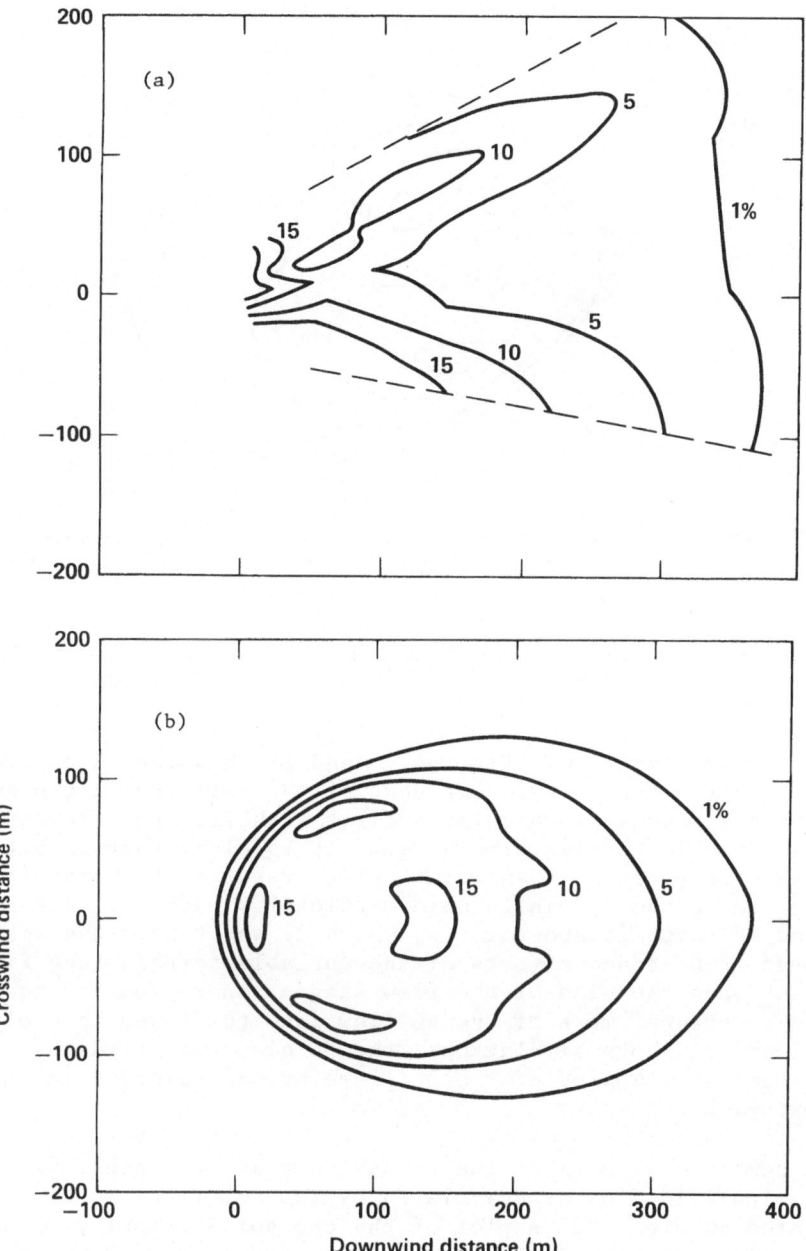

Fig. 8.   Burro 8 contour plots of concentration 1 m above ground
          at t = 180 s: (a) measured; (b) calculated (flat terrain);

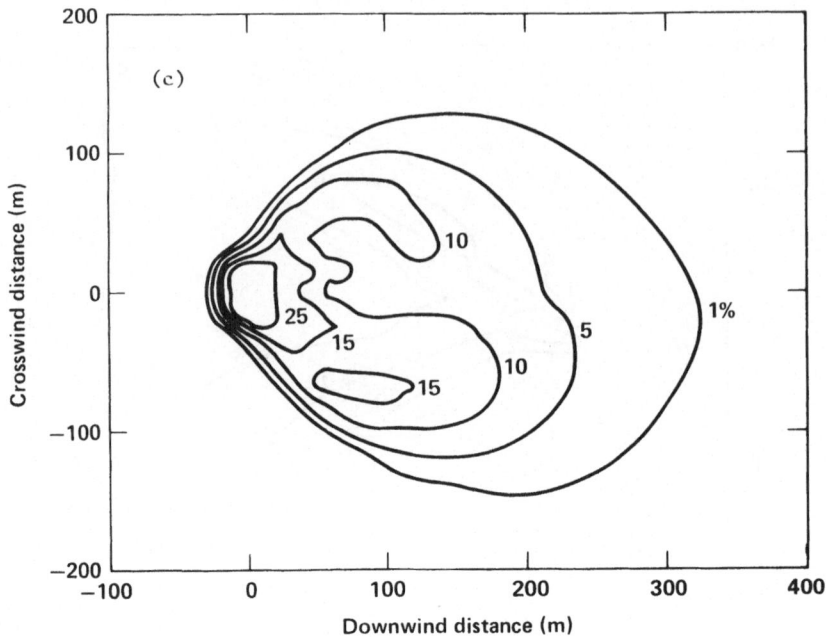

Fig. 8. (Continued); (c) calculated (variable terrain).

surrounding ambient flow. Figures 10 and 11 show the concentration contours and the horizontal component of the velocity in the surface 1 m above the ground as calculated first with flat terrain and then with the variable terrain simulation. It is clear that gravity spreading of the denser-than-air NG cloud results in a significant perturbation to the initially unidirectional windfield. The maximum crosswind velocity is about 1 m/s, which is about half the ambient wind speed. The added effects of the variable terrain (see Fig. 11) are to increase the size of the flow stagnation region in the spill pond and to channel more of the NG flow into the lower lobe of the vapor cloud. In both simulations, the NG cloud was predicted to spread about 40 m upwind of the spill point, as observed in the field experiment.

The combined effects of the crosswind gravity current with the ambient windfield flow as calculated by the numerical model are illustrated in Fig. 12. A plot of the crosswind velocity vectors from the flat terrain solutions for Burro 8 and Burro 9, both with the NG concentration contour superimposed, is given. In both numerical solutions, an outward moving vortex is formed due to gravity spreading of the vapor cloud in the crosswind direction. The maximum crosswind velocity just above the surface is about

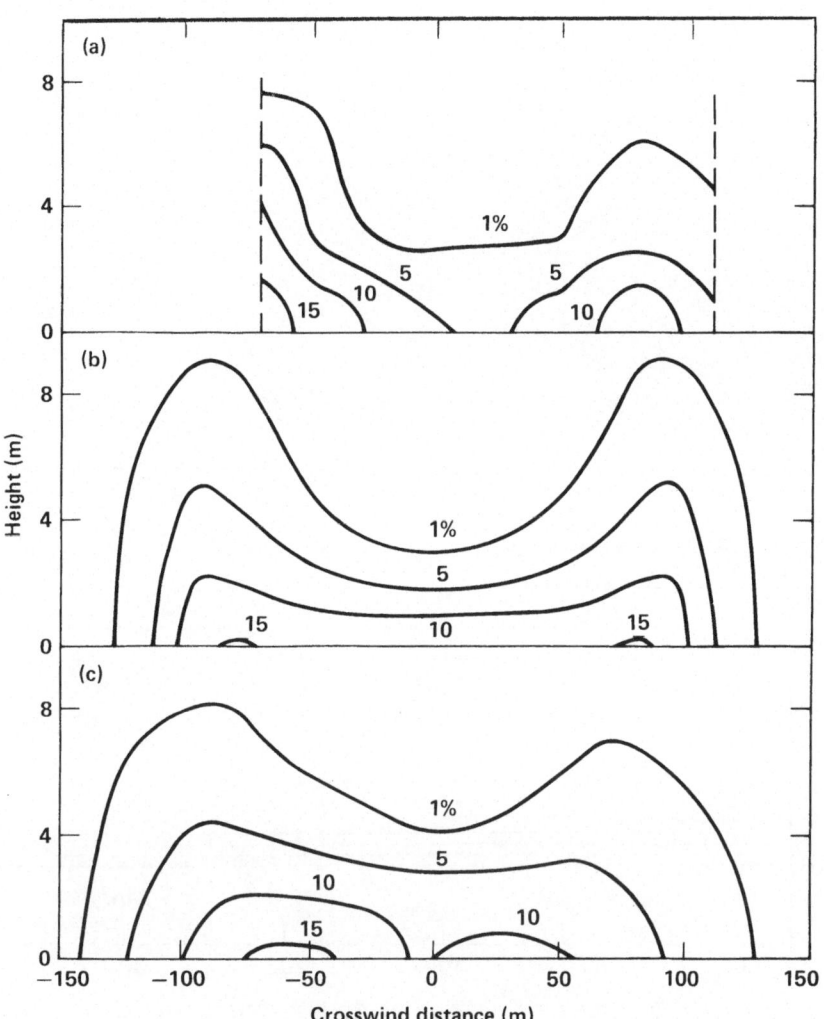

Fig. 9. Burro 8 crosswind contour plots of concentration 140 m
        downwind at t = 180 s: (a) measured; (b) calculated
        (flat terrain); (c) calculated (variable terrain).

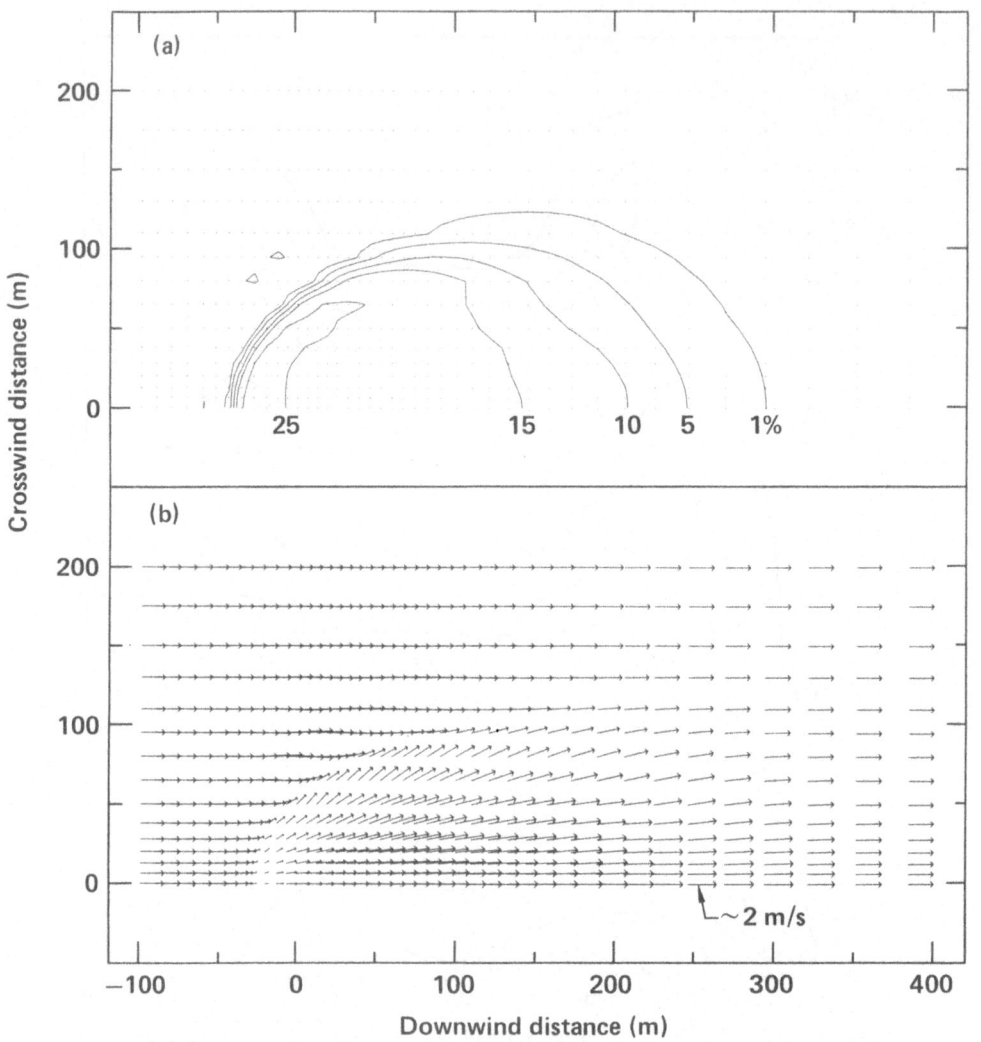

Fig. 10.   Burro 8 calculated results on the plane 1 m above
ground at t = 140 s with flat terrain approximation:
(a) concentration contours; (b) horizontal velocities.

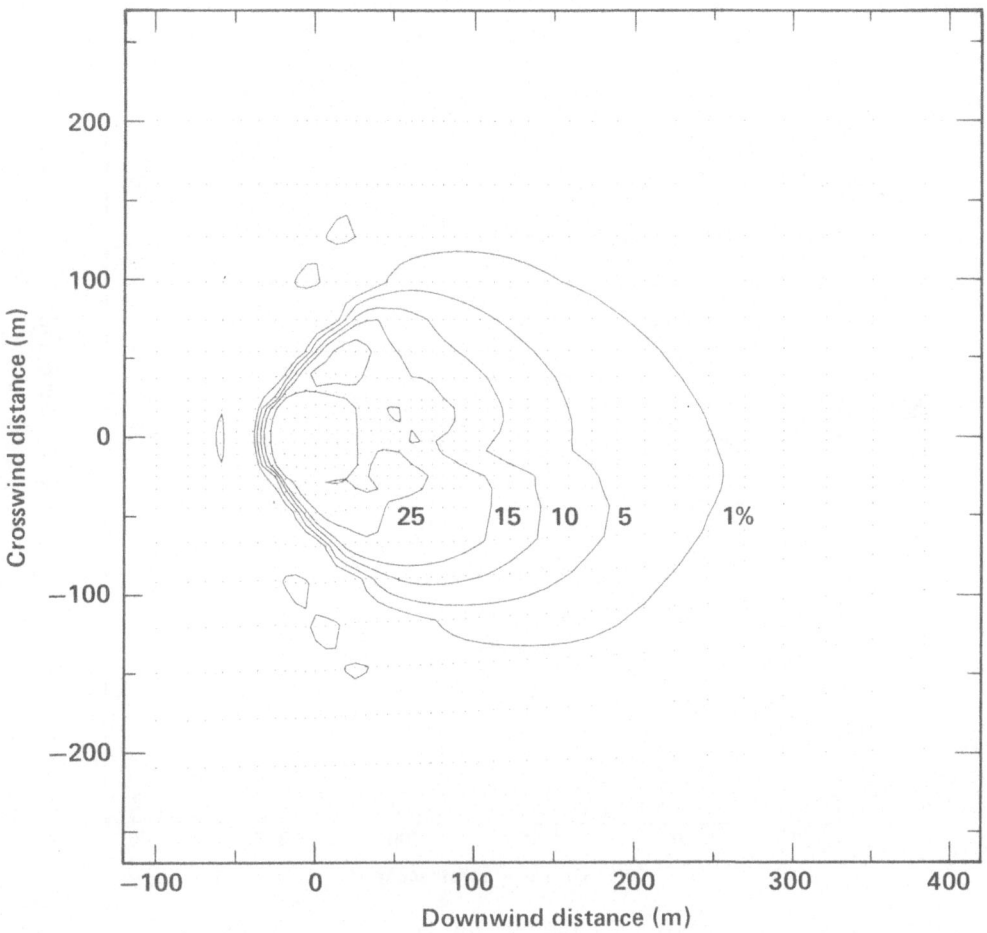

Fig. 11(a). Burro 8 calculated contours of concentration 1 m
above ground at t = 140 s with variable terrain.

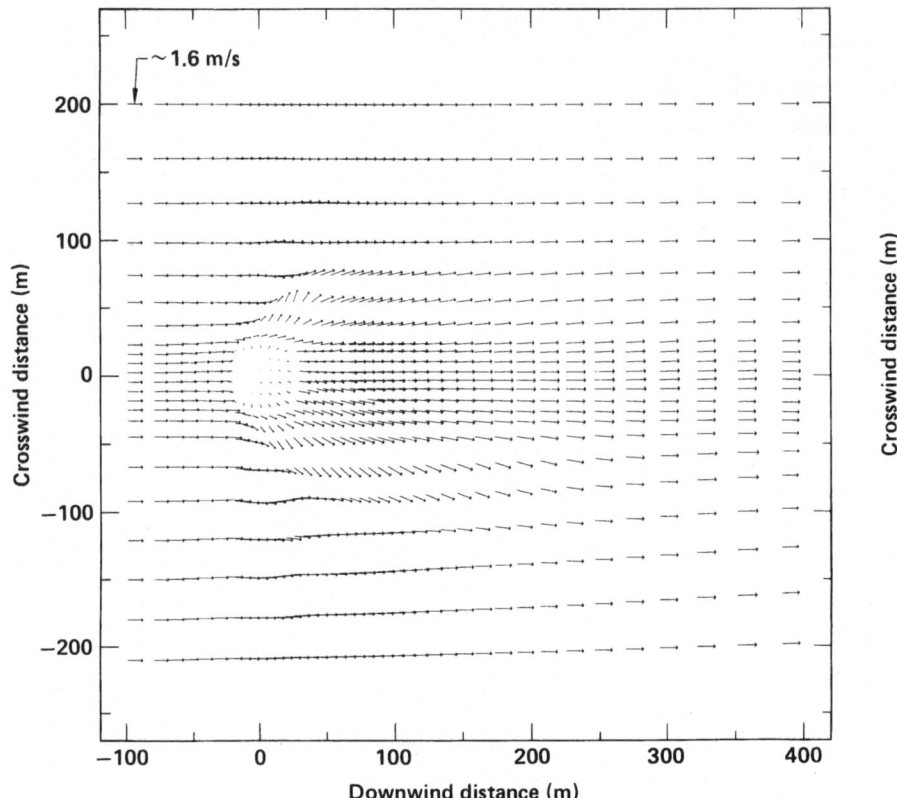

Fig. 11(b).   Burro 8 calculated horizontal velocities 1 m above
ground at t = 140 s with variable terrain.

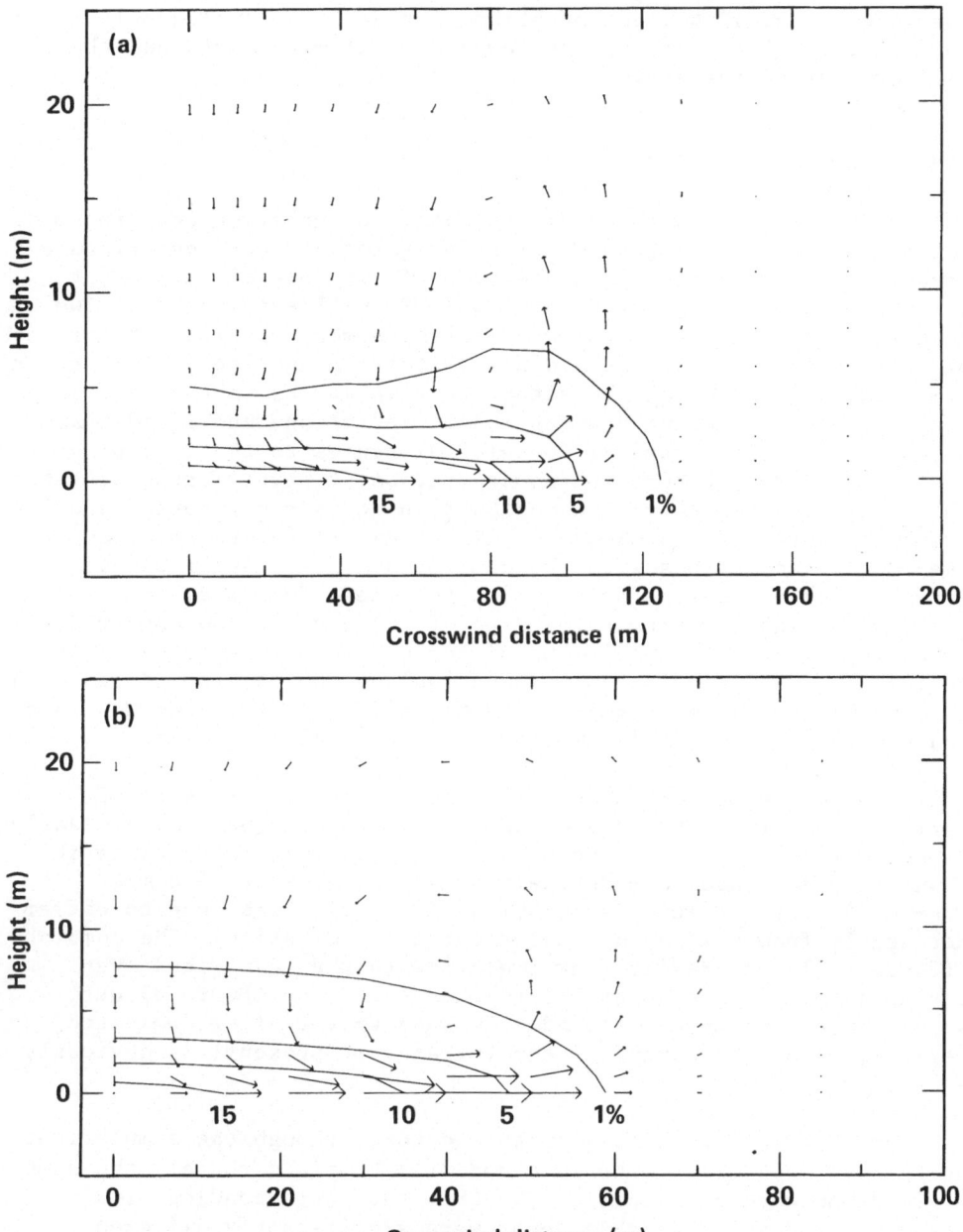

Fig. 12.   Crosswind velocities and concentration contours 140 m
           downwind for (a) Burro 8 and (b) Burro 9, both with
           flat terrain approximation.

0.6 m/s in each case. However, the front of the gravity current
(defined by the 1% NG concentration contour) is elevated only in the
case of Burro 8 because of the lower ambient wind speed and the
smaller eddy diffusivity.

SUMMARY AND CONCLUSIONS

     In this report, a three-dimensional, conservation equation model
for simulating the dispersion of heavy gases has been described and
used to simulate the vapor dispersion of two markedly different
(regarding the role of gravity-flow) LNG spill experiments. Two
numerical simulations of the NG dispersion were carried out for each
experiment. The first assumed a flat terrain and the second used a
numerical simulation of the actual terrain at the test site. In
general, good agreement between model predictions and field measurements, regarding maximum downwind distances to the LFL, time
histories of temperature and concentration at several representative
locations, and concentration contours on certain horizontal and
crosswind surfaces was observed. The overall results obtained in
the model calculations with the simulated actual topography were
shown to correlate much better with the field data and, in
particular, many important features of the vapor cloud observed
under the light wind conditions of Burro 8 were reproduced in the
variable terrain simulation. These include the vortex-induced high
concentration regions resulting in the bifurcation of the NG cloud
and the deflection of the NG cloud due to sloping terrain.

     Through the present numerical simulations, the effects of
variable terrain on the dispersion of heavy gases have been clearly
demonstrated. Even with the relatively mild terrain features at
China Lake and under a moderately high wind speed of $\sim$ 6 m/s
(Burro 9), the resulting vapor cloud dispersion was seen to differ
noticeably from that using a flat terrain simulation. The combined
effects of large gravity-flow (relative to the mean wind) over
variable terrain and under light wind conditions (Burro 8) were
shown to be even more profound. In such gravity-flow dominated
regimes, proper treatment of the terrain, if present, is obviously
necessary.

     Our numerical results also showed that, though the simulations
using flat and variable terrain appeared to yield roughly the same
maximum downwind distances to the LFL, the corresponding cloud
structure and the flammable zones were significantly different.
This suggests that it is generally not sufficient to judge the
performance of a numerical model merely on its accuracy in predicting the downwind distance to the LFL. Any prudent assessment
should also consider the model's ability (or inability) in accurately predicting the cloud structure, including height, width, and

detailed concentration contour levels, and time histories of the field variables.

Although the present model is, by and large, performing reasonably well, further improvements are still desireable in order to model the physics more accurately, especially for cases involving flow regimes such as that of Burro 8. These include better parameterizations of the turbulent mixing, heat transfer from the underlying surface, and an improved vapor generation submodel at the source. Such refinements, of course, would require more systematic numerical studies and the availability of relevant field data. Additionally, for spill scenarios involving relatively high humidity or rain, humidity effects must also be included.

Thus far, predictions of the present model have been compared only with field tests representing a limited range of spill scenarios. Additional comparisons with well-instrumented, field-scale dispersion experiments, including releases of cold, ambient temperature, pressurized, and even chemically reacting dense gases, are obviously needed. Only after being carefully validated can the model be applied with confidence to simulate the atmospheric dispersion of a variety of denser-than-air gases when spilled at rates of practical interest and over a broad range of atmospheric conditions.

ACKNOWLEDGEMENTS

The authors are indebted to all of those who participated in the Burro series of LNG spill experiments. We are particularly grateful to P.M. Gresho for numerous consultations during the development of the numerical model, and to R.T. Cederwall and E.J. Kansa for many helpful discussions regarding the turbulence submodel. We also wish to thank C.D. Upson for his assistance in programming and graphics support. This work was performed under the auspices of the U.S. Department of Energy by the Lawrence Livermore National Laboratory under Contract No. W-7405-Eng-48.

REFERENCES

Businger, J.A., J.C. Wyngaard, I. Izumi, and E.F. Bradley, 1971, Flux-Profile Relationships in the Atmospheric Surface Layer, J. Atmos. Sci., 28, 181-189.

Chan, S.T., P.M. Gresho, R.L. Lee, C.D. Upson, 1981, A Simulation of Three-Dimensional, Time-Dependent, Incompressible Flows by a Finite Element Method, Proceedings of the AIAA 5th Computational Fluid Dynamics Conference, 354-363.

Chan, S.T., P.M. Gresho, and D.L. Ermak, 1982a, A Three-Dimensional
    Conservation Equation Model for Simulating LNG Vapor Dispersion
    in the Atmosphere, Liquefied Gaseous Fuels Safety and Envi-
    ronmental Control Assessment Program: Third Status Report,
    Pacific Northwest Laboratory, PNL-4172, E1 to E17.

Chan, S.T. and P.M. Gresho, 1982b, Solution of the Multi-
    Dimensional, Incompressible Navier-Stokes Equations Using
    Low-Order Finite Elements and One-Point Quadrature, Proceedings
    of the 4th International Symposium on Finite Elements in Flow
    Problems, Tokyo, Japan.

Daley, B. and W. Pracht, 1968, Numerical Study of Density-Current
    Surges, Phys. of Fluids, 11, 1, 15-30.

Dyer, A.J., 1974, A Review of Flux-Profile Relationships, Boundary-
    Layer Meteorol., 7, 363-372.

Dyer, A.J. and E.F. Bradley, 1982, An Alternative Analysis of
    Flux-Gradient Relationships at the 1976 ITCE, Boundary-Layer
    Meteorol., 22, 3-19.

Eidsvik, K.J., 1980, A Model for Heavy Gas Dispersion in the
    Atmosphere, Atmospheric Environment, 14, 769-777.

Engelman, M.S., R.L. Sani, and P.M. Gresho, 1982, The Implementation
    of Normal and/or Tangential Boundary Conditions in Finite
    Element Codes for Incompressible Fluid Flow, Int. J. for Num.
    Meth. in Fluids (to appear).

Ermak, D.L., S.T. Chan, D.L. Morgan, and L.K. Morris, 1982, A
    Comparison of Dense Gas Dispersion Model Simulations with Burro
    Series LNG Spill Test Results, J. of Haz. Material, Vol. 6, Nos.
    1-2.

Gresho, P.M., C.D. Upson, S.T. Chan, and R.L. Lee, 1982, Recent
    Progress in the Solution of the Time-Dependent, Three-
    Dimensional, Incompressible Navier-Stokes Equations, Proceedings
    of the 4th International Symposium on Finite Elements in Flow
    Problems, Tokyo, Japan.

Kato, H. and O.M. Phillips, 1969, On the Penetration of a
    Turbulent Layer into a Stratified Fluid, J. Fluid. Mech. 37,
    643-655.

Knox, J.B., 1982, The Modeling of Dispersion of Heavy Gases,
    NATO/CCMS 13th International Technical Meeting on Air Pollution
    Modeling and Its Applications, Ile d'Embiez (Toulon), France.

Koopman, R.P., 1982, Coyote Series for 40-m^3 Liquefied Natural
    Gas (LNG) Dispersion, RPT, and Vapor Burn Tests, Lawrence
    Livermore National Laboratory Report UCID-19211, Rev. 1,
    Livermore, California.

Koopman, R.P., R.T. Cederwall, D.L. Ermak, H.C. Goldwire, Jr.,
    J.W. McClure, T.G. McRae, D.L. Morgan, H.C. Rodean, and J.W.
    Shinn, 1982, Analysis of Burro Series 40-m^3 LNG Spill
    Experiments, J. of Haz. Materials, Vol. 6, Nos. 1/2.

Lee, R.L., P.M. Gresho, S.T. Chan, and C.D. Upson, 1981, A
    Three-Dimensional, Finite Element Model for Simulating
    Heavier-than-Air Gaseous Releases over Variable Terrain,

NATO/CCMS 12th International Technical Meeting on Air Pollution
    Modeling and Its Applications, Menlo Park, California.
Ogura, Y., and N. Phillips, 1962, Scale Analysis of Deep and
    Shallow Convection in the Atmosphere, J. Atmos. Sci., 19,
    173-179.
Tennekes, H., 1973, A Model for the Dynamics of the Inversion above
    a Convective Boundary Layer, J. Atmos. Sci., 30, 558-567.
Zeman, O., 1979, Parameterization of the Dynamics of Stable
    Boundary Layers and Nocturnal Jets, J. Atmos Sci., 36, 792-804.
Zeman, O., and H. Tennekes, 1977, Parameterization of the Turbulent
    Energy Budget at the Top of the Daytime Atmospheric Boundary
    Layer, J. Atmos. Sci., 34, 111-123.

DISCUSSION

P.K. MISRA   Could you explain the criteria
you use for comparison of your model with single
field experiments ? It seems to me that due to the
turbulent nature of the problem you need to have several releases under similar conditions to make a
meaningful comparison of model results, specially
for mean variables.

S.T. CHAN   In theory, due to the turbulent
nature of the problem, several (or even more) releases under similar conditions should be conducted in
order to obtain useful mean values of the field variables. In practice, however, this is not feasible.
Firstly, it is extremely difficult, if not impossible,
to have similar (meteorological) conditions during
such field tests (unlike laboratory tests, we have
virtually no control over the meteorological conditions). Secondly, even when similar conditions exist,
field tests are usually quite expensive and repeating them can easily become economically un-
affordable.
Therefore, a practical, yet reasonable approach is to
process the data (with turbulence components) by using
a moving-time average. Currently we are using a 10-
second moving average, which appears to have worked
satisfactorily in that it is long enough to smooth out
short wavelength (much less than cloud width) fluctuations, but short enough to preserve cloud meander.

D.R. BLACKMORE   I observe from the film of
                Burro 8 that the bifurcation only occurred late in
                the evolution of the cloud, shortly after the source
                was turned off. Do you think this cessation of the
                source of LNG could have contributed to the bifurcation?
                Do you think that the decreasing windspeed during this
                spill could also have contributed to the bifurcation ?

S.T. CHAN   It is unlikely that termination
                of the source could have contributed to cloud bifur-
                cation because the momentum change associated with
                turning off the source is conceivably quite insigni-
                ficant. Another reason is that, should your hypothesis
                be true, we should have observed cloud bifurcation in
                the Burro 9 experiment also, because this test had
                comparable spill rate and spill duration. In fact, we
                observed no apparent cloud bifurcation during the
                entire Burro 9 experiment. The decreasing in wind
                speed during the Burro 8 experiment, however, could
                have contributed to the cloud bifurcation, because
                it makes the existing gravity. How became more domi-
                nant and hence the cloud bifurcation could have been
                enhanced.

H. VAN DOP   Is there any-more than academic-
                use in detailed modeling of heavy gas dispersion,
                when terrain properties are of that paramount impor-
                tance for the dispersion process?

S.T. CHAN   If enough physics and adequate
                numerics are included in a code to account for the
                terrain properties, detailed modeling of heavy gas
                dispersion can indeed be very useful in practical
                applications. For cases involving significant ter-
                rain, its proper treatment is probably mandatory in
                many applications. While we do not claim to have
                tackled the terrain problem completely, we believe
                our present model is quite useful for practical ap-
                plications, not just of academic interest.

DIFFUSION OF HEAVY GASES

F. Fanaki

Atmospheric Environment Service

Downsview, Ontario, Canada, M3H 5T4

INTRODUCTION

During the past few years there has been a growing interest in utilizing heavy gases as energy sources. These gases are usually found in remote areas. In order to bring the gases to market, they are liquified (e.g. liquified natural gas LNG and liquified petroleum gas LPG), transported and stored in refrigerated pressurized tanks. Because this type of fuel is both flammable and toxic, an accidental release has serious consequences.

Until recently there have been few investigations of such accidental releases. However, some studies have been carried out to estimate the spread of the toxic, flammable clouds from accidental spills of such fuels (see e.g. Segeant and Robinett 1968, Burgess et al 1970 and 1972, van Ulden 1974, Germeles and Drake 1975, Eidsvik 1978 and Picknett 1978 and 1981). An excellent review of this subject is presented by Eidsvik 1980.

Eidsvik (1980) describes the spread of the instantaneously released heavy gas cloud as a process which occurs in three phases. The first phase is characterized by a vortex ring at the rim of the cloud. The vortex ring contains a significant portion of the released gas. In the intermediate phase, because of gravity, the dense gas continues to spread radially. The radius of the cloud is larger than its height. The final phase is passive; its diffusion is governed by atmospheric conditions and topography. It is believed that during this phase the cloud is diluted considerably to a non-hazardous concentration. The transition between each phase is not well defined.

There are other factors as important as the properties of the gas influencing the behaviour and dispersion of the released gas. These factors are wind speed, wind direction, topography and thermal stability of the atmosphere. Under unfavourable conditions, the released gas may reach populated regions with a hazardous concentration. The meteorological phenomena responsible for the unfavourable conditions occurs quite frequently in a cold climate such as in the Canadian winter.

This paper describes a pilot field study to investigate the dispersion of heavy gas under typical summer conditions and will be repeated for the winter season.

EXPERIMENTAL APPARATUS AND PROCEDURE

The dispersion study was conducted in a large grass covered field on the grounds of the Woodbridge Research Station, Ontario, Canada in July 1982. The relative location of the heavy gas source, meteorological towers and sampling equipment is illustrated schematically in Figure 1. During a typical study, heavy gas was released from a circular source at the ground. At the same time, measurements of vertical temperature and wind profiles were made. Concentrations of the released heavy gas were monitored at the ground and at 10 cm above the ground for different downwind distances.

During an accidental release of heavy gas the source takes on different configuration. It ranges from a spilled pool on the ground to a flammable cloud from an erupted pressurized tank. The

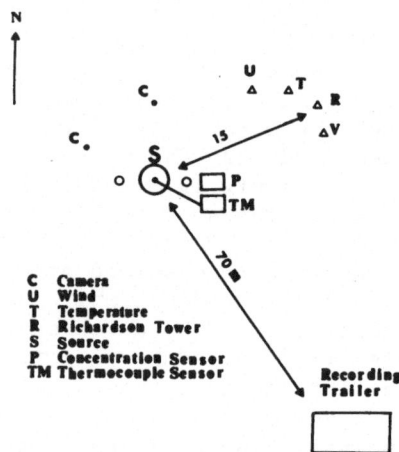

Fig. 1.   A schematic diagram of the layout of the experimental study area at Woodbridge, Ontario.

mode of the release in most cases is instantaneous and in the form
of a puff. In this study an instantaneous type source was simulated. A topless and bottomless cylinder made of insulated plastic
was used to trap the heavy gas at the source location (Fig. 1).
The cylinder was 0.71 in diameter and 0.74 m in height and hung
from specially designed towers (4 m) with an array of pulleys that
were used to lift the cylinder instantaneously from above. The
gas released took the shape of a right-cylinder with its base at
the ground. The gases used in this study were CO_2 and NO_2 obtained
from a fire extinguishing device and liquid nitrogen. Upon release
the gas was cooled to $-15°C$ and it had a density of about 1.5 times
that of air. Gas concentration was measured at the source and at
different downwind distances using a Meloy*. The concentration
sensor was operated just before release and continued to operate
until the heavy gas was completely dispersed.

In order to determine the rise and behaviour of the heavy gas
plume, a photographic technique was applied. This technique is
simple, economical and provides a permanent record at any given
instant on the shape of the puff. Before each release the wind
direction was determined and two cameras were set up. The axis of
one of the cameras was perpendicular to the wind direction and that
of the other camera was parallel to the direction of the wind. Due
to the rapid change of the plume structure many photographs were
required. A series plan and side views of the released puff were
obtained during each experiment. The cameras were operated man-
ually as soon as the gas was released (1 frame sec^{-1}). A thermocouple mounted on the ground at the source was used to monitor the
puff temperature before release.

Micrometeorological data including wind speed, wind direction
and temperature were measured close to the spill area in the experimental field.

The field was flat and uniformly rough, covered with grass
(1-3 cm in height) for at least 200 m in all directions. Trees
(5-10 m), shrubs and cornfields were scattered around the site.
The nearest tall building was 1 km away. The highest building
close to the source was the observation trailer (3 m) at a downwind
distance of 70 m.

The meteorological instruments were mounted on three ($\approx$3 m)
towers and were located close to the spill sites.

----

*The Meloy (model SA285 FPD) is usually used to monitor
SO_2, however the one used in this study measured CO_2 as
well, but not in absolute values.

$$U = \frac{U_*}{k} \; (\ln \frac{Z}{Z_o} - \psi(\frac{Z}{L})) \tag{1}$$

where U_* is the friction velocity, k is von Karman constant (=0.4) and Z_o is the roughness height. $\psi(\frac{Z}{L})$ is a correction function of stability and L is the Monin-Obukhov scaling length. For unstable conditions L is related to the Richardson number (Ri) by the following

$$\frac{Z}{L} = \frac{Ri}{1-5.2Ri} \tag{2}$$

Here Z is an effective height $= \sqrt{Z_1 Z_2}$ where Z_1 and Z_2 are two levels of observations. For Ri<o, ψ is given (Hicks 1976) by:

$$\psi(\frac{Z}{L}) = \exp \; (0.032+0.448 \; (\ln\frac{-Z}{L}) - 0.09 \; (\ln \; \frac{-Z}{L})^2) \tag{3}$$

The values of important micrometeorological parameters that were extracted from the wind and temperature measurements are listed in Table 1. Values of Z_o and U_* ranged from 1.1 to 1.5 cm and from 0.18 to 0.3 m sec^{-1} respectively. In all the runs of this study the atmosphere was unstable (Ri<o).

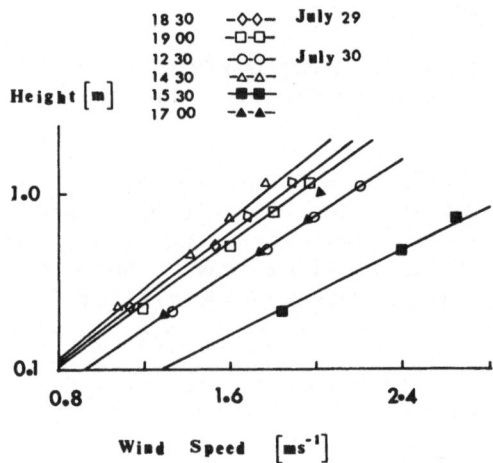

Fig. 2.   Change of wind speed as a function of height.

Wind speeds were measured with three-cup anemometers, mounted at 6 levels; 0.23, 0.49, 0.77, 1.17, 1.69, 2.67 meters above the ground (U, Fig. 1). In addition, a vane anemometer at 2.64 m level was used continuously to monitor wind speed and wind direction (V, Fig. 1). Temperature measurements were made using aspirated linear thermistor probes at heights 0.24, 0.5, 0.78, 1.2 and 1.7 on the T tower (Fig. 1). Wind and temperature data were computer processed and analyzed at the site, yielding wind and temperature profiles in a tabular form and were averaged over 5 min.

A third tower (R Fig. 1), to measure the bulk Richardson number (Ri), consisted of two three-cup fast-response anemometers and a Delta-T Sonde Sensor. This is a high resolution system developed by AES. The Delta-T sensor measured the temperature difference $(T_2 - T_1)$ between the two anemometer levels (Z_1, Z_2). It consisted of two identical fine-wire thermopiles referenced to one another and held at a fixed location by a 1 m long solid rod. The Delta-T output signal, is modulated and transmitted to a UHF receiver connected to a tape recorder. The tape recorder is used in order to permit the subsequent computer analysis of the data. The Delta-T sensor has a time constant of 3×10^{-3} sec and $\pm 0.01°C$ accuracy for temperature differences.

The miniature cup anemometers were located next to each thermopile. They were calibrated in a wind tunnel. They have a low starting speed (<0.1 m sec^{-1}), a small response length <0.5 m and a damping rate of 0.5. The output from each anemometer was stored on magnetic tape for computer analysis. The Richardson numbers were measured in this study at an effective level given as $\sqrt{Z_1 Z_2} = \sqrt{0.62 \times 1.62} \approx 1$ m above the grass.

EXPERIMENTAL OBSERVATIONS AND DISCUSSION

Releases of heavy gas were performed during two days of the four day field study. The observed 15 min average wind speed and direction at 2.6 m level during the releases were examined as function of time. During the study period wind speed at the 2.6 level varied from 0.5 to 2.6 m sec^{-1} and averaged 1.8 m sec^{-1}. This average signifies a relatively calm condition. Wind direction was typical of the site for the month of July (i.e. northeast-southwest).

Profiles of wind speed and temperature were constructed from the tower data. Only those runs that coincided with the heavy gas releases were analyzed. The observed wind speeds (U) were plotted in Fig. 2. The results indicate that U can be described by the relation (see e.g. Panofsky 1963),

Table 1. Micrometeorological Characteristics of the
Boundary Layer During the Heavy Gas Release Study

DATE	TIME	U_* (m sec^{-1})	Z_o (cm)	(Ri) 1 m	$(\frac{Z}{L})$ 1 m	Stability
July 29	1830	0.19	1.5	−0.3	−0.12	Unstable
July 29	1900	0.19	1.2	−0.3	−0.12	Unstable
July 30	1230	0.21	1.1	−2.1	−0.18	Unstable
July 30	1430	0.18	1.5	−0.7	−0.15	Unstable
July 30	1530	0.30	1.3	−0.6	−0.14	Unstable
July 30	1700	0.21	1.1	−1.0	−0.16	Unstable

GAS DISPERSION

Under calm conditions and immediately after release of the gas (CO_2 or NO_2), the puff took the shape of a cylinder. Shortly thereafter, the puff collapsed at the center and started to spread radially to form a disc with a raised annular rim. Fig. 3 shows a photograph of the cloud puff 1 sec after release. The width of the rim varied as the puff expanded. Initially, the width was about 1/3 the puff's radius and decreased gradually. It appeared to move in a highly turbulent fashion involving considerable mixing with ambient air. In a few seconds (4-5 sec) the rim collapsed and the puff expanded as a flat disc. This marked the end of the first phase and the beginning of the second phase. Finally the puff became passive and dispersed at a rate depending on Ri and the atmospheric turbulent intensity. Similar observations have been obtained by Picknett (1981).

Under low wind conditions the dispersion of the puff was different. The vortex rim observed under calm conditions was absent. During the first phase of dispersion the puff took an irregular shape where its upper envelope was torn by rising filaments. These filaments mixed rapidly with the air above.

One expects that it would be difficult to identify the three phases of dispersion under such conditions. Initially, the puff dispersed radially at a faster rate than along the direction of the wind. Later, however, the puff expanded more rapidly downwind than crosswind.

A sample of concentration measurements at two levels Z=0 and Z=10 cm is shown in Fig. 4. As expected the concentration is highest at ground level. As the puff moved downwind the concentration decreased.

Fig. 3. Photograph of the puff shortly after release.

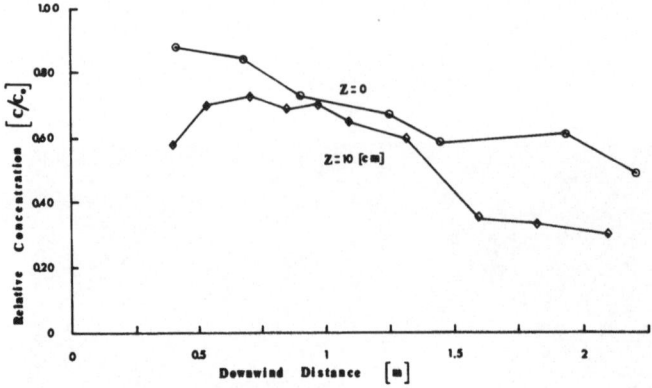

Fig. 4.   Change of relative concentration at different
          downwind distances (C_0 = source concentration).

The time variation of the concentration is presented in Fig. 5
(a and b) where two traces of concentration are given for two different dispersion phases. In the initial phase (Fig. 5a), the trace
shows a sharp increase at the rim of the puff and relatively gradual decrease towards the rear of the puff. At a later stage, the
profile of concentration shows a gradual increase with time (Fig.
5b). It has a relatively broader shape with lower maximum concentration. The traces from other runs have shown similar types of
fluctuation but with a large level of variability in the level of
the peak concentration.

Several models based on experimental observations have been
developed to describe the spread of the dense cloud (see e.g.
van Ulden 1974, Eidsvik, 1978, Rosenzweig, 1980, Fay, 1980). It
appears that each author of these models has interpreted the measurements differently (see e.g. Murphy, 1974 and Fay, 1973).
Moreover, the existing models are sensitive to a large number of
parameters including source type, meteorological variables, type
of heavy gas and the changes of the dispersion coefficients.

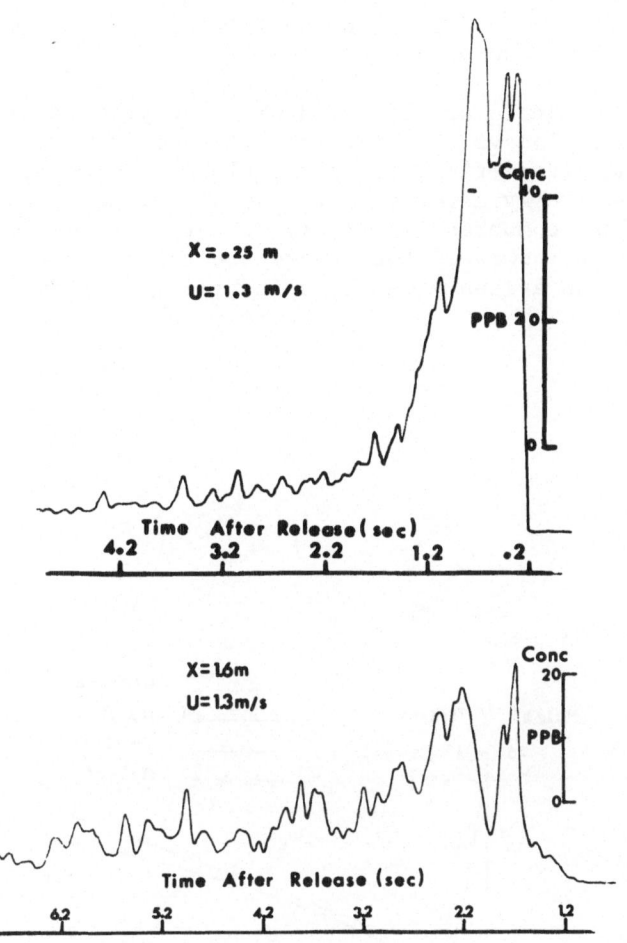

Fig. 5. Time history of the changes of concentration.

In general however, the puff radius R and height H are related to time, t, (see e.g. van Ulden 1974 and Raj 1982) by

and

$$R^2 \ \alpha \ \ t$$

$$H \ \ \alpha \ \ \frac{1}{\sqrt{t}}$$

Jacobsen and Fannelǿp (1982) have shown that $R\alpha(a+bt)^{2/3}$ where a and b are constants depend on the initial radius, initial height of the puff and the wind speed.

The above models are compared with the present measurements of R in (Fig. 6). The values of H are not used in this comparison because visual identification of the cloud becomes more difficult with time, generally after 4 sec. The second model (Jacobsen and Fannelǿp (1982) compares relatively better with the observation. One should note, however, that there are too few data points to provide concrete evidence on the suitability of these models.

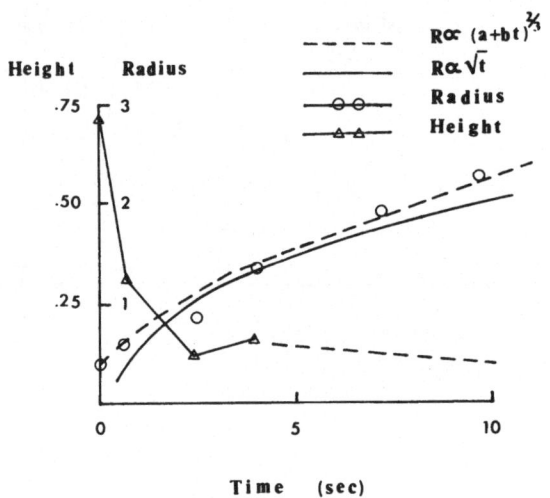

Fig. 6.   Variation of R and H with time.

CONCLUDING REMARKS

Several experiments releasing an instantaneous puff of heavy gas have been carried out under a limited range of atmospheric conditions. In spite of this limitation, characteristic features on the dispersion of heavy gas have been obtained. In particular, the identification of different (stage-phases) after release. One drawback that has been recognized is the limitation that arises due to the reduction in size of the source.

An exact model of the actual accidental spill of LNG, for example, on ground cannot be achieved with small scale sources such as the one described here since the boiloff rates and evaporation of the gas are difficult to model with a small source. This problem has been ignored in this study but will be examined in detail in future work.

The experiments have provided information on the formation, expansion and dilution of the heavy gas cloud. The cloud's behaviour differs with atmospheric conditions. The cloud at its earliest stage is highly turbulent especially at the rim where mixing with the ambient air is strong. At a later stage, the top layer of the cloud is irregular, here mixing is governed by atmospheric conditions.

Since experimental works on actual spills of toxic chemicals are prohibitively expensive and, in some cases, impossible, these simulation experiments are of great value. Attention, however, must be paid when translating the result of the field test to an actual spill.

REFERENCES

Burgess, D, Murphy, J.N. and Zabetakis, M.G., 1970, "Hazards associated with the spillage of liquefied natural gas in water." U.S. Bureau of Mines Rep. No. 7448.

Burgess, D., Boirdi, J. and Murphy, J., 1972. "Hazards of spillage of LNG into water." U.S. Bureau of Mines MIPR Z 70099-9-1239.

Eidsvik, K.J., 1978. "Dispersion of heavy gas clouds in the atmosphere." NILVOR32/78, Lillestrøm, Norway.

Eidsvik, K.J., 1980. "A model for heavy gas dispersion in the atmosphere." Atmos. Envir. 14, pp. 769-777.

Fay, J.A., 1973. "Unusual fire hazard of LNG tanker spills." Com. Sc. and Tech., 7:47-49.

Fay, J.A., 1980. "Gravitational spread and dilution of heavy vapour clouds." In: 2nd Int. Symp. on stratified flows, T. Carstens and T. McClimans, eds. Norwegian Institute of Technology, June 24-27.

Germeles, A.E. and Drake, E.M., 1975. "Gravity spreading and
    atmospheric dispersion of LNG vapour clouds." 4th Symp.
    on Transport of Hazardous Cargoes by Sea and Inland Water-
    ways." Jacksonville, Florida, pp. 519-539.
Jacobsen, O. and Fannel∮p, T.K., 1982. "Gravitational spreading
    of heavy gas clouds." Part II Lecture series von Karman
    Institute for Fluid Dynamics, Belgium, Mar. 8-12.
Murphy, J.M., 1974. "Comments on draft environmental impact
    statement". Final Environmental Impact Statement, Vol. 2.
    U.S.A.
Panofsky, H.A., 1963. "Determination of stress from wind and
    temperature measurements." Quart. J. Roy. Meteorol. Soc.
    89, p. 85.
Picknett, R.G., 1978. "Field experiment on the behaviour of dense
    clouds." Report Ptn. 1L 1/54/78/1, Chemical Defence Est.
    Porton Down, Salisbury Wilts, England.
Picknett, R.G., 1981. "Dispersion of dense gas puffs released in
    the atmosphere at ground level." Atmos. Envir. 15, pp. 509-
    525.
Raj, P.K., 1982. "Heavy gas dispersion - a state of the art
    review of the experimental results and models." Lecture
    series von Karman Inst. for Fluid Dynamics, Belgium,
    Mar. 8-12.
Rosenzweig, J.J., 1980. "A theoretical model for the dispersion
    of negatively buoyant clouds." Ph.D. Dissertation, Mass.
    Inst. of Techn.
Segeant, R.J., and F.E. Robinett, 1968. "An Experimental Inves-
    tigation of the Atmospheric Diffusion and Ignition of Boil-
    off Vapors Associated with a Spillage of Liquefied Natural
    Gas." Report No. 08072-7, Prepared for American Gas
    Association, Inc., New York.
van Ulden, A.P., 1974. "On the spreading of a heavy gas released
    near the ground." 1st Int. Loss Prevention Symp., The Hague/
    Delft Elsevier, Amsterdam, pp. 221-226.

DISCUSSION

E. RUNCA   Does the set of dimensionless
            parameters and functions you have shown depend on the
            size of the heavy gases clouds ?

F. FANAKI   Yes, consider for example the
            Froude number which is an important parameter for
            such study. The Froude number is given as $\frac{gl}{U^2}$ where
            l is a scaling length, U is the wind speed and g is
            the acc. due to gravity; l in this case may be re-
            placed by 1/3 (i.e. the volume of the spilled gas).
            Through the use of the Froude number it is possible
            to translate the results of experiment with small

spilled volume, like ours, to large scale one provided the Fronde number similarity is retained.

T. MIKKELSEN ·   In fig. 6 in your presentation,
your last data point of the height indicates that the
cloud has begun to grow after the slumping phase.
Why do you then extrapolate the data points with a
decreasing function ?

F. FANAKI   The reason for that type· of
extrapolation is that the puff in many occasions was
still slumping. However, I believe at a later time
the puff will be passive. Its behaviour will be
governed only by atmospheric variables and its rise
will depend on the magnitude of the atmospheric
vertical motion.

ENTRAINMENT THROUGH THE TOP OF A HEAVY GAS CLOUD,

NUMERICAL TREATMENT

Niels Otto Jensen and Torben Mikkelsen

Physics Department
Risø National Laboratory
DK-4000 Roskilde, Denmark

INTRODUCTION

After termination of the slumping phase of a heavy gas cloud, the height slowly starts to grow again. In Jensen (1981) it was shown that this cannot be the result of entrainment through the side wall but must be due to entrainment through the cloud top. Further, conservation equations for the density excess, $\Delta\rho$, of the cloud relative to ambient air as well as for temperature difference, ΔT, between cloud and surface were given. For a cold cloud, qualitative arguments were derived to the effect that the density-jump Ri-number, controlling entrainment, would vanish faster than the temperaturejump Ri-number, controlling the general level of incloud turbulence, thus predicting that vertical growth should be an accelerating function of time in this case.

The present paper gives a full numerical treatment of the problem, and vertical growth of clouds initially of equal $\Delta\rho$ but different ΔT is compared. In evaluating the effect of ΔT it is as in Jensen (1981) assumed that the wind speed advecting the cloud along is large enough to make the assumption of a constant surface temperature valid. In case of very low wind speeds and in cases where the release is rather like a plume, a heat budget for the soil would have to be taken into account. The two following paragraphs discuss other assumptions made in the model and the two last paragraphs recapitulize the system of equations which is being solved and discuss the results obtained.

IN CLOUD MECHANICAL TURBULENCE

The values of the advection speed of the cloud and the friction velocity u_{*i} inside the cloud was not given any discussion in Jensen (1981). We will here offer some remarks in order to justify the assumptions made.

The absolute size of the advection speed is not considered important in the present context provided it is large enough for the reason mentioned above, as we are only interested in the development of the cloud relative to its center of mass. Hence the advection speed of the cloud is put equal to the wind speed at some arbitrary level. However, its possible variation may have some implication for u_{*i}.

When the gas cloud results from a pressurized liquid blow-up, the initial entrainment of air during the cloud formation may be 10 times the pure gas volume (van Ulden, 1974; Kaiser, 1979). Thus the horizontal momentum of the cloud will be dominated by the momentum of the ambient air, whereby the average velocity of the cloud will be

$$\bar{u} = \frac{1}{h} \int_{z_0}^{h} \frac{u_*}{\kappa} \ln z/z_0 \, dz = \frac{u_*}{k} (\ln \frac{h}{z_0} - 1). \tag{1}$$

Thus the wind shear Δu over the top of the cloud will be small and of order u_*/k. Hence, the pressure force on the cloud as a result of action from the ambient air will be of order

$$P = \frac{1}{2} \rho_a (\frac{u_*}{k})^2 2rh , \tag{2}$$

which already for $r > 2h$ is less than the frictional force exerted on the cloud top and on the surface in general.

The small shear u_*/k over the top of the cloud will drive some entrainment, but this development is complicated by the slumping of the cloud which intensifies the shear on the upwind edge and reduce it or perhaps reverse it on the down wind edge of the cloud. Also, because the momentum of the cloud initially is determined as the average over the dept h_0, the cloud moves faster than the surrounding air while slumping. This effect will tend to eliminate or reverse the average shear over the cloud top. Further the slumping itself generates turbulent velocities of the order of

$$v_* = \kappa \sqrt{g'h}/\ln(h/z_0), \tag{3}$$

which however is seen to vanish rapidly as h decreases. When the initial slumping phase is terminated and the height starts to increase again all the above effects have vanished, and it may be fair to assume continuity of stress over the cloud top, i.e.

$$u_{*i} = \sqrt{\frac{\rho_a}{\rho}} \ u_* \tag{4}$$

As the cloud at this stage is further diluted, such that $(\rho_a/\rho)^{\frac{1}{2}}$ is close to unity, and because the only use of u_{*i} in the model is to estimate the rate of entrainment caused by mechanical turbulence, which is not that precisely known anyway, we do simply put $u_{*i} = u_*$ and neglect the complicated conditions in the slumping phase.

ENTRAINMENT EQUATION CONSTANTS

Using the turbulent kinetic energy budget, it can be argued that the top-entrainment velocity is of the form (Jensen, 1981)

$$u_{ew} = \frac{a(u_*/\sqrt{e})^3 + b}{c + Ri} \ \sqrt{e} \tag{5}$$

where the entrainment Richardson number is

$$Ri = \frac{g\Delta\rho}{\rho \ e} \ h \tag{6}$$

Further we will assume that the turbulent kinetic energy simply may be written as the sum of mechanically and convectively produced turbulence:

$$e = w_*^2 + u_*^2 \tag{7}$$

Equation (5) has three limiting forms in which it can be calibrated against experiment.

First in strong convection, $u_*/w_* \ll 1$, whereby $\sqrt{e} \approx w_*$, eqs. (5), (6) become

$$\frac{u_{ew}}{w_*} = \frac{b}{c+Ri} \ , \qquad Ri = \frac{g \ \Delta\rho}{\rho \ w_*^2} \ h \ ,$$

which at a sufficiently large Ri-number (Ri >>c) is equivalent to the formula given in Deardorf (1980) as a fit to convection tank experiments. According to these we have to choose b = 0.25.

Similarly, but in the opposite limit with no convection,
$\sqrt{e} = u_*$, whereby eqs. (5), (6) become

$$\frac{u_{ew}}{u_*} = \frac{a+b}{c+Ri} \, , \qquad Ri = \frac{g}{\rho} \frac{\Delta\rho}{u_*^2} \, h$$

which at a sufficiently large positive Ri-number is equivalent to
the formula given by Kato Phillips (1969) as a fit to tank experiments in which the outer layer was stabely stratified with a uniform density gradient. The comparison gives a+b = 2.5 or a = 2.25.

Finally in the limit of small Ri-number and no convection we
would like to obtain the passive entrainment relation $u_{ew} \simeq u_*$,
which then determines c to be approximately 2.5.

RECAPITULATION OF THE MODEL

For a given set of initial conditions ($\Delta\rho$, ρ, ΔT, h, r, c_p)
plus u_* and $u(z,z_0)$, a repeated updating of the equations involved
are made through small time steps. All relevant equations are
given in Jensen (1981). The procedure is as follows:

- Calculation of the temperature jump Ri-number Ri =
  $(g/T_o) \, (\Delta T/u^2)h$.

- Iterative solution of the surface similarity profiles to
  determine the corresponding value of h/L.

- Calculation of surface heat flux Q = $c_H \Delta Tu$ where c_H =
  $\kappa^2/(\ln(z/z_0) - \psi_m)(\ln(z/z_0) - \psi_h)$.

- Calculation of w_*, and the turbulent kinetic energy e.

- Calculation of the density jump Ri-number Ri = $(g/\rho)(\Delta\rho/e)h$.

- Calculation of entrainment velocity u_{ew}.

- Calculation of entrainment volume during a small time step
  and up date of V.

- Calculation of dr from the slumping equation: dr = $\sqrt{g'h}$ dt,
  and update of r.

- Calculation of new value of h through $V/2\pi r^2$.

- Calculation of c_p for the gas mixture (the mass weighted
  average of c_p for the pure gas and for air).

– Calculation of $d(\Delta T)$ through the heat budget. Update of ΔT.

– Calculation of $d(\Delta \rho)$ through the mass budget. Update of $\Delta \rho$.

With the new value of ΔT a new temperature-jump Ri-number is calculated according to the first point above and the sequence is repeated.

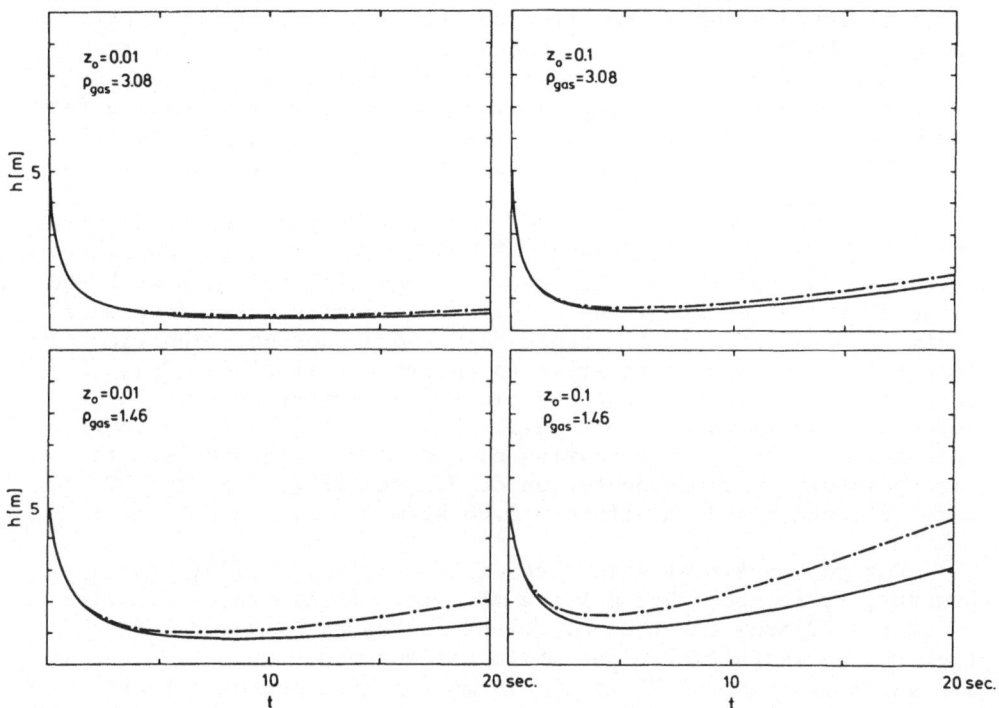

Fig. 1   Diagrams of the height of the cloud h as function of
         time after release t for two different values of the
         surface roughness z_0 and for two different values of
         the initial density difference $\Delta \rho$. Stipled curve is
         for the cold cloud ($\Delta T = 50$ C°) whereas the solid
         curve represents the isothermal cloud ($\Delta T = 0$).

MODEL RESULTS

Typical model results are shown in fig. 1. On each diagram, h(t) is shown for an isothermal cloud ($\Delta T=0$) and for a cold cloud. For the latter we chose $\Delta T=50$ K as representative for typical conditions. The difference between the diagrams is in the choice of $\Delta \rho$ and z_0. All other initial conditions are held constant:

$$u = 5 \text{ m/s}, \quad h = 5 \text{ m}, \quad r = 5 \text{ m}, \quad c_p = 481 \text{ J/kg}^\circ\text{C}$$

(a value which correspond to chlorine gas). Two different initial densities are used: $\rho = 3.08$ kg/m^3 corresponding to pure chlorine gas, $\rho = 1.46$ kg/m^3 corresponding to an initial dilution of 1:10 during cloud formation. For each ρ, results are shown for two different surface roughnesses: $z_0 = 1$ cm corresponding to a smooth grass field and $z_0 = 10$ cm corresponding to a surface covered with small bushes and shrubs. Choice of larger values of z_0, which would be interesting in relation to releases in built up areas, would not be realistic as roughness elements as large or larger than the depth of the cloud would act as individual barriers and obstacles rather than through a single collective roughness effect. The effect of barriers and obstacles on the spreading of dense gases will be subject to future experimental investigations.

In discussing the results of the present model we first note that the temperature difference ΔT has dropped to a few degrees by the end of the time domain shown. Regarding the case with the largest initial density, the calculations confirm the often made claim that the effect of a cloud being cold does not significantly change its development relative to an isothermal cloud (Raj, 1982). The explanation is that the entrainment Ri-number remains large because of small entrainment while the turbulence driving temperature-jump is lost due to heating of the cloud. This result is contradictory to the expectation of Jensen (1981). At t = 20 sec, $h \simeq 0.6$ m and $r \simeq 35$ m with $\rho = 3.08$ kg/m^3.

For the more realistic case of the smaller initial density, however, it is seen that a large ΔT effect is present. Thus is h (at t = 20 sec) for the cold cloud increased by about 50% relative to the isothermal cloud while the corresponding cloud radii are about the same (~ 20 m). This means that concentrations in the cold cloud are reduced by about 30% compared to the isothermal case. The reason for this enhanced dilution is that the entrainment Ri-number, or density jump over the cloud top, is sufficiently small to allow the convective turbulence caused by ΔT to drive some entrainment.

Compared to the example above with larger $\Delta \rho$, the reduction in the initial density alone causes concentrations to drop about 30% while the addition of the ΔT effect causes a further reduction of a similar amount as mentioned above.

Regarding the effect of surface roughness, it is seen that this is a quite significant parameter, both in the large $\Delta\rho$-case and the small $\Delta\rho$-case. In the former the increase in h at t = 20 sec from the smooth to the rough case is a factor of 2.6 and in the latter the increase is a factor 2.3. These factors are relatively insensitive to whether the cloud is cold or isothermal.

These ratios are significantly larger than the neutral limit $u_*(z_0 = 0.1\ m)/u_*(z_0 = 0.01\ m)$ which with u (5 m) = 5 m/s is about 1.6. At t = 200 sec these ratios have decreased to 2.0 and 1.8 respectively which indicates that density effects are still present at this stage of cloud development.

The possibility of parameterizing the various effects discussed above is presently being investigated.

LIST OF SYMBOLS

a	constant
b	constant
c	constant
c_H	heat transfer coefficient $c_H = Q/\Delta Tu$
c_p	heat capacity of cloud mixture
e	a measure of turbulent kinetic energy
g'	reduced gravity = $g\Delta\rho/\rho$
h	height of cloud. Subscript o signifies initial value
L	Monin-Obukhov length = $u_*^3/\left[\kappa(g/T_0)(Q/\rho c_p)\right]$
Q	surface heat flux from ground to cloud
r	radius of cloud
Ri	Richardsons number. Two different numbers are used in this context: one based on ΔT, u; and one based on $\Delta\rho$,e.
t	time since start of slumping
T	absolute temperature. Subscript zero denotes reference value
ΔT	temperature difference between the ground and the gas cloud
u	wind speed
$\bar{u}$	average velocity of cloud
Δu	difference between $\bar{u}$ and wind speed above cloud top
u_*	friction velocity in the ambient flow = $\sqrt{\tau/\rho_a}$. Subscript i denotes the u_* value inside the cloud
u_{ew}	entrainment velocity at cloud top
V	cloud volume
w_*	convective velocity scale = $((g/T_0)Qh)^{1/3}$
z	height above the ground
z_0	surface roughness length
κ	von Kårmåns constant $\simeq 0.4$

ρ   density of gas cloud. Subscript a denotes density of
      ambient air
Δρ   density difference between the cloud and the ambient air
τ   surface shear stress exerted by the wind
ψ   stability correction function. The functional form and the
      proper references are given in Jensen (1981).

REFERENCES

Deardorff J.W.(1980) Progress in understanding entrainment at
      the top of a mixed layer. In: Workshop on the Planetary
      Boundary Layer. Ed. by J.C. Wyngaard (Amer. Meteor. Soc.,
      Boston, Mass.) 322 pp.
Jensen N.O. (1981) Entrainment through the top of a heavy gas
      cloud. In: Air Pollution Modelling and Its Application I.
      Ed. by C. De Wispelaere (Plenum Press, New York) 747 pp.
Kaiser G.D. (1979) Examples of the successful application of a
      simple model for the atmospheric dispersion of dense, cold
      vapours to the accidental release of anhydrous ammonia from
      pressurised containers, SRD 150, UK Atomic Energy Authority.
Kato H., Phillips O.M. (1969) On the penetration of a turbulent
      layer into a stratified fluid, J. Fluid Mech., 37, 643-655.
Raj P.K. (1982) Heavy gas dispersal - A state-of-the-art review
      of the experimental results and models. In: Heavy Gas
      Dispersal, lecture series 1982-04, March 8-12, Von Karman
      Institute for Fluid Dynamics, Rhode-Saint-Genese, Belgium.
van Ulden A.P. (1974) On the spreading of a heavy gas released
      near the ground. In: Loss prevention and safety promotion
      in the process industry. Ed. by C.H. Buschmann (Elsevier,
      New York) 425 pp.

DISCUSSION

J. KNOX   Models of a simpler
structure are needed to estimate global proper-
ties of cloud height and radius; however,
decisionmakers will need 3D-pollutants distributions for assessments which the simpler models
predicting global properties cannot produce.
The rate of advance of a combustion front in
ambient vapour cloud depends on the field of
turbulence in the time dependent heavy gas
boundary layer.

T. MIKKELSEN   The objective of the simple
model presented has been to demonstrate the sig-
nificant difference between top and front
entrainment in a box model of a heavy gas cloud.
I agree that such a simple model cannot give
sufficient information on the structure of the
turbulent field for combustion front propaga-
tion etc.

MAPLIN SANDS EXPERIMENTS 1980: DISPERSION RESULTS
FROM CONTINUOUS RELEASES OF REFRIGERATED LIQUID
PROPANE AND LNG

J.S. Puttock, G.W. Colenbrander*
and D.R. Blackmore

Shell Research Ltd.
Thornton Research Centre
P.O. Box 1, Chester CH1 3SH

*Shell Research BV, Koninklijke/
 Shell-Laboratorium, Postbus 3003
 1003 AA Amsterdam, Netherlands

SUMMARY

     In 1980, a series of spills of up to 20 m^3 of LNG and
refrigerated liquid propane onto the sea was performed by Shell
Research Ltd. at Maplin Sands in the south of England. Both
instantaneous and continuous releases of liquid were made, and some
were ignited. Gas concentration and temperature were monitored from
an extensive array of floating pontoons up to 650 m downwind, and
meteorological profiles were obtained at two locations.

     Results from the continuous spills are presented; these were
performed at liquid flow rates of up to 5.8 m^3/min, and in wind
speeds from about 2 to 10 m/s. A comparison is made of the
measurements with predictions from the dense gas dispersion model
HEGADAS II. The propane measurements provide data on dense gas
dispersion without the effects of large temperature difference and
low molecular weight, which are present for LNG. In general, the
model agrees well with the propane data, but is found to be
conservative for LNG dispersion.

INTRODUCTION

     The Maplin Sands experiments were performed by Shell Research
Ltd. in the summer of 1980. The aim was to study the dispersion
and combustion of releases of dense flammable gases, and for this

353

purpose 34 spills of liquefied gases onto the sea were performed.
The materials used were refrigerated liquid propane and liquefied
natural gas in quantities up to about twenty cubic metres. A
description of the programme is given in reference 1. Dispersion
results from the continuous releases are presented here (instantaneous releases of liquid were also performed); the emphasis is on
LNG since some examples of propane releases have been presented in a
previous paper[2].

   The site chosen for the releases was on an area of tidal sands
with a typical slope of 1/1000 on the north side of the Thames
estuary in England. Experiments were conducted during offshore
winds for reasons of safety. The point of release was 350 m offshore and, when possible, spills were performed at high tide. So
that spills could also be performed at low tide, a 300 m diameter
dike was constructed to retain the sea water around the spill point.
There was a maximum 0.75 m change in level at the offshore edge of
the dike. Behind the 5 m high sea wall was flat farmland.

   In the early experiments the liquid was released from the open
end of a vertical pipe of 0.15 m diameter. In four cases (trials 9
to 17) this was 2 to 3 m above the water surface; subsequently it
was lower. In three cases the pipe end was below the surface.
Later on, the end of the spill pipe was flared out in a vertical-
axis cone with a horizontal plate below it at the water surface.
The liquid emerged from the slot between the cone and the plate with
negligible vertical momentum. Details of these changes are given in
Tables 1 and 2.

   Instruments were deployed on 71 floating pontoons initially in
the layout shown in Figure 1. However, the pattern of deployment
was changed for the propane spills; more pontoons were concentrated
in the near-field, particularly in the direction covered by the prevailing westerly wind.

   There were about 360 instruments in the array. Mounted on a
standard pontoon, in addition to combustion instruments, were three
gas sensors (usually at 0.5 to 0.9 m, 1.4 m and 2.4 m above the sea
surface) and one thermocouple close to the lowest gas sensor. The
gas sensor is a device based on measurement of the heat loss from a
filament under free convection. Two special pontoons each had ten-
metre masts, six gas sensors, one thermocouple and two three-axis
sonic anemometers. Two further sonic anemometers were deployed
elsewhere.

   Another two special pontoons were devoted to meteorological
measurements. These provided vertical profiles of temperature and
wind speed up to ten metres, together with measurements of wind
direction, relative humidity, insolation, water temperature and wave
height.

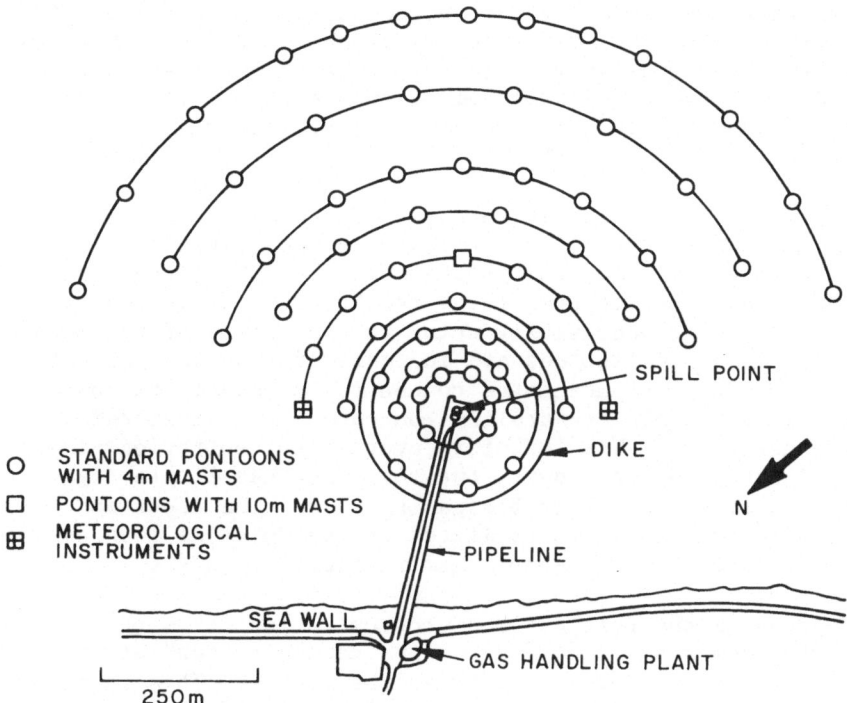

Fig. 1   The layout of the instrument pontoons as used for most of
          continuous LNG spills.

Signals from the instruments were sampled ten times per second,
digitised on the pontoons and relayed by cable via multiplexers to
the computers onshore.

The spills were photographed from three locations, two land-
based towers providing orthogonal views, and a helicopter overhead.
At each location still photographs and video recordings were taken.

METHOD OF ANALYSIS

A problem is posed in these continuous spills by the meandering
of the plume. The array of measuring stations (Figure 1) was set
out as described to cover any offshore wind direction. Restriction
to a smaller range of wind directions would have further reduced the
number of spills that could be performed in the time available. The
steady-state plumes in all but low winds were so narrow that they
could pass completely between sensor stations, although rarely for a
whole experiment because of variations in wind direction. Thus any
sensor might see the plume only intermittently and there were times
when the plume did not pass over any station at a given distance.

If model predictions are to be compared with the data, it must be decided exactly what gas concentrations the models are intended to predict. The long-term average concentration at a fixed location downwind of the source is reduced by meandering of the plume. However, meandering of the plume does not change the maximum concentration inside the plume.

The value of particular interest in flammable gas plumes is the maximum concentration in the plume at a given distance from the source, whatever its angular position may be. This concentration can conveniently be referred to as the "centreline concentration", though it does not necessarily occur at the centre of the meandering plume. If model predictions are in terms of mean concentrations, then the predicted plume-centre concentration should be compared with the mean of this "centreline concentration", measured in the experimental plume. The long-term average concentration at any fixed point may be much lower, but that fact has little significance for the assessment of flammable gas hazards. The crosswind variation in concentration predicted by the model is then assumed to be taken relative to a moveable centreline.

To measure the relevant mean concentration it is necessary to obtain measurements of the "centreline concentration" at each instant, and subsequently to average the values. This is possible only if there were several sensor stations across the plume at each distance, requiring perhaps an order of magnitude increase in the number of sensor stations. Averaging the intermittent signal at any sensor does not give the required results. It is more useful to take the maximum observed concentration, since the "centreline concentration" would be observed as the "centre" of the plume meandered over a sensor. For this reason, maximum values are used in the analysis below. The sensor stations were close enough that the plume centreline rarely failed to pass over one for at least part of the spill time. In fact, there were usually sensors in the plume for a large part of the spill time. Then the maximum signal obtained can be assumed to be above the mean of the "centreline concentration" as defined.

Model predictions

To describe atmospheric dispersion of dense gas emitted by area sources at ground level, the HEGADAS II model has been developed by Shell Research.[3] This can be described as an advanced "similarity" model in the terminology used by Havens.[4] The steady-state version is appropriate to the experiments described here; a time-dependent version also exists.

The model assumes, crosswind, a flat-topped concentration profile with Gaussian edges. The flat part can be eroded so that the profile becomes completely Gaussian far downwind. Gravity spreading can increase the width of the plume. Heat transfer from the underlying surface is not included in the basic model. The interaction of the assumed vertical profile

$$C \propto \exp \left((z/S_z)^{1+\alpha} \right)$$

with a power-law velocity profile $(U \propto z^{\alpha})$ gives an advection velocity which is a function of the height parameter S_z. The vertical entrainment u_e is given by

$$u_e/u_* = (1 + \alpha) \frac{k}{\Phi(Ri_*)},$$

where α is the exponent in the wind velocity profile and k the Von Karman constant. $\Phi(Ri_*)$ is a function of the bulk Richardson number Ri_*. For stably stratified conditions this function has been arranged to be compatible with the results of laboratory two-layer experiments[5] and wind tunnel data.[6] A good fit with these experimental data was obtained using

$$\Phi(Ri_*) = 0.74 + 0.24 \, Ri_*^{0.7} + 1.2 \, 10^{-7} \, Ri_*^3 \qquad Ri_* > 0.$$

For unstably stratified conditions $(Ri_* < 0)$, we arranged $\Phi(Ri_*)$ to be compatible with the results of the Prairie Grass tracer dispersion experiments (as given by Horst[7]). Given the roughness length z_o, the Monin-Obukhov length L and the friction velocity u_* we used Businger's relation for the surface layer temperature profile[8] to obtain the value for the density difference $\Delta \rho$ when evaluating the bulk Richardson number Ri_*. A good fit between the Prairie Grass data and the HEGADAS II predictions was obtained using

$$\Phi(Ri_*) = 0.74/(1 + 0.65(-Ri_*)^{0.6}) \qquad Ri_* < 0$$

as shown in Figure 2. This Figure also shows a good agreement for stably stratified atmospheric conditions (positive Monin-Obukhov length L) where we used the $\Phi(Ri_*)$ function as given for $Ri_* > 0$.

HEGADAS II has been used to simulate the spills described in this paper, using the mean values of meteorological parameters measured during the period of steady release for each trial. The measured air temperature profiles show a nearly neutral atmospheric stability for all trials. One would expect, therefore, very similar values of the ratio of the friction velocity u_* and the 10 m wind speed U_{10}, for all trials. From the sonic anemometer signals we have calculated u_* for each trial (by definition $\sqrt{(-u'w')}$, where u' and w' are the fluctuating wind-speed components in horizontal and vertical direction respectively). The resulting values of u_*/U_{10}

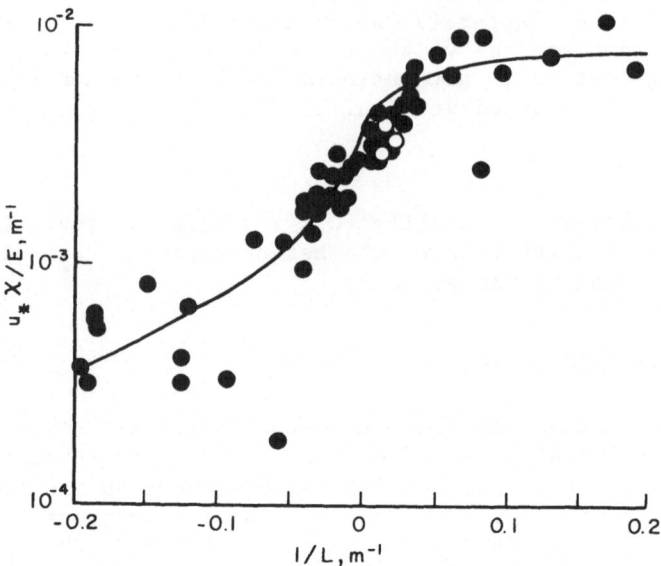

Fig. 2   Measurements 400 m downwind of the source in the
         Prairie Gass experiments (●), as a function of
         atmospheric stability, and the the model prediction
         (curve). The crosswind-integrated concentration χ is
         normalised by the source strength E and the friction
         velocity u_*.

vary widely between the trials. At the time of writing the wind
velocity data are being analysed more thoroughly. It appears that
the readings do not always satisfy the condition of statistical
stationarity during the averaging period used to calculate u'w'.
Moreover, some of the averaging periods used are too short to give
reasonably accurate values of this cross-correlation. The more
detailed analysis of the weather data, currently being undertaken,
will result in more accurate estimates of u_*/U_{10} and atmospheric
stability. However, the mean value of u_*/U_{10} = 0.034 found so far
is consistent with the estimated surface roughness at the Maplin
Sands site and is used for all HEGADAS II simulations; the
atmospheric stability is taken to be neutral for all runs.

Two examples

     We present here some of the detailed results from two LNG
spills. Similar examples of propane spills have been given in
reference 2.

     Spill 29 was a release of LNG which was kept steady at 4.1 m^3/
min for 225 s. The mean wind speed during this period was 7.4 m/s.

Fig. 3   Spill 29.

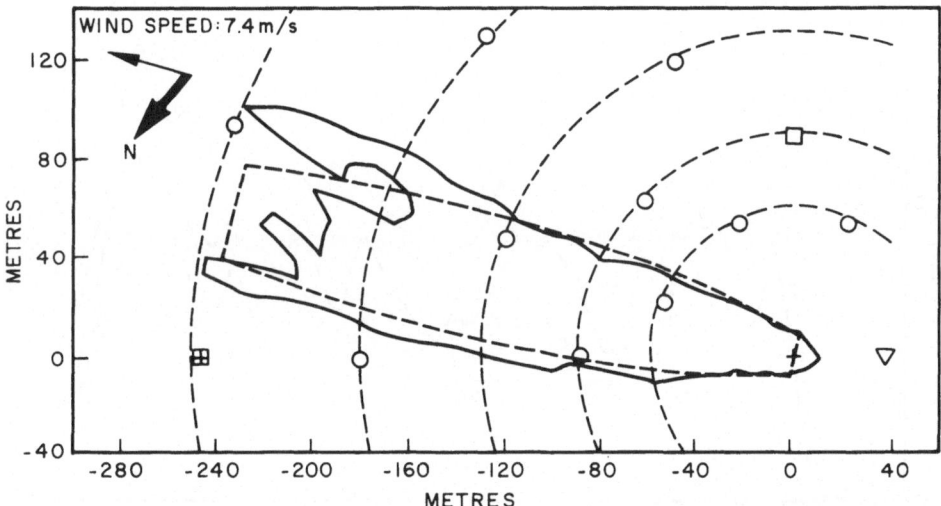

Fig. 4   A plan view of the spill 29 plume, derived from Figure 3,
          together with the limits of the visible plume predicted by
          HEGADAS II using the ambient relative humidity measured at
          10 m height.

Figure 3 is a photograph of the fully-developed plume. The strongly visible white plume extended generally to some 250 m although wisps of cloud travelled further, occasionally as far as 500 m from the source.

The cloud outlines from overhead still photographs such as Figure 3 have been input to the computer using an image analyser. After perspective correction, the visible outline can be plotted and compared with the HEGADAS II model prediction (Figure 4). The curve plotted from the model is the ground level contour of 5.1% which is the concentration at the limit of visibility for the ambient relative humidity of 52%, as determined by an adiabatic mixing calculation.

Gas concentrations at three levels at each sensor station can be plotted as in Figure 5. The signals have been smoothed using a three-second moving average to eliminate any high frequency noise present. The gas sensor itself has an exponential time constant of about 1.1 s (2.5 s to 90% response).

The maximum concentration observed at each sensor can be obtained and then plotted as a function of distance from the source. The model predictions of mean concentration are shown in Figure 6 for comparison.

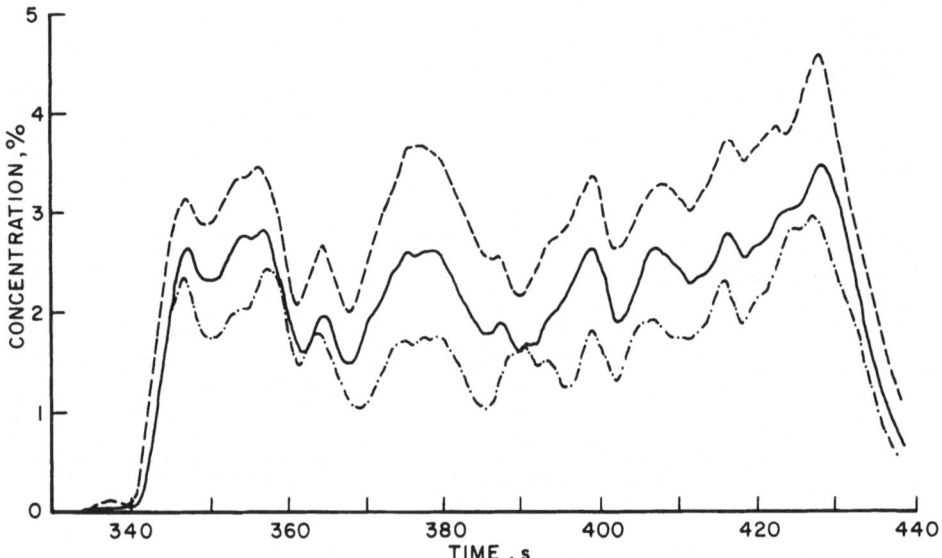

Fig. 5   Gas concentrations measured in spill 29 at 180 m from the
          source, -90° from the array axis, at heights ----0.9 m,
————— 1.3 m and—·—·· 2.3 m.

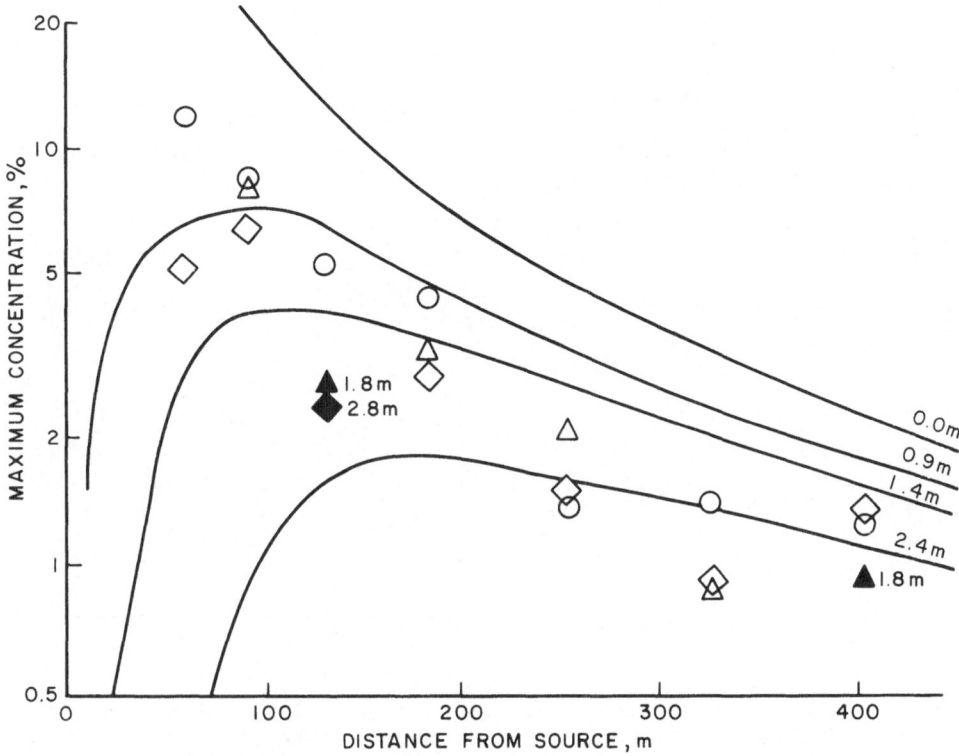

Fig. 6   Maximum concentrations (smoothed using 3s average)
          measured at each radius for spill 29, at heights ○ 0.9 m,
          △ 1.2-1.4 m, ◇ 2.2-2.4 m.   The curves are HEGADAS II
          predictions of mean concentrations.

Fig. 7   Spill 15.

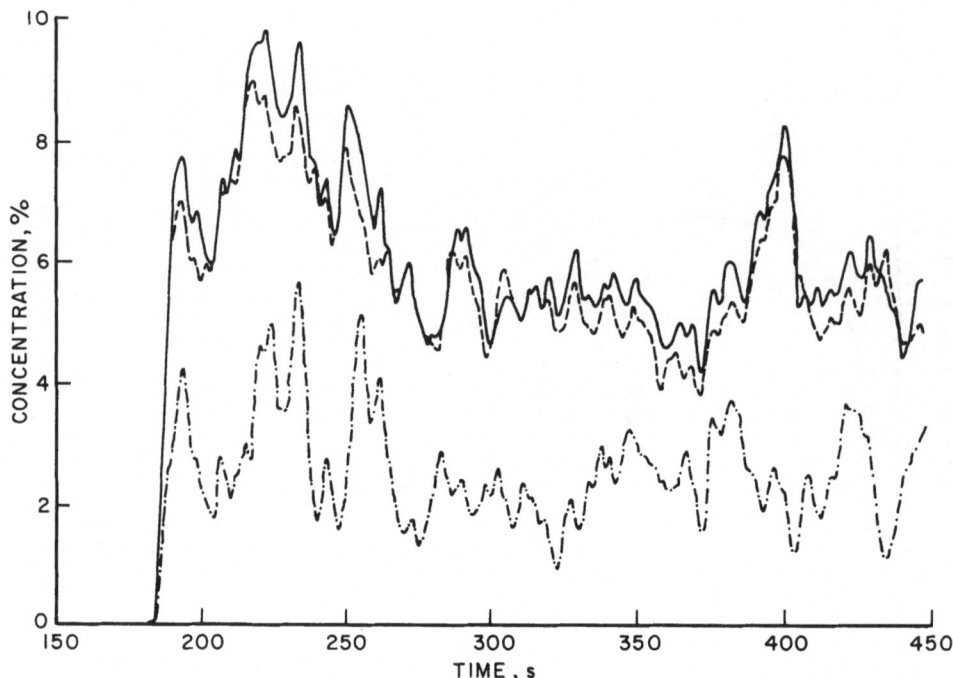

Fig. 8   Gas concentrations measured in spill 15 at 58 m from the
          source, -23° from the array axis, at heights of---0.9 m,
          ——1.3 m, and —·—·2.3 m.

Spill 15, by contrast, was a release of 2.9 m^3/min of LNG
performed in a low wind (3.6 m/s) and at much higher relative
humidity (88% at 10 m height). The result was a wider plume,
clearly affected by gravity spreading, which was visible to a
greater distance (Figure 7): the connected visible plume extended to
600 m and wisps could be seen 800 m or more from the source. The
end of the spill pipe on this occasion was elevated 2 or 3 m, rather
than close to the surface.

An example of gas sensor data is shown in Figure 8. In
Figure 9 the maximum observed concentration is plotted against
distance from the source for the three sensor heights.

RESULTS

Details of all eleven continuous propane spills and thirteen
continuous LNG spills are given in Tables 1 and 2. The period of
time for which the flow was controlled at the stated rate is shown,
and the wind speed quoted is the mean over this period. In order to
compare the strength of density effects in the spills, an initial
Richardson number is calculated, as defined in reference 1.*

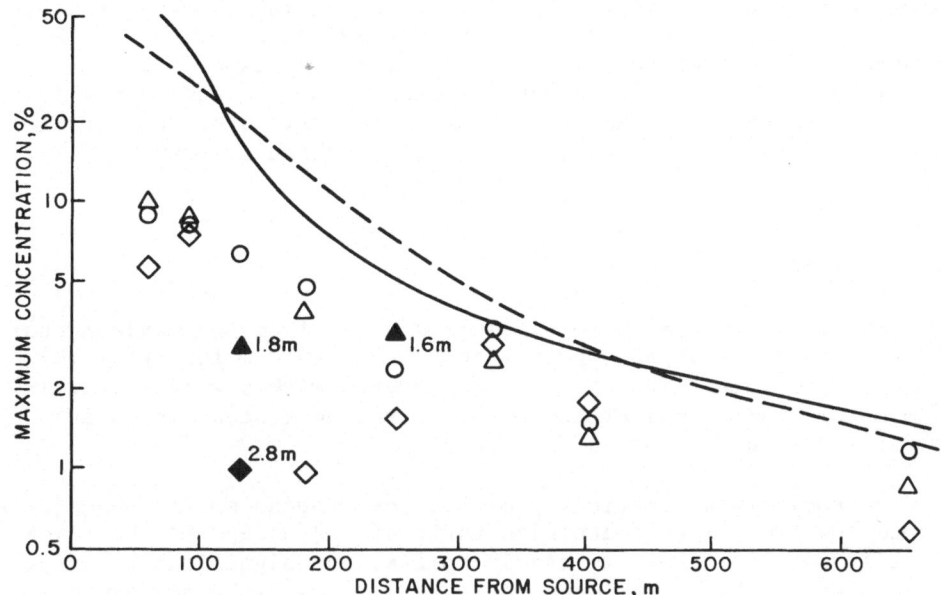

Fig. 9   Maximum concentrations (smoothed using 3 s average)
          measured at each radius for spill 15, at heights:
          O 0.9–1.0 m, △ 1.3–1.5 m, ◇ 2.2–2.4 m.   HEGADAS II
          predictions at the surface: —————"pool evaporation
          and — — — —"jet" evaporation.

Seven of the plumes were ignited, but four of these provided
useful dispersion data in the period before ignition. The
usefulness of the data from several other spills was limited for
reasons noted in the tables. Nevertheless, LFL** distances were
obtained, as described below, from the maximum concentration data
for seventeen of the spills.

For each spill, results were plotted as in Figures 6 and 9. A
smooth curve, estimated by eye, was drawn through the data points
for the lowest sensor level. Taking into account the scatter of the
data, two more curves were drawn to estimate upper and lower

---

* $Ri_o = g'H_o/u_*^2$, where $g' = g (p_g-p_a)/p_a$, p_a and p_g are the
    density of the air and the cold gas, $H_o = Q/UD$, Q is the volume
    flow of gas from the source, and D is the steady liquid pool
    diameter calculated from the liquid flow rate and an evaporation
    rate of 2×10^{-4} m^3/m^2/s.

** LFL = lower flammable limit, 2.1%v for propane, 5%v for LNG.

extremes of concentration. From the intersection of the first curve with the LFL concentration level an observed distance to the LFL was obtained. The other two curves provided an estimate of the uncertainty in this value. A better estimate of the confidence interval for this parameter may result from an analysis of the experimental variability of the sensor calibration, which is currently in progress.

Propane data

The LFL distance so obtained is derived from the maximum concentrations observed at approximately 0.9 m (0.6 m for spill 54) above the surface. This has to be compared with a value derived from model predictions of the mean centreline concentration at ground level, as listed in Table 1.*

An analysis of vertical profiles for propane spills suggests a ground level/0.9 m concentration ratio of 1.2 except in the low wind of spill 54; data from two steady spills, not significantly affected by meander, imply a peak/mean ratio of 1.4. Details are given in reference 2. Thus the observed peak concentrations at 0.9 m might be expected to exceed the predicted mean concentrations at ground level by a factor of $1.4/1.2 \simeq 1.2$. The actual ratios in the region near the LFL for propane (2.1%v) are listed in Table 1. Excluding spill 54, the results seem to lie around the 1.2 value, although scattered 30–40% in either direction. Spill 54 is the exception.

In two of the propane spills (42 and 52) the liquid was released below the surface of the water in a downwind jet. Most of it probably evaporated below the surface and the result was a stream of gas reaching the surface with considerable vertical momentum and jetting up into the air. We believe that the subsequent slumping of the gas, from the several metres height attained in this way, is the cause of the bifurcation of the plume observed in spill 42. In the higher wind speed of spill 52 the effect was not strong enough to produce visible bifurcation; but for both these spills the LFL distance was significantly lower than for comparable above-surface releases.

---

* The more obvious approach would seem to be to compare data obtained at 0.9 m with model predictions at 0.9 m. However, because the vertical profile adopted in the model is somewhat arbitrary in the part of the plume where gravity effects play a dominant role, the experimentally observed vertical profile has been introduced to translate the 0.9 m observations to ground level. In this way the most important feature of the model is tested, namely its ability to predict the maximum concentration at a given distance, which is at ground level.

Table 1. Details of continuous propane spills. LFL distances were obtained from plots of maximum concentration against distance; in two cases long-term concentrations could also be used.

Spill no.	Rate m³/min	Duration of steady flow, s	Wind speed, m/s	Ri_0	Observed LFL distance peak, z = 0.9 m, m	Predicted LFL distance mean, z = 0 m, m	Concentration ratio (far field) peak, z = 0.9 m / model, z = 0 m	Comments
42	2.5	180	3.7	100	220 ± 35	315	0.50 ± 0.10	Underwater release
43	2.3	330	5.5	28	215 ± 20	245	0.79 ± 0.10	Pipe-end just above surface
45	4.6	330	~2	800	-	675	-	Pipe-end just above surface. Wind very unsteady and non-uniform
46	2.8	360	8.1	10	245 ± 35	225	1.18 ± 0.23	
47	3.9	210	5.6	35	235 ± 25	320	0.59 ± 0.09	
49	2.0	90	6.2	18	285 ± 25	210	1.7 ± 0.3	Ignited
50	4.3	160	7.9	13	210 +50 -25	280	0.61 +0.17 -0.09	Ignited
51	5.6	140	6.9	22	-	335	-	Ignited. Plume centre missed sensors when steady
52	5.3	140	7.9	14	200 ± 30 (mean 130 ± 30)	315	0.45 ± 0.08	Underwater release
54	2.3	180	3.8	90	450 ± 70 (mean 400 ± 100)	295	2.0 ± 0.6	(Data from lowest sensors at z = 0.6 m)
55	5.2	150	5.5	40	-	370	-	At edge of array

LNG data

When the LNG spills are examined, a different picture emerges. The results are summarised in Table 2. These include, for several spills, data from an extra model run. If the LNG from a high jet were to evaporate before hitting the water surface it would obtain its latent heat of vaporisation from air entrained into the jet. The plume resulting from this, which we may call "jet" evaporation, would be considerably colder than that from LNG which obtained its latent heat from the water, and then entrained the same amount of air. (The latter may be called "pool" evaporation). The resulting difference in density would affect the dispersion. For the first three LNG spills, when the source was a jet 2 to 3 m from the surface, there must have been at least an element of "jet" evaporation. In fact, the plume in spill 15 was found to be colder, at a given concentration, than predicted in the model calculation, whereas it would be expected to be warmer because of the neglect of heat transfer in the calculation. (Compare Figure 10, discussed below.) Thus for these three spills, both "pool" and "jet" calculations have been performed to cover the range of possibilities for the evaporation. The calculated LFL distances are given in Table 2.

It can be seen from Table 2 that there is reasonable agreement between model and data at Ri_o = 3, where density effects are not important. However, as Ri_o increases, there is a strong trend to greater disagreement, with the model overpredicting the LFL distance by about a factor of four for spill 12. Additionally, the measured peak values may well be more of an overestimate of the mean centreline LFL distance than for propane; the gas is less dense in its later stages and appears to give large peak/mean ratios, and the plumes are higher, resulting in shallower vertical concentration gradients.

Some overprediction is to be expected since the HEGADAS II model does not include the effects of heat transfer from the water surface. Indeed the effects may be expected to be strong in some circumstances for LNG, since heat transfer increases the chance of the plume becoming slightly buoyant, rather than denser than the ambient air. In the absence of confidence in the correct form of heat transfer relations, no such allowance has yet been included in the model. However, we have tried the effect of incorporating fairly standard relations for free- and forced-convection heat transfer similar to those used by, for instance, Eidsvik.[9] The results are encouraging. In spills 12 and 15, the measured LFL

Table 2. Details of the continuous LNG spills.

Spill no.	Rate m^3/min	Duration of steady flow, s	Wind speed, m/s	Relative humidity at 10 m %	Ri_0	Observed LFL distance peak, z = 0.9 m, m	Predicted LFL distance mean, z = 0 m "pool" evap.	"jet" evap.	Concentration ratio (far field) peak, z = 0.9 m model, z = 0 m	Height of source above surface[a]	Comments
9	1.6	300	8.9	84[b]	3	135 ± 15	125	130	1.12 ± 0.20	2-3 m	
12	0.7-1.1	340	1.5	91	400	63 ± 4	200	295	0.072 ± 0.010	2-3 m	Wind speed from 1 to 2 m/s.
15	2.9	285	3.6	88	50	165 ± 15	250	305	0.60 ± 0.11	2-3 m	
17	2.8	170	≈6	68[b]	11	–[c]	230	215	–	2-3 m (perforated drum)	Ignited. Photographic data only.
27	3.2	160	5.5	53	16	190 ± 20[c]	260	–	0.68 ± 0.11	Close to surface	Ignited.
29	4.1	225	7.4	52	7	140 ± 15	250	–	0.44 ± 0.10	"	
34	3.0	95	8.6	72	4	150 ± 20	195	–	0.73 ± 0.12	"	
35	3.8	135	9.8	63	3	175 ± 25	205	–	0.82 ± 0.15	"	
37	4.1	230	4.7	66[b]	–	–	310	–	–	Below surface	Buoyant plume.
38	5.8	25	4.3	61[b]	40	–	385	–	–	Close to surface	Ignited. Steady flow too brief for useful dispersion data.
39	4.7	60	4.1	63[b]	50	130 ± 20[c]	355	–	0.25 ± 0.07	"	Ignited.
56	≈2.5	80	5.1	83	17	110 ± 30	210	–	0.37 ± 0.12	Cone at surface	Plume only briefly over sensors. Lowest sensors at z ≈ 0.6 m.
57	4.3	190	3.7	93[b]	60	–	295	–	–	Cone at surface	Plume missed sensor array.

Notes: (a) Source was open end of downward pointing pipe except where stated.

(b) Derived from observations at meteorological station 3 km. distant.

(c) Furthest extent of flame for spill 17:140 m; spill 27:130 m; spill 39:130 m

distance then lies between the predictions given by the assumption
of total "pool" evaporation and of total "jet" evaporation. But for
spills in high wind speeds and low relative humidity (trials 29, 34
and 35) incorporation of heat transfer has a negligible effect on
the predicted LFL distance; it appears that the increased vertical
mixing is offset by the loss of horizontal spreading due to gravity.
For spills in moderate wind speeds and low relative humidity (trials
27, 38 and 39) incorporation of heat transfer slightly improves the
predicted LFL distance but not enough for a good agreement with the
observations. Model runs show that the level of humidity has a pronounced effect on the calculated LFL distances. When combined with
heat transfer from the water surface, high humidities cause the
plume to become buoyant over an extended distance, resulting in
enhanced vertical entrainment.

Indications exist from measurements during trial 29 at 0.7 m
above the water surface that at this level the humidity was higher
(60-73%) than the 52% observed at 10 m and used in the initial model
run. For a relative humidity up to 60% the calculated LFL distance
turns out to be almost independent of humdity; above that value the
distance drops to 200 m for 73%, and 135 m for 90%.

This example may serve as an indication of how sensitive the
results of LNG vapour dispersion calculations may be to small
changes in heat content of the plume, in this case due to the supply
of latent heat by water vapour condensation.

A full assessment of the applicability of any heat-transfer
relations requires a check on whether the model, as amended, can
simultaneously correctly predict the measured gas concentration and
temperature in the plumes. Such an assessment has not yet been
completed. The data from each thermocouple at Maplin can be plotted
against the measurements from the coincident gas sensor as in
Fig. 10. These results are from spill 56, with the conical pipe-
end, and show that the plume indeed had a greater heat content than
could be accounted for by the entrained air. The water content of
the plumes also requires study, since it appears that they remained
visible to lower gas concentrations than would be deduced using the
humidity measurements at 10 m height.

The one underwater release of LNG resulted in a plume which
initially appeared similar to the other LNG plumes. But after less
than a minute it became totally buoyant, leaving the surface about
30 m from the source and passing above all the sensors. (The change
may have been due to warming of the pre-cooled pipe, allowing a
small amount of flashing in the LNG and so increasing the jet
velocity from the end of the pipe.) It thus appears that not only
could the LNG be evaporated under water, but also that the resulting
gas was warmed sufficiently to be buoyant on emerging from the
water.

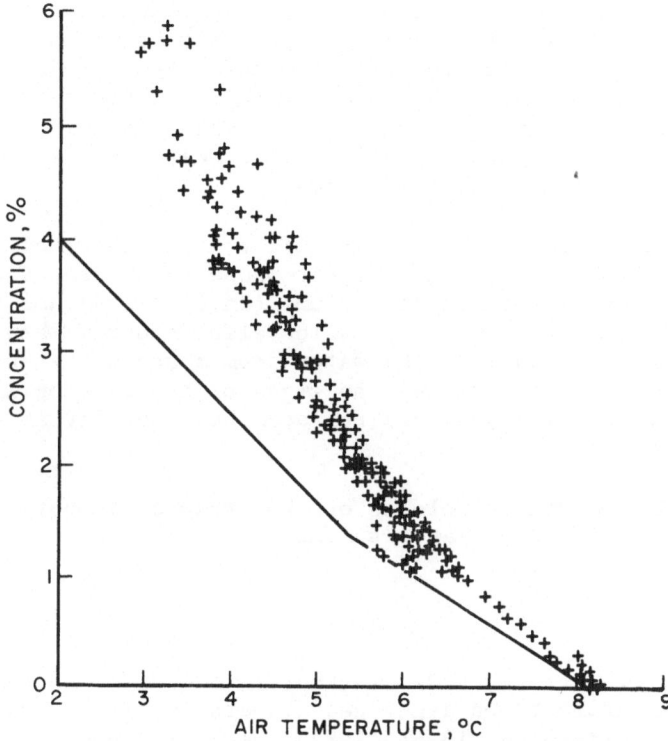

Fig. 10   Concentration against temperature measurements at 130 m
           from the source in an LNG spill. The line shows the
           relation given by an adiabatic mixing calculation. Data
           obtained mainly during a period of increased spill rate
           at the end of spill 56.

CONCLUSIONS

1. For the well-defined propane spills with good data, the
    maximum distance to which flammable gas was found to travel
    was, with one exception, between 0.75 and 1.35 times the
    mean LFL distance predicted by HEGADAS II (concentrations
    within ±50% of the expected value). The exception was a
    spill where the discrepancy was about twice as great,
    although with more uncertainty.

2. With LNG, the data similarly confirmed the model
    predictions for spills at high wind speeds, where density
    effects were unimportant. However, for other spills the
    model consis-tently overpredicted the LFL distance, the
    discrepancy roughly increasing with increasing initial
    Richardson number to about a factor of four at the lowest

wind speed. Incorporation of standard heat-transfer relations into the model appears to give much improved agreement with the experimental data, provided account is taken of the high ambient humidity close to the water surface. Further analysis of the heat and water content of the plumes is necessary before heat transfer relations may be applied with confidence.

3. Underwater release of liquid propane resulted in a strong jet of gas up into the air whose subsequent slumping produced a bifurcated plume (in low wind) and increased dilution. Some bifurcation also occurred when LNG was jetted down into the water from a short distance above the surface. There were no signs of bifurcation in spills where the liquid was released horizontally at the water surface.

4. An underwater release of LNG produced a buoyant plume which rose from the water surface.

REFERENCES

1. J.S. Puttock, D.R. Blackmore and G.W. Colenbrander, Field experiments on dense gas dispersion, in J. Hazardous Materials, 6: 13-41 (1982); also in "Dense Gas Dispersion", ed. R.E. Britter and R.F. Griffiths, Elsevier, Amsterdam (1982).
2. J.S. Puttock, G.W. Colenbrander and D.R. Blackmore, Maplin Sands experiments 1980: dispersion results from continuous releases of refrigerated liquid propane, in Proc. Symp. "Heavy Gases and Risk Analysis", Frankfurt, May 1982 (publ. D. Reidel, Dordrecht), to appear.
3. G.W. Colenbrander, A mathematical model for the transient behaviour of dense vapour clouds, 3rd Intl. Symp. "Loss Prevention and Safety Promotion in the Process Industries", Basle, September 1980.
4. J.A. Havens, A review of mathematical models for prediction of heavy gas atmospheric dispersion, in Proc. Symp. "The Assessment of Major Hazards", Manchester, April 1982 (publ. I. Chem. E., Rugby).
5. L.H. Kantha, O.M. Phillips and R.S. Azad, On turbulent entrainment at a stable density interface, J. Fluid Mech. 79: 753-768 (1977).
6. J. McQuaid, Some experiments on the structure of stably stratified shear flows, Technical Paper P21 (1976) Safety in Mines Research Establishment, Sheffield, UK.
7. J.W. Horst, Lagrangian similarity modelling of vertical diffusion from a ground-level source, in Proc. 2nd Symp.

"Turbulent Shear Flows", Imperial College, London, July 1979.

8. J.A. Businger, in "Workshop on Micrometeorology", ed. D.A. Haugen, American Meteorological Society, Boston, Mass. (1973).

9. K.J. Eidsvik, A model for heavy gas dispersion in the atmosphere, Atmospheric Environment, 14:769-777 (1980).

DISCUSSION

P.K. MISRA   The 'peak' concentrations observed in your experiment reflect single realisations in a turbulent flow and hence in principle should be random variables. This renders your comparison of model results with the 'peak' concentrations somewhat questionable. You should probably attempt to obtain mean concentrations and compare the model results with them ; and have a different model to predict the peak values.

J.S. PUTTOCK   I would repeat our suggestion that long-term mean concentration at a fixed point is not a quantity of interest in the dispersion of flammable gases, and not even what (mean concentration) models should be attempting to predict. The relevant mean concentration, that at the "centre line" of the meandering plume, cannot be obtained from the data except in a few cases. We have therefore attempted to estimate a typical ratio of peak (using three-second-averaged data) to mean centre-line concentration ; this is discussed more fully in ref. 2. In any case the ratio seems to be small compared with some of the differences discussed in this paper.
The fact that we are dealing with a random variable can explain some of the scatter found in the measured model ratio for the propane data.
We are not aware of the existence of practical models to predict probability distributions of peak concentrations.

H. VAN DOP   In view of your problems to resolve the cross-wind profiles, would it not have been better to increase the number of sensors per circular array ? (and given the amount of sensors is constant, to have skipped one or two arrays)

J.S. PUTTOCK   The sensor array was designed
              for both continuous releases and instantaneous
              releases, which can give very different cloud shapes.
              I think we had a reasonable compromise. If we had
              doubled the number of sensors in the circular arcs,
              this would probably not have been enough to provide
              measurements of mean "centreline" concentration for
              many of the narrow plumes ; and the absence of data
              at half of the radial distances would have been a
              significant loss.

                                    In examination of slide of
              spill 29, bifurcation of the plume was noted and is
              perhaps due to gravity effects.

J.S. PUTTOCK   We have described in the paper
              the difference observed between propane releases
              from an open pipe underwater and from the conical
              pipe-end with splash-plate. The underwater releases,
              from which the gas jetted up to a considerable height,
              produced bifurcated plumes with strong vortices at
              the edges. Since we saw no evidence of bifurcation
              when the liquid was delivered gently onto the surface
              of the water, we feel that the edge vortices and
              bifurcation are a result of the slumping of the gas
              from a significant height. Spill 29, shown in fig. 3,
              was a release above the water surface from the open
              pipe-end which again resulted in some evaporation
              under the surface and upward jetting ; the result was
              a mildly bifurcated plume.

J. KNOX   The concentration time series
              data, uneffected by plume meanders show high frequen-
              cy content of several seconds; this corresponds to
              frequency of gravity waves for such systems.

J.S. PUTTOCK   The concentration time-series
              shown consist of data smoothed by a three-second moving
              average; so the shortest time-scales remaining are of
              the order of a few seconds. I do not know whether the
              influence of gravity waves can be distinguished from
              the effect of ambient fluctuations here.

J. KNOX   Underwater release data, sho-
              wing + buoyant behaviour, brings up the question of
              effect of rain on LNG vapour cloud behaviour.

J.S. PUTTOCK   There would presumably be a
small warming effect, but it is difficult to estimate
its extent. In addition, as we have shown, the high
humidity during a rainstorm would tend to enhance
LNG dispersion.

F.B. SMITH   Your experiments are in conditions of flow from land to sea, and it is likely the
turbulence, the stress, the heat flux and the wind
speed are not in equilibrium with the underlying surface. Do you feel this affects the behaviour of the
plumes ?

J.S. PUTTOCK   The fetch over sand and water
was 300 m. when the wind was directly offshore
(which occurred rarely) and more than this when the
wind was at an angle to the shoreline. So we feel
that there would not be a strong effect on the low
plumes at this distance from the shore. However, in
our continuing examination of the meteorological
data we are alert to the possibility of such effects.

D.R. MIDDLETON   Have you tried to use the
integral of the concentration probability distribution from lower to upper flammable units as a parameter for model comparisons with data ? This was
used by British Gas in jet flow as a criterion for
ignition probability.

J.S. PUTTOCK   No. We hope to undertake a
more detailed examination of the fluctuations of
concentration, with a consideration of ignition
probabilities, at a later stage of the analysis.

4: REMOTE SENSING AS A TOOL FOR
   AIR POLLUTION MODELING
   Chairmen: J. Irwin
              A. Venkatram
   Rapporteurs: G. Schayes
                 C. de Wispelaere

CORRELATION SPECTROMETRY AS A TOOL FOR MESOSCALE AIR POLLUTION

MODELING

N.D. van Egmond, D. Onderdelinden and H. Kesseboom

National Institute of Public Health
P.O.Box 1
Bilthoven - The Netherlands

INTRODUCTION

In the discussion on the merits of complex versus simple air pollution models, generally two criteria are considered:
- the physical credibility of the model, and
- the accuracy of model results in a comparison with measurements.
Both criteria are interrelated in the sense that a more detailed description of physical processes should result in an increase of model accuracy for situations in which these processes are relevant. It is rather inefficient to design complex models which only can be validated against overall parameters such as long term average concentrations; one should find a rational optimum between measuring and modeling effects. In plume modeling a widespread use is made of both mobile techniques, measuring ground level concentration profiles, and remote sensing methods like lidar and correlation spectrometry. An example of the application of correlation spectrometry in mesoscale studies will be given in this paper. The measurement data, vertically integrated concentration profiles (so called gasburden), are used for validation of a mesoscale model generating spatial gasburden profiles, which are directly comparable to the measured profiles.

THE MESOSCALE MODEL

The mesoscale PUFF model was described earlier in detail by Van Egmond and Kesseboom (1982). The lagrangian model describes the transport and dispersion over the 400 x 400 km surroundings of The Netherlands. The model is based on the transport of a large number of gaussian puffs with horizontal concentration distribution

$$C(r) = f \frac{M}{2 \, \Pi \, \sigma_r^2} \exp(-r^2/2 \, \sigma_r^2) \qquad (1)$$

where   r   the distance from the puff-centre
           σ_r the standard deviation (puff width)
           M   the mass represented by the puff
           f   a shape factor for the vertical profile over the surface
             layer, resulting from dry deposition at the surface
           h   height of the layer modelled by the puff.

The degree of sophistication is adapted to the routinely available input data:
- wind speed and direction at 40 stations at 10 m height and 5 stations (tv-towers) at levels between 150 and 300 m.
- solar radiation measured at 3 stations
- diurnal profile of mixing height measured by an acoustic sounder.
    In the model three atmospheric layers are distinguished:
- mixing layer, assumed to be constant during the night and increasing during daytime. The vertical concentration profile in this layer is assumed to be gaussian close to the source. The vertical dispersion σ_z is computed in the usual way, where stability class is derived from surface layer parameters.
- surface layer of 50 m, being the lowest part of the mixing layer. From wind and radiation data the surface layer parameters Obukhov length L and friction velocity u_* are derived. Together with deposition resistances near the surface, the concentration gradient f over the surface layer in quasi steady state can be obtained.
- reservoir layer, in which the pollution emitted by high sources is transported. This layer is situated above the mixing layer. Every puff is divided in a mixing layer and a reservoir layer part; during the morning increase of mixing height (fumigation), material is transported from the reservoir part to the mixing layer part.

Horizontal disperion is modelled by increasing σ_r according to

$$\sigma_r^2{}_{t + \Delta t} = \sigma_{rt} + 2 \, K_H \, \Delta t + 2 \, K_S \, \Delta t \qquad (2)$$

Herein K_H is the apparent horizontal diffusivity and K_S the additional diffusivity resulting from wind shear over the mixing or reservoir layer. The apparent horizontal diffusivity K_H is time dependent and given by

$$K_H(t) = \overline{v_M^2} \int_0^t R_L(t) \, dt \qquad (3)$$

$\overline{v_M^2}$ is the crosswind turbulence as derived from measured standard deviation of wind direction and from wind speed.
$R_L(t)$ is the lagrangian correlation function, assumed to be a negative exponential.

Shear dispersion is modelled as an additional 'dispersion' K_S. Given the wind shear S in rad/m over the extend Z of the vertical pollutant profile, a crosswind shear displacement $\Delta Y_s = Z\, S\Delta t$ will result after travel time Δt. The equivalent K_S then follows from

$$\frac{\overline{dy_s^2}}{dt} = S^2 \int_o^t \overline{z(t)z(t')}dt' \quad \text{(where the averaging is performed over all possible correlated z values at times t and t'):}$$

$$K_S\,(t) = \frac{1}{2}\frac{\overline{dy_s^2}}{dt} = \frac{1}{2}\,\sigma_z^2\,S^2\,t \qquad (4)$$

In this approximation $\sigma_z^2 = \overline{z^2} - \overline{z}^2$ gives the vertical extend of the plume with an upperbound at complete mixing $\sigma_z^2 = h^2/12$, i.e. the second momemt at a constant concentration over layer depth h. Given the approximative character of equation 4 the coefficient '1/2' will be considered as a parameter k_S to be determined from comparison of model and measurement results.

The model computes hourly average SO_2-concentration fields after a initialization period of 24 hours. The concentrations are presented as maps over the 400 x 400 km^2 area as constructed from concentrations at a set of (network)measurement stations and as spatial profiles of concentration and vertically integrated concentration (gasburden) along the route traversed by mobile measurement systems.

MEASUREMENTS: CORRELATION SPECTROMETRY

Two sources of measurement information are available for validation of the model:
- A network of 100 baseline grid stations over The Netherlands measuring a.o. hourly average SO_2-concentrations
- Mobile systems, measuring a.o. SO_2-(ground level) concentrations and -gasburden by means of a Barringer correlation spectrometer (Cospec IV). The mobile measurements are made over traverses of several hundreds of kilometers, more or less perpendicular to the overall wind direction, and take about three hours. Time variability is not taken into account.

The correlation spectrometer is used in the passive, upward looking mode. The total amount of SO_2 above the moving vehicle is obtained by comparing the measured diffuse skylight spectrum with the SO_2 absorption spectrum. The measurement procedure is delineated in figure 1. By integration of the measured gasburden profile and additional information on wind speed, the SO_2-flux over the traverse (or source strength) can be determined.

Resulting from light scattering at lower atmospheric levels on molecules and in particular aerosols, under- and overestimation of the true gasburden level occur. As higher SO_2-levels in mesoscale

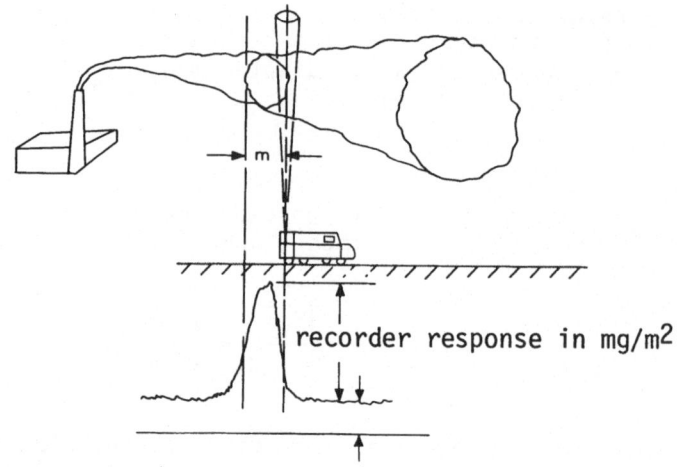

Fig. 1.   Measurement procedure for passive, upward looking corre-
           lation spectrometry.

plumes are mostly associated with high aerosol concentrations overestimation of gasburden is expected.
        Computations with single scatter models by Millan (1980) and
Van der Meulen (1982) suggest that overestimations of the order of
10% can occur under a wide plume. Due to the small elevation of the
sun in the time of the year considered, the effective optical path
length might increase considerably depending on the measure of "lower
fraction" of the sky light at ground level. This lower fraction, i.e.
the rays from the sun scattered inside the pollution layer into the
detector, can amount upto some 30% in the case considered. The sun's
elevation being of the order of 20°, than gives an optical path
length about a factor three higher than the assumed vertical path
length for this fraction of 30%. As the remaining "upper fraction"
of 70% has an efficiency of one, the overall efficiency will be
3x30% + 1x70% = 160%. Apart from a loss of spatial resolution this
implies a (maximum) overestimation of 60%.
        Nevertheless the gasburden profiles provide information which
ease the use of the model by creating a validation possibility in
between emission and ground level concentration. The use of mobile
remote sensors has the following advantages over the fixed station
concentration measurements:
- the emission position, relative source contribution and (with lower
  accuracy) absolute source strength can be determined,
- the gasburden levels are not affected by very local disturbances
  (such as the traffic) and are thus more representative for meso-
  scale pollutant patterns,
- the extremely high spatial density allows validation and, if necessary, updating of model parameters.

These features will be illustrated by the following example.

Example

To demonstrate the combined use of the PUFF-model and the mobile Cospec-measurements, the concentration field for February 3rd
1982 - 13.00 hours was computed and presented in figure 2. At a
stable south easterly anti-cyclonic circulation with relatively
high wind speed (~ 6 m/s) and low mixing depth (350 m) the mesoscale
plumes arising from source areas in The Netherlands, Belgium and
Germany dominate the 400 x 400 km^2 model area. In the same figure
the concentration field as measured by the 100 station network is
given. Further the spatial concentration and gasburden profiles are
presented, both measured and modelled; the profiles all correspond
to the measured concentration profile which is plotted over the modeled concentration field in the lower left part of the figure.

The modeling results presented in figure 2 apply to a $K_S = 0$
in equation 2. In figure 3 the corresponding results according to
equation 4 is presented ($k_s = \frac{1}{7}$). The following conclusion can be
drawn from these combined results:

1. The measured SO_2-gasburden seems to be unrealistically high. The
   modelled flux over the traverse amounts 150 tons/hour as derived
   from known emission figures in the upwind part of the model area.
   As the discrepancy between measured and modelled gasburden profile
   cannot fully be ascribed to the above mentioned optical overestimation, it is suggested that background SO_2 from outside the
   model region contributes to the measured gasburden profile.
2. On the southern part of the traverse two dominating plume zones
   are computed. This corresponds to the shape of the measured gasburden profile (figure 3). The southerly (left) plume does not
   affect ground concentrations. In the model the respective emission is erroneously allocated to the mixing layer; actually the
   plume appears to have penetrated through the inversion layer.
3. The correlation between network and model results is rather insensitive for variation of the shear parameter k_S.

For $K_S = 0$ (fig. 2), $K_S = \frac{1}{2} \sigma_z^2 s^2 t$ (fig. 3) and $K_S = 2 \sigma_z^2 s^2 t$

the correlations between measured (network) and modelled concentrations are respectively 0.48, 0.52 and 0.57. The comparison of measured and modelled gasburden profiles (fig. 2 and 3) allow a more
accurate determination of the coefficient in equation 4.

Comparison of measured and modelled concentration profiles
(fig. 3) supports the theoretical value of $k_s = \frac{1}{2}$. The extra (shear)-
diffusion suggested from the gasburden profile should be interpreted
carefully as the broadening of the plumes (loss of resolution) at
least partly might be due to the above mentioned optical effects.

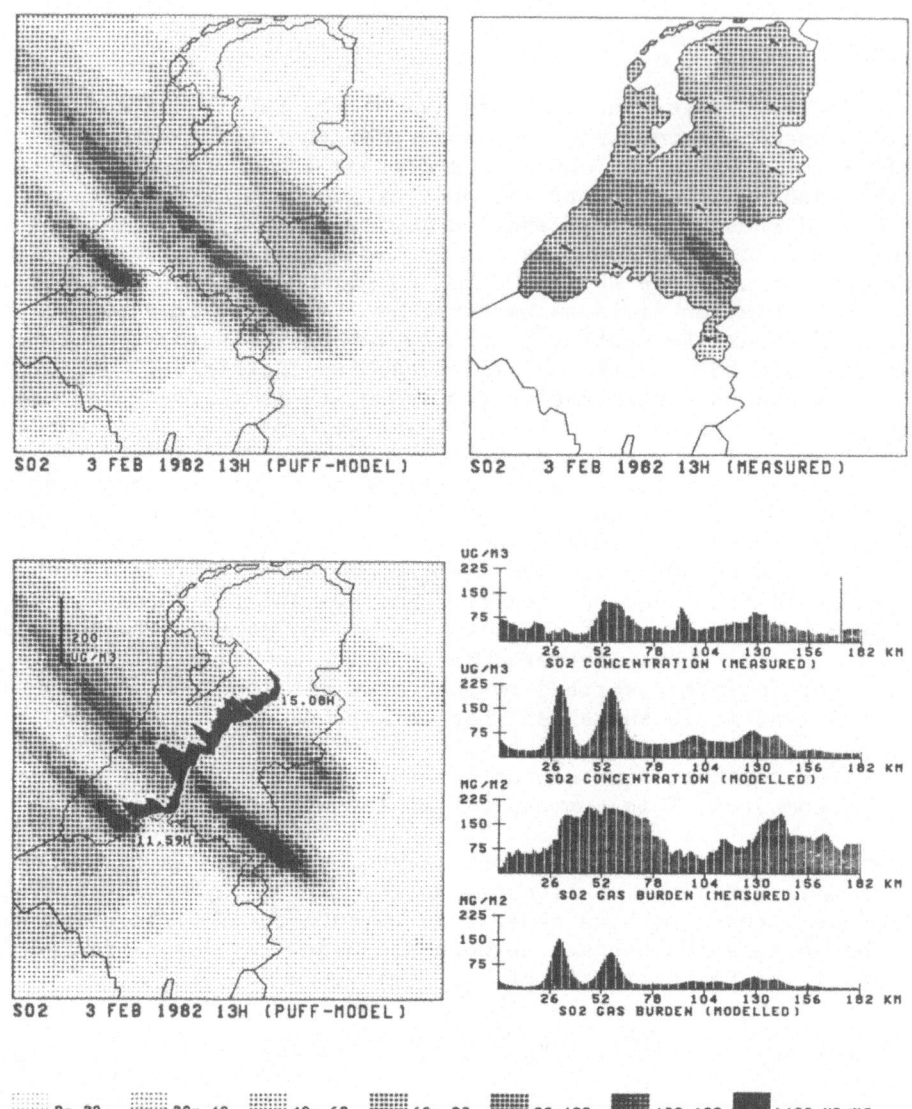

Fig. 2.   Modelled and measured (network and mobile traverse) SO_2
           concentration patterns; modelled and measured (mobile
           system) concentration and gasburden profiles. Additional
           wind shear disperion not modelled ($K_S = 0$).

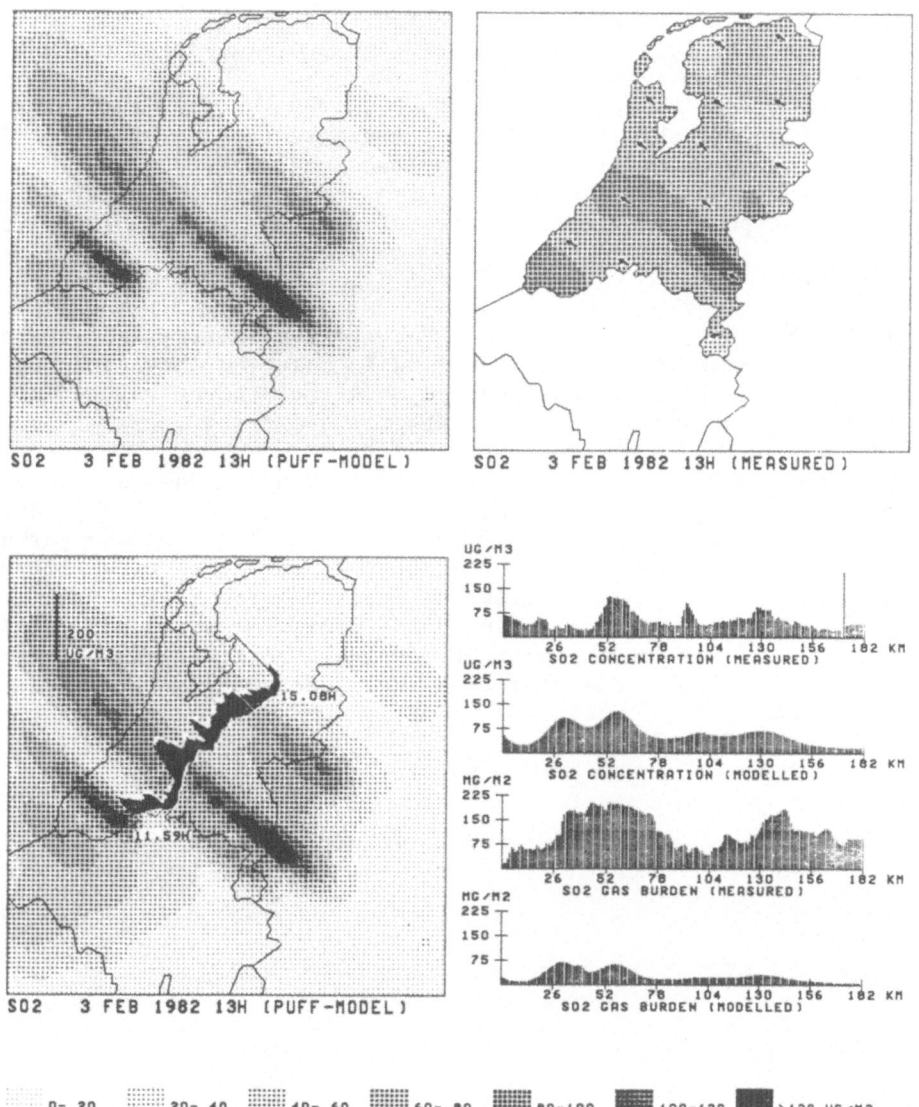

Fig. 3.   Modelled and measured (network + mobile traverses) SO₂-
          concentration patterns; modelled and measured (mobile
          system) concentration and gasburden profiles. Additional
          wind shear dispersion modelled according to equation 4.

REFERENCES

Egmond, N.D.van and Kesseboom, H., 1982, Mesoscale air pollution
        dispersion models B: Lagrangian PUFF-model, and compari-
        son with Eulerian GRID-model, Atm.Env. (accepted for publi-
        cation).
Meulen, A.van der and Onderdelinden, D., 1982, Optical pathlengths
        of zenith sky light in passive remote sensing of air pollu-
        tion, Atm.Env. (accepted for publication).
Millan, M.H., 1980, Remote sensing of air pollution. A study of some
        atmospheric scattering effects. Atm.Env. 14, pp. 1241-1253.

DISCUSSION

M. WILLIAMS   Could you enlarge upon the pro-
        blems, if any, involved in matching the averaging
        times of measured and modelled concentrations in rela-
        tion to model validation.

K. VAN EGMOND   Discrepancies due to averaging
        time were not considered so far, although this could
        be easily done by deriving concentration and gasbur-
        den levels from the model at the appropriate times.
        As the variability of concentration fields within the
        two hour duration of the mobile measurements is limi-
        ted for the current mesoscale applications no signi-
        ficant discrepancies are expected.

DOPPLER ACOUSTIC SOUNDING:

APPLICATION TO DISPERSION MODELING

Michael Chan, Ivar Tombach and Paul MacCready

Environmental Programs Division
AeroVironment Inc.
Pasadena, CA 91107

INTRODUCTION

The use of acoustics to study atmospheric properties is well-established. A review of atmospheric effects on acoustic signals can be found in Thomson (1975) and the history of acoustic sounder development is given by Gaynor (1982). However, the use of Doppler acoustic sounders in air pollution applications, as pointed out by Gaynor (1982), has gained acceptance only very recently. Doppler acoustic sounders with proven reliability have become commercially available only in the past decade and, even now, only a limited number of units are routinely being used to collect wind data.

This paper describes the measurement capabilities of one commercially available Doppler acoustic sounder and suggests how Doppler acoustic sounder data can be used in atmospheric dispersion modeling studies.

METEOROLOGICAL INPUT DATA REQUIREMENTS OF AIR POLLUTION DISPERSION MODELS

A very wide spectrum of mathematical air pollution dispersion models is now available, ranging from the simple Gaussian model designed for a single source on smooth terrain to models dealing with long-range transport over complex topography, taking into account chemical transformation, radioactive decay, and various scavenging phenomena. Furthermore, these models can be deterministic, stochastic, or adaptive (Drake, 1981). All these models, however, require only two types of meteorological information in order to determine the transport and dispersion of pollutants: the mean wind to describe the transport and some aspect of turbulence to relate to the dispersion.

Although various assumptions can be made to describe the mean and turbulent parts of the flow field, there are obvious benefits to the use of observed data. Air pollution dispersion models usually take the observed mean wind information at plume level or extrapolate and/or interpolate from data observed at other locations to arrive at the mean wind at the level of interest. That value is then applied directly to the transport term of the dispersion equations. One assumption regarding the mean wind is the use of an exponential law (DeMarrias, 1959) to account for the increase in wind speed with height. However no adjustment is made for changes in wind direction with height. This assumption might be acceptable in simple dispersion problems, such as emissions from a single stack in smooth terrain. However, it fails in coastal or complex terrain applications where strong wind shears and flow reversals are observed in the vertical wind profiles (Keen et al., 1979; Whiteman, 1981; Barr and Clements, 1981). This assumption also fails in strongly stable conditions in flat terrain.

The treatment of turbulence in the governing set of equations may be in the form of a well-mixed volume (Lettau, 1970), semi-empirical diffusion coefficients (Gifford, 1976), eddy diffusivities (Shir and Bornstein, 1977), Lagrangian statistics (Lamb, 1978), or higher-order turbulence closure approaches (Mellor and Yamada, 1974). Lagrangian statistics and higher-order turbulence closure approaches require the knowledge of the turbulence structure, such as the Reynolds stress, or the dissipation rate of kinetic energy. The most common approach, however, is the use of semi-empirical diffusive coefficients, which are estimated in various ways. These coefficients can be derived from the second moment of the wind field (turbulence intensity) as long as we assume that low diffusivity, which is concerned with a Lagrangian frame of reference, can be related to the turbulence intensity, which is established in an Eulerian frame of reference. For horizontal diffusion, a relation can be satisfactory which depends only on the time or space scales.

For vertical diffusion, the relation between diffusion and intensity depends additionally on a stability factor or classification (see the Gifford curves). The phenomenon underlying this relation is that an upward moving eddy which contributes a certain amount to the observed vertical turbulence produces more diffusion in unstable conditions than in stable conditions. (In the unstable case, it rises higher before merging its material with the surrounding air).

DESCRIPTION OF A DOPPLER ACOUSTIC SYSTEM

Continuous measurement of wind and turbulence throughout the atmospheric boundary layer (and, therefore, at plume height) has not been possible with commercially available equipment until recently. One such instrument capable of these measurements is the Doppler acoustic system (DAS) manufactured by AeroVironment Inc. A study of the applicability of that system to diffusion modeling has just been completed (MacCready and Worden, 1982).

System Description

The sensing element of the AeroVironment system consists of three adjacent pencil-beam antennas. Figure 1 shows one version of the system. One of the antennas is tilted 30° from the vertical and is aligned to point toward true north. Another is tilted 30° from the vertical and is aligned to point toward the east, while the third antenna points vertically.

These antennas are operated sequentially and in the monostatic mode. The system employs a standard driver feeding downward into a tuned horn, which spreads sound out to a 1.8-m parabolic reflector from which a pencil beam is emitted upward. The sound pulse is transmitted at 150 to 200 watts, a duration of 180 ms and a frequency of 1500 Hz.

To reduce sidelobes for transmitting or receiving, the assembly is housed in a large, heavy-walled acoustic enclosure. The enclosure interior is lined with a 5-cm thick layer of sound absorption material. The top edges of the enclosures are equipped with Thnadners™, which are comb-like structures which acoustically feather the edges and thereby greatly decrease diffraction.

The return signals are processed through a computer and displayed on a facsimile chart. The processed data are stored on digital tape as well as printed on a line printer. Figure 2 shows the data processing system.

Principle of Operation

The theory of using acoustics to determine atmospheric structures has been addressed by Little (1969) and Thomson (1975). Because the AeroVironment Doppler acoustic system operates in the monostatic mode, scattering of acoustic signals is due primarily to the thermal structure constant, C_T^2. The motion of scatterers causes a shift in the frequency of the returning acoustic signal. The system detects this shift, measures the motions of the scatterers along the beam, and, hence, is able to calculate the wind component in that direction.

Fig. 1.   The three-component Doppler acoustic system on a trailer.

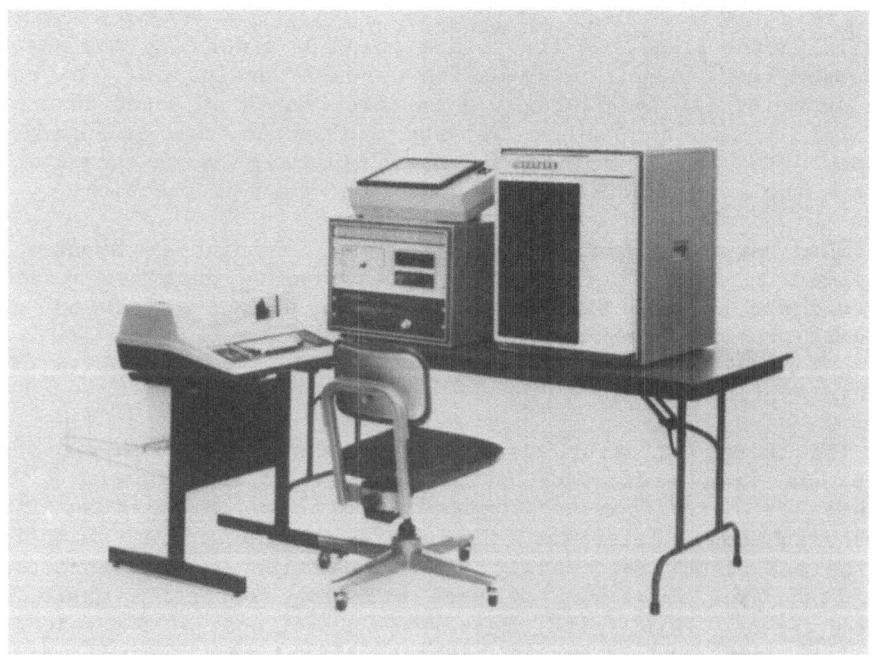

Fig. 2.   The data processing system for the DAS.

In actual operation, the DAS is detecting the motions of scatterers in a volume bounded by the area common to the transmitter/receiver beam and the range gate. The range gate of the system is 30 m in the vertical. The spectrum of frequency shifts is examined for acceptance or rejection based on a signal-to-noise criteria. By curve fitting to the integrated spectrum, the peak Doppler shift is determined, and by processing the return signals from all three antennas, we obtain the wind velocities along the three antenna beams. From these three velocities, the mean wind speeds in the horizontal and vertical, the horizontal wind direction, and the second moments of the along-beam components can be calculated.

It would not be appropriate to use data generated by this system to compute cross-correlations, since the three beams are sensing velocities sequentially and also probing different volumes of air.

Comparing the output parameters of this system with the meteorological input requirements of air pollution dispersion models, one finds that the parameters satisfy the requirements of most models, except some higher-order turbulence closure type models that need cross-correlations of velocities.

PERFORMANCE OF THE SYSTEM

The accuracy of the system was tested in August-September 1979 in Boulder, Colorado. This test was part of the Boulder Low-Level Intercomparison Experiment (Kaimal et al., 1980). Outputs from the system were compared against data collected by the 300-m instrumented tower of the Boulder Atmospheric Observatory as well as data collected by instrumented aircraft.

The tower was instrumented at eight levels (10, 22, 50, 100, 150, 200, 250, and 300 m) while the DAS provided wind information at 33.3-m intervals, beginning at 66.7 m. Thus, comparisons were possible from only the 50-m and higher levels. In addition, since the measurement heights of the two systems were not identical, some adjustments were made to allow intercomparison. These included comparing the DAS observations at 67 m with the tower observations at 50 m; comparing the average of the two DAS range gates at 133 and 167 m against the 150-m tower data; and comparing the average of the two DAS range gates at 233 and 267 m against the 250-m tower data.

The performance of the acoustic system is comprehensively discussed in MacCready and Worden (1982). The following summarizes their findings.

Accuracy of the Mean Velocity Values

Data from both systems were averaged over 20-minute intervals. For 308 cases for each horizontal axis (N-S and E-W), the horizontal component comparison showed the means agreeing within 0.05 m/s, the RMS differences being 0.82 m/s for E-W and 0.85 m/s for N-S, and the correlation for each component being 0.97. For 244 vertical cases, the means matched to 0.1 m/s and the RMS difference was 0.34 m/s. The correlation coefficient was 0.32.

Without more information, we cannot know how much of the variance between the tower and DAS data comes from instrumental inaccuracies and how much from meteorological variability. However, even if all the variance is due to the instrument, the instrument still appears adequate for providing the key inputs for diffusion study. However, additional investigation indicates that the main contribution to the variance was meteorological variability, not instrumental errors.

The problems with comparing DAS and tower data are that the sampling periods were not identical (DAS was operated for the last 17 minutes of each 20-minute sampling interval), and the horizontal separation between the DAS and tower was over 400 m. One comparison case emphasizes the problem (see MacCready and Worden, 1982). For three consecutive 20-minute periods, the DAS gave rather similar velocity component profiles, and for the first two periods, the DAS profiles agreed well with the tower profiles. However, for the third period, the tower top observation showed a wind about 9 m/s different from earlier observations (and temperature gradient observations indicated a cold wedge aloft prevailed at the tower). We have no reason to suspect a measurement error in either the DAS or tower. We are thus confronted with determining DAS errors of the order of 0.1 m/s in the presence of meteorological aberrations two orders of magnitude larger.

Balser and Netterville (1981), other participants in the intercomparison experiment, also felt that the meteorological variability was responsible for much of the variance between the tower data and that from their Doppler acoustic sounder. Their system was sited 300 m from the tower. During a two-hour period when the transport wind was steady and blowing from the tower to their system, they were able to show that imposing a two-minute lag on the tower data resulted in a good correlation between the vertical wind components from their system and the tower.

The vertical profiles of mean wind data from the AeroVironment DAS and the tower showed the same general features and a smooth continuity with altitude. Figure 3 shows a comparison of these vertical profiles. If the differences between the DAS and the tower data were due to random instrumental errors in the DAS, the difference would destroy the continuity of the DAS profiles. Therefore, the main difference is caused by variations in the meteorology at the tower and in the volume sampled by the DAS. The continuity of the profiles would be destroyed if the measurement errors in U and V for the DAS exceeded 0.1 m/s and errors in W exceeded 0.02 to 0.05 m/s.

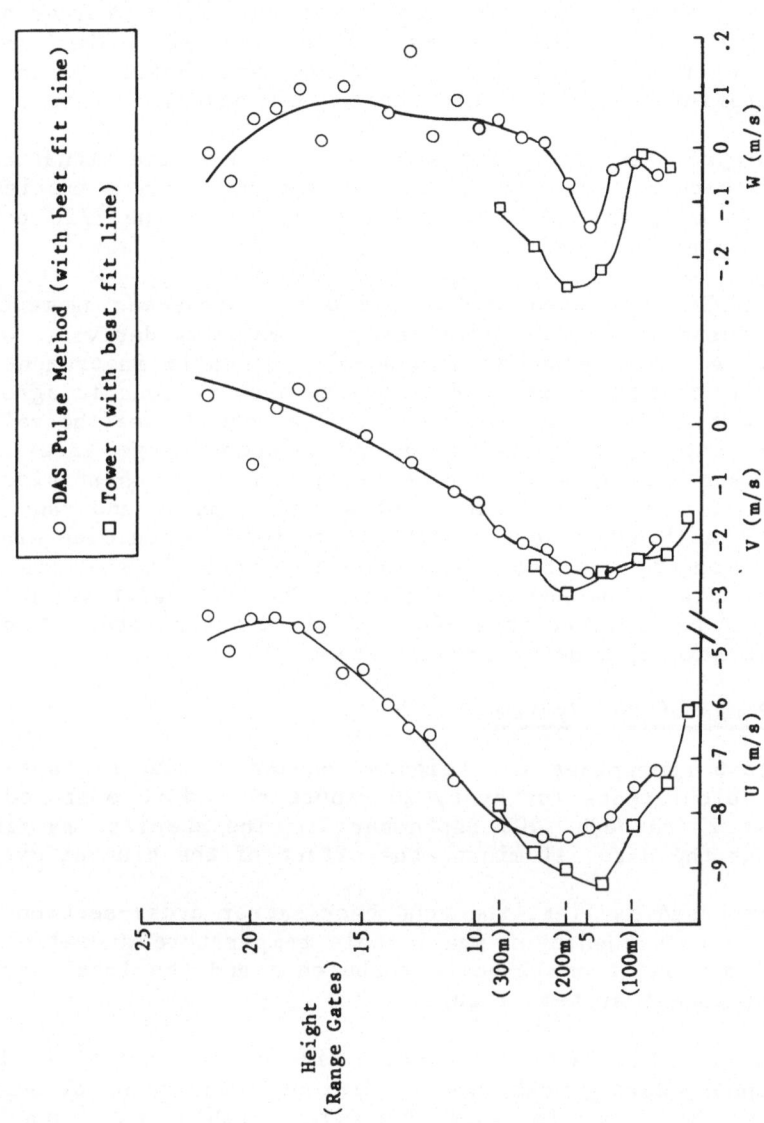

Fig. 3. Height profiles of mean components for a weakly stable early evening case (8/30, 1900).

Accuracy of the Turbulence Calculations

Two turbulence parameters, σ_w and ϵ, calculated from the DAS data were compared with those from the tower data.

Figure 4 shows a comparison of σ_w at 300 m. σ_w values as low as 0.09, 0.10, 0.12, and 0.15 ms^{-1} were observed. The true σ_w cannot be negative, so in these cases the contribution to apparent σ_w from instrument errors cannot exceed the observed values.

Height profiles of σ_w for an evolving convective situation are shown in Figure 5. The continuity of the pulse data implies DAS errors of no more than 0.1 ms^{-1}, and portions of the 1120 profile suggest a number half as large.

The DAS-Boulder tapes were analyzed by a differencing method to see if the turbulence dissipation rate, ϵ, could be derived. Gaynor (1977) had used the method with a Doppler acoustic instrument and, for carefully selected cases, was able to show reasonable agreement between tower and Doppler system data. The method uses the velocity differences between adjacent or second-adjacent range gates. For the DAS-tower comparison, the technique proved unsatisfactory. Evidently, the minimum resolution of a single pulse and range gate for the DAS, although fine for statistics yielding the mean wind and the second moment, was too large when the variable was the small difference between nearby velocities. The DAS relative accuracy would have to be improved three-fold to yield satisfactory ϵ observations for stable, weak turbulence cases.

Altitude Range of the System

Figure 6 summarizes the heights reached by DAS for speed and direction calculations for every 20-minute period it operated from 0000 August 27 through 2400 September 2. The abscissa is time of day, so that the data illuminate the effect of the diurnal cycle.

As discussed earlier, the echo backscatter cross-section coefficient is a consequence of small-scale temperature anomalies, and hence, a function of small-scale turbulence and the local vertical gradient of potential temperature.

As shown on Figure 6, the poorest time of day, for altitude, is late afternoon, during weak wind conditions. The onset of negative heat flux at the ground is associated with a stable lapse rate in the lowest regime and decoupling of the air above from the surface-generated turbulence. The turbulence aloft decays, and velocity and potential temperature gradients are very weak.

In summary, at Boulder at night, the DAS reached 1400 m once and 1,000 m often; during late afternoon it reached 150 to 800 m, averaging about 500 m.

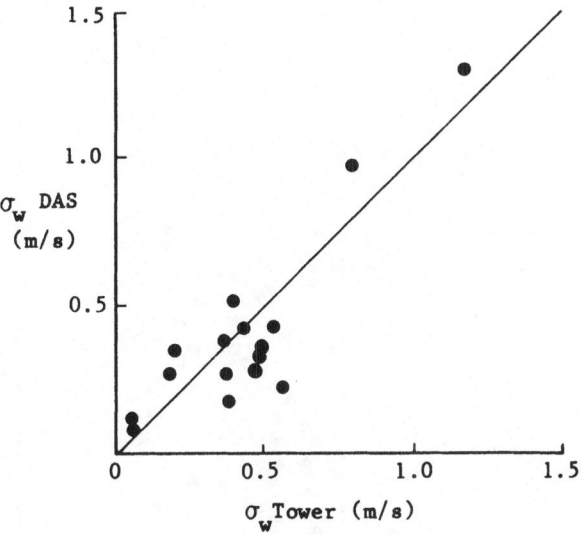

Fig. 4. σ_w DAS versus σ_w tower, 300 m.

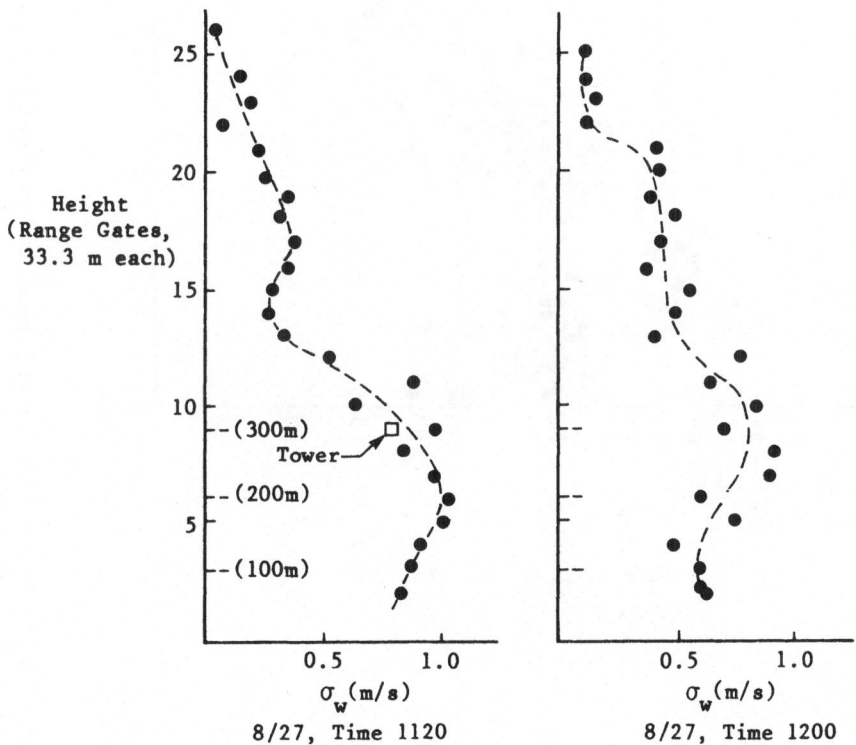

Fig. 5.   Height profiles of σ_w for an evolving convective situation.
Dashed lines are eyeball best fit to pulse method points.

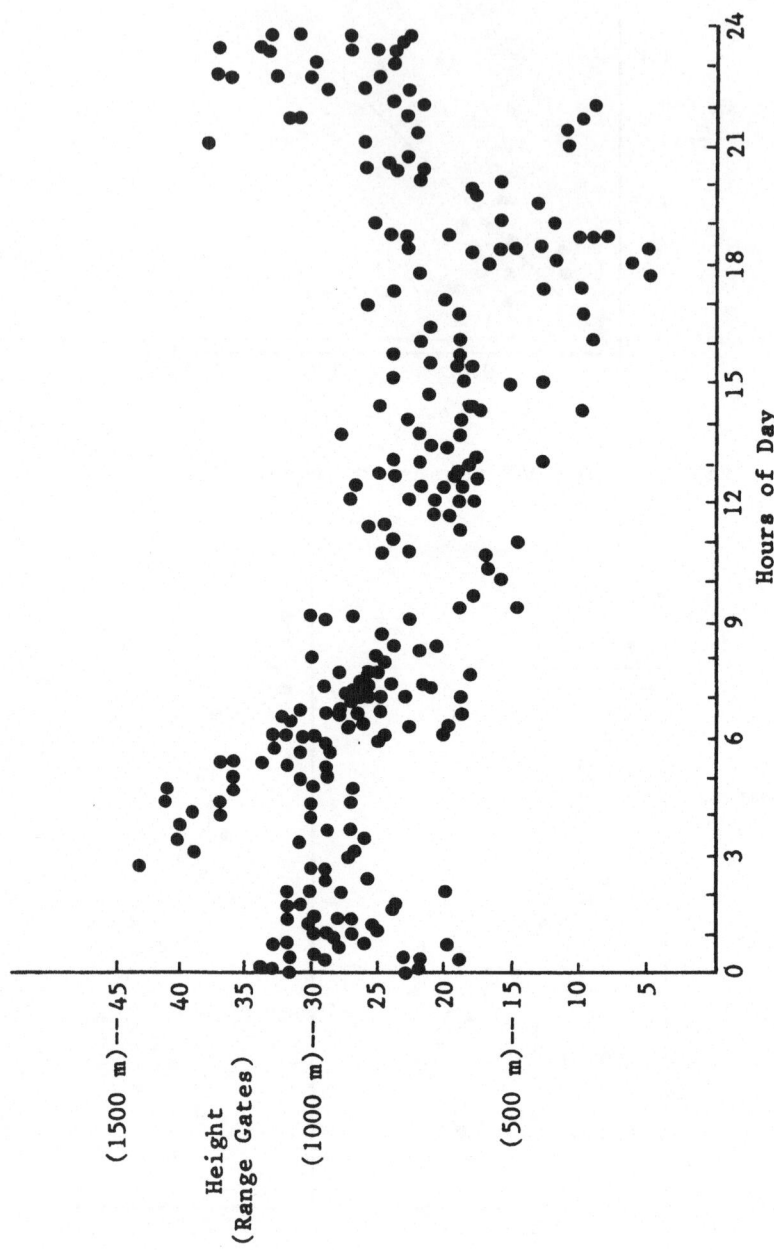

For the 53 official cases on seven days of DAS-tower intercomparison measurements, the tower top height of 300 m was reached 98% of the time, 400 m 92%, 500 m 77%, 750 m 45%, and 1,000 m 13%.

Summary of DAS Performance

From the Boulder Low-Level Intercomparison Experiment, the performance of the DAS can be summarized as follows:

1. For 20-minute averages, with good echoes, mean vertical component relative accuracy is in the range of 0.02 to 0.05 m/s, mean horizontal component accuracy (ignoring vertical correction) is better than 0.1 m/s, and sigma accuracies are better than 0.1 m/s and perhaps even 0.05 m/s for small sigmas.

2. The 95% and 5% data recovery percentiles correspond approximately to 300 and 1000 m, with percent versus altitude varying linearly between these heights.

3. $\epsilon^{1/3}$ calculations have errors too large to make this a useful DAS measurement. A three-fold improvement in resolution/relative accuracy between pulse pairs would permit the DAS to be useful for $\epsilon^{1/3}$ measurements.

APPLICATION OF DAS TO MODELING STUDIES

The greatest benefit of the DAS to the modeling community is the availability, on a continuous basis, of mean wind and turbulence information at a reasonable cost. The only other commercially available means of obtaining these data is by an instrumented tower. The cost of installing and operating an instrumented tower over 100 m high, however, is formidable. Upper-air wind information can be obtained by pilot-balloon or rawinsonde releases. But these do not provide continuous information; rather, they give a snapshot of the vertical wind profile.

The precision of DAS wind velocity component measurements, as presented in the last section, should be adequate for most air pollution dispersion modeling purposes. As discussed earlier, all models need mean wind information to predict the transport of pollution. The DAS provides the mean wind data at the effective stack height, the level at which such information is most important. The other measurements provided by DAS are the along-beam turbulent components. From these, we can obtain directly σ_w, the standard deviation of vertical velocity, and indirectly σ_θ, the standard deviation of wind direction (or equivalently, σ_v, the horizontal crosswind component). The techniques for computing σ_θ have been described by MacCready and Worden (1982). The report also discusses ways of deriving stability category information from DAS observations, particularly σ_w profiles. A rapid fall-off of σ_w with height denotes stability and the top of the mixing region; an increase of σ_w with height is consistent with instability. From σ_θ, the lateral plume spread, σ_y, can be obtained. From σ_w and stability classification information, the vertical plume spread, σ_z, can be obtained.

DAS should be used for the following types of modeling studies:

1. Impact Assessment. Data collected with a DAS can be used to predict expected exceedances of certain pollutant levels or as a climatological data base to determine frequency distributions of impacts.

2. Model Validation Studies. Vertical wind (mean and turbulence) profiles can be used to evaluate the accuracy of a model. Using actual wind data rather than relying on extrapolation would reduce the number of variables necessary to test the accuracy of a model.

3. Supplementary Control Systems. DAS provides real-time continuous information on wind profiles that can be used to activate supplementary control systems.

4. Emergency Response Decisions. In the event of accidental releases of radioactive or toxic substances, the availability of continuous upper-air wind data would be essential to modeling tasks, the results of which are the basis of emergency response decisions.

ACKNOWLEDGEMENT

The funding for this study was provided by the Electric Power Research Institute.

REFERENCES

Balser, M., and D. Netterville, 1981, Measuring wind turbulence with Doppler-acoustic radar, J. Appl. Met., 20:27.

Barr, S., and W.E. Clements, 1981, Nocturnal wind characteristics in high terrain of the Piceance Basin, Colorado, presented at the 2nd Conference on Mountain Meteorology, November 9-12, Steamboat Springs, Colorado.

DeMarrias, G.A., 1959, Wind speed profiles at Brookhaven National Laboratory, J. Appl. Met., 16:181.

Lamb, R.G., 1978, A numerical simulation of dispersion from an elevated point source in the convective planetary boundary layer, Atmos. Envir., 12:129.

Drake, R.L., 1981, Evaluation of models for prediction of photo-chemical secondary pollutants, Battelle Pacific Northwest Laboratories, Richland, Washington.

Gaynor, J.E., 1977, Acoustic Doppler measurement of atmospheric boundary layer velocity structure functions and energy dissipation rates, J. Appl. Met., 16:148.

Gaynor, J.E., 1982, Present and future use of sodars in air quality studies by industry and government, presented at the 75th Annual Meeting of the Air Pollution Control Association, June 20-25, New Orleans, Louisiana.

Gifford, F.A., 1976, Turbulent diffusion typing schemes: a review, Nuclear Safety, 17:171.

Kaimal, J.C., H.W. Boynton and J.E. Gaynor, 1980, The Boulder low-level intercomparison experiment, NOAA/ERL, Boulder, Colorado.

Keen, C.S., W.A., Lyons, and J.A. Schuh, 1979, Air pollution transport studies in a coastal zone using kinematic diagnostic analysis, J. Appl. Met., 16:181.

Lettau, H.H., 1970, Physical and meteorological basis for mathematical models of urban diffusion processes, in: Proceedings of the Symposium on Multiple-Source Urban Diffusion Models, Publ. No. AP-86, U.S. Environmental Protection Agency, Washington, D.C.

Little, C.G., 1969, Acoustic methods for the remote probing of the lower atmosphere, Proc. IEEE, 57:571.

MacCready, P., and J. Worden, 1982, Doppler acoustic sounding: observational inputs to pollution dispersion models, AeroVironment Inc., Pasadena, California.

Mellor, G.L., and T. Yamada, 1974, A hierarchy of turbulence closure models for planetary boundary layers, J. Atmos. Sci., 31:1791.

Shir, C.C., and R.D. Bornstein, 1977, Eddy exchange coefficients in numerical models of the planetary boundary layer. Atmos. Envir., 12:1297.

Thomson, D.W., 1975, Acdar meteorology: the application and interpretation of atmospheric acoustic sounding measurements, presented at the 3rd Symposium on Meteorological Observations and Instrumentation, February 10-13, Washington, D.C.

Whiteman, C.D., 1982, Breakup of temperature inversions in deep mountain valleys, point observations, J. Appl. Met., 21:270.

DISCUSSION

P. MISRA   There seems to be a decrease
                  in maximum σ_w between 11 a.m. and 12 a.m. in one of
                  your slides, whereas the mixed layer height increased
                  by almost a factor of two in the same period. I
                  should have expected an increase instead. Please com-
                  ment ?

M. CHAN   The σ_w data presented are ave-
                  raged over 17 minutes. The increase in the peak va-
                  lue is within the accuracy of the instrument. One
                  explanation might be that thermals were more organi-
                  zed during that period, thus the σ_w at that level was
                  reduced. The important point, however, is that there
                  was an overall increase in σ_w throughout a deeper layer
                  indicating the growth of the mixed layer.

F. FANAKI   I have noticed the speaker did
                  not comment on the limitations of the system in par-
                  ticular during emergency use. One serious limitation,
                  I believe, is the system requires a noiseless envi-
                  ronment. Would the speaker like to comment on how to
                  overcome this problem ?

M. CHAN   The Doppler acoustic radar does
                  not need an absolutely quiet environment to operate.
                  It has been used routinely in airports and urban
                  areas. However, noise does reduce the altitude range
                  of the system as well as its performance. Research
                  is going on to develop better electronics to improve
                  the signal to noise ratio.

J. KNOX   In regard to use of acoustic
                  sounder for emergency, a reactor accident involving
                  entrainment failure and a steam-jet existing during
                  accident leads to a noisy environment; the proposed
                  use of acoustic sounder in such a noisy environment
                  is doubtful. Whereas, laser systems operating on a
                  Doppler principe, for instance, could produce use-
                  ful 3D wind data for input to models for transport
                  and dispersion of release.

M. CHAN   Even though accidents in a
                  nuclear reactor can result in a noisy situation, the
                  presence of a Doppler acoustic system could still be
                  extremely useful.

This is because one needs 3-D meteorological data
to exercise dispersion models immediately following
the accident and data collected continuously by the
Doppler acoustic system prior to the accident can be
used to forecast wind conditions as well as to model
dispersion. The use of a Doppler laser system is a
good idea but I am not sure such a system that is
reliable is commercially available whereas the Doppler
acoustic system is.

PLUME MODELING FROM METEOROLOGICAL DOPPLER RADAR DATA

Henri Sauvageot[*] and Assad Emile Saab[**]

[*]Centre de Recherches Atmosphériques
65300 Lannemezan - France

[**]Direction des Etudes et Recherches - Electricité de
France - 6 quai Watier - 78400 Chatou - France

1. INTRODUCTION

To compare the results of modeling with full scale situations
in air pollution problems, it is necessary to measure accurately the
successive states of the local atmosphere ; from this point of view
remote sensing techniques are useful so far as they are able to probe
great atmospheric volumes with the necessary spatial and temporal
continuity.

Remote sensing techniques used in air pollution problems were
almost exclusively restricted up to now to lidar and sodar. Lidar is
employed to obtain some data on the tridimensional distribution of
some air pollution concentration (for example aerosol concentration)
and for air motion measurements (see for example Teissier du Cros,
1977 ; Kunkel et al., 1980). Nevertheless lidar is inefficient in
the optically opaque or poorly transmissive media. Sodar is used
because it enables to characterize easily the dynamic and thermodynamic structures of the low atmospheric layers : wind profiles,
height of stratified layers, convection in clear air, etc... (Beran
and Hall, 1974). But sodar is generally restricted to observing along
fixed directions around the vertical. In the particular case of
isolated pollution sources, the above quoted limitations for lidar
and sodar techniques apply but with more disadvantage because
the measurement problems for plumes are more acute. Indeed
frequently local atmospheric perturbations are rapidly evolving and
cannot be treated from space and time average measurements.

Radar was almost never used for air pollution studies. The
reason is that radar is mainly known as a tool for observations and

measurements of reflectivity and velocity fields in regions of the atmosphere where precipitations are present because precipitating particles are the main scattering cause for the electromagnetic waves of meteorological radar. In fact many other possibilities exists (see for example Battan, 1973 or Sauvageot, 1982).

The object of this paper is to discuss the ability of radars to provide observations and measurements useful for air pollution modeling, especially in the case of an isolated pollution source as a plume. The problem of radar tracers is firstly examined, then the determination by radar of the characteristics of environmental mean field and plume structure is discussed.

2. THE RADAR TRACERS

Radar measurements depend on the presence in the atmosphere of a tracer convenient to the wanted observations. Tracers can be natural or artificial. In fact the tracer problem comes up in a slightly different way for observing the mean field or an isolated pollution source as a plume. In the following section a short synthesis on this point is given (for more details, see Sauvageot, 1982).

2.1. The natural tracers

The natural tracers generally used for meteorological radar observations are the precipitating hydrometeors (rain, snow, hail) ; non precipitating hydrometeors (droplets or tiny ice crystals in clouds) are close to the limit of radar detection (cf. section 4).

For the usual meteorological radar wavelengths (between 3 and 10 cm), the hydrometeor size to wavelength ratio is small (except for hail) so that radar backscattering is situated in the Rayleigh scattering region. In these conditions, the backscattered intensity is proportional to the inverse of the fourth power of the radar wavelength and to the sixth power of the scatterer diameter. The radar reflectivity η (defined as the average radar backscattering cross section of the target by unit volume) is given by the equation :

$$\eta(\lambda) = \frac{\pi^5}{\lambda^4} |K|^2 \frac{1}{V} \sum_V D^6$$

where :
$\sum_V$ is the summation for the radar pulse volume V

D is the equivalent diameter of the scatterer

$|K|^2$ is the dielectric factor of the scatterer

Then a radar is very sensible to large hydrometeors and the sensibility of detection increases for decreasing wavelengths.

Hydrometeor detection by radar enables to obtain an estimate of some physical characteristics of the target (precipitation rate, liquid water content, etc...) and also many data on air motions.

Yet, other natural tracers are available in the atmosphere : insects and turbulent fluctuations of air refractive index.

Always present in dry atmosphere at positive temperatures, insects generally have a proper motion relative to the air very slow. So their motion relative to the ground can be used as an approximative air velocity measurement. It was also shown that their spatio-temporal distribution exhibits a close dependency on the thermodynamic structure of the atmospheric boundary layers (Campistron and Sauvageot, 1974 ; Campistron, 1975). However, on account of geographic, seasonal and diurnal variations of his distributions, this tracer, is only occasionally usable (except at low latitude).

Turbulent fluctuations of air refraction index (n) are an important and nearly universal radar tracer. In a turbulent mixing process, the air parcels rapidly displaced keep temporarily their identities; the pressure undergoes a continuous equilization with the environment but the potential temperature and the specific humidity are preserved. The result is to create inhomogeneities in the refractive index field. The stronger the turbulence and the sharper the initial gradient in temperature and humidity, the stronger will be the refractive index inhomogeneities.

Detection of such fluctuations are accessible to high power, high sensibility radars using wavelengths equal to or larger than about 10 cm.

It is shown that, for a well developped turbulence and if the half radar wavelength is included in the inertial subrange, the radar reflectivity η is proportional to the structure constant of the refractive index C_n^2 (see Ottersten, 1969 among others) and we have for a unit volume :

$$\eta(\lambda) \simeq 0.38 \ C_n^2 \ \lambda^{-1/3}$$

This tracer is almost always available, the atmosphere never being at rest. From the detection of this tracer, data are obtained on the turbulence intensity, on many dynamic and thermodynamic structures of small and medium scales (stratification, waves of diverse kinds, convective cells, etc...) and also obviously on the mean and turbulent air motions.

Although radars able to detect the turbulent fluctuations of n are rare, this technique is rapidly evolving. It offers a real and important possibility owing to the permanence of the tracer.

2.2. Artificial tracers

When no natural tracer (detectable by the used radar) is available in the atmosphere, a possibility remains which is the use of artificial tracers. The most currently employed artificial radar tracers for atmospheric observations are resonant passive electromagnetic dipoles calles chaff. These dipoles are made of metallised glass fibre cut to a length around the half radar wavelength. Their terminal fall velocity is smaller than 0.3 m s^{-1}. Consequently chaff displacements follow reliably the atmospheric motions.

For N dipoles by unit volume, randomly oriented and homogeneously distributed in the radar pulse volume, the radar reflectivity value is given by the relationship :

$$\eta(\lambda) = 0.18 \ N \ \lambda^2$$

Then, radar reflectivity is proportional to the number of dipoles by unit volume since all scatterers have same radar backscattering cross section. However, chaffs are not really randomly oriented, owing to aerodynamic effects, they tend to fall horizontally.

Chaff can be used to trace air motions characterizing the environmental mean field or to visualize by radar any air flow particularities. Chaffs are particularly convenient for plume study (see section 4).

3. BASIC RADAR DATA

The main quantities measured by a coherent (Doppler) radar are :

- the radar reflectivity η which is proportional to the number and radar backscattering cross section of the scatterers situated inside the pulse volume. η is deduced from the measurement of the average received power $\overline{P}_r$ from the radar equation :

$$\overline{P}_r = C \ \eta/r^2$$

where C is a constant depending on the radar technical characteristics and r is the radar-target distance ;

- the mean radial Doppler velocity $\overline{V}$;

- the radial Doppler velocity variance $\sigma_v{}^2$;

- the radial Doppler velocity spectrum $S(V)$.

The first three parameters are the first three moments of the Doppler spectrum:

$$\overline{P}_r = m_0$$

$$\overline{V} = m_1/m_0$$

$$\sigma_v^2 = (m_2-m_1^2) / m_0$$

with $m_n = \displaystyle\int V^n \ S(V) \ dV$

Then knowing $S(V)$, it is possible to calculate $\overline{P}_r$, $\overline{V}$ and σ_v^2. However these three quantities can be obtained directly, without explicit spectrum determination, from estimators working in real time.

With conventional (non Doppler) radar, mean velocity measurements are not possible but the other quantities can be obtained.

4. DETERMINATION OF MEAN FIELD CHARACTERISTICS

The useful data for modeling concern the physics and dynamics of the local atmosphere. The informations obtained from radars are qualitative and quantitative ; their usefulness depend on the peculiarities of each case.

4.1. Physical structure

The data provided by the radar are:

- the tridimensional structure of clouds and precipitation : size and spacing of the cells, some informations on the nature of the hydrometeors at each level. Particularly the height of the clouds and the convective field top is determined ;

- the reflectivity field. Under some conditions relative to the knowing of the hydrometeor nature, reflectivity field can be converted into physical quantities describing the scattering medium : precipitation rate R, water content M or characteristic diameters of the particles such as median volume diameter D_0 ;

- from Doppler spectra measured at the vertical of the radar, accurate precipitation size distributions are obtained with their incidental anomalies and their temporal variations.

4.2. Dynamic structure

From Doppler velocities, vertical air velocities can be determined at the vertical of the radar by correcting the Doppler values for the terminal fall velocity of the tracer in still air.

The other parameters describing the environmental dynamic field can be measured by different methods. When two Doppler radars (or more) are available, the tridimensional velocity field is obtained by combining the radial velocity measurement of each radar.

With only one Doppler radar, dynamic characteristics of the environmental field can be determined from azimuthal scanning at constant elevation. This method called VAD (Velocity - Azimuth - Display) enables to calculate (Browning and Wexler, 1968) :

- the vertical profile of the horizontal wind in velocity and direction ;

- the vertical profile of the mean vertical air velocity ;

- the horizontal divergence : $\dfrac{\partial \overline{u}}{\partial x} + \dfrac{\partial \overline{v}}{\partial y}$

where $\overline{u}$ and $\overline{v}$ are the mean wind velocity along the two perpendicular horizontal direction x and y ;

- the stretching deformation : $\dfrac{\partial \overline{u}}{\partial x} - \dfrac{\partial \overline{v}}{\partial y}$;

- the shearing deformation : $\dfrac{\partial \overline{u}}{\partial y} + \dfrac{\partial \overline{v}}{\partial x}$.

Taking into account the Doppler velocity fluctuations along the VAD scanning circles, different informations concerning the field of turbulence are attainable (Wilson, 1970) :

- an estimate of the horizontal turbulent cinetic energy ;

- an estimate of the turbulent field isotropy ;

- estimates of the momentum fluxes : $\overline{u'w'}$, $\overline{v'w'}$, $\overline{u'v'}$.

5. DETERMINATION OF PLUME CHARACTERISTICS

Among the various encountered plume conditions, the main two cases are dry plumes and wet plumes.

5.1. Physical structure

Most dry plumes are constituted of gaseous pollution, dry aerosols or dust with size smaller than ten μm. Radar detection without artificial tracers is not possible in this case.

Dry plumes including larger particles and wet plumes with hydrometeor sizes larger than several ten μm are detectable with millimetric radar. Then the physical structure of the plume can be analysed from the measurement of the radar reflectivity distribution. The interpretation of the data as physical quantities characterizing the scattering medium (such as volumic, massic or numerical concentration of scatterers) is possible only if relations between the radar reflectivity and the physical quantities are available.

For the plumes, in most cases, no general relation is usable.
Each case is a particular one. Then it is indispensible to set up a
specific relation, if possible, for each plume. This work needs to
have sufficient knowledges on the properties of the scattering medium (size distribution, dielectric factor, etc.) to calculate a
relation. Such knowledes can be obtained from simultaneous (in time)
and concordant (in space) measurements of the scattering medium properties by instrumented aircraft and of the radar reflectivity by radar.

The case of wet plumes developed from the same physical processes
as natural clouds has to be considered separately. An example of that
is the cloudy condensation associated to thermal plumes when the
physical properties of the convective air (humidity and condensation
nucleis) are similar to those of the environmental air (fig. 1). In
this case, only the convection cause is artificial, the microstructure

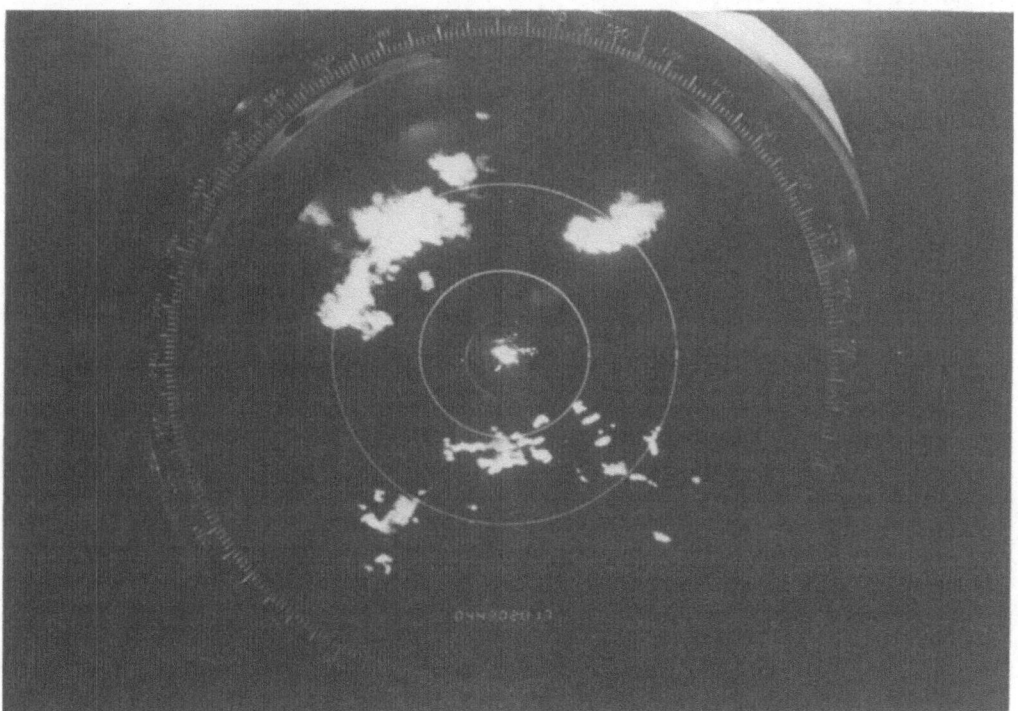

Fig. 1. PPI radar scope photograph showing : in the upper right part
        the radar echo of an artificial thermal plume merging in a
        partly artificial cloud, and in the upper left part natural
        cloud echoes. The echoes in lower half part are ground clutter.
        Range markers are 2 km from each other. The observation was
        made at 1540 GMT on 27 June 1978 during the COCAGNE experiment
        with a millimetric radar (see Sauvageot et al., 1982).

of the associated cloud is nearly the same as in natural clouds and
the same relations can be used. Inversely, the plumes associated
with wet cooling towers in presence of anomalous droplet size distributions (large "primage" droplets) cannot be treated like natural
clouds, particularly for the radar reflectivity.

For example in wet plumes with granulometric distributions close
to natural clouds and droplets smaller than 150 µm, mean radar
reflectivity factor is approximately related to liquid water content
M and mean volume diameter D_0 by the relationship (Sauvageot and Omar,
1982).

$$Z = 0.068 \ M^{1.94}$$

$$Z = 3.6 \times 10^{-7} \ D_0{}^{3.16}$$

where :
$$M = \frac{\pi}{6} \rho \int_{D_{min}}^{D_{max}} N(D) \ D^3 \ dD$$

D_0 is defined by :

$$\frac{M}{2} = \frac{\pi}{6} \rho \int_{D_{min}}^{D_0} N(D) \ D^3 \ dD$$

Z is in $mm^6 m^{-3}$, M is in $g \ m^{-3}$ and D_0 is in µm

N(D) dD is the number of droplets with diameter between D and d + dD.

5.2. Geometry and dynamic structure

A plume observed with a radar from a distance of some kilometers
is a target relatively narrow in so far as the observations are
made with small amplitude beam scans. Obviously, in such conditions,
the only measurement of the Doppler velocity of the plume tracers
in the aim direction do not enable to determine unambiguously the
velocity field. Using simultaneously two radars would permit to
reach a complete observation. When a single radar is available, it
is necessary to have some complementary data.

In the case a plume undetectable directly by radar, observations
can be done from artificial tracers released in and around the plume
source and used as lagrangian tracers (fig. 2). This technique is
particularly convenient for thermal plume ; indeed in this case the
tracers permit to visualize the entire volume occupied by the convective air (Sauvageot et al., 1982). Then, observing with a conventional
radar rapidly scanning the plume by sectoral scans the motions of
the chaff bundles discontinuously released from the heat source
permits to obtain :

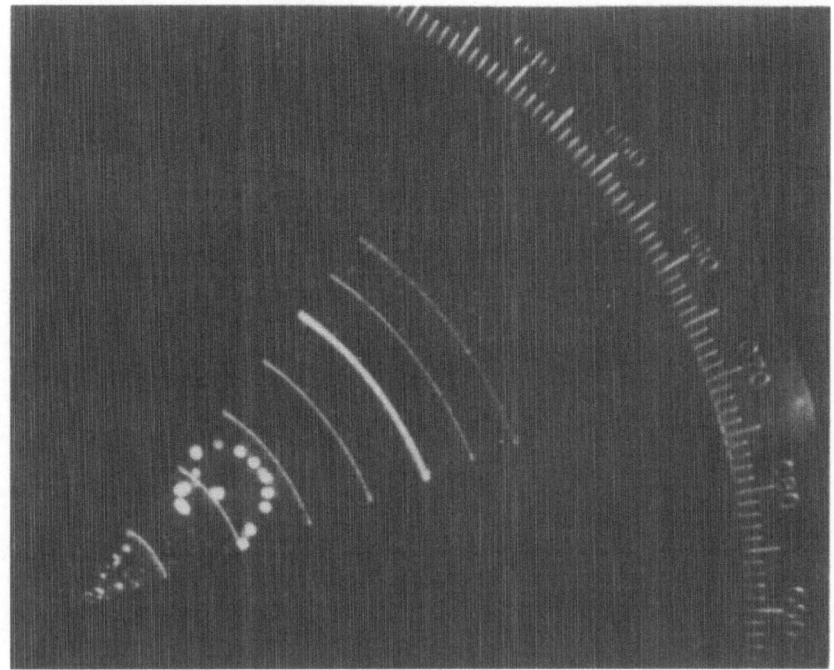

Fig. 2. PPI photograph showing a ring of chaff bundles released
around a thermal plume from a light airplane in order to
visualize the environmental circulation. The echo located
near the centre of the ring in the one of the thermal plume
(also due to chaff backscatter). The observation was made
on 19 October 1978 at 1151 GMT during the COCAGNE experiment.
The range markers are 2 km from each other.

- an accurate determination of plume boundary ;

- a measurement of the maximum height of the plume ;

- measurements of the turbulence inside the plume.

In addition the observation of the spatio temporal evolution
of the chaff bundles across the plume and the calculation of the
statistical moments of the chaff distribution give (Sauvageot et al.,
1982).

- an objective determination of the plume axis (fig. 3) ;

- an estimate of the mean air velocity along the plume ;

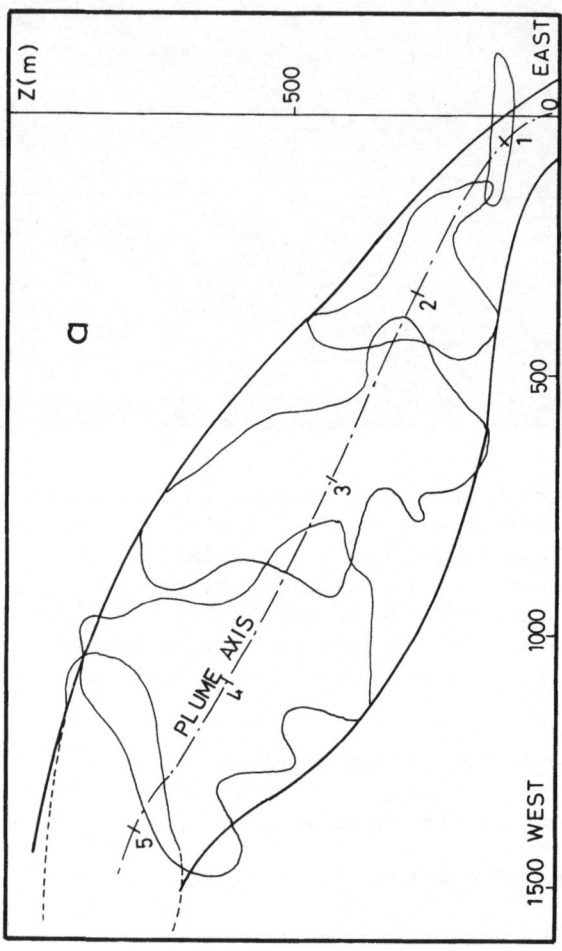

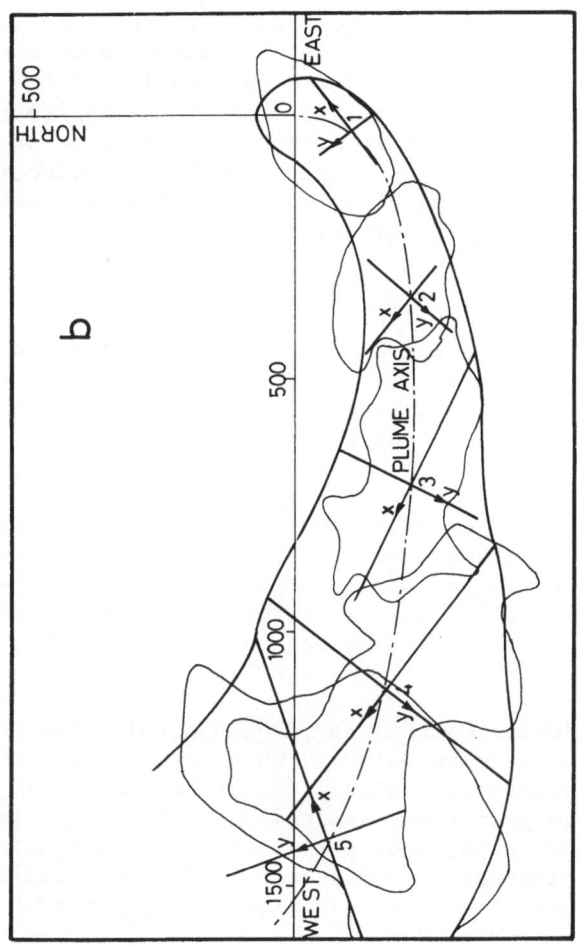

Fig. 3. Example of the evolution of a chaff bundle in a thermal plume on 8 June 1979 projected in vertical (a) and horizontal (b) planes. All distances on the vertical and horizontal axes are in meters. The thin lines show the contours of the chaff bundle for five successive positions (numbered 1-5) observed between 1211:30 and 1214:50. The jet axis was determined from the position of the chaff bundle centroids ; x and y are the main horizontal axes of the dispersion ellipsoïds. The observations was made during the COCAGNE experiment.

- an estimate of the axes of the dispersion ellipsoïds (fig. 3).

When the plume is detectable by radar (without chaff), outlines
and turbulence inside the plume can be directly obtained. An approximate determination of the velocities in the plume can be done
only from the observation of the motions of particularities in the
plume reflectivity distribution (relative maximum or minimum).

For dry or wet plumes, a major question is the interactions with
environmental clouds. As regard to this problem, the usefulness of
the radar is to provide data concerning the environmental mean field
(see section 4) and to permit an accurate observation of the interaction zone. Sometimes, however, the efficiency of the observations decre:
ses when plume echoes are diluted among natural clouds or precipitation echoes. In this case, it is partly possible to overcome the
difficulty with discontinuous chaff release in order to create obser-
vable modifications in the reflectivity field.

6. SUMMARY AND CONCLUSION

The usefulness of meteorological Doppler radar as a tool for
air pollution modeling has been discussed. The problem of the radar
tracer, which is of particular interest, was firstly examined. Several
types of tracer can be used : the hydrometeors, the turbulent fluctuations of the air refractive index and the artificial tracers (chaff).

When a tracer is available, the measurements of the basic radar
parameters permit to deduce many quantitative and qualitative results
for the characterization of the meteorological environment : tridimensional fields of reflectivity, mean air motions and turbulence,
precipitating particle size distributions and cloud or precipitation
water content.

Radar is of particular concern for isolated pollution source
as cooling tower and power plant plumes. When artificial clouds or
interactions between plumes and natural clouds are concerned, millimetric radar permits to gather statistically significant data on
reflectivity (i.e. on microphysical parameters such as liquid water
content, mean volume diameter, etc...) in natural and totally or
partly artificial clouds and to monitor different aspects of the local
meteorological field perturbations.

In dry plume, chaff techniques permits in almost all atmospheric
conditions, an accurate determination of lateral limits and maximum
height of the plume. In addition air velocity and some other parameters concerning airflow conditions in the plume can be measured.

REFERENCES

Battan, L.J., 1973 - Radar observation of the atmosphere. The University of Chicago Press.

Beran, D.W. and F.F. Hall, 1974 - Remote sensing for air pollution meteorology. Bull. Amer. Meteor. Soc., 55, 1097-1105.

Browning, K.A. and R. Wexler, 1968 - A determination of kinematic properties of a wind field using Doppler radar. J. Appl. Meteor., 7, 105-113.

Campistron, B., 1975 - Characteristic distributions of angel echoes in the lower atmosphere and their meteorological implications. Boundary Layer Meteorology, 9, 411-426.

Campistron, B. et H. Sauvageot, 1974 - Sur l'évolution matinale de la distribution dans l'espace des échos radar en air clair. C. R. Ac. Sc., Paris, Série B, 479-482.

Kunkel, K.E., E.W. Eloranta and J.A. Weinman, 1980 - Remote determination of winds, turbulence spectra and energy dissipation rates in the boundary layer from lidar measurements. J. of Appl. Meteor., 37, 978-985.

Ottersten, H., 1969 - Atmospheric structure and radar backscattering in clear air. Radio Sci., 4, 1179-1193.

Sauvageot, H., 1982 - Radarmétéorologie. Edition Eyrolles, Paris, 296 p.

Sauvageot, H., G. Lafon and M. Oruba, 1982 - Air motion measurements in and around thermal plume using radar and chaff. J. of Appl. Meteor., 21, 656-665.

Sauvageot, H. et J. Omar, 1982 - Sur la mesure de la microstructure des nuages (to be published).

Teissier du Cros, F., D. Renaut et J.C. Pourny, 1977 - Le lidar en météorologie. La Météorologie, 6, 29-59.

Wilson, D.A., 1970 - Doppler radar studies of boundary layer wind profile and turbulence in snow conditions. Prep. 14th Radar Meteor. Conf., Amer. Meteor. Soc., Boston, 191-196.

DISCUSSION

H. HASENJAEGER   Could and especially designed
           electromagnetic radar become a competitive instrument
           to the Doppler acoustic sounder ?

H. SAUVAGEOT   In principle, the information
           provided by Sodar could be obtained by electromag-
           netic radar. However, to make the radar competitive,
           one needs to reduce its price and its operational
           cost.

OBSERVATIONS AND PREDICTIONS OF INVERSION PENETRATION

BY A BUOYANT INDUSTRIAL PLUME

P.R. Slawson and G.A. Davidson

Dept. of Mechanical Engineering
University of Waterloo
Waterloo, Ontario, Canada N2L 3G1

ABSTRACT

Observations of the penetration of elevated inversions by a plume from a tar sands plant in northern Alberta, Canada are compared with the predictions of a numerical plume rise model. Time-mean plume behaviour was recorded simultaneously by ground-based photography and aircraft monitoring of sulphur dioxide concentration. Supporting data on source conditions as well as atmospheric temperature and wind profiles were also obtained concurrently with the plume measurements.

The results of the comparison indicate that reasonable predictions of the position and the relative penetration of the buoyant plume into the inversion are possible with a simple one-dimensional plume model, given the measured atmospheric wind and temperature profiles.

INTRODUCTION

Chimney plumes are often trapped completely or in part within or below an elevated inversion, and subsequent rapid erosion of the inversion can lead to local high ground level pollutant concentrations. This particular dispersion condition, often referred to as inversion break-up and fumigation, has been identified as important to the dispersion meteorology of the tar sands area of northern Alberta. The degree to which a given chimney plume either penetrates or is trapped by an elevated inversion depends on the inversion height, the lapse rates below, within and above the inversion and on the source height, momentum flux and buoyancy flux. Thus,

415

given the inversion climatology for a particular site, it may be possible to 'design' source conditions for a chimney such that some fraction of the expected elevated inversions could be completely penetrated and excessive concentrations at ground level could be averted.

Measurements of the rise and spread of the plumes from the Suncor plant in the tar sands area of Alberta, Canada were collected during four seasonal periods in 1977. Both ground-based photography and an instrumented fixed wing aircraft were used to document the position and shape of the plume in space. Supporting source and meteorological data in the form of wind and temperature profiles through the plume rise region were also recorded. The wind field data was obtained by double theodolite tracking of pilot balloons every fifteen minutes, while the temperature data was obtained from hourly minisonde releases. The purpose of this paper is to present some observations of the position of the plume relative to various elevated inversions, and to compare these observations with the predictions of a simple integral model of plume rise and growth.

PLUME MODEL

The model used for plume rise and growth is based on the integral form of the conservation equations of mass, momentum and heat applied to a plume elemental disc across which top hat profiles of all variables are assumed (Davidson and Slawson[1]). The plume is considered to rise and grow in two phases, an initial buoyant phase where self-generated turbulence is deemed responsible for the entrainment of ambient air into the plume, and a second phase where atmospheric turbulence is primarily responsible for plume growth. In stable air, a maximum plume rise is reached during the initial phase of plume rise, and at this point the atmospheric phase is assumed to begin.

The governing equations for plume rise and growth in the initial buoyant phase are given by

$$\frac{d}{dt} (\rho_p R^2 V) = 2\rho_a RVv_e \tag{1}$$

$$\frac{d}{dt} (\rho_p R^2 Vv_x) = 2\rho_a RVUv_e \tag{2}$$

$$\frac{d}{dt} (\rho_p R^2 Vw) = gR^2 V(\rho_a - \rho_p) \tag{3}$$

$$\frac{d}{dt} [gR^2 V(\rho_a - \rho_p)] = -\rho_a N^2 R^2 Vw \tag{4}$$

where each symbol is defined in the Nomenclature. In addition, the hydrostatic law,

$$\frac{d}{dt} (\rho_p T_p) = -\frac{7}{2} \gamma_{ad} \rho_a w \tag{5}$$

the kinematic relations,

$$\frac{ds}{dt} = V \tag{6}$$

$$\frac{dz}{dt} = w \tag{7}$$

$$\frac{dx}{dt} = V_x \tag{8}$$

$$V^2 = V_x^2 + w^2 \tag{9}$$

and an equation for the entrainment velocity given by Ooms[2]

$$v_e = \alpha \left| V - \frac{v_x U}{V} \right| + \beta \left| \frac{U v_x w}{V^2} \right| \tag{10}$$

completes the equation set. Optimization of trajectory predictions over the large set of plume rise data collected at the Suncor site yielded the best fit entrainment constants as $\alpha = 0.15$, $\beta = 0.63$.

In the application of the model, an average value for ambient air density ρ_a, wind speed U and the square of the Brunt-Vaisala frequency N^2 must be calculated for each plume cross-section at each step forward in the numerical integration procedure. Integration is then performed using a Runge-Kutta technique with error check by step-size halving. The maximum (and assumed final) rise of the plume is reached when the vertical plume velocity w first becomes zero. Here the atmospheric phase is assumed to begin and plume growth is described by a Gaussian plume model where the standard deviations or sigmas are specified by empirical curves such as those of Pasquill or Briggs[3]. The USNRC stability classification[4] and the average lapse rate across the plume at its final rise are employed to choose a stability class and hence the appropriate empirical sigma curve. Following Slawson[5], the rise and atmospheric phases of plume growth are matched by equating the circular area of the plume cross-section at the final rise to that of an ellipse with major and minor axes a multiple of the lateral and vertical standard deviations. This is equivalent to establishing a virtual point source at the final rise altitude of the plume.

OBSERVATIONS AND MODEL PREDICTIONS

Six plume cross-sections recorded during elevated inversion
conditions were selected from the data set collected at the Suncor
site during 1977 for comparison with model predictions. The
observed (solid line) and predicted (broken line) plume altitudes
and depths for these six cases are shown in Figures 1 and 2, along
with corresponding profiles of ambient wind and temperature. The
observed plume altitude and depth was obtained from aircraft trans-
ects of the plume at a known downwind distance from the chimney
source (indicated in brackets on the plume depth lines in each
figure). Table 1 lists the source conditions (at stack top) for the
powerhouse for each observation period, which are used as input for
the model predictions. The powerhouse stack has an exit diameter
of 5.8 m and a height of 107 m.

In all of the observations collected, considerable lateral
wind shear existed over the plume rise region. This wind shear
results in a significant lateral dispersion of the plume which is
not adequately described by the present Gaussian model. Figure 3
illustrates the typical shape of a plume cross-section as identified
by the 10% concentration isopleth at only 2.3 km downwind from the
source. Here the vertical extent of the plume cross-section covers
some 200 m, while the lateral extent covers 2800 m. Since significant lateral wind shear accompanies most elevated inversions at this
site, one must adequately model the resulting lateral plume spread
in order to partition the relative fractions of the plume cross-
section that is either below or trapped within an elevated inversion.

On inspection of Figures 1 and 2, it is evident that the simple
numerical integration model places the plume centre-line very close
to that observed. Also, the vertical extent of the plume depth
about that centre-line is also reasonably well predicted. However,
as discussed above, the model in its present form tends to underestimate the lateral extent of the plume by a factor of 1.5 to 2.

CONCLUSIONS

A simple integral model of plume rise employing numerical integration of the one-dimensional conservation equations allows the incorporation of complex profiles of wind and temperature. Reasonable
predictions of plume rise and vertical extent can be obtained, even
in the presence of elevated inversions and severe kinks in the wind
profile. Averaging ambient profiles of wind and temperature over
the depth of a plume element in order to obtain a top-hat representation thus appears to provide sufficient detail on atmospheric structure to allow reasonable model predictions.

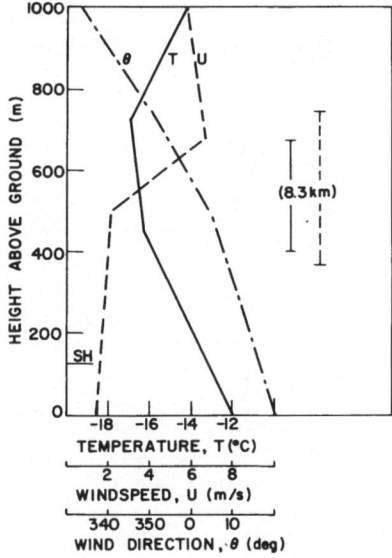

January 25 (1415-1520)

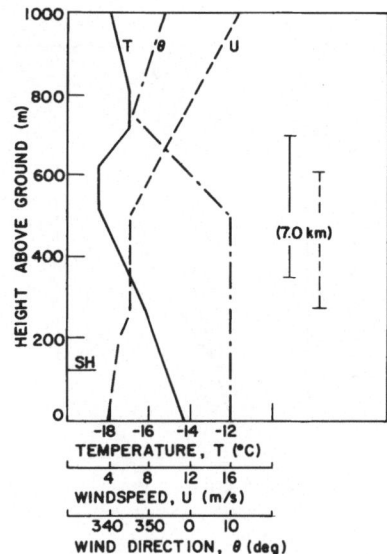

January 26 (1140-1235)

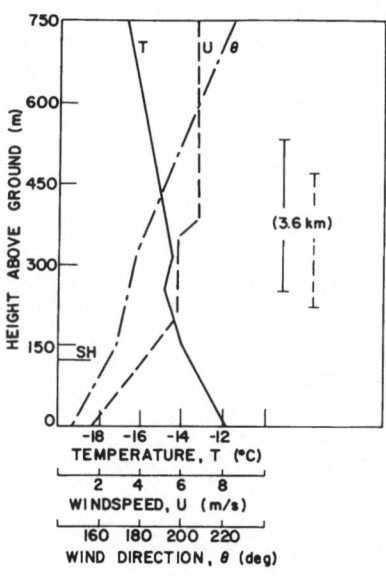

January 31 (1353-1442)

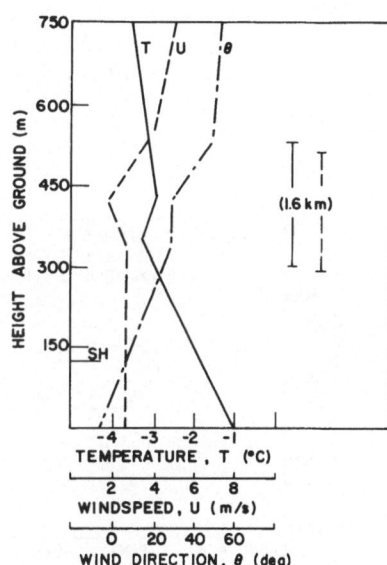

March 26 (0918-1002)

Fig. 1.   Observed and predicted plume positions and depths with
         ambient profiles of temperature and wind for three obser-
         vations in January and one in March, 1977.

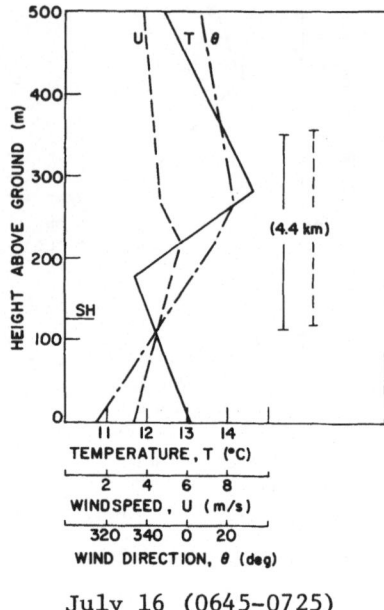
July 16 (0645-0725)

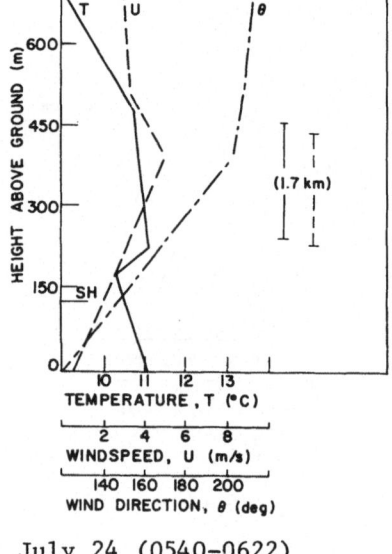
July 24 (0540-0622)

Fig. 2. Observed and predicted plume positions and depths with
         ambient profiles of temperature and wind for two observa-
         tions in July, 1977.

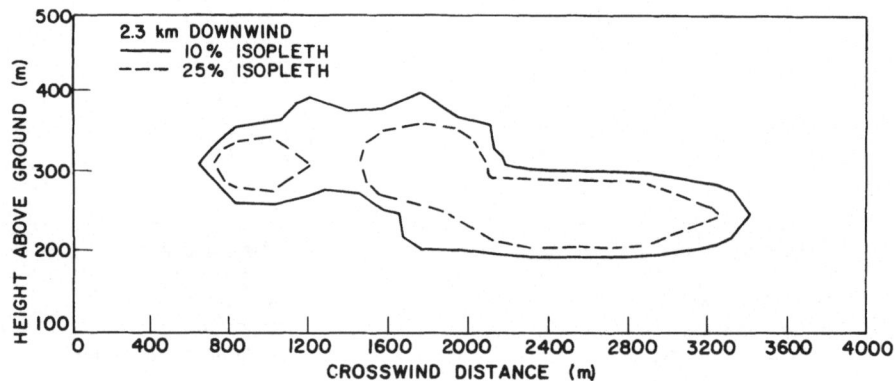

Fig. 3. Observed plume cross-section at 2.3 km from the source on
         July 15, 1977 at 0610 hrs showing the ten and twenty five
         percent isopleths of the maximum observed concentration.

Table 1. Powerhouse exit velocities and temperatures

Date	Time	w_o (m/s)	T_{po} (^{o}C)
Jan. 25	1415–1520	18.4	280
Jan. 26	1140–1235	18.5	280
Jan. 31	1353–1442	22.0	307
Mar. 26	0918–1002	18.5	280
July 16	0645–0725	17.1	274
July 24	0540–0622	16.5	284

REFERENCES

1. G. A. Davidson, P. R. Slawson, A Comparison of Some Plume Dispersion Predictions with Field Measurements, Proc. 11th Nat. Int. Tech. Meeting, Amsterdam, 1980.
2. G. Ooms, A New Method for the Calculation of The Plume Path of Gases Emitted by a Stack, Atmos. Environ. 6: 899 (1972).
3. F. A. Gifford, Turbulent Diffusion Typing Schemes – A Review. Nuclear Safety, 17:68 (1976).
4. U.S. Nuclear Regulatory Commission, Proposed Revision 1 to Regulatory Guide 1.23 – Meteorological Programs in Support of Nuclear Plants. U.S. NRC Office of Standards Development, 34 pp., 1980.
5. P. R. Slawson, Observations and Prediction of Natural Draft Cooling Tower Plumes at Paradise Steam Plant, Atmos. Environ. 72: 1713 (1978).

NOMENCLATURE

g	gravitational acceleration
N	Brunt-Vaisala frequency
R	plume radius
s	plume trajectory length
T	temperature
t	downwind travel time
U	windspeed
V	plume velocity
v_e	entrainment velocity
v_x	downwind component of V
w	vertical component of V
x	downwind plume centreline coordinate
z	vertical plume centreline coordinate
α	entrainment constant
β	entrainment constant
θ	wind direction

Subscripts

a   atmospheric property
o   plume property at the source
p   plume property
s   stack property

ACKNOWLEDGEMENTS

    The authors would like to express their appreciation to
Gordon J. Hitchman for his assistance in preparing this report and
also to Colleen Janer for typing the manuscript. This research
has been funded through grants awarded to the authors by the
Natural Sciences and Engineering Research Council of Canada.

DISCUSSION

G. RESELE   If I understood you correctly,
            you calculated the plume axis and the depth and pre-
            dicted the position of the upper and lower edge of
            the plume by assumming a symmetric plume. I am asto-
            nished about the quality of your predictions for in-
            versions, where you expect an assymmetric plume,
            which is compressed at the upper part. Can you com-
            ment on that ?

P.R. SLAWSON   The plume cross-section is
            circular in the initial phase to the final rise point
            in stable air.
            An equivalent area ellipse is matched to the plume
            circular cross-section at the final rise point, where
            the major and minor axes are proportional to the sig-
            mas (standard deviation) as given by a set of empi-
            rical sigma curves. The subsequent plume depth is
            compared then with the observed plume depth. No at-
            tempt has been made to accurately predict the shape
            of the plume cross-section.

F. FANAKI   The area of the study is cha-
            racterized by multi-inversion layers; is your model
            capable to handle such condition ?

P.R. SLAWSON   Some of the cases that I have
            chosen do have multilayers and the model appears to
            position the plume rather well, relative to those
            layers and the observed plume.

F. FANAKI   Were the entrainment coefficients calculated or observed experimentally ? Do you think two entrainment coefficients should be used one when the plume is under the inversion and one when the plume is in the inversion ?

P.R. SLAWSON   From the apparent success of this model on these few results, I do not believe it is necessary in order to satisfactorily place the plume relative to the observed inversions.

ACOUSTIC SOUNDER DATA AS METEOROLOGICAL INPUT

IN DISPERSION ESTIMATES

H. Gland

Direction des Etudes et Recherches

EDF - Chatou - France

INTRODUCTION

The SODAR technique is very interesting for EDF because it is now the only one able to determine, on a routine basis, the vertical distribution of wind velocity and direction and to give information on the vertical thermal structure and the turbulence in the first five hundred meters of the atmosphere.

The tests of qualification of a monostatic Doppler SODAR delivering numerical information began in 1979.

Since 1979, measurements have been carried out over long periods of time and for different sites (JUMEAUVILLE, LYON-SATOLAS, SAINT-ALBAN, CREYS-MALVILLE). They enabled us to test the performances and the operational qualities of the system : fiability rate greater than 90 %, informations available up to 500 meters in more than 50 % of the cases, up to 200 meters in more than 90 %.

Many comparisons have been carried out with conventional measurement techniques in order to test the representativeness of wind measurements (comparison with radar or meteorological tower measurements) but also to establish relationships between SODAR data (echo intensity profiles, velocity fluctuations profiles) and the vertical thermal structure (comparisons with radio-soundings or meteorological tower measurements).

Good correlations were obtained between the shapes of echo intensity profiles and the vertical temperature profiles and also between the vertical velocity standard deviation and the temperature lapse rate [1], [2], [3].

These results enable us to consider now the direct use of SODAR data in dispersion modeling.

HOW DOES THE SODAR OPERATE

The functioning of the SODAR is based on the back-scattering of an acoustic wave by turbulent cells present in the atmosphere whose dimensions are around one-half the emitted wave length, i.e. around 10 centimeters for the frequency of 1600 Hz.

The three-dimensional detection is obtained by means of three antennas oriented in three different directions, each functioning in turn as transmitter and receiver.

For each sample, the SODAR transmits, along each beam axis, a short acoustic signal ($\simeq$ 100 milliseconds), after which back-scattered signal is monitored for a period of 6 seconds (each antenna being therefore interrogated every 18 seconds).

The signal processing is based on a zero-crossing method coupled with a signal-to-noise correction and invalidation technique.

Each information is digitized every 6 milliseconds corresponding to a spatial resolution of about one meter.

The amplitude of the back-scattered signal gives the intensity of the echo.

The frequency of the signal undergoes a frequency shift (Doppler effect) used to compute the displacement velocity of the target cells along the axis of the antennas, and further the u, v, w components of the wind.

Average wind and echo intensity are computed for 40 or 50 meter thick layers over integration periods of 15 minutes.

A micro-computer operates the system as a whole. At the end of each integration period a "measurement unit" is recorded on cassette. Each "measurement unit" gives for the different levels the following information :

- Intensity of the echo (average and standard deviation).

- Mean velocity and direction of the horizontal wind.

- Mean vertical velocity of the air and its standard deviation (σ_w).

In the near future, the lateral-wind-direction standard deviation (σ_θ) for the same sampling time of 15 minutes will also be measured.

USE OF SODAR DATA IN DIFFUSION MODELING

Models we are interested in, are simple operational ones such as climatological dispersion models or gaussian plume or puff models using stability categories to characterize the dispersion capacity of the atmosphere.

The immediate advantage of the SODAR is to provide, on a continuous mode, a measurement of the wind direction and velocity near the level of the release height, even for very tall chimneys.

The mixed layer depth and the stability parameters (leading to the gaussian dispersion parameters σ_y and σ_z) can be infered from the echo intensity profiles and the measurement of σ_w and σ_θ as we shall discuss now.

1. Detection of Stable Layers

A lot of comparisons have been carried out between vertical back-scattered echo profiles and thermal profiles.

In most cases, the presence of stable layers is shown by the appearance of a peak in the reflectivity at the same altitude as these layers, with a very rapid decrease in the intensity of the back-scattered signal at higher levels. In the case of an elevated temperature inversion, the increase in the intensity of the echo is very sudden (figure 1a).

In the case of inversions beginning at ground-level, or very thick inversions, a regular increase in reflectivity is noted from the base of the stable layer (figure 1b).

There is no evident connection between the value of the peak reflectivity or the form of the pattern of echo intensity and the characteristics of the temperature inversions (size, thickness).

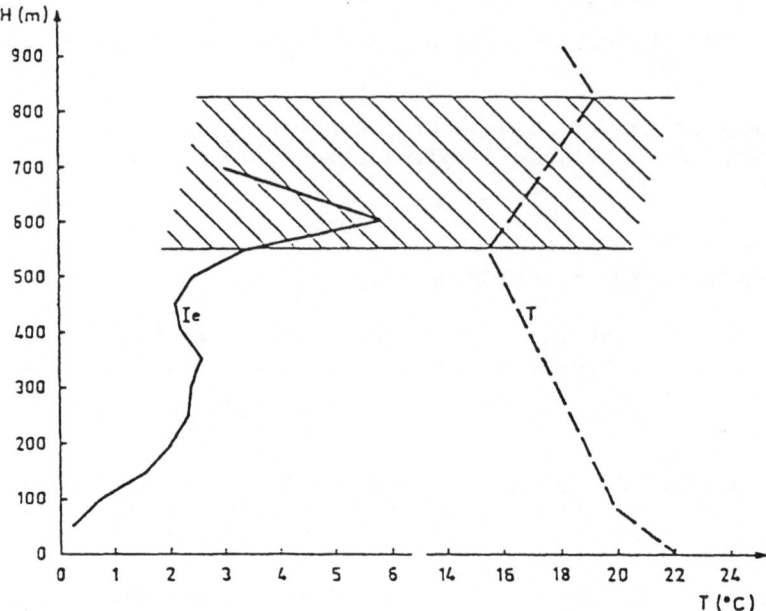

(1.a)   12 September 1980 (12.00 GMT) - Elevated temperature inversion

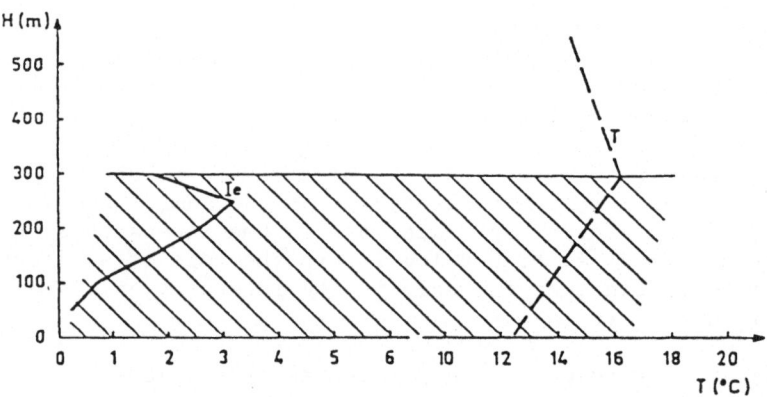

(1.b) 1 October 1980 (12.00 GMT) - Temperature inversion between the
ground and 300 meters.

Figure 1.   Verticle contours of reflectivity Ie (arbitrary units)
and corresponding temperature soundings.

In the case of a neutral or adiabatic atmosphere, which corresponds to a temperature lapse of less than $-0.5\,°C/100$ meters the intensity profiles of the echo are much smoother, and do not show any significant "irregularities" (figure 2.a.b).

Finally, it will be noted that a temperature inversion does not necessarily constitute a barrier for the back-scattering of the acoustic signal : data are often obtained above the stable layer, especially in the case of nocturnal radiation inversions, and that enable to detect several zones of stability at different elevations (figure 2c).

So, the shape of the vertical profile of the echo allows a good detection of the inversion layers. We can see, for example in figure 3, the evolution as a function of time, of a stable layer detected near the ground and then aloft (the points correspond to the maximum of the SODAR echo ; the solid lines are the thermal profiles issued from radio-sounding.

2. A Stability Indication

σ_w is the standard deviation of the sample of vertical velocities measured every 18 seconds over an interval of 15 minutes. These statistics integrate the turbulent motions of the atmosphere at various levels, especially those caused by convection. For this reason, σ_w is particularly sensitive to diurnal variation.

When analysing the average values of this parameter at various hours of the day, we observed that σ_w remains quite often below 45 cm/s during nocturnal periods (see for example figure 4) while in the middle of the day, variations between 65 and 100 cm/s, depending on the season, are observed.

We have therefore attempted to find a critical value of σ_w that is capable of separating cases of "stable" atmosphere (frequent at low levels during the night) from cases of "neutral or instable" atmosphere (generally observed during the day), as it is done with temperature soundings, on both sides of the threshold :

$$\Delta T/\Delta Z = -0.5\ °C/100\ m$$

Between the ground and 400 meters, the σ_w values shown in table 1 were obtained : they vary from 43 to 51 cm/s.

By adopting a critical value of 45 cm/s for all levels, the atmospheric stability is defined in the following manner :

$\sigma_w < 45 \Rightarrow$ stable atmosphere

$\sigma_w > 45 \Rightarrow$ neutral or unstable atmosphere

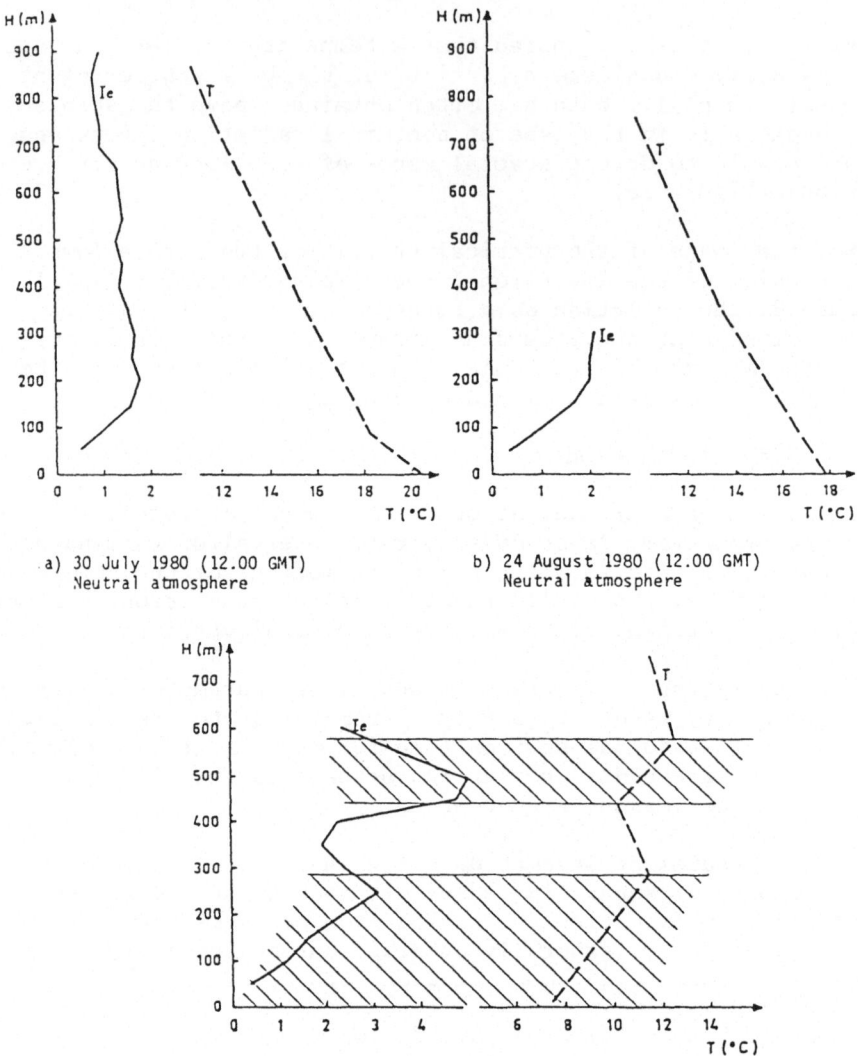

a) 30 July 1980 (12.00 GMT)
Neutral atmosphere

b) 24 August 1980 (12.00 GMT)
Neutral atmosphere

c) 25 August 1980 (06.00 GMT) - Multiple temperature inversions.

Figure 2.   Vertical patterns óf reflectivity Ie (arbitrary units)
and corresponding temperature soundings.

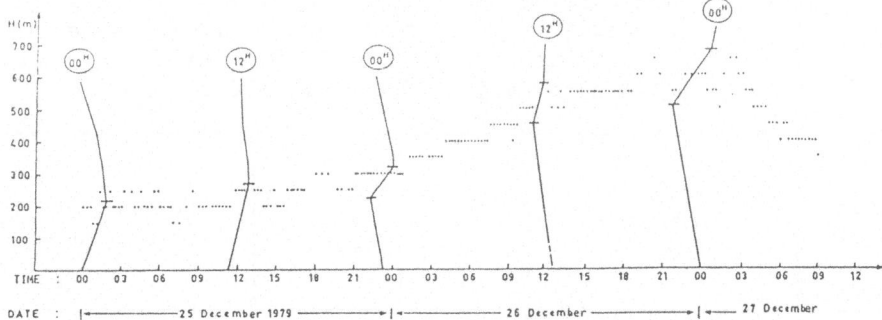

Figure 3. Continuous surveillance of the thermal structure of the lower atmosphere: Detection of temperature inversions using digital data provided by the Sodar every quarter-hour.

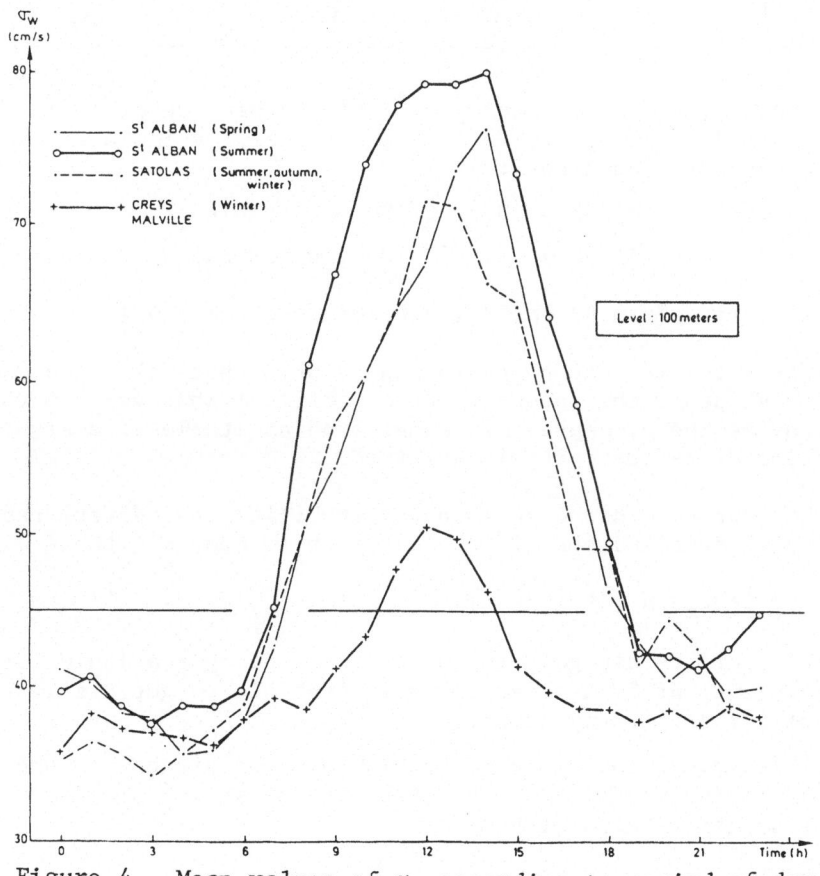

Figure 4. Mean values of σ_w according to period of day.

Table 1.   Values of σ_w separating the "stable" and the
"neutral" or "unstable" classes in the same
proportions as the -0.5 °C/100 m criteria
(SATOLAS : July to December 1980).

Level (m)	Critical value of σ_w (cm/s)	Correct evaluation in percent of the cases (with σ_w = 45) (%)	Number of cases studied
50	48	67	681
100	43	72	743
150	45	66	670
200	43	66	634
250	46	68	583
300	43	67	530
350	45	64	473
400	51	63	402

Table 1 shows, for each level, the percentage of evaluations
thus carried out which conform to the classification made from
the temperature soundings :

$$\Delta T/\Delta Z > -0.5 \text{ °C/100 m} \Rightarrow \text{stable}$$

$$\Delta T/\Delta Z < -0.5 \text{ °C/100 m} \Rightarrow \text{neutral or unstable}$$

It varies from 63 to 72 % according to the level.

These results can be performed by using both the σ_w parameter
and the shape of the intensity echo profile : whatever the value
of σ_w may be the atmosphere is considered as stable if a significant peak of reflectivity is observed.

The correspondence between the stability thus determined
from SODAR data and the $\Delta T/\Delta Z$ evaluation is shown in table 2.

. <u>Applications</u> :

1/ The stability category in two classes, being available for
each level of 50 meters, makes it possible to get easily the
depth of the mixing layer.

2/ Statistics giving the distribution of wind directions and
velocities according to the stability category (stability
roses) can be established.

Then, they can be used to initialize statistical models using only two stability categories (or dispersion conditions classes) widely used for mean annual concentrations estimates (or impact studies).

Table 2. Correspondence between the stability categories "stable" and "neutral or unstable" determined from $\Delta T/\Delta Z$ and SODAR data (σ_w and significant reflectivity peak combined).

Level (m)	Frequency of equivalent evaluations made by radiosonde and SODAR	
	In general (%)	In most unfavorable conditions (worst dispersion conditions) (%)
50	71	77
100	74	86
150	67	73
200	66	72
250	70	68
300	69	73
350	70	73
400	69	63

3. PASQUILL-GIFFORD Stability Classes Determined by SODAR Data

In the standard gaussian plume or puff models, the horizontal and the vertical spread can be considered separately, in a so called "split sigma method" [4], [5], [7].

Using SODAR data, the stability category for the vertical spread can be infered from σ_w and the stability class for the horizontal spread from the standard deviation of the wind azimuth angle σ_θ.

3.1. Stability Category for the Vertical Spread

We started from the relationship between PASQUILL-GIFFORD categories and the temperature lapse rate $\Delta T/\Delta Z$, described by the US Nuclear Regulatory Commission, and we searched the values of σ_w separating the entire sample of σ_w in the same proportions as the $\Delta T/\Delta Z$ criteria.

The correspondence is shown in table 3.

Table 3.   Correspondence between stability categories, $\Delta T/\Delta Z$
and σ_w (SATOLAS, July-December 1980).

Stability	PASQUILL-GIFFORD category	$\Delta T/\Delta Z$ (°C/100 m)	σ_w SODAR (cm/s)
Extremely unstable	A	< -1.9	> 123
Moderately unstable	B	-1.9 to -1.7	109 to 123
Slightly unstable	C	-1.7 to -1.5	104 to 109
Neutral	D	-1.5 to -0.5	45 to 104
Slightly stable	E	-0.5 to 1.5	45 to 24
Moderately stable	F	1.5 to 4.0	24 to 18
Extremely stable	G	> 4	< 18

The distribution of the stability categories determined by
the σ_w or $\Delta T/\Delta Z$ technique is shown in table 4.

Table 4.   Stability categories determined from SODAR
(σ_w) and radio-sounding ($\Delta T/\Delta Z$) measurements
(SATOLAS : level 100 m).

$\Delta T/\Delta Z$ \ σ_w	A+B+C	D	E	F	Total
A+B+C	4	14	0	0	18
D	10	145	61	12	228
E	3	62	119	43	227
F+G	1	4	33	19	57
Total	18	225	213	74	530

These two tables are issued from 530 comparisons made at
100 meters at SATOLAS (July-December 1980) between simultaneous
measurements of σ_w (SODAR) and $\Delta T/\Delta Z$ (sounding).

In table 4, categories A, B and C have been got together due
to lack of unstable cases at this level.

The total class agreement is 54 % (at higher level, 200 meters for example, we found 51 %).

We used the σ_w threholds indicated in table 3 to determine the stability categories for a nuclear power plant site, CREYS-MALVILLE, where continuous σ_w (SODAR) and $\Delta T/\Delta Z$ (80 meters meteorological tower) measurements were performed for a period of four months (November 1980-February 1981). In this method, whatever the value of σ_w may be, the atmosphere is considered as "stable" if a significant peak of reflextivity exists at the same time at the studied level.

The results of the comparison are shown in table 5.

Table 5.   Correspondence between stability categories determined by σ_w (SODAR) and $\Delta T/\Delta Z$ (meteorological tower). CREYS-MALVILLE - Level : 80 meters (November 1981-February 1982).

$\Delta T/\Delta Z$ \ σ_w	A+B+C	D	E	F	Total
A+B+C	1	60	85	16	162
D	14	219	356	86	675
E	2	209	555	206	972
F	3	92	300	207	602
Total	20	580	1296	515	2411

The total class agreement is 41 %.

But it can be noted that both SODAR and $\Delta T/\Delta Z$ data lead to similar stability distributions and, in particular, show an important proportion of stable cases (E and F) during the test period (winter) for this typical valley site.

3.2. Stability Category for the Horizontal Spread

No comparison have yet been carried out between the standard deviation of horizontal wind direction measured by the SODAR and other σ_θ measurement techniques, because such improvement on our SODAR is too recent.

In a first time, we plan to use the classification scheme recommended by the US Nuclear Regulatory Commission (table 6) ; the class limits will possibly be modified as soon as new data will be available, especially for higher elevations.

Table 6

Stability category	σ_θ (degrees)
A	> 23
B	23 to 18
C	18 to 13
D	13 to 8
E	8 to 4
F	4 to 2
G	< 2

CONCLUSION

Comparison between SODAR data and conventional measurements have been carried out on various sites and for various long periods since 1979.

Good correlations were always observed between wind measurements. Relationships were also found between the echo intensity profiles, the vertical wind velocity fluctuations and the vertical thermal structure of the low atmosphere.

For each level, a stability index can be determined by using both the shape of the echo intensity profile and the value of σ_w.

The method was experimented on two different sites to determine the PASQUILL-GIFFORD stability categories : the stability classes showed similar distributions when either the SODAR data or the vertical lapse rate technique was used.

Although this information can be used for determining both the vertical and the horizontal spread, we plan to use it only for the vertical spread ; the horizontal one will be obtained from the horizontal wind fluctuations measured recently by our SODAR : this last point is still under development.

REFERENCES

[1] GLAND H.
     Wind Measurements using Doppler Acoustic Sounding :
     Comparison with Measurements by RADAR.
     EDF Report : HE/32-80.6 - 1980.

[2] GLAND H.
     Experiments with a Doppler Acoustic Sounder.
     EDF Report : HE/32-80.24 - 1980.

[3] GLAND H.
     Qualifying test on a three-dimensional Doppler SODAR
     (SATOLAS : July-December 1980).
     EDF Report : HE/32-81.9 - 1981.

[4] MAC CREADY P. - WORDEN J.
     Doppler Acoustic Sounding : observational Inputs to Pollutant
     Dispersion Models.
     EPRI-EA - 2219 - 1982.

[5] SAGENDORF J.F. - RAY DICKSON C.
     Diffusion under low windspeed, inversion conditions.
     Third Symposium on Atmospheric Turbulence, Diffusion and
     Air Quality.
     RALEIG - 1976.

[6] US Nuclear Regulatory Commission
     On-site meteorological programs. Regulatory Guide 1.23.
     Safety guide 23 - 1972.

[7] SEDEFIAN L. - BENNETT E.
     A comparison of turbulence classification schemes.
     Atmospheric Environment, Vol. 14.

[8] SCHUBERT J.F.
     A method for using acoustic sounder categories to determine
     atmospheric stability.
     4th Symposium on Turbulence Diffusion and Air Pollution.
     RENO - 1979.

DISCUSSION

A. VENKATRAM   What is the rationale for trans-
              lating direct measurements of turbulent intensity σ_{ω}
              to the crude measure of stability classifications ?

H. GLAND   The main reason is that we use
              to work in our company with some models based on the
              P.G. classification scheme and that having at our
              disposal an increasing number of sodar data (issued
              from different power plants) we need to find quickly
              a way for a direct use of these data in operational
              dispersion models.

SIMULATION OF SULPHUR DIOXIDE MASS-FLOW OVER MILAN AREA

S. Sandroni[1], M. de Groot[1], G. Clerici[2], S. Borghi[3]
and L. Santomauro[3]

[1]Commission of the European Communities, Joint Research
    Centre - Ispra Establishment, I-21020 Ispra, Italy

[2]A.R.S., Milano, Italy

[3]Osservatorio Meteorologico di Brera, Milano, Italy

AIM OF THE STUDY

The evaluation of the air pollution impact on a large urban
area from industries partially or totally surrounding it, is a cause
for concern in many European cities. The problem is particularly
important in Milan, a city with about 2 million inhabitants and an
area of about 190 km^2, located at the centre of the Po Valley in an
industrial region. The continental climate characterized by frequent
and prolonged temperature inversions, much fog during the winter and
winds scarcely exceeding 2 $m.s^{-1}$, leads to pollution build-up.

The air quality level and its trend in the urban or industrial
areas are usually deduced from a network of fixed monitoring
stations, the number and distribution of which depend mostly on
financial and local policy factors. Local authorities may have a
source inventory at their disposal from which, depending on the
day and the meteorological forecast, the air quality level over
24 hours can be predicted (Finzi et al. 1980). Beside the well-
known advantage of a continuous control, a monitoring network
gives only a part of the information about the transport of pollution
coming from remote regions and in particular on the relative contribution of the surrounding industrial areas to the urban air
quality level.

   The city of Milan has a high density of residential and commercial premises with extensive building development, while industries are mainly confined to the Northern (NW to NE) and South-
Eastern outskirts (refineries, power plants, chemical and metal-
lurgical factories). In the past pollution levels were notoriously
high: a decade ago the yearly mean concentrations were about
790 ug/m^3 for SO_2 and 280 ug/m^3 for smoke and particulates (Concawe,
1976). Early measurements showed that during inversion periods SO_2
ground-level concentration(glc) frequently exceeded 1500 ug/m^3, the
threshold of taste and smell. A systematic study of measurements
made during the 1969/70 winter heating season and regression analysis
of the temperature-concentration relation showed that SO_2 pollution
was mostly associated to low level emissions from domestic heating
equipment. Because of the low winds, industry seemed to contribute
scarcely to the high pollution in the city's centre, but the
prevailong wind distribution suggested that any industrial development to the West could produce a significant contribution.

   The Ministerial Decree Law 615 of July, 1966 defined zones
in Italy where special measures should be taken to control air
pollution. Milan is in zone B which specifies the unrestricted use
of kerosene and gas oil but limits the use of light fuel (3% S max.)
to furnaces rated above 500,000 kcal/h and heavy fuel (4%S) to furnaces above 1,000,000 kcal/h. In comparison with the past, the actual pollution level has decreased sharply (the actual fuel consumption is given in Table 1); nevertheless, in connection with its specific climate
and location, Milan area needs particular attention for the air
quality level.

NEW APPROACHES TO THE PROBLEM

   The problem of pollution emission in the Milan area as well
as pollution transport towards it, coming from the industrial
regions, has been tackled in two ways:

(1) experimentally, by evaluating the input and output mass-flow
    by a remote-sensing technique. Measurements have been performed
    along the motorways around the city while collecting at the
    same time the necessary meteorological information. Similar
    measurements have recently been made in the Ghent industrial
    area during the 5th CEC Campaign (final report in preparation);
    and
(2) by comparing the experimental data with those produced by
    already tested codes as well as by an analytical model. For
    the modellistic approach, the emission inventory has been
    assembled not only for urban sources (domestic heating) but
    also for isolated sources located inside and outside the area
    under investigation. While an inventory of urban emissions

Table 1. Fuel Consumption in Milan in the Cold
Season (October, 1980 - March, 1981)[1]

Heavy fuel (4 % S)	300,000 tonnes
Light fuel (2.7 % S)	200,000 "
Gas oil (0.7 % S)	1,500,000 "
Methane and ass. (0.0 % S)	

Data accuracy $\pm$ 10 %
Calculated average SO_2 emission   14.5 tonnes/h

[2]Centro Riscaldamento Combustibili Liquidi,
Milano.

was already available (Gualdi et al., 1979), a tedious work
was necessary to collect information on location, characteristics and emission of isolated sources external to the area, at
least for those having an emission rate higher than 0.5 tonnes/h
and to evaluate their impact on the urban air quality level
(Clerici et al., to be published).

This study has been extended to some typical days of the year
with different stable meteorological situations. In the following
sections the experimental procedure and the modellistic approach
are described. Reference is made in particular to surveys performed
on May 20th (1), August 12th (2) and September 24th (3), 1980;
January 14th (4), February 18th (5) and November 25th (6), 1981.

EXPERIMENTAL AND MODEL

Remote-Sensing Measurements

For an experimental evaluation of pollution transport over the
city, surveys were made along the motorways around the urban area
with a mobile laboratory equipped with remote sensors and point
analysers. An absorption correlation spectrometer (COSPEC III or V,
Barringer Research) as remote sensor looked up into the atmosphere
through a window in the roof of the van and measured the integral
vertical burden of SO_2 using the diffused skylight as a source
(Hoff and Millan, 1981); SO_2 glc was simultaneously measured by
a Bendix 8302 monitor. The measurements were performed while the
van moved along the motorways. On board, simultaneous information
about the time and van position were stored together with the
measured data on a tape via a desk-computer (Hewlett-Packard 9825)
at a sampling resolution of 100 m. The tape was analysed later and
the measured burden and glc plotted as a function of the travelled
distance or directly on a map (Sandroni et al., 1982). The measured
SO_2 levels were essentially attributed to industrial emission to

which domestic heating was added in the cold season. The contribution
of diesel traffic emission could be neglected as traffic congestion
hours were avoided and the van speed was not too low. Furthermore,
the overall survey duration should be kept as short as possible as
the arbitrary zero level of the cospec signal may show a considerable
drift over the day, due to spectral changes in the UV skylight
background. In our case the entire survey (80km) took about 2 h,
usually from 11.30 to 13.30 LMT, which can be considered a reasonable
time interval for this type of measurements.

The evaluation of the pollutant mass-flow required information
on the wind field over the depth of the polluted layer. The incoming
and outgoing mass-flows have been calculated by integration of the
individual burdens projected perpendicularly to the average synoptic
wind direction and multiplied by the average wind speed (de Groot
et al., 1982).

Local Meteorological Conditions

As mentioned previously, the climate of the Po Valley is
characterized by low winds. Synoptic information on wind, temperature and pressure were collected by a network of meteorological
stations of the Aeronautica Militare (AM) distributed as in Fig. 1.
On the local scale, data were available from Laboratorio Provinciale
Igiene e Profilassi and from the AM-station at Linate Airport, near
the survey circuit (Fig. 2).

Important aspects affecting the atmospheric field and the
meteorological parameters are : (a) the urban heat island, (b) the
air mass circulation, and (c) the height of the mixing layer. The
urban heat island and the air mass circulation are associated phenomena described by the temperature and wind fields (Santomauro,
1971). The perturbation induced by the heat island may be neglected,
if one selects a sufficiently large survey circuit, as we did. The
mixing height may be evaluated from the vertical temperature gradient obtained by a radiosonde routinely launched at 12.00 GMT from
Linate AM-station. Three types of temperature stratification near
the ground were observed during the days quoted:

(a) an underlying unstable layer, with $\partial T/\partial Z < 10^{-2}$ °C/m (cases
    (1), (2) and (3));
(b) an approximately neutral layer under a more stable one (cases
    (5) and (6));
(c) a very stable layer characterized by a temperature inversion
    ($\partial T/\partial Z > 0$) (case (4)).

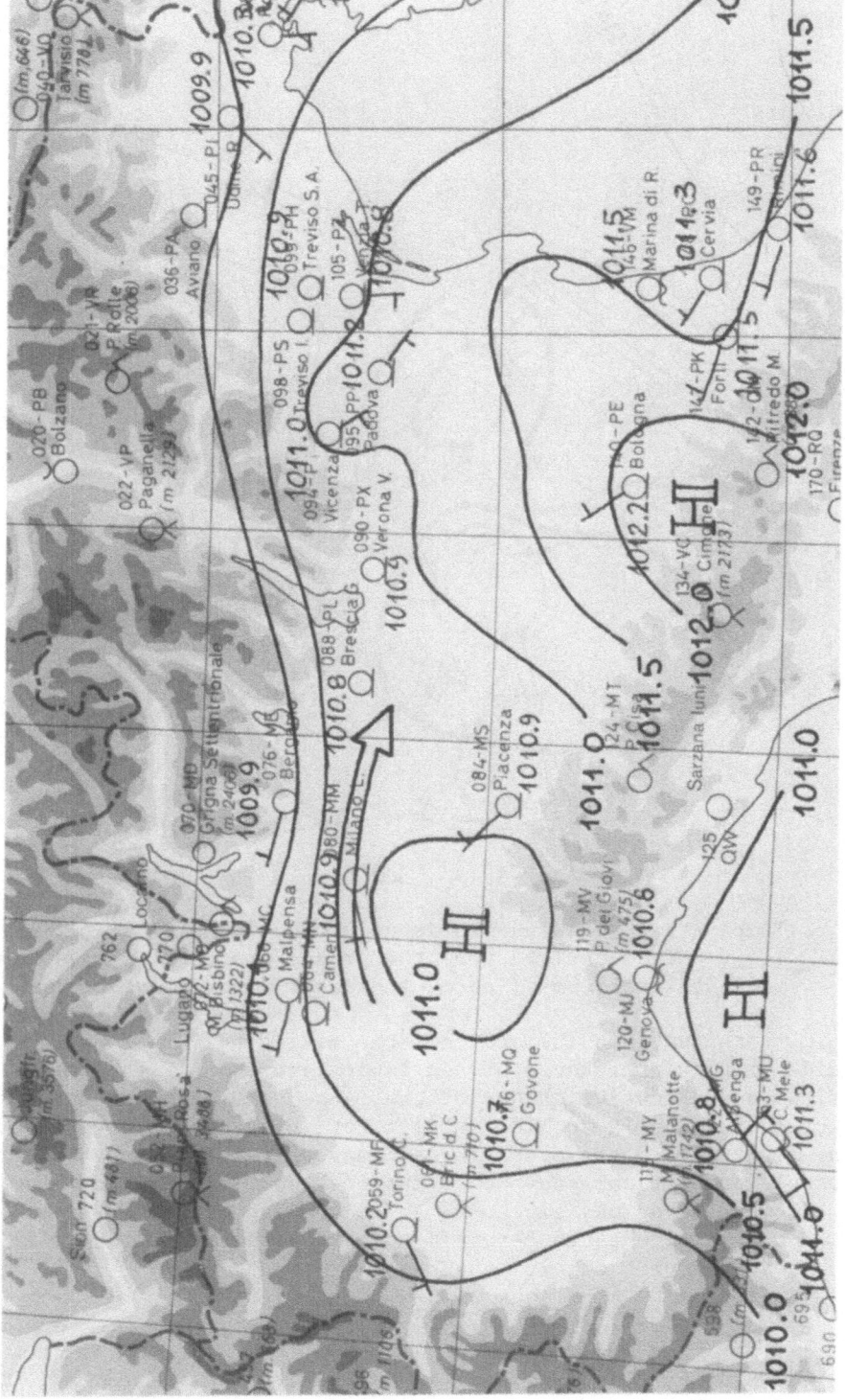

Fig. 1. Ground-level baric field over N-Italy on May 20th, 1980, at 12.00 GMT. The isobars are given in mbar at a 0.5 mbar resolution. The arrow indicates the dominant wind circulation.

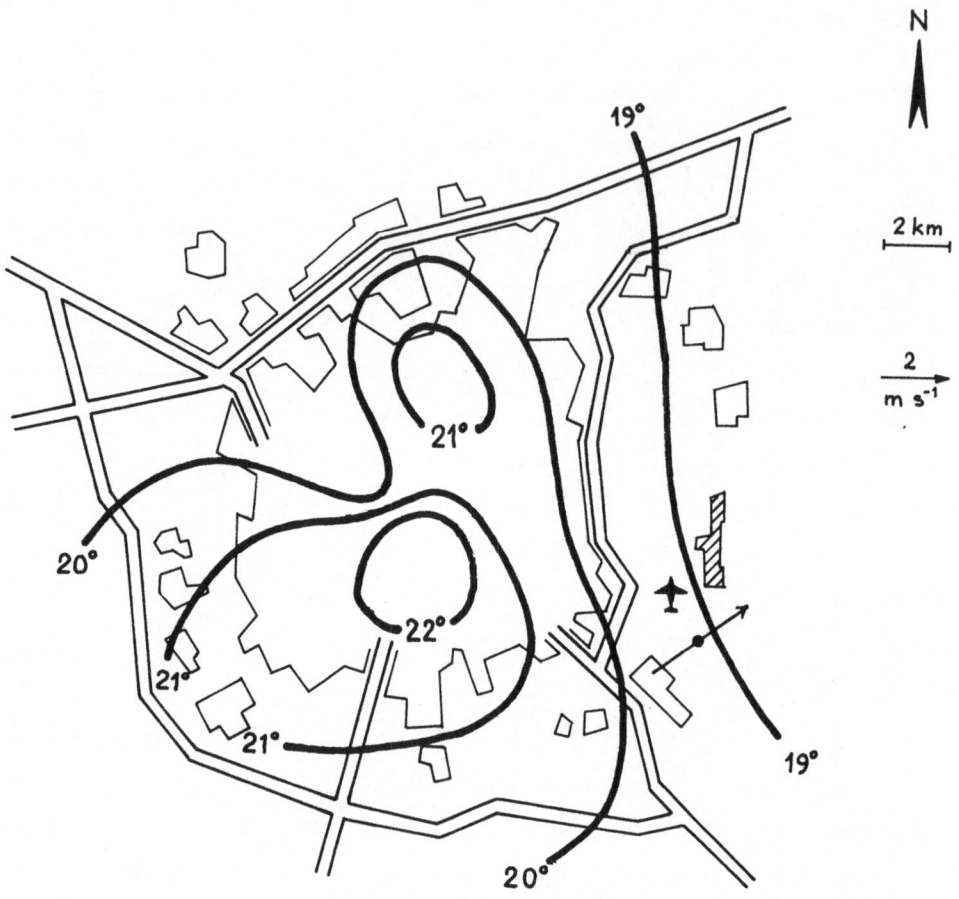

Fig. 2. Temperature and wind fields at ground level on 20-5-80,
        12.00 GMT as derived from the ground stations. Dark lines
        indicate isohterms ; arrow indicates wind direction and speed.

        In the situations (b) and (c), the mixing height is assumed to
coincide with the thickness of the lower layer, as it is limited by
the first significative change of the lapse rate. Such an assumption
is particularly valid for (c), whilst for (b) a possible vertical
mixing might affect the overlying air masses more stable and less
favourable to vertical convective motions. For (a), the instability
of the lowest layer may give rise to convective motions, whose
extension in altitude can be evaluated by classical methods of thermodynamical analysis of air masses.

As an example, let us summarize the meteorological analysis
associated with the survey (1). On May 20th, 1980, the measurements
around the city were performed from 10.30 to 12.30 GMT counter-
clockwise, the sounding at Linate-station at 12.00. The synoptic
situation over Northern Italy given in Fig. 1 showed that the area
was affected by Westerly winds. The veering wind from SW observed
at Linate was associated with the heat island indicated in Fig. 2
(a temperature gradient of $\sim 3°C$ was observed between the centre
and the rural outskirts). Such a superposition of a cyclonic (i.e.
counterclockwise) rotation on a rectilinear synoptic motion has
produced the observed complex wind field. For the evaluation of the
mixing height, the vertical temperature profile recorded at Linate,
has been transferred to the Herlofson diagram (Fig. 3) in which
areas are equivalent to energy per unit mass. If an air particle
lifts because of its thermic instability without heat exchange
with surrounding masses (adiabatic lifting), its vertical motion
is over when the two areas $P_1P_2P_3A$ and $ABCP_4$ are equivalent, P_1 and
C being the initial and final points of the vertical displacement

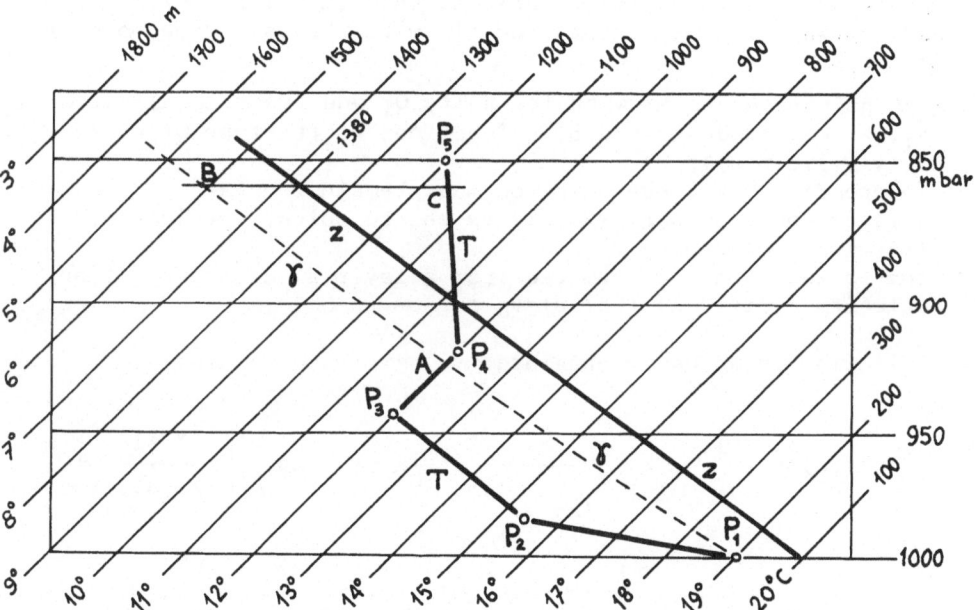

Fig. 3. Graphical evaluation of the mixing height for survey (1) :
        Herlofson's diagram T-log p.

(Haltiner and Martin, 1957). In the case considered (Fig. 3), H
was found to be $1380 - 104 = 1276$ m.

For the experimental evaluation of mass-flow and for the
modellistic approach, wind parameters have been averaged over the
mixing height previously evaluated. For (1) a wind direction of
280° and a wind speed of 3.5 m.s.$^{-1}$ have been used. As mentioned
in a preceding paper (Sandroni et al., 1982) the mixing heights
evaluated via Herlofson's diagram agreed with the values calculated
according to van Egmond et al. (1978) from experimental burden and
glc data.

Mathematical Model

For the evaluation of the SO_2 burden integrated along the z-
direction, the mean concentration $c(x,y,z)$ has to be determined.
The classical atmospheric diffusion equation may be expressed as

$$u \cdot \frac{\partial c}{\partial x} = \frac{\partial}{\partial y} \left(K_y \frac{\partial c}{\partial y}\right) + \frac{\partial}{\partial z} \left(K_z \frac{\partial c}{\partial z}\right) - \lambda c \tag{1}$$

where u is the mean wind speed, K_y and K_z the horizontal and vertical coefficients of turbulent diffusion and λ the kinetic constant
of a 1st order reaction which includes the removal of pollutant by
rainout or washout. By integrating (1) with the following boundary
conditions:

(a) the point-sources M_i with intensity Q_i and known localization x_i,
   y_i, z_i may be described by δ-functions of the type $Q_i/u \cdot \delta(x-x_i)$.
   $\cdot \delta(y-y_i) \cdot \delta(z-z_i)$;
(b) independent of y, the solution must remain regular;
(c) a layer localized at $z=H_m$ limits the turbulent vertical mixing
   (for $z=H_m$, $\partial c/\partial z=0$);
(d) the pollutant may be removed from a layer z_a at a deposition
   velocity v_d, i.e. $K_z (\partial c/z)=v_d \cdot c$;

the following equation is obtained :

$$c(x,y,z) = \frac{1}{2V\sqrt{\pi\beta}} \sum_{i=1}^{M} \frac{Q_1}{(x-x_1)^{1/2}} \ U(x-x_1) \ \exp\left\{-\left[\frac{(y-y_1)^2}{4\beta(x-x_1)} + \right.\right.$$

$$\left.\left. +\lambda(x-x_1)\right]\right\} \cdot \sum_{n=1}^{\infty} A_n\{\cos_n z + B_n \sin_n z\} \cdot$$

$$\cdot \exp\left[-C_n(x-x_1)\right] \tag{2}$$

where U is the Heavyside function (equal to zero if $x < x_i$ or equal to 1 if $x > x_i$), γ_n are the eigenvalues of the transcendent equation

$$\tan_n = N/\gamma_n \qquad n = 1,2,\ldots,\infty. \qquad (3)$$

The coefficients A_n, B_n, C_n are defined by

$$B_n = N/\gamma_n \qquad C_n = \gamma_n^2 \qquad (4)$$

$$A_n = \frac{\cos\gamma_n z_{si} + B_n \sin\gamma_n z_{si}}{1/2(1+B_n^2)+(1-B_n^2)\ \sin\ 2\gamma_n/4\gamma_n + B_n \sin^2\gamma_n/\gamma_n} . \qquad (5)$$

The numerical solution of eq. (3) and the application of eq. (4) and (5) give the values of A_n, B_n and C_n whose substitution in eq. (2) provides the mean concentration of the pollutant.

The accuracy of the analytical solution described above has been checked for M = 1 (i.e. for a single point-source) by comparing it with a numerical solution based on collocation methods (Clerici, to be published). Recently these methods have been applied by Wengle (1982) in the solution of a diffusive problem for which the analytical solution was previously known, obtaining about the same results. Therefore, the numerical solution of eq. (1) being in good agreement with the analytical solution (2), one can conclude that (2) is fully satisfactory for our purposes (Clerici et al., to be published).
A similar equation is derived for area sources (Clerici et al., to be published).

Apart from the complex nature of eq. (2) (and of the corresponding one for area sources), the z-dependence appears only through trigonometric functions of the Fourier expansion. The integral burden of SO_2 along the z-direction is consequently given by the equation

$$C_i\ (x,y) = \int_0^{H_m} C_t(x,y,z).dz \qquad (6)$$

where H_m is the mixing height and C_t, the total concentration in a point of given coordinates x,y,z, represents the sum of the concentrations related to point- and area-sources. The integral (6) may be calculated in an explicit way through the evaluation of integrals of the type:

$$F_{1p} \sum_{n=1}^{\infty} G_{np}^{1} \int_{0}^{H_m} \cos\gamma_n z.dz \qquad F_{2p} \sum_{n=1}^{\infty} \int_{0}^{H_m} \sin\gamma_n z.dz$$

$$F_{1a} \sum_{n=1}^{\infty} G_{na}^{1} \int_{0}^{H_m} \cos\gamma_n z.dz \qquad F_{2a} \sum_{n=1}^{\infty} G_{na}^{2} \int_{0}^{H_m} \sin\gamma_n z.dz$$

$$(7)$$

where the x- and y-dependent functions are included in F_{1p}, G_{np}, F_{2p}, G_{np}^{2}, F_{1a}, G_{nd}^{1}, F_{2d}^{2}, G_{na}^{2} only for point- and area-sources, respectively. If ℓ indicates the closed line in each point of which the vertical burden of SO_2 was measured, the net mass-flow of SO_2 above the area defined by the line is given by:

$$\Phi = \oint_{\ell} C.I(\ell).\bar{u}.\vec{n}.d\ell . \qquad (8)$$

For the solution of eq. (8) one can approximate ℓ by a closed pentagonal broken line (Fig. 4). If no other source is included in the approximation, the estimated value for (8) remains the same.

RESULTS AND DISCUSSION

The comparison between measured and calculated burdens for cases (3), (4), (5) and (6) is shown in Figs. 5 and 6.

The accuracy of the experimental technique is influenced by several factors such as the light-source and its stability in time, the atmospheric scattering, the background value, the calibration and some instrumental factors. The instrumental response associated to atmospheric scattering effects has been analysed by Millan (1980). The fluctuations of UV-light intensity associated with solar elevation can be neglected if surveys are performed in a reduced time interval around noon, as in our case. The calibration of the sensor was made according to the procedure adopted in the CEC-campaigns (Sandroni and de Groot, 1980), by using some standard cells. As background, whose evelation represents a crucial point (final report of the 5th CEC-campaign at Ghent, in preparation), the burden value was taken, measured at Ispra, a non-polluted area, before starting and after each survey. The intrinsic accuracy of the technique, as evaluated from static in-field calibration, is of the order of 10 ppm-metre.

To validate the experimental procedure, specific surveys were performed. Case (2) refers to Mid-August time in which industrial activity is absent and traffic reduced to a minimum. During

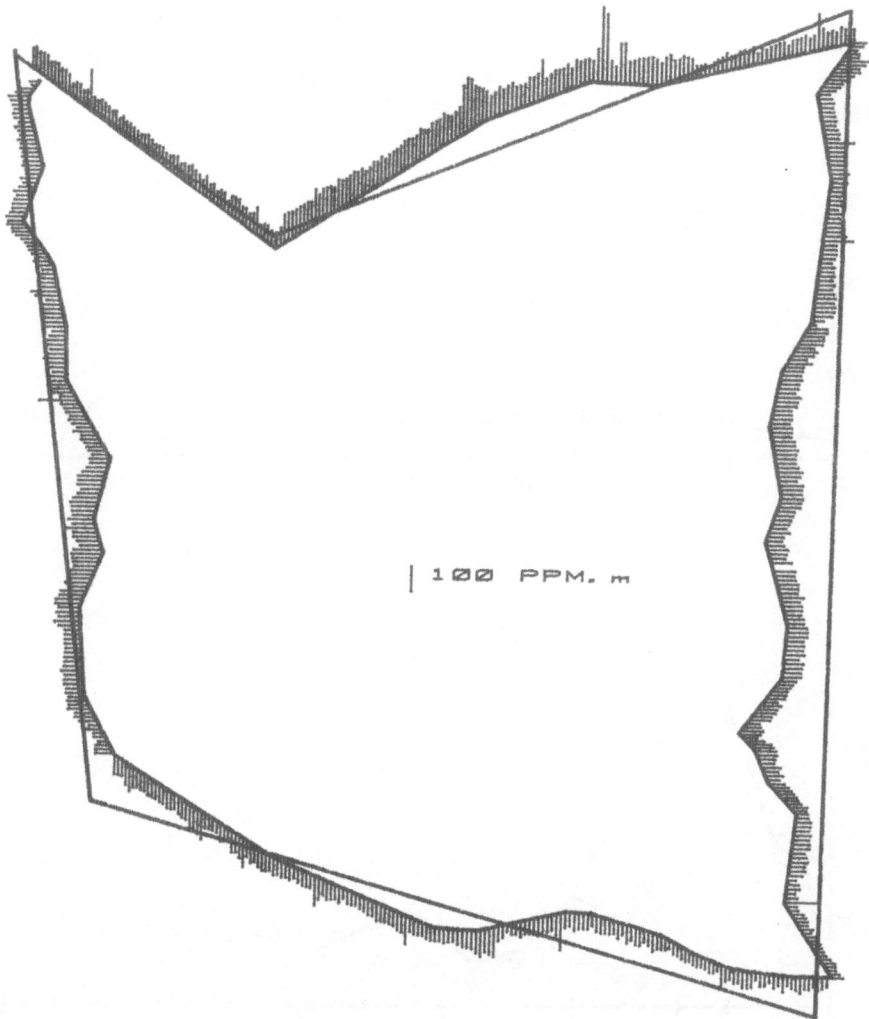

Fig 4. Survey road and approximated pentagonal line.
Burden data refer to (1).

that survey, average burden of 5-10 ppm-metre and glc of 20 ppb
SO_2 were measured. Case (5) refers to a survey performed twice
consecutively from 11.30 to 15.00 LMT: the measurements performed
during the first run have been confirmed within 15 ppm-metre, while
the meteorological conditions remained about unchanged.

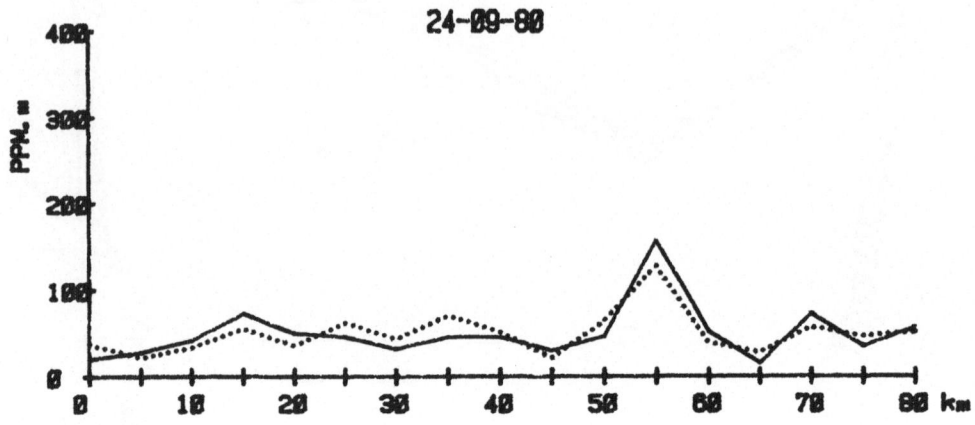

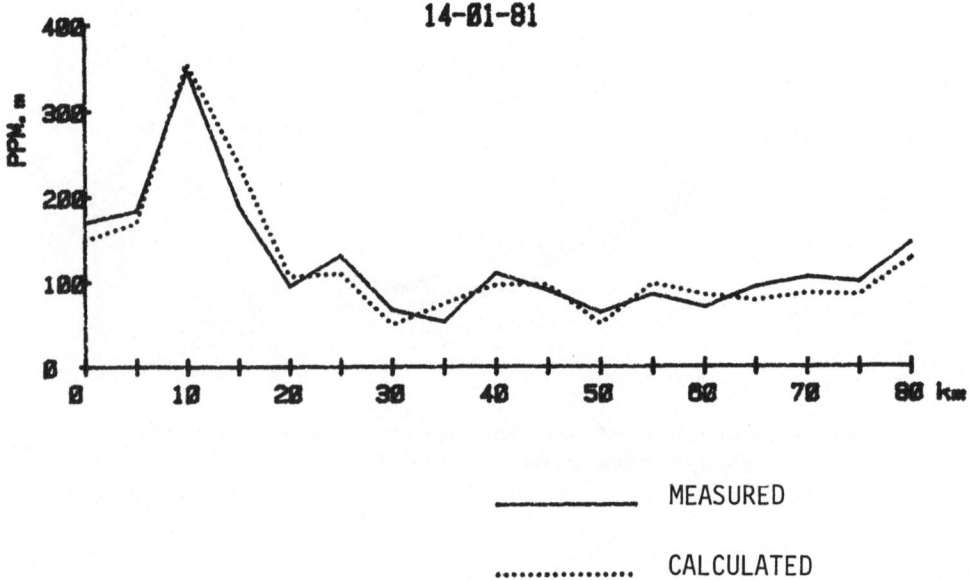

Fig. 5. Comparison between measured and calculated SO_2 burdens for cases (3) and (4).

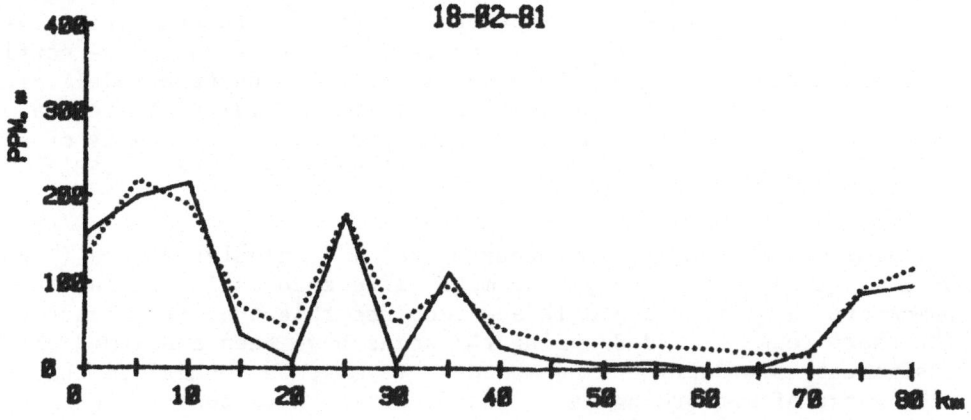

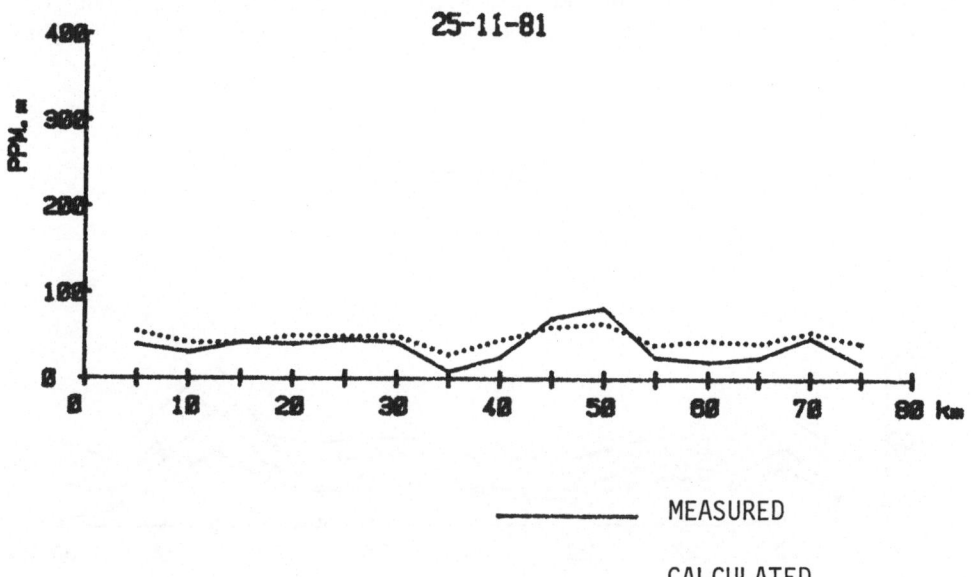

Fig. 6. Comparison between measured and calculated SO2 burdens for
        cases (5) and (6).

For the modellistic approach two classical codes (ATDL and PLUME)
have been tested in parallel to the analytical model previously described. As is known, the classical models give only ground concentration data, i.e. one has to assume a uniform vertical distribution

of the pollutant, which is erroneous. In general these models give
values lower than an analytical model, and that is once more confirmed
in these cases. In Fig. 7 the data gathered by the tree models are
compared to experimental values for (1). The validity of the analytical
model could be verified only in an indirect way on the basis of a comparison with a numerical solution as cited above (Wengle, 1982;
Clerici, 1982).

        The comparison between experimental and calculated data (Figs.
5-7) is quite satisfactory. The mean difference is lower than 20
ppm-metre, a quite acceptable accuracy for this type of measurements.
The sharp peaks recorded during the scans have been confirmed as their
assignment to specific strong sources. It must be emphasized that a
wide range of meteorological situations (which represent about 70 %
of cases) and air quality levels was analysed; for instance, (4)
refers to a mean temperature of 0°C and glc exceeding locally 1 ppm.

        From the experimental data at 100 m spatial resolution, input
and output mass-flows have been evaluated. Table 2 summarizes for the
individual cases the meteorological parameters, the mixing heights,

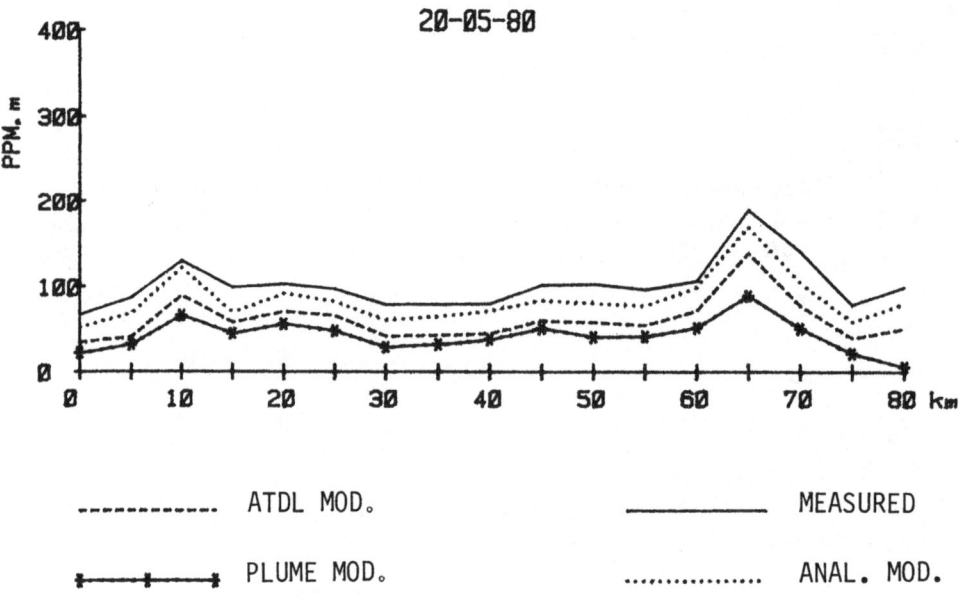

Fig. 7. Comparison of data gathered by ATDL, PLUME and the analytical
        model for (1).

Table 2 – Sulphur Dioxide Mass-Flow Above Milan Area.

Date	Tempera-ture ($^\circ$C)	Mixing Layer (m)	Av. Wind Dir. ($^\circ$N)	Av. Wind Speed (m$\cdot$s^{-1})	ϕ_{in} (t.h^{-1})	ϕ_{out} (t.h^{-1})	$\phi_{out}-\phi_{in}$ (t.h^{-1})
20.05.80	+19	1300	280	3.2	40.9	32.8	- 8.1
12.08.80	+25	1950	50	1.6	0.6	0.5	- 0.1
24.09.80	+24	850	90	1.7	22.0	17.8	- 4.2
14.01.81	0	350	90	1.0	20.5	39.0	+19.5
18.02.81	+ 2	870	90	1.5	8.3	40.5	+32.2
25.11.81	+12	210	290	5.0	27.2	40.2	+13.0

mixing heights, the input and output mass-flows and the corresponding mass-balance for the Milan area. The results confirm
what was expected, i.e. the urban area is a pollutant sink in the
warm season and an emitter in the colder one. A detailed analysis
on the absolute values would require a more complete analysis of
the meteorological situations for the previous days.

CONCLUSIONS

The mass-flows of sulphur dioxide above the Milan area in
different stable meteorological situations have been successfully
simulated by an ad-hoc developed analytical model. It has allowed
to validate experimental data gathered by a mobile laboratory
equipped with a remote sensor. Input data for the model are (a)
the emission inventory of point- and area-sources located inside
and outside the area, (b) synoptic and local meteorological information, and (c) glc and mixing heights. Local problems such as
the urban heat island may be overcome if one chooses an external
survey circuit not influenced by the phenomenon and for which
synoptic wind data can be used.

By combining experimental data with the model one can extend the range of information such as the contribution of single
sources to local urban air quality level.

No limitation is foreseen for the application of the described
procedure to other complex areas, provided that sufficient input
dat are available.

REFERENCES

Clerici, G.C., 1982, Analytical model for calculation of the ver-
    tical burden of a pollutant (to be published).
Clerici, G.C., Sandroni, S., 1982, Results to be published.
Concawe, 1976, Characteristics of urban air pollution: sulphur
    dioxide and smoke levels in some European cities, Report
    4/76.
de Groot, M., Cerutti, C., Sandroni, S., 1982, A mobile unit for
    mapping of atmospheric pollution, EUR 8260 en.
Final Report of the 5th CEC Campaign on Remote Sensing of Air
    Pollution at Ghent, June 1981, to be published.
Finzi, G., Fronza, G., Spirito, A., 1980, Multivariate stochas-
    tic model of sulphur dioxide pollution in an urban area, J.
    Air Poll. Control Ass., 30:1212.
Haltiner, G.J.,Martin, F.L., 1957, "Dynamical and physical
    meteorology", McGraw Hill, N.Y., 21-45.
Hoff, R.M., Millan, M.M.,1981, Remote SO_2 mass flux mea-
    surements using Cospec, J. Air Poll. Control Ass., 31:381.

Millan, M.M., 1980, Remote sensing of air pollution. A study of some
    some atmospheric scattering effects, Atmos. Environ. 14 : 1241

Sandroni, S., de Groot, M., 1980, Intercomparison of SO_2 and NO_2
    remote sensors at the 4th CEC campaign, Turbigo, 1979, Atmos.
    Environ., 14:1331.

Sandroni, S., de Groot, M., Borghi, S., Santomauro, L., 1982, Air
    pollution mass-flow over Milan area, Atmos. Environ., 16:1271.

Santomauro, L., 1971, Analysis of wind trend in an urban area
    situated on flat land, 51st Annual Meeting of the Am. Met. Soc.,
    San Fransisco.

van Egmond, N.D.,Tissing, D., Onderdelinden, D., Bartels, C., 1978,
    Quantitative evaluation of the mesoscale air transport, Atmos.
    Environ., 12:2279.

Wengle, H., 1982, Numerical solution of advection diffusion problems
    by collocation methods, Proceedings of the IIIrd GAMM Conf. on
    Numerical Methods in Fluid Mechanics, Part 2, 295, J. Wiley &
    Sons, N.Y.

DISCUSSION

A. VENKATRAM   Why did you use the diffusion equation
            rather than the Gaussian equation to calculate the
            vertical concentration profile?

G. CLERICI   The use of Gaussian equation requires
            the knowledge with a great accuracy of the dispersion
            coefficients, σ_x, σ_y, σ_z.
            Analytical solution of diffusion equation contains instead
            only the ratio K_y/K_z represented by parameter β. All
            the remaining part of the solution is in a dimensionless
            form, so the possibility to introduce errors and their
            propagation is limited.

I. TROEN   In your diffusion equation (1) it is
            not clear to me how you specify the turbulent diffusivities
            K_y en K_z. Could you comment on the specification on these
            quantities in your analytical model?

G. CLERICI   Diffusion equation (1) is put in a
            dimensionless form, so the turbulent diffusivity K_y and K_z
            in the solution are present only through the ratio $\beta =$
            K_y/K_z. The value of this parameter depends upon meteo
            conditions, but some different test performed by ranging
            this value do not exhibit remarkable modifications on the
            net mass flow.

G. SCHAYES   The 5th EEC Campaign in Ghent, where
                  a similar technique with at least 6 Cospecs were used,
                  showed a large discrepancy between individual measurements
                  making SO$_2$ fluxes estimation very difficult.

S. SANDRONI   The means and the information available
                  at the 5th CEC Campaign held at Ghent in June 1981 were
                  different from those in Milan. At Ghent several mobile
                  units made simultaneous surveys in an area for which the
                  hourly emissions of the single sources were available.
                  In Milan surveys have been performed by only one mobile
                  unit ; furthermore the emission was available on a seasonal
                  base. We have limited the possible sources of errors by
                  (i) a reduced measurement time (2 hours, around noon),
                  (ii) an accurate calibration procedure, and finally (iii)
                  a background procedure. In contrast to Ghent case, we
                  could operate in stable meteorological situations. In
                  our case, we noticed that the estimated SO$_2$ mass-flows
                  agreed, as order of magnitude, to the fuel consumpution
                  inventory.

TETROON FLIGHTS AS A TOOL IN

ATMOSPHERIC MESO-SCALE TRANSPORT INVESTIGATION

W. G. Hübschmann, P. Thomas and S. Vogt

Kernforschungszentrum Karlsruhe
Hauptabteilung Sicherheit/Projekt Nukleare Sicherheit
D-7500 Karlsruhe

1. INTRODUCTION

Increasing attention is being paid to the meso-scale transport and dispersion of airborne pollutants. The reasons are the growing emission of pollutants, and their impact to the oecosystem over hundreds of kilometers.

Several meso-scale models call among other parameters for knowledge of the dispersion parameters - the standard deviations of the Gaussian plume equation - and of the trajectories of the pollutant plumes.

Vertical and horizontal dispersion parameters have been determined up to about 10 km downwind as a function of the stability class and emission height by tracer experiments by various insti- tutions, see e. g. /TH81a, TH81b/. For meso-scale calculations, extrapolation of the locally determined diffusion parameter curves is not adequate without proof, and direct experimental evidence of the plume behaviour is only scarce. Tetroon flights can provide more information on atmospheric dispersion in the range up to 100 km. The horizontal dispersion parameter can be derived from radar tracked tetroon flights.

Estimates of plume trajectories have usually been based on wind data obtained at fixed points (Eulerian network). Atmospheric diffusion however depends on the movement of individual air parcels (Lagrangian consideration). Although Eulerian wind data are suitable for diffusion estimates over short times and distances, there are important difficulties inherent in this technique for longer durations and distances: On its transport, a plume of pollutants encounters

457

changes of wind-speed and -direction, that are difficult to deduce
from an Eulerian network. Tetroon flights are supposed to provide
Lagrangian-like data for the meso-scale.

2. EXPERIMENTAL TECHNIQUE

The balloon used has a tetrahedral shape and is commonly
referred to as tetroon (see Fig. 1). It is manufactured from
polyester dyed red film with a skin thickness of 51 μm. Its mass
and volume is 470 g and 1 m³, respectively. It is inflated with
helium and ballasted for the desired altitude. Once a tetroon has
been released, it will rise until its buoyancy equilibrium is reached.
Then the tetroon will float at an isopycnic surface. The inflation
and ballasting procedure is described in /VO82/.

Tracking of the tetroon was performed in most cases by an
MBVR-120 (Mobiles Ballon Verfolgungs Radar) leased from the German
Army Services. This radar is seen on the rear of Fig. 1. Its
characteristics are compiled in Tab. 1.

The antenna with its gears, motors and the Gunn-oscillator is
mounted on a 3 tons trailer pulled by a 5 tons truck. The radar
electronic and the control panel is housed in the truck.

After the ballasted tetroon has been launched it is tracked by
an optical target searching device connected to the antenna. Each
movement of the searching device is followed by the antenna. Once
the echo of the tetroon is seen on the video-display of the control
panel, the radar is switched to the automatic tracking mode.

During automatic tracking the following informations are
printed:

- time after release of the tetroon,
- distance between radar and tetroon,
- elevation angle,
- azimuthal angle.

For short flights the tetroon is equipped with a passive
corner reflector made of wooden bars and aluminum coated paper
(see Fig. 1). The radar receives the surface echoes of the
reflector and uses them for automatic tracking. Due to the small
height above ground of the reflector, the so called ground clutter
disturbs the tracking with increasing distance. To get rid of the
ground clutter a transponder (transmitter-responder) is used
instead of the reflector. The transponder (Fig. 2) receives the
radar pulses and responds at a slightly different frequency. The
transponder characteristics are listed in Tab. 1. To track the
transponder the radar is run in its secondary mode: The receiving

Table 1.   Characteristics of Radar and Transponder.

	Radar	Transponder
Dimension		
Antenna	2.1 m Ø	
Total		15.4x8.1x8.1 cm³
Mass		410 g
Gain of the antenna	39 dB	5 dB
polarization	linear, vertical	
Frequency	9375 Hz	9200 Hz
Peak power	75 kW	180 mW
Pulse repetition frequ.	250 Hz	250/1000 Hz
Pulse length	0.5/2.0 µs	0.5/1.0 µs
Scanning frequ.	30 Hz	
Range	>50 km	
Operating time		>5 h

Fig. 2. Photo of a transponder.

Fig. 1. Photo of a tetroon with a corner reflector and the MBVR in the rear.

chain of the radar is tuned to the frequency of the transponder.
Now the radar is able to discriminate the ground clutter which has
the same frequency as the radar pulses emitted.

The finder of a transponder is asked to send it back to KNRC
(Karlsruhe Nuclear Research Center). About 60 % of the valuable
transponders are returned and can be used several times after
exchange of batteries. Beside we know the place of landing and
time of finding, which gives more information about the complete
tetroon trajectory.

3. TETROON FLIGHTS PERFORMED

The data of our tetroon campaigns are compiled in Tab. 2,
along with the stability classification which prevailed during the
flights. The mean flight level varies from 250 m to 1200 m above
ground, although all tetroons had been set to a level of 400 m.
The differences between the desired and the actual flight levels
are partly due to inadequate ballasting, which was difficult on
account of high wind velocities near ground.

The campaigns have been performed at three different sites:
The Witthoh is a flat hill (860 m) and higher then the hills in the
surrondings of more than 40 km. It is located 20 km northwest of the
Lake Constance.

Dorum is a small village at the coast of the Northern Sea
situated about halfway between Bremerhaven and Cuxhaven. In the
environment of Dorum the PUKK-campaign (Projekt zur Untersuchung
des Küstenklimas) was performed, in which we participated among
more than ten other organisations. In this project the difference
of the structure of the atmospheric boundary layer on both sides
of the coast-line was investigated in the meso-scale.

Three tetroon campaigns were performed in the vicinity of the
KNRC situated in the Upper Rhine Valley. At the KNRC a 200 m high
meteorological tower is operated /DI76/, giving detailed information
about atmospheric conditions at this site. The radar was operated
at two different sites which are chosen dependent on the wind
direction forecast: during NE winds at the east side of the Rhine
valley, during SW winds at the west side of the Rhine valley.

4. EVALUATION OF TETROON FLIGHTS

4.1 Vertical Motion

Fig. 3 shows the height profile of a tetroon flight plotted in
dependence on the travel time. The profile of the terrain passed
by the tetroon is indicated in solid black. The tetroon,released

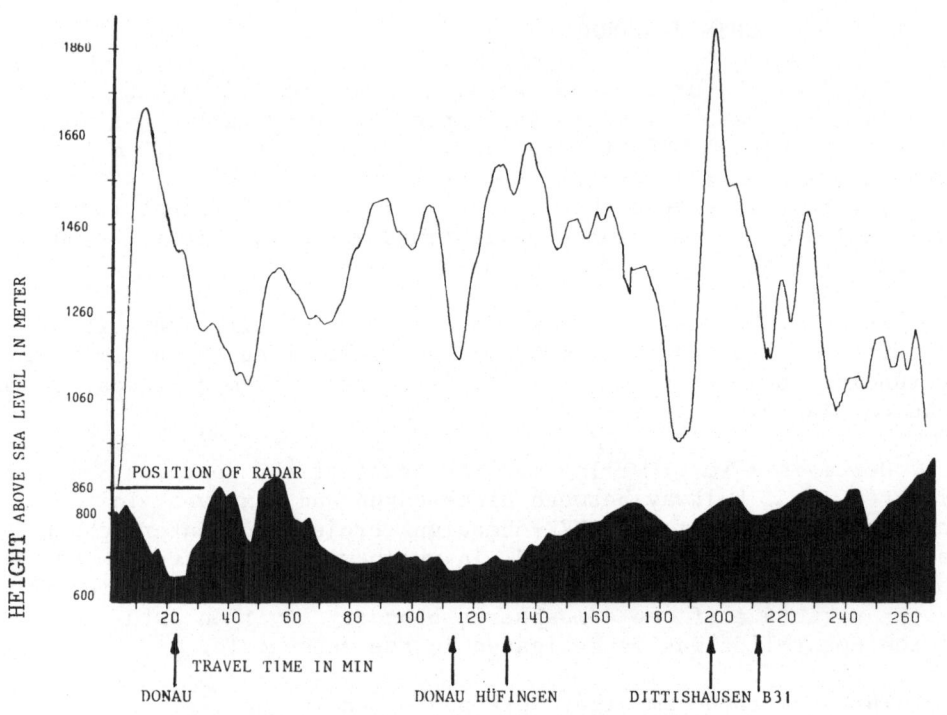

Fig. 3.   Profile of a tetroon launched from Witthoh on Aug. 25, 1981,
          12:00 MET.

at the 25th August 1981, 11:00 Mean European Time (MET) from
Witthoh showed strong vertical motions of up to 500 m above and
below the mean flight level. Sometimes the vertical velocity of
the tetroon exceeded its horizontal one and the upward velocity
is generally greater than the downward one (up to 2.7 m/s vertical
upward velocity at 190 min flight time in Fig. 3). During this
flight the temperature gradient measured with a structure sonde
was about -1.1 K/100 m. Synoptic observation (wind speed and
cloudiness) at noon indicated very unstable stratification.

Fortak /FO80/ experienced similar vertical motions ($\sim$ 3 m/s)
in the Upper Rhine Valley 600 m above ground during a measuring
campaign with a motor glider. Fortak characterized the meteorological situation by calm high pressure and pronounced convective
activity. Angell /AN61/ experienced vertical velocities up to 4 m/s
with tetroons over a desert in the USA.

The contour of the flight (Fig. 3) was Fourier analysed to
determine the period of vertical oscillations. Two peaks in the
spectral density of the vertical velocity revealed periods of
15 and 27 min.

4.2 Horizontal Dispersion Parameter

As shown in /SL68/, it is possible to estimate lateral
diffusion from successive tetroon flights. This technique of
estimating the atmospheric diffusion is adequate as far as the
diffusion of a continuous point source is considered.

Panofsky and Brier /PA58/ showed that the best estimate of
the variance of the tetroon trajectories (σ_y^2) is given by:

$$\sigma_y^2 = (N-1)^{-1} \sum_{i=1}^{N} (y_i - \bar{y})^2 ,$$

where N is the number of trajectories, y_i is the position of
tetroon No. i on the y-axis, and $\bar{y}$ is the mean position of the
N Tetroons on the y-axis.

The square root of this lateral variance is the horizontal
dispersion parameter σ_y in the Gaussian dispersion formula.

According to this procedure and to Fig. 4, circles with
different radii are drawn and the distances between points of
intersection of the trajectories with these circles are measured
yielding σ_y-values at individual downwind distances x.

The results of several flight series are plotted in Fig. 5

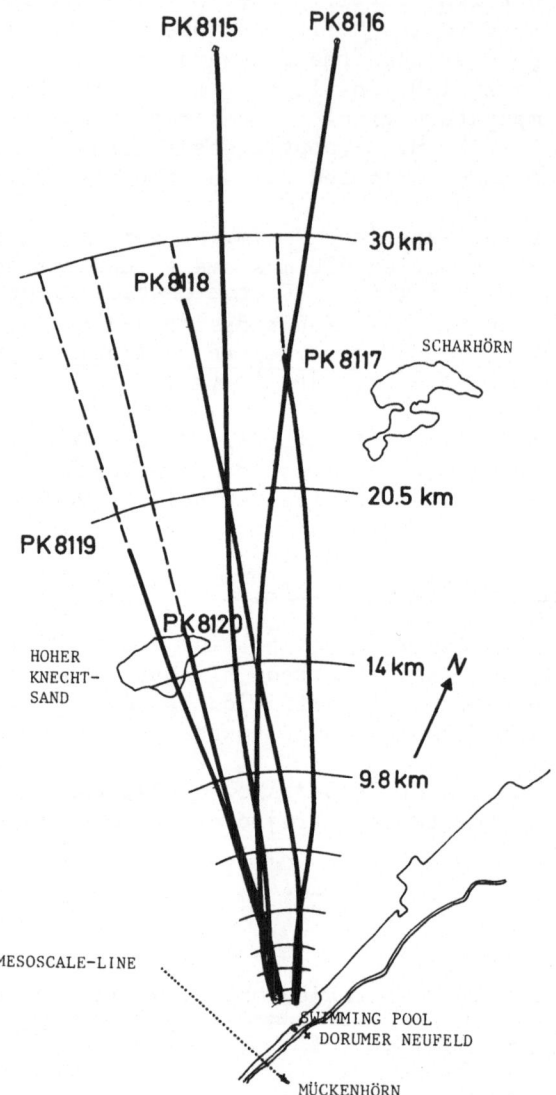

Fig. 4. Trajectories of tetroon flights on
         Oct. 2, 1981,
           ----: extrapolated trajectories.

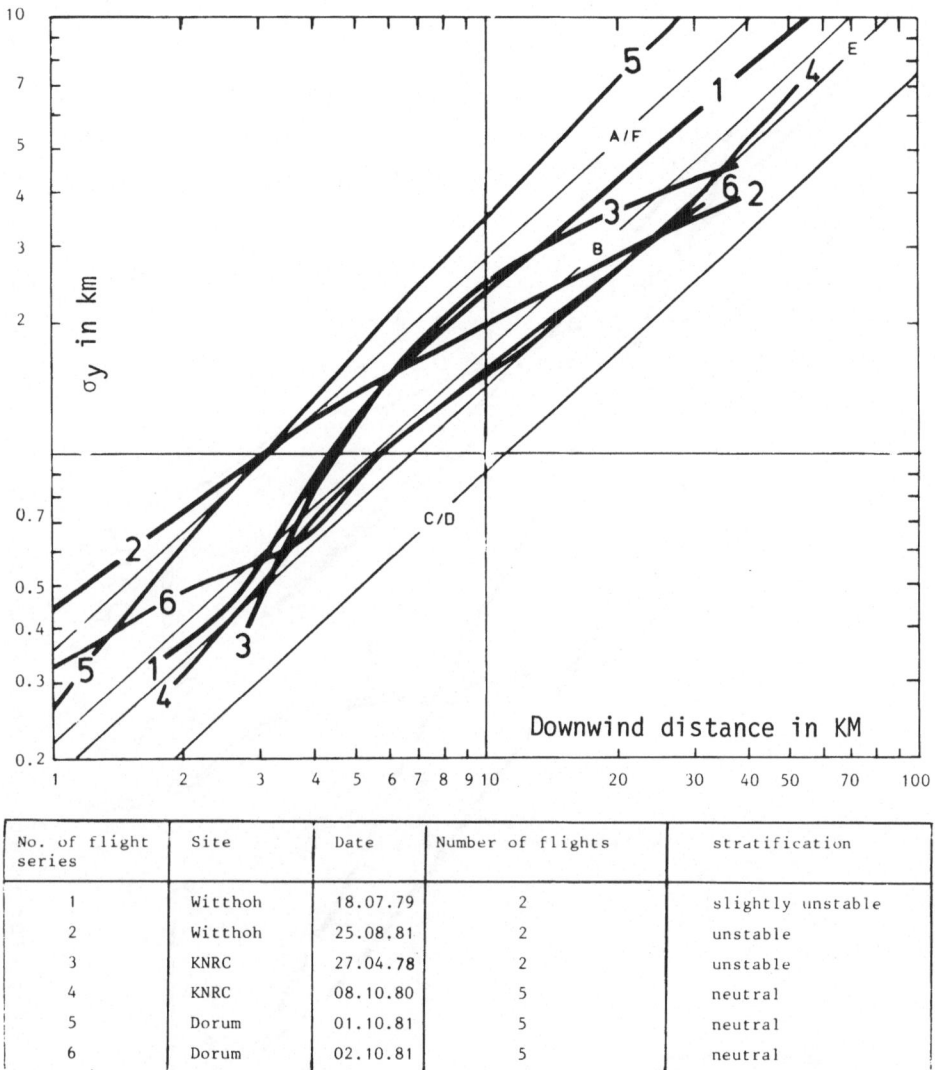

No. of flight series	Site	Date	Number of flights	stratification
1	Witthoh	18.07.79	2	slightly unstable
2	Witthoh	25.08.81	2	unstable
3	KNRC	27.04.78	2	unstable
4	KNRC	08.10.80	5	neutral
5	Dorum	01.10.81	5	neutral
6	Dorum	02.10.81	5	neutral

Fig. 5. Horizontal Dispersion Parameter σ_y.

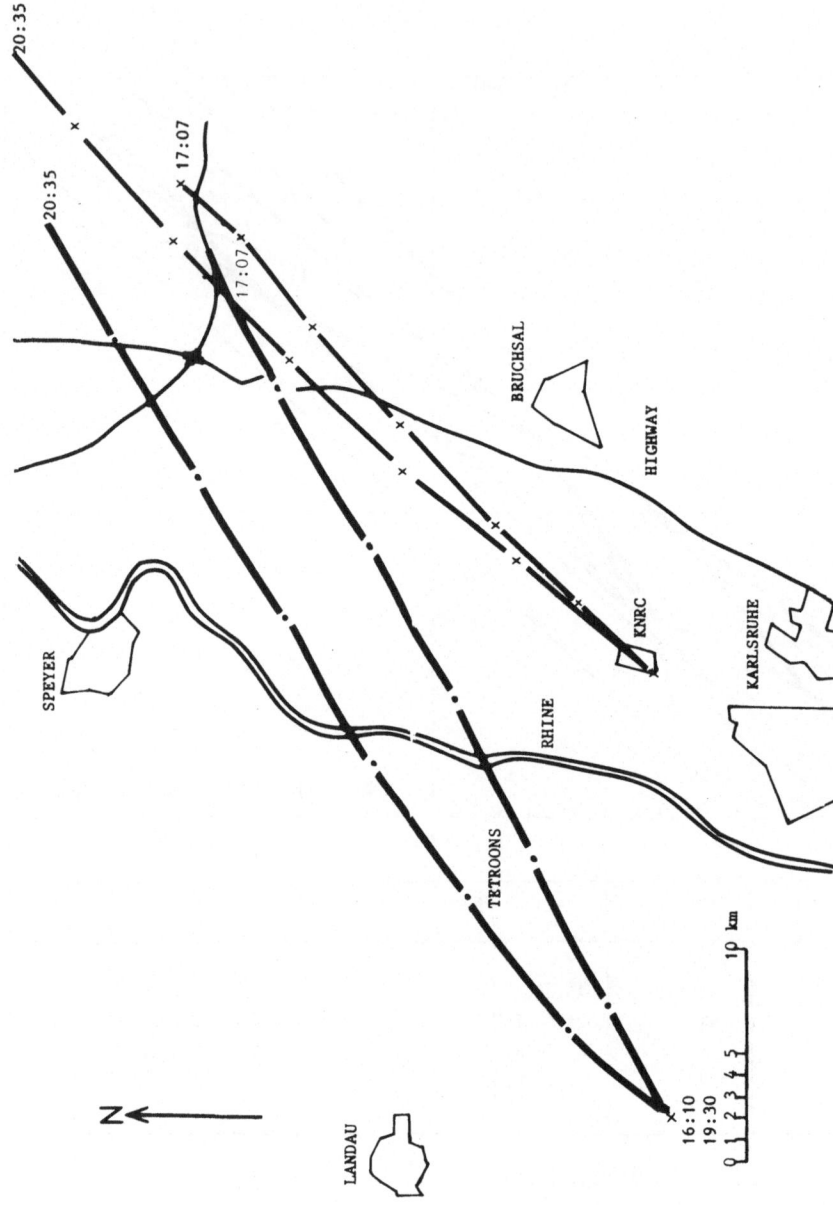

Fig. 6. Trajectories from tetroon flights on Oct. 8, 1980 and from tower wind data.

in a double logarithmic scale in dependence on the downwind
distance. The σ_y-curves are an arbitrary best fit to the individual
σ_y/x-pairs. The thin lines represent the results of tracer
experiments performed at KNRC with emission heights of 160 m and
195 m /HU80/. These tracer experiments have been performed up to
source distances of 10 km; beyond 10 km the curves are extrapolated.

        From Fig. 5 the following conclusions can be drawn:

- The σ_y-curves deduced from tetroon trajectories are well comparable
  to those from tracer experiments.
- The σ_y-curves of individual tetroon campaigns scatter hardly. in
  contrast to σ_x-curves of individual tracer experiments which
  differed considerably.
- There is not yet a significant dependence of σ_y on atmospheric
  stability.

4.3 Comparison of Observed and Calculated Trajectories

        Trajectories are calculated normally from the windfield which
is derived from as many measuring stations as are available. In
the simplest form, the trajectory is drawn using the wind measurements of only one site. The tetroon course is a test of the
accuracy of the so calculated trajectory. This method is being
applied to the compaign of Oct. 8, 1980.

        The trajectories of two tetroons released west of the Rhine
at 16:10 and 19:30 MET are plotted in Fig. 6. The mean flight
level was 1000 m and 850 m, respectively. The trajectories derived
from wind measurements taken at the top of the 200 m high tower of
the KNRC are added. The tower is situated 21 km east of the tetroon
release point. The wind measurements are available as mean values
over 10 min. To indicate the timing and to label the trajectories
the times of beginning and end of each trajectory is indicated in
MET. Spacings between dots or crosses, respectively, represent
10 min intervals.

        The following effects can be observed:

- The tower trajectories are shifted counter-clockwise as compared to
  the tetroon trajectories due to the channelling effect of the
  Upper Rhine Valley which runs in the NNE direction. This effect
  is more pronounced at the lower tower measuring level than at
  the tetroon flight level. The shift angle is about 11° and 16°,
  respectively.
- The divergence of the tower trajectories is smaller. This might
  be due to the difference between Eulerian and Lagrangian method
  of observation.
- The tetroon transport velocity is underrated by wind data, meas-

Table 2.   Campaigns with radar tracked tetroons.
Witthoh: Highland in Southern Germany
KNRC: Upper Rhine Valley
Dorum: North Sea Coast

Date	Site	Number of flights		Flight level in m	Distance tracked in km	Atmospheric stability
		Transponder				
		Without	With			
25.08.77	Witthoh	2		500– 750	14–24	slightly unstable
25.04.–27.04.78	KNRC	6		350– 700	10–29	slightly unstable
16.07.–18.07.79	Witthoh	4	2	300– 850	6–65	slightly unstable
08.10.80	KNRC	5		250–1000	17–49	neutral
24.08.–26.08.81	Witthoh		4	450–1000	45–94	unstable
29.09.–06.10.81	Dorum	27	2	250–1200	11–51	neutral
12.05.–17.05.82	KNRC		8	300–1000	46–83	unstable

ured at the tower level. This can be accounted for by considering the vertical profile of the wind velocity.

5. PERSPECTIVE TO FUTURE WORK

Previous tetroon flight series and their evaluation indicated the possible gain of this technique for meso-scale modeling. Further studies and improvements are necessary to draw full advantage of this experimental tool.

A fair comparison of observed and constructed trajectories is based on wind measurements at different sites and up to the tetroon flight level. This is being planned for the PUKK-campaign, see Tab. 1, where extensive wind measurements have been performed throughout the coastal area.

The σ_y-evaluation from tetroon flight series will give more reliable results, if two or more tetroons are launched and tracked simultaneously. As it is too costly and needs too much effort to engage more than one radar, it is presently being planned, to enable the radar to perform multiple tracking, but further development is still necessary.

Due to the numerous technical problems that are difficult to overcome tetroon experiments need much time for development and management. But as some of the difficulties are being solved, we are hopeful to have provided a valuable tool for meso-scale modeling.

6. REFERENCES

/AN61/ Angell, J. K., Pack, D. H.: Estimation of Vertical Air
        Motions in Desert Terrain From Tetroon Flights, Monthly
        Weather Rev. pp. 273-283 (1961).

/DI76/ Dilger, H.: Das meteorologische Meßsystem des Kernforschungs-
        zentrums Karlsruhe, KFK 2347 (1976).

/FO80/ Fortak, H., Lindemann, C.: Abwärmeprojekt Oberrheingebiet,
        Erste Phase des Projekts 1976-1979, Editor: Umweltbundesamt
        (1980).

/HU80/ Hübschmann, W., Nester, K., Thomas, P.: Ausbreitungsparameter
        für Emissionshöhen von 160 m und 195 m, in Jahresbericht 1979
        der Hauptabteilung Sicherheit, pp. 182-184, KfK 2939 (1980).

/PA58/ Panofsky, H. A., Brier, G. W.: Some Applications of Statistics
        to Meteorology. Pennsylvania State Univ., University Park
        Penn. (1958).

/SL68/ Slade, D. H., (Editor): Meteorology and Atomic Energy.
       TID-24190 (1964).

/TH81a/ Thomas, P., Dilger, H., Hübschmann, W., Schüttelkopf, H.,
        Vogt, S.: Experimental Determination of the Atmospheric
        Dispersion Parameters at the Karlsruhe Nuclear Research
        Center for 60 m and 100 m Emission Heights, Part 1:
        Measured Data, KfK 3090 (1981).

/TH81b/ Thomas, P., Nester, K.: Experimental Determination of the
        Atmospheric Dispersion Parameters at the Karlsruhe Nuclear
        Research Center for 60 m and 100 m Emission Heights, Part 2:
        Evaluation of Measurements, KfK 3091 (1981).

/VO82/ Vogt, S., Thomas, P.: Investigation of Meso-scale Atmospheric
       Transport by Means of Radar Tracked Tetroons, to be published
       in "Beiträge zur Physik der Atmosphäre".

DISCUSSION

M.W. CHAN   What is the sensitivity of σ_y
to the number of tetroon trajectories ?

P. THOMAS   Until now we have evaluated
six flight series comprising two to five trajec-
tories per σ_y-curve. In these six series no significant dependance of σ_y on the number of trajectories
has been found.

P.J.H. BUILTJES   Can you say something about
the averaging time connected with the values of σ_y?

P. THOMAS   The averaging time is the time
interval between the first and the last tetroon
lauchings in one series. The number of trajectories
used to determine σ_y varied from 2 to 5. The duration
of one tetroon flight was 1 to 2 hours, in seldom
cases more than 2 hours. The averaging time is between 2 and 4 h in the σ_y-values presented.

B. SMITH   Have you been able to use
your tetroon data to calculate Lagrangian spectra
for the vertical and crosswind components and compare these with corresponding Eulerian spectra derived from wind measurements collected on the tower at
Karlsruhe ?

P. THOMAS   Until now we have not performed this evaluation, but we will do it in the near
future for the vertical components measured with
vector vanes on our tower at three levels.

H. VAN DOP   During its flight the tetroon
is subject to strong vertical motion. Aren't you
measuring in fact vertical wind shear rather than
horizontal dispersion ?

P. THOMAS   I do not think so. If we
measured vertical wind shear when the balloon under-
goes vertical motion, there would also be horizontal oscillations, which we did not observe. The
horizontal component of a trajectory varies only
in a slow and monotone manner.

Crossings of trajectories of the same series are seldom. The vertical motion, which I showed by way of example, was a very pronounced one : Motions like this were not observed frequently during other flights.

T. WARNER   Have you used the cases when 5 tetroons were launched, to calculate the σ_y for all possible pairs of tetroon flights ? Have these σ_y's been significantly different from one another?

P. THOMAS   The method of calculating σ_y from all possible pairs of tetroon flights, has been used by us in some cases only. No significant differences have been found as compared to the method mentioned in my contribution. But this we will still investigate in more detail.

M. WILLIAMS   You said, in response to an earlier question, that the averaging time of the σ_y's was 1-3 hours. Can you explain how you derived this value ?

P. THOMAS   We tracked a tetroon until it was lost by the radar. Depending on atmospheric conditions for the radar beam, on wind speed and on the height of the tetroon above ground, this was between one and two hours. As we evaluated two to five trajectories to determine σ_y, the averaging time of the σ_y's lies between 2 and 4 hours. The time of 1 to 2 hours refers to the duration of one trajectory.

DISCUSSION

M.W. CHAN   What is the sensitivity of σ_y
                    to the number of tetroon trajectories ?

P. THOMAS   Until now we have evaluated
        six flight series comprising two to five trajec-
        tories per σ_y-curve. In these six series no signifi-
        cant dependence of σ_y on the number of trajectories
        has been found.

P.J.H. BUILTJES   Can you say something about
        the averaging time connected with the values of σ_y ?

P. THOMAS   The averaging time is the time
        interval between the first and the last tetroon
        launchings in one series. The number of trajectories
        used to determine σ_y varied from 2 to 5. The duration
        of one tetroon flight was 1 to 2 hours, in seldom
        cases more than 2 hours. The averaging time is be-
        tween 2 and 4 h in the σ_y-values presented.

B. SMITH   Have you been able to use your
        tetroon data to calculate Lagrangian spectra for the
        vertical and crosswind components and comparese these
        with corresponding Eulerian spectra derived from wind
        measurements collected on the tower at Karlsruhe ?

P. THOMAS   Until now we have not perfor-
        med this evaluation, but we will do it in the near
        future for the vertical components measured with
        vector vanes on our tower at three levels.

H. VAN DOP   During its flight the tetroon
        is subject to strong vertical motion. Aren't you
        measuring in fact vertical wind shear rather than
        horizontal dispersion ?

P. THOMAS   I do not think so. If we mea-
        sured vertical wind shear when the balloon undergoes
        vertical motion, there would also be horizontal oscil-
        lations, which we did not observe. The horizontal
        component of a trajectory varies only in a slow and
        monotone manner.

Crossings of trajectories of the same series are seldom. The vertical motion, which I showed by way of
example, was a very pronounced one: Motions like
this were not observed frequently during other flights.

T. WARNER   Have you used the cases when
5 tetroons werd launched, to calculate the σ_y for
all possible pairs of tetroon flights ? Have these
σ_y's been significantly different from one another ?

P. THOMAS   The method of calculating σ_y
from all possible pairs of tetroon flights has been
used by us in some cases only. No significant differences have been found as compared to the method
mentioned in my contribution. But this we will still
investigate in more detail.

M. WILLIAMS   You said, in response to an
earlier question, that the averaging time of the
σ_y's was 1-3 hours. Can you explain how you derived
this value ?

P. THOMAS   We tracked a tetroon until it
was lost by the radar. Depending on atmospheric conditions for the radar beam, on wind speed and on the
height of the tetroon above ground, this was between
one and two hours. As we evaluated two to five trajectories to determine σ_y, the averaging time of the
σ_y's lies between 2 and 4 hours. The time of 1 to
2 hours refers to the duration of one trajectory.

5: DISPERSION MODELING INCLUDING
   PHOTOCHEMISTRY
   Chairmen: F. B. Smith
            J. Knox
   Rapporteurs: B. Silvertsen
               B. Misra

PROGRESS IN PHOTOCHEMICAL AIR QUALITY SIMULATION MODELING

J.H. Shreffler, K.L. Schere, and K.L. Demerjian[1]

Environmental Sciences Research Laboratory
U.S. Environmental Protection Agency
Research Triangle Park, NC 27711 U.S.A.

INTRODUCTION

Over the past 15 years we have witnessed the development of atmospheric dispersion models which include increasingly complex approximations of photochemistry. Unlike existing Gaussian or analytic schemes, the models simulate atmospheric chemical reactions, some of which act at very rapid rates, and therefore involve time steps much shorter than are generally regarded as necessary for transport and dispersion of inert species. The computational demands rise rapidly with the number of chemical species considered, and it is not unusual to see present-day models for urban and regional scales having computer simulation speeds comparable to the real-world events. Computer restrictions have tended to engender decisions to diminish the spatial resolution of the models in order to increase the number of species and reactions treated.

We have recognized for a number of years that within the class of existing or conceived photochemical air quality simulation models (PAQSM's) there are only several basic approaches. The United States Environmental Protection Agency (U.S. EPA) in the mid-1970's reviewed the various urban scale models which were in existence and chose three, embodying distinct approaches, for further refinement, development, and evaluation. In the case of the EPA, the ultimate goal was to provide regulatory tools for control of photochemical pollutants, O_3 in particular. The models

---

[1]All authors are on assignment from the National Oceanic and
 Atmospheric Administration, U.S. Department of Commerce.

were to be used to predict the problems created by new sources or
the benefits of controlling existing sources. Shortly thereafter,
work also began on a regional-scale (1000 km) photochemical model
to be used initially in the northeastern U.S. Interest in larger
scales derived from a growing awareness that elevated O_3 levels
may be found over large regions and at considerable distances from
urban centers (Wolff et al., 1977). The goal of the regional
modeling effort was again to allow defensible regulatory decisions,
but the complexity of the problem and its immense resource requirements precluded study of several alternate approaches as was done
with the urban models.

The purpose of this paper is to describe the PAQSM's emerging
from research and development projects of the U.S. EPA. The presentation will not attempt a detailed discussion of chemical mechanisms or numerical schemes. Rather, emphasis will be placed
on the models' basic structures, data requirements, computer requirements, and problems encountered in applying them. Also, we
will describe the two major field programs which have been conducted by EPA to support testing and evaluation of the models.

COMPONENTS OF MODEL SIMULATIONS

PAQSM's may differ in complexity and therefore data and resource requirements through deliberate choices of the builders.
For example, restriction to a simple box-model makes spatial resolution of emissions and winds unnecessary. Furthermore, data
requirements will vary greatly depending on the goal of the simulations. If model development and evaluation are planned, then the
data needs are extensive. On the other hand, if the model is
already viewed as a reliable operational tool, data needs are
greatly reduced as climatological scenarios may be used for
worst-case and average events. In the latter case, predicted
pollutant levels are only checked for reasonableness against previous records and not against specific observations. Photochemical
air quality simulations share similar attributes, and requirements,
regardless of the particular model used. Generally, all PAQSM's
are logical frameworks which synthesize information from meteorology, sources, and air quality and predict the photochemical
pollutant concentrations consistent with expressions for governing
physical laws. The simulation is therefore an interplay of four
equally important components.

PHYSICAL LAWS: Physical laws are expressed as mathematical equations governing atmospheric motions (transport and diffusion) and
the chemical kinetics. The equations enforce mass conservation
and promote appropriate chemical transformations in a manner
consistent with theoretical considerations and observational data.

Kinetic mechanisms generally have been developed by comparing numerical predictions to concentrations resulting from carefully controlled smog chamber experiments. It is important to note that equations of both the meteorological and chemical aspects of the problem are approximations. For example, mixing rates are usually difficult to prescribe, and hydrocarbon (HC) species must be lumped in terms of reactivity classes.

SOURCE EMISSIONS: Pollutant emissions are an important component in the application of a PAQSM and must be compiled on spatial and temporal scales comparable to the resolution inherent in the model structure. Photochemistry involves a strong diurnal cycle which dictates a temporal resolution on the order of 1-h in the emissions inventory. Inventories usually are divided into the source types of area, point, and line. Area sources include many diffuse, population-related sources such as those associated with space heating or residential automotive traffic. Point sources are large, identifiable sources such as power plants, and line sources refer to major automotive thoroughfares. A typical inventory for a large city may include several hundred point sources. Unless the model has the unusual capacity to treat line sources, they may be incorporated into the area source inventory. For PAQSM's the emphasis is usually on developing emissions for HC, NO_x, and CO, although mechanisms generally include other species such as SO_2 and SO_4. The HC emission is broken into reactivity classes consistent with the kinetic mechanism (e.g., olefins, paraffins, aromatics, formaldehyde, and other aldehydes). Also, the NO_x must be appropriately split between NO and NO_2.

METEOROLOGICAL FACTORS: Meteorological factors account for the transport and diffusion of pollutants and influence reaction rates in the chemical mechanism. These data are introduced into the model from observations and are not predicted. Important parameters include the wind field, inversion height, atmospheric stability, solar radiation, water vapor, and temperature.

AIR QUALITY MEASUREMENTS: Ambient concentration measurements are needed to set initial conditions, boundary conditions, and as evaluation data. Photochemical simulations usually begin with an initial set of observations used to assign concentrations of important O_3 precursors, HC and NO_x, within the modeled domain. As an alternate strategy, the model could generate its own initial state by allowing sufficient time for emissions to establish suitable concentration levels. However, this strategy is usually impracticable when considering other aspects of the problem such as ventilation rates, especially on the urban-scale. Boundary conditions refer to the concentrations assigned at the extremities of the modeled region, both in the horizontal and vertical. Concentrations at the upwind boundary of a grid-model domain are of obvious impor-

tance. Moreover, the concentrations aloft may also play a role
as they are entrained during the diurnal growth of the mixed layer.
Ambient data must have a spatial and temporal resolution at least
comparable to the model output if evaluation of the model is to be
attempted.

URBAN MODELS

    The U.S. EPA currently has interest in three urban models
embodying distinctly different approaches and levels of complexity.
Those models are:

    Photochemical Box Model (PBM) - a single cell Eulerian model
    constructed by EPA.

    Lagrangian Photochemical Model (LPM) - a multi-level parcel
    model originally developed by Environmental Research and
    Technology, Inc.

    Urban Airshed Model (UAM) - a multi-level, Eulerian grid model
    originally developed by Systems Applications, Inc.

The versions of the models at EPA have been structured to easily
use urban data compiled during the Regional Air Pollution Study
(RAPS) and have been subject to continuing modifications. The
RAPS will be discussed in a later section. In this section, the
basic frameworks and requirements of the models will be surveyed.

Photochemical Box Model

    The Photochemical Box Model (PBM) is a single cell Eulerian
air quality model whose purpose is to simulate the transport and
chemical transformation of air pollutants in smog prone urban
atmospheres. The model's domain is set in a variable-volume,
well-mixed reacting cell where the physical and chemical processes
responsible for the generation of O_3 by its HC and NO_x precursors
are mathematically created. These processes include the transport
and dispersion of pollutant species through the cell, the injection
of primary precursor species by emission sources, and the chemical
transformation of the reactive species into intermediate and secondary products. They are schematically illustrated in Figure 1. In
a typical application of the model, the horizontal length scale of
the single cell is about 20 km and the vertical scale is time-
varying, equivalent to the depth of the mixed layer. The model
domain is centered on the city such that the area encompasses most
of the major emissions sources. Source emissions are assumed to
be distributed uniformly across the surface face of the cell.

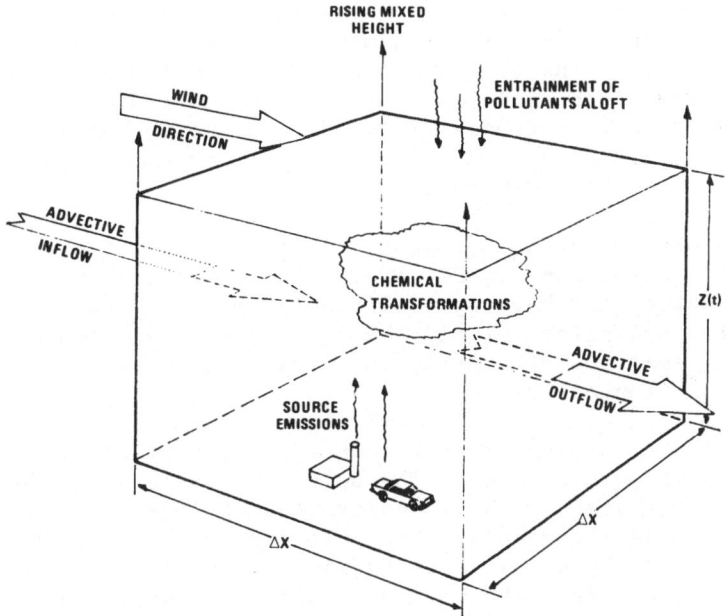

Fig. 1.   Schematic of the domain and processes in the Photochemical
          Box Model.

While the data requirements of the PBM are not particularly
rigorous, three data preprocessors must be executed prior to each
model simulation. The first accesses the 1-h average air quality
and meteorological data base forming the initial concentrations of
participating species, determining the 1-h average wind vectors
that guide transport through the model domain, and choosing the
hourly updated concentration of species at the upwind boundary of
the box. The first preprocessor also forms similar spatial averages of concentrations of monitored pollutant species for comparison with model predictions.

A function of the second preprocessor is to determine 10 min
average values of total solar radiation from pyranometer measurements. From these values the time-varying photolytic rate constants can be determined through a series of empirical and theoretical relationships. The temporal resolution of these data is
greater than for most other data because of the rapid chemical
reactions which are associated with these rate constants. This
preprocessor also calculates the diurnal growth pattern of the
mixed layer, or the depth of the model domain. To perform this
calculation, the morning minimum and afternoon maximum mixed layer
heights are provided by the user. These heights were determined
in the RAPS application by studying the vertical structure of the
temperature and moisture from radiosondes. Releases occurred

throughout the day at 6-h intervals beginning 1 h before sunrise.
The model fits a piece-wise linear function between the specified
minimum and maximum depths that simulates the actual mixed layer
growth as observed by acoustic sounder, lidar, and radiosonde
techniques. Ten-minute averages of mixed layer height are formed
for the PBM; photolytic rate constants for the corresponding time
period are integrated through this depth.

The third preprocessor is responsible for forming hourly
emissions source terms for CO, NO, NO_2, and non-methane HC. Even
though the emissions may have spatial resolution, all source emissions, both area and point, within the model domain are summed to
provide a total emission rate for each primary pollutant species
at every hour of simulation.

The PBM may be executed when the data files from the three
preprocessors have been created. The only other relevant information needed at this time is the concentration of O_3 at the top
boundary of the modeling domain, that pollutant having been transported into the area by winds aloft during the night. This concentration is determined from the O_3 measurements at the far upwind
surface monitoring sites after the nocturnal temperature inversion
has eroded and the air aloft mixes down through the atmosphere to
the surface. Concentrations of other pollutants at the top of the
modeling region are assumed to be negligible. The simulation is
typically started at 0500 CST and continues through 1700 CST.

The PBM contains a chemical kinetic mechanism with 36 reactions and 27 reactive species (Demerjian and Schere, 1979). The
set of equations describing the rates of change of the concentrations of these species is numerically solved at time steps on the
order of 10 min. From these solutions the model determines the
1-h average predicted concentration of each modeled species.

Lagrangian Photochemical Model

The Lagrangian Photochemical Model (LPM) was developed by
Environmental Research and Technology, Inc. of Westlake Village,
California and adapted under contract with EPA for use with the
RAPS data base. (The LPM is essentially identical to the general-
use model named ELSTAR.) The LPM envisions a portion of the atmosphere as an identifiable parcel which can be tracked from early
morning to the late afternoon. As the parcel moves over the
various emissions sources, pollutants are assimilated, vertically
mixed, and subjected to photochemical reactions in the presence
of solar radiation (see Figure 2). The LPM is attractive relative
to grid models in that it is fairly simple to execute and uses a
moderate amount of computer time. On the other hand, the LPM

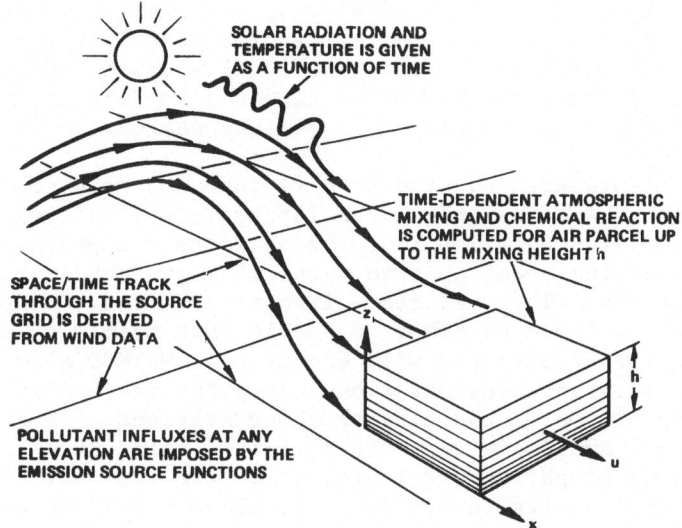

Fig. 2. Schematic of the modeling concepts of the Lagrangian
Photochemical Model.

calculates concentrations only within a parcel and not over a
complete spatial field.

The LPM is executed using a series of program modules. They
are METMOD, EMMOD, and KEMOD, sequentially performing calculations
on meteorology and air quality, emissions, and photochemistry. The
input and running procedures described by Lurmann et al. (1979) have
been generally followed, although some modifications were deemed
necessary as more experience with the model was acquired (Lurmann,
1980; 1981).

The first step in setting up a simulation is to determine the
starting point of a parcel so that it will arrive at a specified
point at an assigned time. A backward trajectory can be generated
by the METMOD. The parcel trajectory is usually determined by
$1/R^2$ weighting of winds from the closest three stations. However,
experience with wind data suggests that even closely situated
stations can show large, unexplained differences in wind vectors
from time to time. Reliance on a single anomalous station, if the
parcel has a close approach, can create an erratic trajectory. To
eliminate the possibility of such vagaries, it may be prudent to
compute a single resultant wind vector from the wind station network
and assign it to all stations for a given hour. The resultant gives
a general movement of the air mass over the region.

Because of the difficulty in solving the chemistry set with minimal solar radiation, the start time of a parcel (which is on the hour) must be at least 10 min past local sunrise. Once the start position is set, the METMOD is run in a forward-trajectory mode until 1800 CST or the parcel leaves the region.

Mixing heights are computed from radiosondes released approximately 1 h before sunrise and at 6-h intervals thereafter. The temperature and wind profiles from the balloons are processed automatically by the model to give vertical eddy diffusivities throughout the day. The trajectory information from METMOD is stored and is available to EMMOD to enable that module to obtain the emissions that the parcel will encounter. METMOD also stores the observed pollutant concentrations along the trajectory using a $1/R^2$ weighting scheme on the closest three stations. These observations are available for setting initial conditions in the parcel and later comparisons with the model predictions. In most cases, the parcel starts in a relatively clean rural environment, and levels of HC, NO, and NO_2 are assumed to decrease with height according to an assigned formula. On the other hand, ozone is depleted near the ground at night; thus, the initial O_3 increases from the surface to a value at 400 m which is equal to the observed 1000-1200 CST surface concentration upwind of the city (Shreffler and Evans, 1982).

The EMMOD creates a record of the emissions entering the parcel as it traverses its trajectory. These emissions are used by the KEMOD to simulate the photochemical reactions which will take place as a function of time, yielding concentrations of 39 chemical species at 30-min intervals. A total of 65 reactions are modeled.

Urban Airshed Model

The Urban Airshed Model (UAM) is a three-dimensional (3-D) grid-type, or Eulerian, PAQSM developed by Systems Applications, Inc. (SAI) of San Rafael, California. The structure of the model consists of a latticework array of cells (see Figure 3), the total volume of which represents an urban-scale domain and in which the physical and chemical processes responsible for photochemical smog are mathematically simulated. These processes include the advection of pollutant species through the modeling domain, the species entrainment from aloft by a growing mixed layer, the diffusion of material from cell-to-cell, the injection of primary source emissions into the modeled volume, and the chemical transformations of reactive species into intermediate and secondary products. The horizontal dimensions of each cell are constant but the heights of the cells vary throughout a model simulation as the depth of the mixed layer in the UAM changes accordingly.

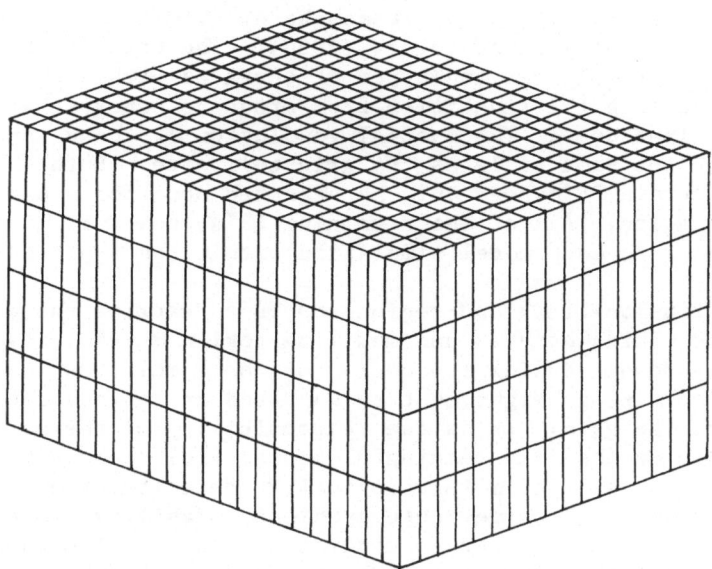

Fig. 3. Schematic of the grid domain used in the Urban Airshed
         Model.

In typical applications, the area modeled is about 60 x 60
km, and each individual cell is 4 km on a side in the horizontal.
Vertically, there are four layers of cells in total; the bottom
two layers simulate the mixed layer and the top two represent the
region immediately above the mixed layer. The 3-D grid model is a
sophisticated type of PAQSM and provides both spatially and temporally resolved concentration predictions. Thus, the UAM attempts
to estimate the 1-h average observed concentration of a pollutant
species at each monitoring site within the model domain.

The package of computer programs constituting the UAM actually
contains 12 data preprocessing programs as well as the simulation
model. The data requirements for applying the model are rather
intensive. The preprocessors (PP's) access surface-based, hourly
air quality and meteorological data base, the upper air pibal and
radiosonde data, and the source emissions inventory for the necessary parameters, and process the parameters as required by the
simulation model. A brief description of the PP's follows in order
to convey a sense of the UAM data requirements.

The chemistry PP sets up the rate constants and other parameters related to the chemical kinetic mechanism within the model.
This mechanism, named Carbon Bond II, is a state-of-the-art set of
chemical reactions describing the NO_x-HC-O_3 species interactions

within a photochemically active atmosphere. The mechanism was
developed at SAI by Whitten et al. (1980). The terrain PP describes the spatially varying types of surfaces within the modeling
domain, reflecting the transition from urban to suburban and finally to rural land uses. The spatial resolution of these land forms
should be on the same scale as the grid cell size. Both the chemistry and terrain PP's need only be executed once for a particular
model application. All of the remaining PP's, however, must be
executed prior to each model simulation within the application.

The diffusion-break and region-top PP's describe the variations in the mixing heights and modeling region depths, respectively, over the domain during the course of the simulation. Hourly
values of the mixing height must be supplied at representative
locations on the grid. It is the responsibility of the user to
specify these depths in a meaningful manner from upper air soundings
or equivalent data. The meteorological PP describes the temporal
variation of vertical temperature gradient, stability class, atmospheric pressure, water vapor concentration, and the NO_2 photolysis rate constant. Values for these parameters do not vary
spatially in the model. The top-concentration PP specifies the
concentrations of principal species at the top of the modeled
region throughout the simulation. In most cases these concentrations will be close to the clean air background values for these
species, although substantial concentrations of O_3 can result
from advection over the region. Its value is determined from the
O_3 measurements at the far upwind surface monitoring sites after
the nocturnal temperature inversion has eroded and the air aloft
mixes to the surface.

The air quality, temperature and wind speed PP's all require
data from a surface monitoring network. Hourly averages of observed
species concentrations are objectively interpolated across the model
grid by the air quality PP to produce a field of initial concentrations for the UAM. Typically this initial field is applied near
sunrise. As the density of monitoring locations increases, the
reliability of the objective interpolation does likewise. The
temperature PP produces gridded fields of surface temperature at
each hour of model simulation. These are required for both the
wind and point source PP's. Finally, the wind PP assimilates all
available surface and upper air wind speed and direction measurements and produces a gridded field of u and v wind components at
each of the model's four vertical levels. This objective wind
field analysis routine was developed by Anderson of SAI (Killus et
al., 1977). Vertical wind velocities are calculated internally by
the UAM from mass continuity considerations. Anderson's interpolation routine smooths the data in such a way so as to eliminate
unrealistic vertical motions. It simulates the mesoscale urban
circulation patterns through the use of the surface temperature

patterns. The boundary concentration PP determines the modeled species concentrations along each of the domain's borders for each hour of model simulation. A vertical concentration gradient is also described.

The last two PP's needed before the model is executed deal with source emissions. The area sources, including highway and line sources, and the point sources are treated separately. Hourly emission rates of all primary pollutants for each grid are calculated. Organic HC emissions must be distributed into particular structural classes.

Each PP generates a data file required by the UAM. Model simulations begin at 0500 CST and continue through 1700 CST. The model numerically calculates the rates of change of species concentrations at time steps on the order of several minutes, and from these determines the 1-h average predicted concentrations.

Status of the Urban Models

The three models discussed in this section are viewed by EPA as being in final form. That is, substantial breakthroughs in chemical mechanisms or the compilation of a clearly superior data base will be needed before revisions would be considered. Although the model preprocessors are geared to easy use of the RAPS data, it would not be difficult to adapt the models to other locations and other forms of data. In the next section the RAPS data set will be described along with some results of model applications. Before proceeding, we review some problems which arose during the testing of the models and look at estimated resource requirements.

Because of its relative simplicity, the PBM is attractive for use where resources are limited and results do not need to be highly detailed. With this model, it should be noted that emissions control strategies are difficult to target on particular sources because of the lack of spatial resolution. Also, the highest O_3 concentrations in the domain, upon which decisions may be more critically centered, are lost in the averaging assumptions. Finally, the simplified wind field (a single vector for each hour) and large size to the Eulerian domain make the model best suited for low-wind or stagnation episodes. These, of course, are usually conditions most favorable to high O_3 concentrations.

Compared to the UAM, the LPM may seem less resource intensive. However, the data requirements are nearly identical. The LPM may be used along a single trajectory, perhaps that trajectory encountering the observed maximum concentration, and therefore use much less computer time. If multiple trajectories are executed to display

concentrations over the region, such as the UAM does in a single
simulation, then computer resources and preparation time will grow
accordingly.

A number of problems have become apparent in the development
and testing of the models. For example, the LPM seems quite sensitive to initial conditions since the parcel retains all pollutants.
Thus, care must be taken to put realistic vertical distributions of
precursors in the parcel at the beginning of the simulation day.
The LPM has five vertical levels in which to distribute the pollutants. As originally conceived the model did not include any lateral
diffusion, and this produced unrealistically high O_3 concentrations
since all emissions were retained in a fixed volume. The model now
allows the parcel to expand laterally in a manner commensurate
with empirically determined diffusion. Whereas this feature considers dilution of a parcel when beyond the high emissions area,
it introduces a problem in setting side boundary condition concentrations representative of the air that is entrained into the
parcel. Our approach has been to set the initial parcel size
about equal to the downtown area of highest emissions (5 x 5 km for
RAPS) and treat the entrained air as having background values of HC
and NO_x. The parcel is viewed as a segment of the urban plume.
However, large power plant sources are liable to present conditions
seriously in conflict with that assumption from time to time. This
example evinces a fundamental truth about all of the PAQSM's; there
is an element of art in their application which is unlikely to be
eliminated without jeopardizing the models' flexibility in treating
different situations.

The UAM as originally constructed suffered from substantial
numerical diffusion. We believe this potential problem should be
thoroughly investigated in any grid model of this sort. The tendency of the UAM to underestimate ozone maxima in our tests (see
results in next section) was suspected of being due to the spurious
diffusion. However, the underestimation was only slightly improved
by a new and accurate numerical scheme, which, incidentally, increased computer time by 15 percent (Schere, 1982).

The choice of which particular model to use in a specific
application involves not only the accuracy of the model but also the
resources required to operate it. The models discussed here have
resource requirements correlated with their level of complexity. In
terms of man-months needed to set up a single day simulation and
computer time expended (minutes of CPU on a Univac 1100/82) the
approximate requirements are:

```
PBM ----------- 0.15 man-month ------------    1 minute CPU
LPM ----------- 0.20 man-month ------------   10 minutes CPU
UAM ----------- 0.50 man-month ------------  110 minutes CPU
```

The initial man-month investment needed to become adequately familiar with the model perhaps exceeds the simulation set-up time by at least a factor of 10.

The three models discussed here have all shown themselves to be acceptable tools for analysis of urban O_3 air quality. The specific configuration of an application along with the quantity and quality of related data and resources available to the user must all be considered in the final selection of a model. For an indication of average O_3 air quality in an urban area under stagnation conditions or as a screening method for a more complex model, the PBM is appropriate. The choice of a trajectory model, such as the LPM, or a grid model, like the UAM, might well be decided by resource requirements or by the number of proposed simulations. In any event, the user of any of these models must have a strong scientific background and exercise extreme care in implementing the air quality simulations.

THE RAPS AND URBAN MODEL EVALUATION

Beside the physical laws expressed in the PAQSM, essential components of air quality simulations include emissions, meteorological and ambient concentration data. In this section we describe an extensive field effort for gathering data on an urban scale and some of the modeling results based on that data.

The Regional Air Pollution Study (RAPS) was carried out in the St. Louis area during 1974-1977. St. Louis is a city of negligible topographic relief in the central U.S.; about 2.3 million people reside in the metropolitan area. Although augmented by a variety of special studies, the principal data base was gathered by a 25 station network comprising the Regional Air Monitoring System (RAMS). Measurements included both meteorological parameters and pollutant concentrations. The purpose of RAPS was to develop a data base for testing numerical air quality simulation models and especially models involving the complex photochemical reactions leading to O_3 formation. The RAPS expenditures totaled about $25 million over a five-year period, and the data base remains the most comprehensive available.

RAMS measurements, recorded as 1-min averages, included wind speed and direction, temperature, vertical temperature difference, solar radiation and concentrations of O_3, CO, HC, NO_x and SO_2. The sampling manifold used by all gaseous monitors had an intake at about 4 m from the ground. Figure 4 shows the layout of the RAMS network. A more detailed description of the RAPS program and the RAMS instrumentation is given by Schiermeier (1978).

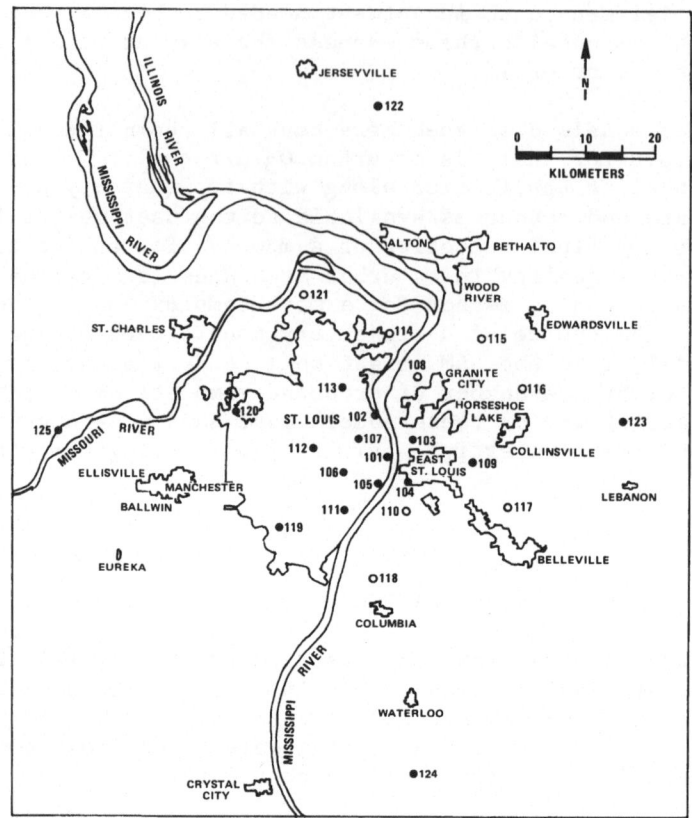

Fig. 4. The St. Louis area with locations of the RAPS stations
        101-125.

     The 1-min average values of all parameters were objectively
screened. The screening excluded values which were null, part of a
calibration set, taken during a period of excessive drift, outside
instrument limits, taken under abnormal instrument status, or
indicating persistence. The remaining data were used to form 1-h
averages, and the 1-h average archive was generally the one used in
model evaluation.

     Overviews of the RAPS O_3 and NO_x data are given by Shreffler
and Evans (1982) and Shreffler (1982). Figure 5 displays a 4-mo
record of ozone taken in mid-1975. The time series are for the
inflow concentration (that from upwind rural areas) and the maximum
1-h concentration recorded at any station in the network. The
locations of the maxima, especially the strong peaks, usually show

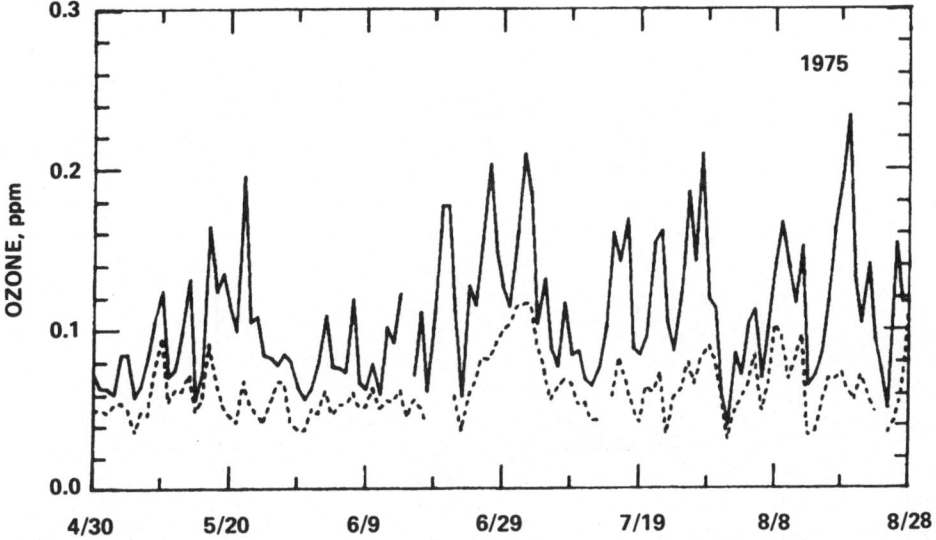

Fig. 5. Time series of the daily maximum 1-hr ozone concentration
(solid) and the daily rural inflow ozone concentration
(dashed) for May – August 1975 as measured in the RAPS.

clear downwind relations to high emissions areas. The differences
between concentrations in the two series may be ascribed to net
enhancement of O_3 levels due to the urban emissions.

In preparation for model evaluation, the entire data base was
examined to establish suitable case-study days. In all, 20 days
were selected for simulations based on high observed O_3 maxima and
adequate availability of data. The maximum observed O_3 on these
days ranged from 160 to 260 ppb.

The three urban models (PBM, LPM, and UAM) were adapted to
access the RAPS data archive, which included a detailed emissions
inventory with resolution to 1 h and 1 km (Littman, 1979). Although precise input requirements differ among the models, effort
was made to supply information in a consistent form for all.
Thus, the simulations were done in parallel, using the same basic
data sets, and the results give a comparison among models as well
as an indication of accuracy against observed concentrations.

As examples of model results, Figures 6 and 7 show observed
and predicted O_3 values for the PBM and LPM on a single day, 1
October, 1976. Stagnation conditions prevailed on that day. For
the PBM, the time series show concentrations within a fixed box
20 km x 20 km over the city center. Maximum and minimum observed

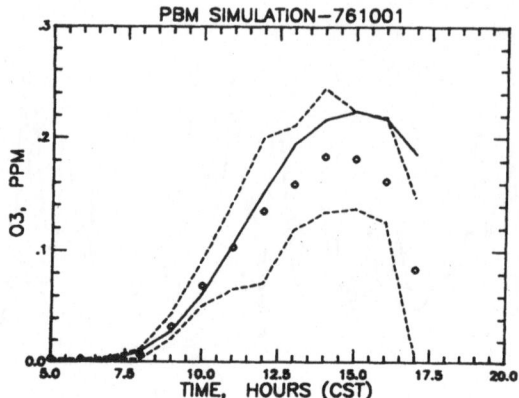

Fig. 6. PBM predicted ozone (solid) compared to measured domain
average (circles) and maximum and minimum within the
domain (dashed). This result is for the 1 October 1976
RAPS simulation.

concentrations over all the stations (13) in the box are given by
the dashed lines. The average observed concentration is given by
circles, and the model prediction is given by the solid line. For
the LPM, the concentrations refer to a moving parcel with initial
horizontal dimensions 5 km x 5 km. Predicted values are given for
two vertical levels in the parcel, the surface (L-1) and about 200 m
(L-3). This particular parcel arrives at the station (102) observing the maximum 1-h O_3 value at the time of that observation
(1400 CST).

The ability of the models to reproduce observed O_3 maxima
over the 20 test days is summarized by statistics given in Table 1.
Complete results of the urban model evaluation are now available in
report form (Schere and Shreffler, 1982).

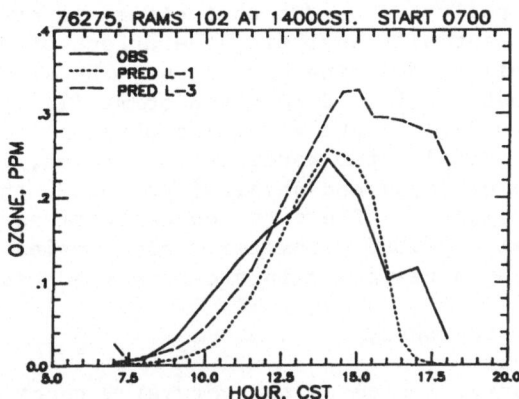

Fig. 7. LPM predicted ozone at the surface (L-1) and 200 m (L-3) compared to the observed ozone. This result is for the 1 October 1976 RAPS simulation.

Table 1. Statistics on residual concentrations (observed minus predicted) of maximum ozone (ppb) from 20 days of RAPS data. The observed maxima ranged 160-260 ppb.

	PBM	LPM	UAM
$\overline{\Delta C}$	-12	5	62
s.d.(ΔC)	39	58	35
$\overline{|\Delta C|}$	29	50	62

A REGIONAL SCALE (1000 KM) MODEL

The U.S. EPA is presently developing a model that can guide the formulation of regional emissions control strategies by estimating the effect of sources on concentrations in remote regions, determining the pollution burden that cities impose on distant neighbors, and eventually analyzing the effect of emissions on acid rain, visibility and fine particles. The utility and credibility of the model will be determined primarily by the extent to which it accounts for all the governing physical and chemical processes. Accordingly, the model is formulated, in principle, to treat all of the chemical and physical processes that are known, or presently thought, to affect the concentrations of air pollutants over several day/1000 kilometer scale domains. Among these processes are (not necessarily in order of importance):

1. Horizontal transport;

2. Photochemistry, including the very slow reactions;

3. Nighttime chemistry of the products and precursors of photochemical reactions;

4. Nighttime wind shear, stability stratification, and turbulence "episodes" associated with the nocturnal jet;

5. Cumulus cloud effects – venting pollutants from the mixed layer, perturbing photochemical reaction rates in their shadows, providing sites for liquid phase reactions, influencing changes in the mixed layer depth, perturbing horizontal flow;

6. Mesoscale vertical motion induced by terrain and horizontal divergence of the large scale flow;

7. Mesoscale eddy effects on urban plume trajectories and growth rates;

8. Terrain effects on horizontal flows, removal, diffusion;

9. Subgrid scale chemistry processes resulting from emissions from sources smaller than the model's grid can resolve;

10. Natural sources of HC, NO_x and stratospheric O_3;

11. Wet and dry removal processes, e.g., washout and deposition.

Of the eleven processes listed above, only the first and last have been treated in any detail in the regional scale models of air pollution developed to date. In fact, a review of these models (see, for example, reviews by Drake and Bass in Henderson et al.

1980) reveals that virtually all of the Eulerian type models are in essence simply expanded urban scale models. They account for the physical processes that are active during daylight hours and within 10 km or so of a source, but they neglect both the processes that are important beyond this distance and those that are active at night.

The U.S. EPA has taken the approach of developing a truly regional model, allowing the processes described above to influence the structure of the model, rather than trying to force the processes into an existing urban structure. The original goal of this work was to develop a specific model of regional scale photochemical air pollution. However, as the work progressed and new developments and ideas continually emerged, the need was seen for a general modeling framework within which the various physical and chemical processes that play important roles could be treated in modular form. This would permit ongoing incorporation into the model of state-of-the-art techniques without the need to overhaul the model each time. The structure and modular form of the Regional Oxidant Modeling System (ROMS) are unique for studying regional scale pollution. Complete documentation of the ROMS is being prepared (Lamb, 1982a, 1982b).

When this model development work was initiated some four years ago, an attempt was made to derive from the observational evidence available at that time an estimate of the minimum vertical and horizontal resolutions necessary to describe regional scale air pollution phenomena. The aim was to arrive at the best compromise between the restrictions imposed upon the model by computer time and memory limitations and the need to describe as accurately as possible all of the governing processes cited above. Careful review was made of the NO, O_3 and meteorological data reported in Siple (1977) by the participants of the 1975 Northeast Oxidant Transport Study. The characteristics of the O_3 distribution described in those reports would require at the very least a three-level model--one level assigned to the surface layer, another level to the remainder of the daytime mixed layer, and an additional layer atop the mixed layer. The top level would be used in conjunction with the mixed layer to account for downward fluxes of stratospheric O_3 as well as upward fluxes of O_3 and its precursors into the subsidence inversion layer above. Material that entered this top layer could be transported by winds aloft to areas outside the modeling region; it could enter precipitating clouds and be rained out of the atmosphere; or it could undergo chemical transformation. Representing the subsidence inversion, where cumulus clouds often form under stagnant high pressure conditions, the top level of the model would be instrumental in simulating the chemical sink effect of heterogeneous (within cloud droplets) reactions among O_3, its precursors, and other natural and pollutant species. Including

cloud effects in the model would be especially important in future
simulation of SO_2 and sulfates.

Having three layers in a model is insufficient in itself to
simulate all relevant phenomena. For example, three layers of
constant thickness are incompatible with the spatial and temporal
variability that the radiation inversion and mixed layer thickness
are known to have. What is needed in the model is three "dynamic"
layers that are free to expand and contract locally in response to
changes in the phenomena they are intended to treat. The model
discussed here possesses this property. Figure 8 illustrates the
vertical structure of this model and the physical phenomena that
each layer is intended to simulate. The surfaces that comprise the
interfaces of adjacent layers in our model are variable in both
space and time in order that each layer can keep track of the
changes that occur in the particular set of phenomena that layer is
designed to describe (summarized in Figure 8). A consequence of
this structure is that the volumes of the grid cells vary in both
space and time. By contrast, in conventional models the grid
network and cell volumes remain fixed and surfaces such as the mixed
layer top move through the grid system. In the following several
paragraphs, we elaborate on some of the phenomena cited in Figure 8
that our model will take into account.

During the day the highest layer shown in Figure 8a represents
the synoptic scale subsidence inversion, which may or may not
contain cumulus clouds. Stratospheric O_3 is transported downward
through this layer and anthropogenic O_3 and its precursors can be
carried into it by cumulus clouds or penetrative convection. The
base of this layer is normally 1 to 2 km above ground level. Below
it pollutants are kept well mixed vertically by turbulent convection. If the winds are strong or the surface heat flux is weak,
wind speed and direction may vary appreciably within the first
several hundred meters above ground. There is usually also a marked
difference in the wind speed and direction between the inversion
layer and the mixed layer below. Over large lakes and along sea
coasts there is frequently a second inversion layer below that
generated by synoptic scale subsidence. This lower inversion is
produced by sea or lake breeze regimes, and it restricts the vertical mixing of pollutants emitted over the water and within several
kilometers inland from the water's edge.

Air drawn into young cumulus clouds originates primarily in
the lower portion of the mixed layer. Fresh emissions of NO_x and
HC can be transported by the vertical currents that feed these
clouds from ground-level to altitudes well above the top of the
mixed layer in one steady, upward motion. In the process little
or no mixing with aged pollutants in the mixed layer occurs. At
night, cumulus clouds usually evaporate, and when they do they

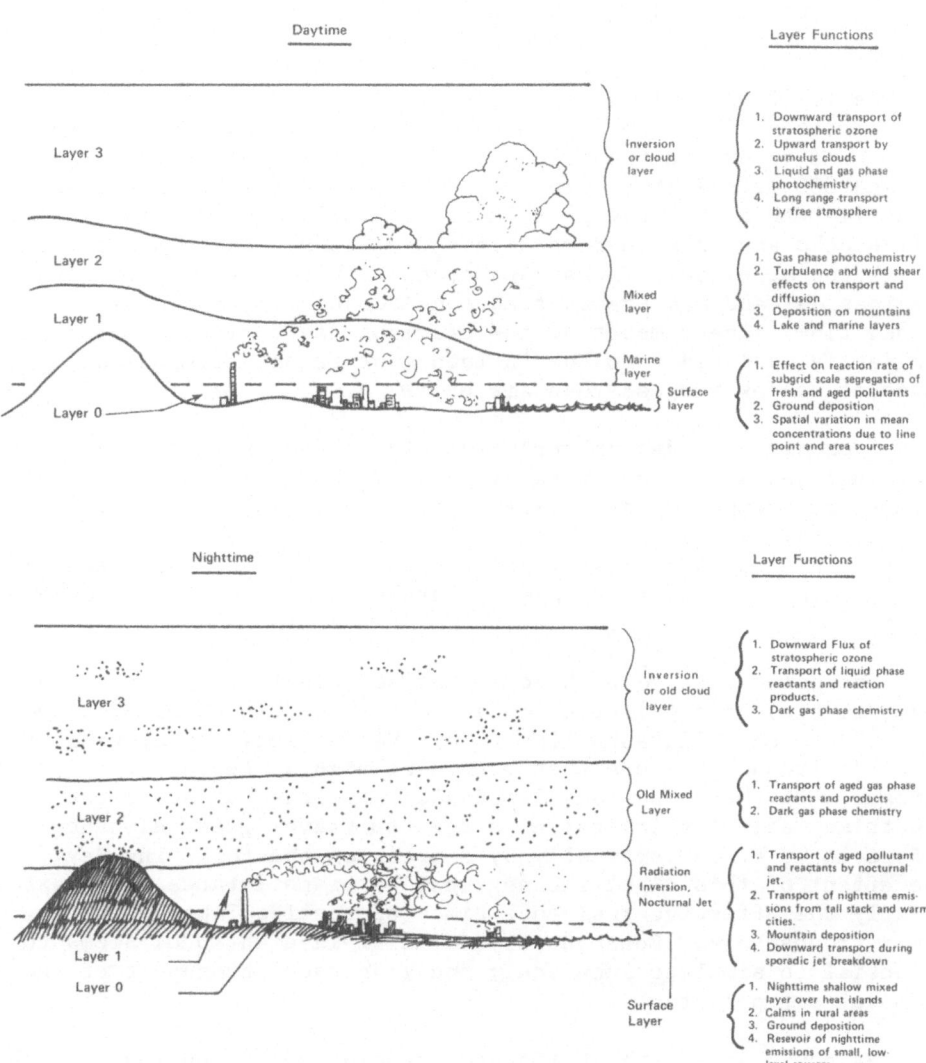

Fig. 8. Schematic illustration of the dynamic layer structure of
the regional model and phenomena each layer is designed
to treat: (a) daytime, (b) nighttime.

leave behind products of liquid phase reactions that can be transported hundreds of kilometers before sunrise.

Dramatic changes occur in the mixed layer at night. With the
onset of surface cooling following sunset, a stable layer of air
forms near the ground that quenches the vertical momentum fluxes
that give rise to frictional drag on the horizontal flow. With
retardation forces eliminated, the wind just above the stable layer
accelerates giving rise to the phenomenon known as the nocturnal
jet. Wind speeds in the core of the jet, which usually lies between
300 and 500 meters above ground, may be 10-15 m/s while at the
same time the air is nearly calm at the surface. Emissions from
tall stacks and from sources within the urban heat island enter the
jet region at night. There they react with aged pollutants from the
previous day and are transported considerable distances by the
strong flow. The remnant of the previous day's mixed layer above
the jet is isolated from the influence of fresh emissions and it
moves at a slower speed than air below.

Sporadic episodes of turbulence in the shear layer beneath the
nocturnal jet are a mechanism by which O_3 and constituents of urban
plumes are brought to ground-level at night. There, deposition on
surfaces and reactions with emissions of small, low-level sources
occur. This sporadic mixing process is perhaps the only mechanism
by which the reservoir of aged pollutants aloft can be depleted at
night.

One point that we wish to emphasize here is that one-layer
regional scale air pollution models are incapable of simulating
the effects on pollutants like O_3 of the vertical segregation of
aged and fresh emissions that occurs at night. Being cut-off from
contact with the ground and fresh NO_x emissions, O_3 above the
nighttime radiation inversion is free to travel great distances
before it is mixed vertically by convection the following day.
The effect of this nighttime segregation of pollutants is to extend
greatly the effective residence times of species like O_3 in the
lower troposphere. Consequently, a multi-layered model seems to be
essential to simulate accurately the long range transport of photochemical air pollutants.

As now planned, the horizontal resolution of the model is about
18 km. The resolution should be as high as possible to mitigate
the effects of subgrid scale concentration fluctuations. A scheme
to treat subgrid chemistry is implemented in Layer 0, adjacent to
the ground. Layer 0 is treated diagnostically in the governing
equations, and also handles surface depletion. In modeling atmospheric processes over 1000 km scale regions, effect of the earth's
curvature must be taken into account. Model equations are transformed into a curvilinear frame in which latitude, longitude and

elevation are coordinates and the basis vectors point north, east
and vertically upward at every point on the earth. We have chosen
this frame because it is a natural one from which transformations
to any rectilinear system are easily performed. Also, is the
frame in which worldwide meteorological data are reported.

There are three basic problems that must be overcome to make
operational a model as large and comprehensive as the one we are
developing.

First, due to the large number of processes that we plan to
treat, our model is rather complicated. In order to alleviate the
problems that this might cause in operating the model and in making
future refinements, we have structured it so that its central core
consists solely of a set of algorithms for solving the coupled set
of generalized finite difference equations that describe processes in
each of its layers. The modeling functions of describing the mixed
layer dynamics; topographic effects on winds; chemistry; cloud
fluxes, etc. will be handled by a set of special processors that are
external to the central model and which feed the model key variables
through a computer file. Within this framework the techniques used
to describe the various physical processes can be altered without
overhauling the model itself. An additional advantage is that
execution times are greatly reduced when several runs of a given
scenario are to be performed in which only one or a few parameter
values are altered.

A second problem is limitations of computer storage capacity.
To simulate air quality over the northeastern United States with the
horizontal resolution we desire, our model has roughly 10^4 grid
points and treats 25 (eventually more) chemical species. Thus, the
concentration variables alone require 250K words of storage and
this is just under the working limit of 260K words of memory on
EPA's Univac computer. To accomodate a model of the anticipated
size we have developed special techniques for handling the modeling
domain in piecewise fashion.

Finally, the empirical data needed to parameterize some of the
physical phenomena cited earlier are not presently available. To
remedy this, EPA initiated project NEROS (Northeast Regional Oxidant
Study) to collect during the summers of 1979 and 1980 the meteorological and chemical data required to formulate the model. A second
goal of NEROS was to gather the data required to perform comprehensive test runs and evaluation exercises of the model (see Clark and
Clarke, 1982; Clarke et al. 1982).

Aircraft sampling during NEROS was designed to gather evaluation data for the regional model. Using a trajectory model and
radiosonde data, flight tracks were directed to obtain Lagrangian

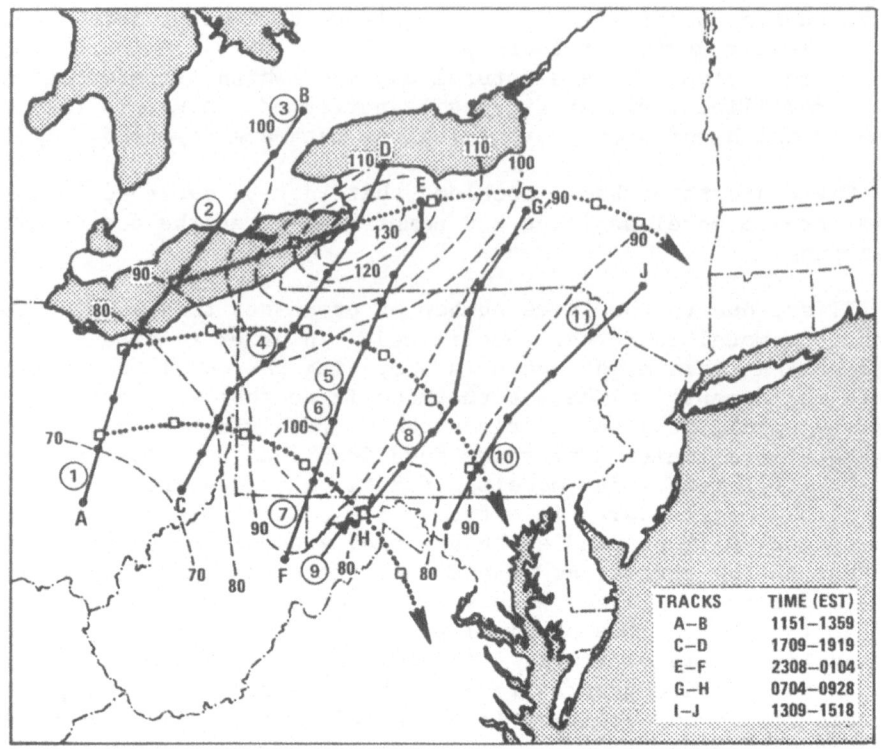

Fig. 9. Aircraft flight tracks, ozone concentrations (ppb), and
air trajectories (6-hr segment given by (□)) for 3-4
August 1979.

sampling of an air mass. Figure 9 shows computed trajectories and
corresponding aircraft flight tracks with O_3 concentration isopleths on one day. Figure 10 gives the O_3 concentrations in a
vertical cross-section along one of the flight tracks (E-F). The
significant O_3 concentrations at the level of the nocturnal jet
are evident in these data taken near midnight.

The model is currently set up with a 60 x 42 array of grid
cells. Figure 11 shows the region of the northeast U.S. being
modeled as well as NO_2 isopleths resulting from a short test run.
Although the model is functioning, substantial development work
must yet be accomplished. Preprocessors must be thoroughly checked
for accuracy, a refined emissions inventory (including biogenics)
must be finished, and the NEROS data base must be analyzed. We

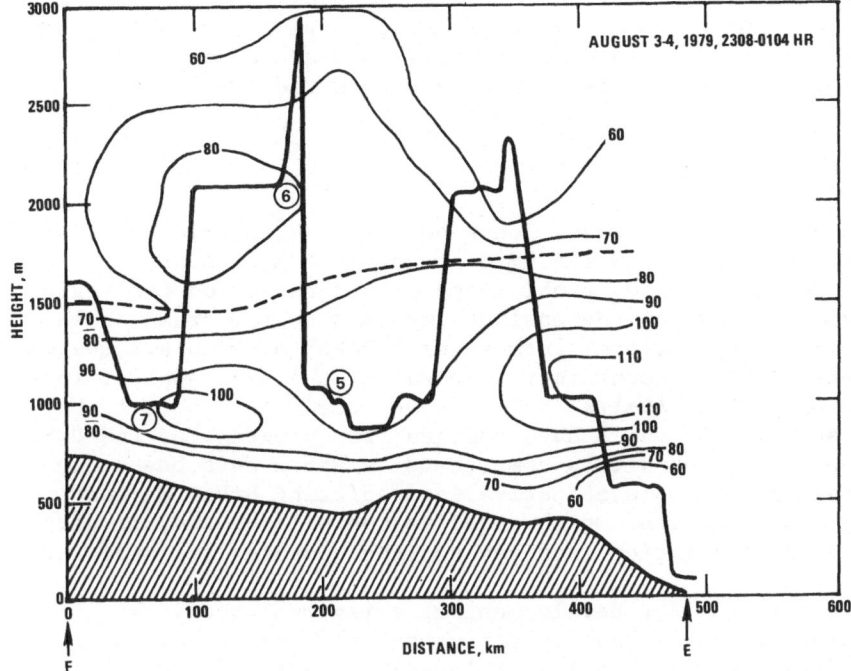

Fig. 10.   Cross-section of ozone concentrations (ppb) for the track
E-F in Fig. 9. Aircraft path is given by solid line,
and the dashed line shows an elevated inversion.

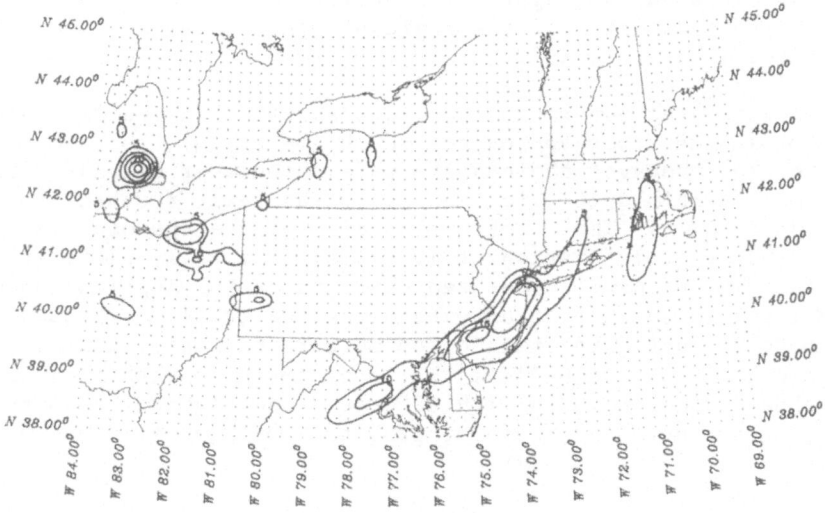

Fig. 11.   The grid of the regional model and NO_2 isopleths from a
test run. The isopleths are for 0800 local time and
reflect sources about urban areas.

believe these tasks can be completed in mid-1983. After that we
would need about another year of model running and evaluation
before the model would be operational and could conceivably be
transferred to other user groups (perhaps mid-1984).

CONCLUDING REMARKS

Over the past decade, focus of research, development, and
evaluation efforts concerning photochemical models has been initially on the urban O_3 problem and more recently on the regional O_3
problem. Because of the earlier emphasis on the urban scale, there
now exist in operational form several PAQSM's which are designed to
simulate urban photochemistry and which have been applied in the
U.S., Europe, and Australia. We have summarized the attributes of
three such models which have emerged from programs of the U.S. EPA
and been evaluated with a comprehensive urban data base. The models
have been shown to be effective in predicting urban O_3 concentrations. The choice of which model to use will depend, in part, on
the spatial resolution required and the computer resources available.

The impetus for development of a regional model resulted from
a growing awareness that constituents of photochemical smog travel
long distances, and control of locally generated O_3 may not be
sufficient to alleviate air quality problems. On the regional
scale, the multi-day nature of long-range transport necessitates
consideration of factors which may be neglected for a one-day
episode on the urban scale. These processes include slow photochemical reactions and segregation of surface and upper layers at night.
Testing and evaluation of the regional model using data of the NEROS
program will begin in 1983.

In confining our review to progress with EPA models, we do not
mean to exclude other approaches which are being considered but
believe that whatever choices are made the number of basic frameworks must remain quite limited due to the large resource requirements of PAQSM's. Thus, refinement of a single approach is preferred over a proliferation of models with slightly different characteristics. We have emphasized that successful application of a
PAQSM involves the use of an extensive, and usually expensive,
data base including information on meteorology, air quality, and
source emissions. We believe that commitment to gathering such
data should stand equally with efforts at developing the mathematical constructs and computer programs which constitute the
PAQSM.

REFERENCES

Clark, T.L., and J.F. Clarke, 1982: Boundary layer transport of NO_x and O_3 from Baltimore, Maryland-A case study. Proceedings of Third Joint Conference on Applications of Air Pollution Meteorology, San Antonio, Texas, American Meteorological Society.

Clarke, J.F., J.K.S. Ching, R.M. Brown, H. Westburg, and J.H. White, 1982: Regional transport of ozone. Proceedings of Third Joint Conference on Applications of Air Pollution Meteorology, San Antonio, Texas, American Meteorological Society.

Demerjian, K.L. and K.L. Schere, 1979: Application of a photochemical box model for O_3 air quality in Houston, Texas. In Proceedings of Ozone/Oxidants: Interactions with the Total Environment II, Houston, Texas, October 1979, Air Pollution Control Association, pp. 329-352.

Henderson, R.G., R.P. Pitter and J. Wisniewski, 1980: Research Guidelines for Regional Modeling of Fine Particulates, Acid Deposition and Visibility, Report of a Workshop held at Port Deposit, MD October 29-November 1, 1979.

Killus, J.P., J.P. Meyer, D.R. Durran, G.E. Anderson, T.N. Jerskey and G.Z. Whitten, 1977: Continued research in mesoscale air pollution simulation modeling: Volume V--Refinements in numerical analysis, transport, chemistry, and pollutant removal. Report No. ES77-142, Systems Applications, Inc., San Rafael, CA 94903.

Lamb, R.G., 1982a: A Regional Scale (1000 km) Model of Photochemical Air Pollution, Part I: Model Formulation. EPA Report (in press), U.S. Environmental Protection Agency, Research Triangle Park, North Carolina.

Lamb, R.G., 1982b: A Regional Scale (1000 km) Model of Photochemical Air Pollution, Part II: Procedures for Model Operations, Validation and Refinement, EPA Draft Report (June 1982), U.S. Environmental Protection Agency, Research Triangle Park, North Carolina.

Littman, F.E., 1979: Regional Air Pollution Study-Emission inventory summarization. Report No. EPA-600/4-79-004, U.S. Environmental Protection Agency, Research Triangle Park, NC 27711.

Lurmann, F., D. Godden, A.C. Lloyd, and R.A. Nordsieck, 1979: A Lagrangian Photochemical Air Quality Simulation Model. Vol. I-Model Formulation, Vol. II-Users Manual. EPA-600/8-79-015a,b (available NTIS).

Lurmann, F., 1980: Modification and Analysis of the Lagrangian Photochemical Air Quality Simulation Model for St. Louis. Environmental Research and Technology, Inc. Document No. P-A095. Westlake Village, CA. 25pp.

Lurmann, F., 1981: Incorporation of Lateral Diffusion in the Lagrangian Photochemical Air Quality Simulation Model. Environmental Research and Technology, Inc. Document No. P-A748. Westlake Village, CA. 32pp.

Schere, K.L., 1982: An evaluation of several numerical advection schemes. Submitted to Atmospheric Environment.

Schere, K.L. and J.H. Shreffler, 1982: Final Evaluation of Urban-Scale Photochemical Air Quality Simulation Models. EPA Report (in press), 249 pp.

Schiermeier, F.A., 1978: Air monitoring milestones: RAPS field measurements are in. Environmental Science and Technology, 12, 664-651.

Shreffler, J.H., 1982: Observations and modeling of NO_x in an urban area. Proceedings of the U.S.-Dutch Symposium on NO_x. Maastricht, The Netherlands, May 24-28, 1982.

Shreffler, J.H. and R.B. Evans, 1982: The surface ozone record from the Regional Air Pollution Study, 1975-1976. Atmospheric Environment, 16, 1311-1321.

Siple, G.W., 1977: Air Quality Data for the Northeast Oxidant Transport Study, EPA-600/4-77-020, Environmental Protection Agency, Research Triangle Park, North Carolina.

Whitten, G.Z., J.P. Killus, and H. Hugo, 1980: Modeling of Simulated Photochemical Smog with Kinetic Mechanisms - Vol. 1. report No. EPA-600/3-80-028a, U.S. Environmental Protection Agency, Research Triangle Park, NC 27711.

Wolff, G.T., P.J. Lioy, G.D. Wight, R.E. Meyers and R.T. Cederwall 1977: An investigation of long-range transport of ozone across the midwestern and eastern United States. Atmospheric Environment, 11, 797-802.

DISCUSSION

J. KNOX   In the context of an envolving
source inventory (wherein errors are being removed
for significant species, O_3, NO_x) one must be very
careful; early model evaluations may <u>not</u> be comparable to model evaluations done later, when the data-
base has a new character.

J. SHREFFLER   I agree. The three urban models I discussed were all executed with the final
source inventory. Over several years, continuing refinements in the inventory resulted in higher total
hydrocarbon emissions and a general increase in their
reactivities. In the models, the final inventory produces maximum ozone concentrations about 20 percent
higher than those generated by the first available
inventory.

F. LUDWIG   Would you comment on how well
you are able to verify the models' basis in physics
and chemistry when you only test its performance
under conditions that produce high ozone concentrations ?

J. SHREFFLER   The higher ozone levels are
precisely the ones of most interest to EPA. As the
maximum ozone concentrations become lower, the simulation generally becomes more dependent on initial
and boundary conditions, and these conditions are
difficult to accurately define. Therefore, I think
that evaluation of the model's ability to simulate
locally-generated ozone would be more difficult
using days with lower concentrations.

E. RUNCA   Natural sources of hydrocarbons can be important in regional modelling of oxidants. Do you plan to include these sources and
how ?

J. SHREFFLER   The inventory now being developed for the regional model will include natural
hydrocarbons. I am not acquainted with the precise
method of determining such emissions, but I know
that fluxes of hydrocarbons have been measured
from various types of vegetation using profile techniques and enclosures. These experimental results
could be related to the gridded land-use classifi-
cations also being developed for the model.

M. MULLER   Do you put the recently known
kinetic rates data on elementary photo-chemical reactions (e.g. the radical .OH reactions) in your photochemical models ?

J. SHREFFLER   In fact, we use mainly the
results of experiments on photochemical reactions
performed in atmospheric simulation chambers. In
terms of the reaction rates, the chemical mechanisms
used in these models represent the state of knowledge
in about 1980.

EVALUATION OF THE PERFORMANCE OF A PHOTOCHEMICAL DISPERSION MODEL

IN PRACTICAL APPLICATIONS

P.J.H. Builtjes(1), K.D. van den Hout(2) and S.D. Reynolds(3)

(1) MT/TNO P.O. Box 342, 7300 AH APELDOORN, the Netherlands
(2) IMG/TNO P.O. Box 214, 2600 AD DELFT, the Netherlands
(3) 101 Lucas Valley Road, San Rafael, Ca. 94903. U.S.A.

ABSTRACT

A photochemical dispersion model has been used to calculate concentrationlevels for an area covering the Netherlands and its surroundings during a given episode. The agreement between measured and calculated concentrations of O_3, and to a somewhat less extent of NO_2, showed to be remarkably good. With this model the influence of traffic emissions, the emissions from industry and the inflow of emissions from outside the area during the episode considered on the existing concentration levels of O_3, NO_2 and NO has been determined. The emissions from outside the area showed to have a large influence on the O_3-concentrations. The influence of industry on O_3- and NO_2-levels is larger than that of traffic.

The model has proven to be a valuable tool in determining the influence of different source categories and control regulations on the concentration levels.

INTRODUCTION

A few years ago the Dutch Ministry of Health and Environmental Protection initiated the start of a project dedicated to the development of an Air Quality Management System, AQMS. The AQMS consists in principle of three modules. The first one is an economic module, which in the framework of different scenarios based on economy development, energy consumption and air pollution abatement measures, produces emission values for the future. The second module contains transport, dispersion, transformation and removal models by which concentrations of

pollutants can be calculated. The third module contains the
social economic and environmental impact assessment, and its
results form the basis for policy decisions.

The AQMS has first been used to establish a policy for SO_2-
abatement (1). The emphasis was restricted to annually averaged
ground-level concentrations of SO_2, as a consequence of different
scenarios. The area considered was approximately 500 x 500 km^2.
The dispersion model used was a combination of a Gaussian plume
model for short distances and a box model for larger distances(2, 3).

Recently, it was decided to develop an AQMS for NO_x. Emission
data for NO_x have been obtained, the scenario system is under development. For the determination of annually averaged concentrations of NO_x, the same dispersion model as for SO_2 will be used
(for preliminary results see (3, 4)). However, in establishing a
policy for NO_x-abatement it is also necessary to pay attention to
the role of NO and NO_2 separately, including their behaviour under
photochemical conditions and their connection with O_3. At this
moment no reliable model seems to exist by which annually averaged
concentrations of NO_2 can be determined (5), although some progress in this direction has recently been made (5, 7).

The concentration levels of O_3, NO and NO_2 under episodic
conditions, however, can be determined in principle by a grid
model. With such a model the contribution of different source
categories to the concentration levels can be obtained. The performance of the model used, the so-called SAI-airshed model, including the results of a sensitivity analysis and of some applications forms the subject of this paper.

DESCRIPTION OF THE DISPERSION MODEL, AND OF THE AREA AND
EPISODE CONSIDERED

The basis of the model used in this study, the SAI-airshed
model, is the diffusion equation for each pollutant considered.
The diffusion equation describes the change with time of the
concentration of a pollutant, the advection by the mean wind
velocity, the turbulent diffusion described in the gradient-type
transport form, the source or sink term of the chemical reaction
and the emission. The equations of the different pollutants are
coupled by the chemical reaction term.

With the emissions of mainly NO, NO_2, SO_2 and hydro-carbons
the model calculates the concentrations of O_3, NO, NO_2, Pan, SO_2,
etc. for every grid for hourly averaged and instantaneous concentrations (averaging time 6 minutes). The meteorological part of
the model uses measured (synoptic) wind data to determine the
wind field. Also the mixing height, atmospheric stability, temperature field and photolysis rate are inputs to the model.

The chemical mechanism used is the so-called Carbon-Bond Mechanism (CBM). The Hydrocarbons are split into four "effective" hydrocarbon groups, according to their reactivity. Uptill now, the CBM-version of 1978 has been used in this study (8) . The numerical solution method used in the SHASTA-scheme. For a more general description of the model the reader is referred to Reynolds (8) .

Two remarks have to be made with respect to the model used in the present study:
The model is constructed in such a way that different sub-models can be distinguished, for example a windfield sub-model, a chemical sub-model, etc. These sub-models can easily be replaced by new models in case more information or knowledge becomes available. So, the model presents a framework into which information, also new information, can be placed and used. An example is the family of chemical mechanisms, up to the CBM-IV version.
The second remark is that, although different sub-models can be used, it is of great importance to keep the sophistication of the different models in balance. In this sense also the accuracy of the several input data should be kept in mind.

The model is being applied to the Netherlands and its surrounding countries. It should be noticed that the distances between industrialized areas in Western Europe are relatively small. Consequently, the degree in which the different countries influence one another with respect to air pollution is large. By applying the model it is desirable to choose the modelling region if possible in such a way that the concentrations in the area under consideration are significantly determined by emissions inside the modelling region. Therefore the modelling region has been taken large enough to include important industrialized areas in Belgium (Antwerp) and Germany (Ruhr area), as shown in Fig. 1.

The model has been applied to the simulation of the photochemical episode of 7 and 8 June 1976. This episode can be considered as a normal type of episode, which type can occur quite frequently during the summer months. During this episode an extension of a high-pressure area centered over the North Sea was drifting North-Eastwards, resulting in weak winds during daytime of June 7 and a South-Easterly flow during the remainder of the episode. Surface wind speeds at June 8 were around 5 m/s. The maximum temperature was around 28°C. Based on the desired resolution and computer capacity the area has been divided into grids of 10x10 km^2; there are five levels in the vertical direction. The size of the vertical cells varies. The ground cell has a height of 50 m, the total height of the modelling region is 2100 m.

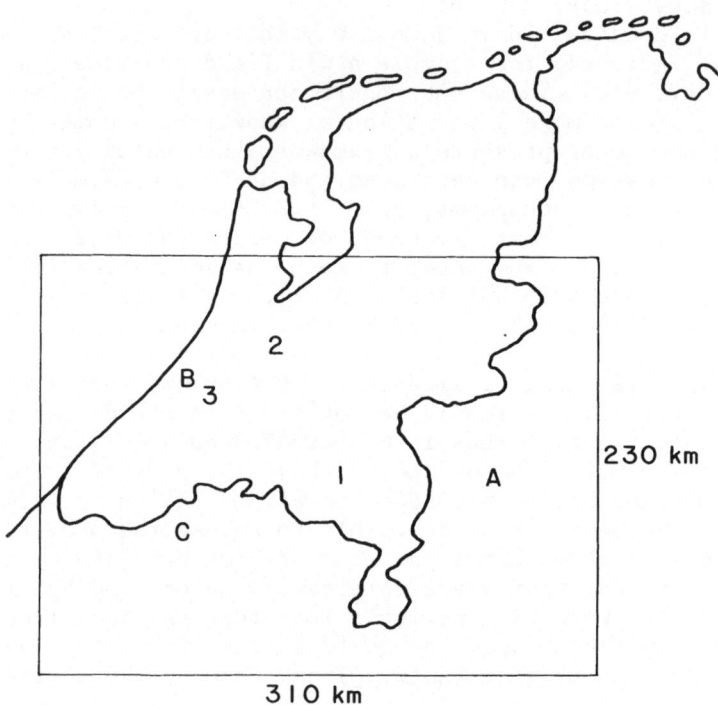

Fig. 1   The modelling area
             A: Rurh srea
             B: Rijnmond area
             C: Antwerpen
             1: Eindhoven
             2: Cabauw
             3: Vlaardingen

Table 1 : Description of calculations

Run 0   :   preliminary calculation, described in ref (9)

Run 1   :   basic run

Run 2   :   boundary conditions of HC divided by 2
            boundary conditions of NO_x multiplied by 2

Run 3   :   boundary conditions of HC multiplied by 2

Run 4   :   average boundary conditions of HC (see table 4)

Run 5   :   emissions of HC and NO_x multiplied by 1.3
Run 6   :   emissions of HC and NO_x multiplied by 0.7

Run 7   :   no emissions whole area, reference run 1
Run 8   :   no emissions whole area, reference run 4
Run 9   :   no traffic whole area, reference run 1
Run 10 :   no traffic whole area, reference run 4
Run 11 :   no local traffic in the Netherlands, reference run 1
Run 12 :   no industry whole area, reference run 1
Run 13 :   no industry in the Netherlands, reference run 1

The emission input has been obtained from the Dutch emission registration, supplemented by German and Belgium information. The initial and boundary conditions input was based on existing outdoor concentration measurements.

A more detailed description of the input has been presented earlier, including some preliminary results of the model calculations (9).

DESCRIPTION OF THE SENSITIVITY RUNS

The first results obtained with the model described in ref. (9), showed a good agreement between calculated and measured O_3-concentrations. The temperal correlation coefficient for all the measuring stations for O_3-data greater than 40 $\mu g/m^3$ was 0.91, the spatial correlation 0.62. The results for NO_2, and especially NO were poor. Also the calculated O_3-data during night time were significantly higher than the measured results. Part of this discrepancy was found to be caused by a coding error, resulting in too strong vertical diffusion. In view of these results, it was decided to carry out a new run, for which run also the photolysis rate is changed, (an increase to a maximum value of 0.45 min^{-1}, instead of 0.38 min^{-1}).

The results for O_3 during daytime with this new run, run 1, were about equal to the results of the previous run, run 0; the nighttime results were different. The calculated averaged maximum O_3-value at 11 Dutch measuring stations for run 1 was 244 $\mu g/m^3$, the measured averaged maximum 259 $\mu g/m^3$; for run 0 this value was 233 $\mu g/m^3$. As an example of the results obtained from run 1, Fig. 2 shows the O_3, NO and NO_2 data at Vlaardingen. The calculation has been carried out for a time period of 26 hours, starting at June 7, 18.00 h. The results are shown from midnight on, at which time the direct influence of the initial conditions has decreased. The calculated O_3-values are in good agreement with the measured ones. The calculated NO-maximum is clearly higher than the measured value. It should be noticed that the calculated value is an averaged value over a grid of 10×10 km^2, which averaging procedure has a large effect on a primary, reactive pollutant like NO. The agreement between the measured and calculated NO-maxima at Delft, at the other NO measuring station, 82 $\mu g/m^3$ and 72 $\mu g/m^3$ respectively, is much better. The agreement between measured and calculated NO_2-data is reasonable, also for the other five measuring stations. The measured average maximum NO_2-concentration at the six stations is 90 $\mu g/m^3$, the calculated value 80 $\mu g/m^3$.

In view of the good agreement in general between measured and calculated concentration levels of the pollutants under consideration, run 1 will be considered as the basic run.

To evaluate the performance of a model it is of primary importance to carry out a sensitivity analysis (10) . Several statistical methods exist to evaluate the results obtained (11). To gain a first insight, the evaluation will be restricted here to the averaged maximum O_3-concentration at the eleven measuring stations, and the NO, NO_2 and O_3-data at the location Vlaardingen, Cabauw and Eindhoven, see Fig. 1.

Table 2 : Ozon-results for the different calculations

AVERAGED MAXIMUM O_3-VALUE AT 11 DUTCH MEASURING STATIONS

Measured	259 $\mu g/m^3$	Run 7	203 $\mu g/m^3$
Run 0	233	Run 8	198
Run 1	244	Run 9	240
Run 2	182	Run 10	223
Run 3	310	Run 11	243
Run 4	225	Run 12	216
Run 5	244	Run 13	242
Run 6	243		

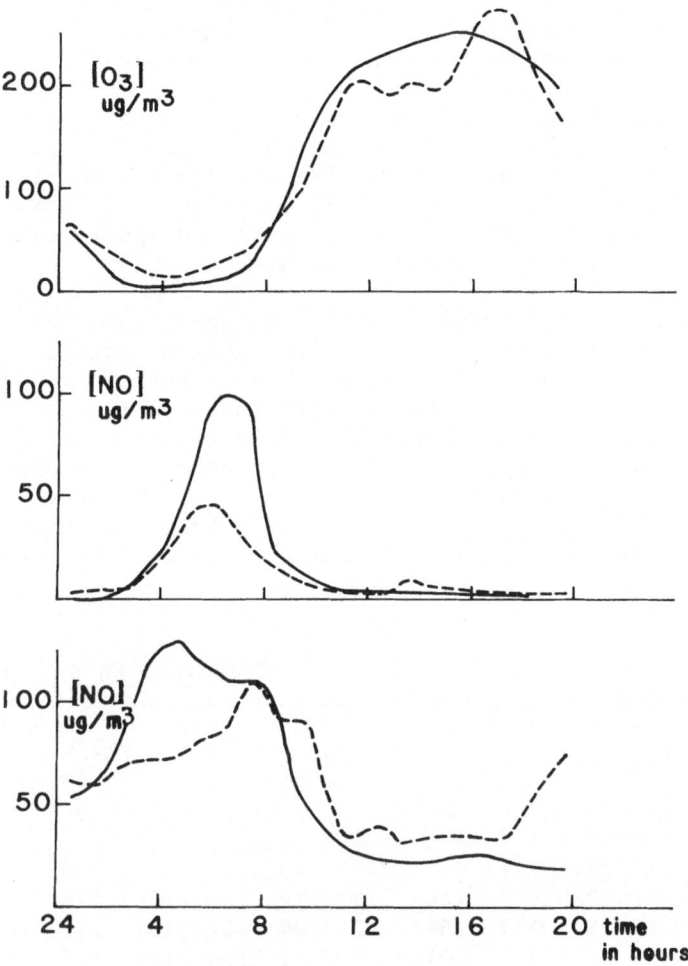

Fig. 2   Concentrations at Vlaardingen
         --------- observed
         _____ calculated, run 1

Vlaardingen is located in the centre of the Rijnmond area, the highly industrialized area around the harbour of Rotterdam. For Dutch standards Cabauw can be considered as a rural station. In its surroundings no industry is located. During this episode, however, the edge of the plume coming from the Ruhr area hits this location. The city of Eindhoven is close to the Dutch-German border, so the city is under the direct influence of the Ruhr area. Vlaardingen is about 100 km down wind of Eindhoven.

In Fig. 3 the above given characterisation of the locations is illustrated by the calculated concentrations for run 1. Next to the results of NO, NO_2 and O_3 the data for the total oxidant level $O_x = NO_2 + O_3$ are given. This oxidant-level is a useful quantity to consider because its temperal and spatial variation is less than that of NO_2 and O_3 separately.

The calculated O_3-maximum at Vlaardingen is somewhat higher than at the other locations; there is a general increase in the O_3-level in the downwind direction. The patterns of NO and NO_2 at Vlaardingen and Cabauw are similar; the situation at Eindhoven is completely different, however. The large values of NO and NO_2 calculated for the latter location are caused by the plume of the Ruhr area. The influence of the Ruhr area is so large that the morning peak shown at the other locations has disappeared completely. The O_x-results show also a slight increase in the downwind direction.

Table 3 : Maximum concentrations at Eindhoven, Cabauw and
          Vlaardingen, (the maxima can occur at different times)

	$O_3, \mu g/m^3$			NO, $\mu g/m^3$			$NO_2, \mu g/m^3$			O_x, ppb		
	E	C	V	E	C	V	E	C	V	E	C	V
Measured	215	277	271	–	–	43	–	–	109	–	–	157
Run 1	213	238	252	145	78	99	156	100	128	121	133	142
Run 2	164	181	177	154	92	104	151	137	135	101	108	109
Run 3	279	305	328	201	73	82	134	107	115	158	165	179
Run 4	202	221	229	145	73	96	156	105	128	114	125	130
Run 5	214	239	249	281	119	132	147	120	129	127	136	145
Run 6	219	240	252	122	48	54	120	75	80	122	131	138
Run 7	192	198	197	1	1	1	11	10	10	101	104	103
Run 8	188	193	192	1	1	1	12	11	10	99	101	101
Run 9	212	236	248	107	45	61	151	81	122	118	129	136
Run 10	200	220	226	107	42	59	151	82	122	112	121	125
Run 11	214	238	252	145	70	90	156	94	128	121	132	140
Run 12	200	213	214	3	12	15	61	45	52	106	114	115
Run 13	213	238	249	145	72	92	156	97	126	120	132	139

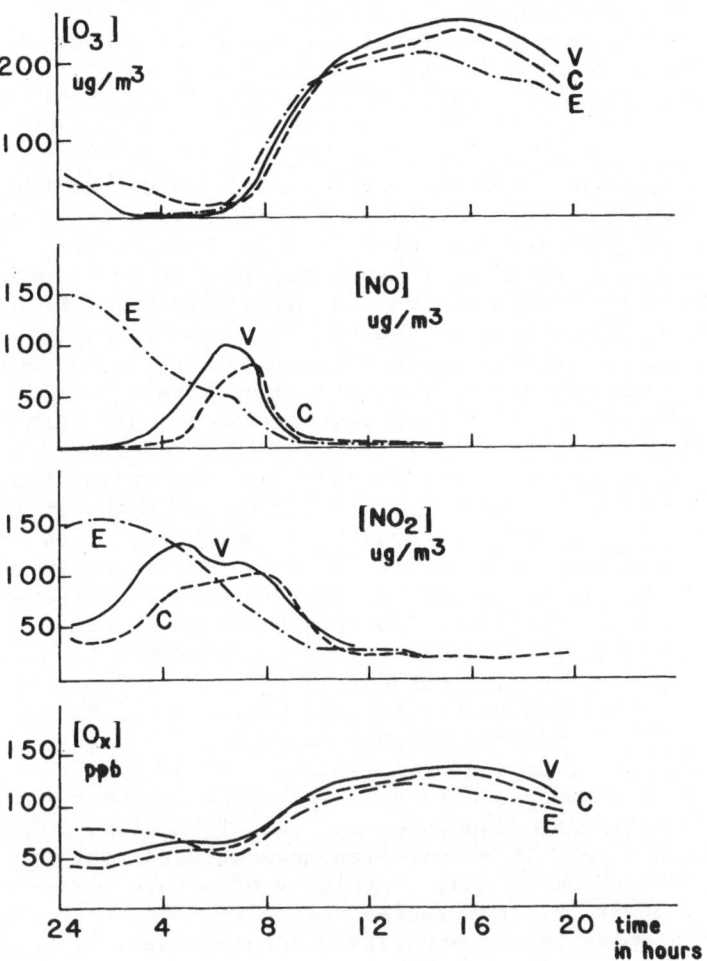

Fig. 3   Calculated values for run 1
--------- Cabauw
-·-·-·-·- Eindhoven
——————— Vlaardingen

In considering these results it should be noticed that the Dutch national ambient air quality standard for the 98-percentile of NO_2 is 135 $\mu g/m^3$; at the moment no standard exists for O_3.

For further comparison the emphasis will be directed to the calculated maximum levels at the three locations. When considering these results it should be kept in mind that these maximum levels can occur at different times, especially at Eindhoven.

Several trajectory runs carried out showed that the principal influence on the calculated concentrations seems to be caused by the boundary conditions for the hydrocarbons (HC). Because very few outdoor experiments for HC were performed during the period under consideration, the uncertainty in the boundary condition for HC is large. The boundary conditions for run 1, the basic run, were based on some measurements in the Netherlands during this episode and on the experience obtained from measurements carried out in following years. The boundary conditions for the first calculated hour, 18.00 - 19.00 of June 7, are given in Table 4. A remark should be made here concerning the ratio of NMHC (non-methane HC) and NO_x. This ratio NMHC/NO_x on a ppmC-basis, has for the boundary conditions which are based on outdoor experiments, a value of 17. For the emissions in the area considered this ratio is about 1.4. This large difference between these two ratios has also been found in an other study, although in that case the difference was somewhat less, for the emission ratio 3, for the outdoor experiments 15 - 21 (12) . The calculated ratio of NMHC/NO_x for the basic run at Cabauw and Vlaardingen has a minimum value of 2 in the morning around 6.00 h, and a maximum value of 14 around 14.00 h. So the calculated ratio, which is of course influenced by both emissions, boundary conditions, dispersion and transformation, shows large changes with time. It should also be mentioned that the uncertainty in the HC-emissions, especially of natural emissions, is quite large. Still, the precise reason for the large difference between the emission and boundary condition ratio is unclear at the moment, and requires further attention.

Three sensitivity runs concerning the HC-boundary conditions have been carried out. For run 2, the boundary conditions for HC are decreased by a factor 2, for NO_x increased by a factor 2. This means that the ratio NMHC/NO_x changes from 17 to 4. For run 3 the boundary condition for HC is increased by a factor 2, for NO_x the value is equal to the basic run. Finally, for run 4 boundary conditions for HC have been used which can be viewed as the average levels found from a set of studies, mainly performed in the U.S.A. (see Table 4). The NMHC/NO_x-ratio in this case is about 9.

Table 4 : Boundary conditions for HC at 18.00 - 19.00, June 7.

	Parafins ppmC	Olefins ppm	Aromatics ppm	Aldehydes ppm
run 1	.090	.001	.006	.008
run 2	.045	.0005	.003	.004
run 3	.180	.002	.012	.016
run 4	.035	.0004	.0034	.015

In Tables 2 and 3 the results from these runs are presented. It is clear from Table 2 that run 1, with the first estimate of boundary conditions, gives the best overall agreement concerning O_3. Changing the boundary conditions for both NO_x and HC (run 2) has a much larger influence on the O_3-results than changing only the boundary conditions for HC (run 4). Increasing the boundary conditions for HC gives a remarkable increase of O_3 (run 3). It seems that the O_3-results are more sensitive to an increase in NO_x than to a decrease in HC; an increase of the boundary conditions of NO_x decreases the O_3-level. From the results of the three stations given in Table 3 it can be concluded that, as anticipated, the influence of a change in boundary conditions is somewhat less pronounced for O_x than it is for O_3. The largest difference between the three stations appears for the levels of NO and NO_2 at Cabauw. The more rural character of this station and consequently its larger sensitivity to changes in boundary conditions is clearly demonstrated, expecially for run 2. Run 3 shows a large sensitivity on the NO- and NO_2-level at Eindhoven. It can be concluded that the behaviour of the three stations is quite similar with respect to O_3 and O_x, but different as regards NO and NO_2. For example, increasing the boundary conditions for HC (run 3) gives an increase of NO at Eindhoven and a decrease at Cabauw and Vlaardingen, and a decrease of NO_2 at Eindhoven and Vlaardingen and an increase at Cabauw. This is also related with the different times at which the maxima occur.

Sensitivity runs have also been made in respect of the emissions. The overall uncertainty in the emissions has been estimated to be approximately 30%. Runs 5 and 6 show the effect of an increase, respectively a decrease in the emissions of both NO_x and HC by 30%. The results show that there is hardly any change with respect to O_x and O_3. Regulations regarding both NO_x and HC-control by decreasing their emissions with the same percentage will have no effect on the maximum O_3-level.

Of course there is a clear effect on NO-concentrations, and to a less extent on NO_2-levels.

From the results of the sensitivity runs it can be concluded that run 1, taking into account the estimates made in this run concerning required input data such as boundary conditions for HC, gives the best overall agreement with the measurements and can be therefore considered as the basic run.

DESCRIPTION OF THE APPLICATION RUNS

When making regulations to improve air quality it is of great importance to know the relative contribution of different source-categories to existing concentration levels. First the influence of the transport of pollutants from outside the area, reflected by the boundary conditions, has been investigated. Calculations have been carried out with no emissions inside the area, and with the HC-boundary conditions of the basic run (run 7), and of run 4 (run 8). It should be remarked that for all the application runs the initial and boundary conditions have been left unchanged. The results of the runs are given in Table 2 and 3, and in Fig. 4.

The most remarkable result is that with no emission in the area on the average still about 85% of the O_3-maximum value remains. This means that this amount of O_3 is caused by emissions from outside the area considered. These emissions are reflected by the boundary conditions of NO, NO_2, NMHC and O_3. It can be seen from Table 3 that the maximum O_x resulting from the inflow is about 100 ppb. With no emissions the NO- and NO_2-levels decrease to about their boundary conditions at the start of the calculations. The influence of the emission inside the area on the PAN-concentration is larger than on the O_3-levels. For the three locations the maximum mean PAN-concentration of the basic run is about 12 $\mu g/m^3$; for run 7 this value is about 6 $\mu g/m^3$, a decrease of about 50%.

Of great importance for possible regulations is the relative influence of traffic versus industry. Run 9 calculates the concentrations when there was no traffic emissions in the whole area, with the boundary conditions of the basic run. The boundary conditions of run 4 have been used in run 10.

In this case there is hardly any decrease in O_3- or O_x-level. There is, on the average, a decrease in the NO-level of about 35% and in the NO_2-level of about 10%. As shown in Fig. 4, the decrease for Cabauw is much larger, for NO_2 about 20%. It should be mentioned that, although there is no traffic emission, there is still an NO- and NO_2- peak in the morning. This peak is caused by earlier emissions above as well as below the mixing

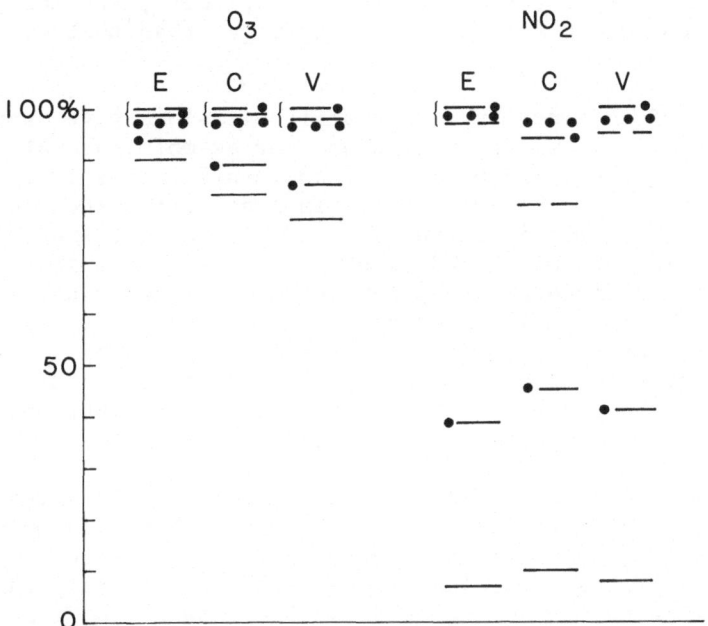

Fig. 4   Results of the application runs
         100% = run 1, basic run
         C = Cabauw, E = Eindhoven, V = Vlaardingen

───────────── run 7 no emissins whole area
─ ─ ─   run 9 no traffic whole area
───────────● run 11 no local traffic Netherlands
●──────────── run 12 no industry whole area
●   ●   ● run 13 no industry Netherlands

height. Consequently, the "morning traffic peak" as shown in Fig. 2 for example is mainly caused by stack emissions emitted sometime before.

Run 12 is the complementary run in which there is no emission from industry in the whole area. There is, on the average, a decrease in the O_3-level of 10% and in the O_x-level of 15%. The NO-level decreases by about 90%, the NO_2-level by about 60%. The results show that for these kinds of episodes, the influence of industry on concentration levels is larger than that of traffic.

The sum of the industry and traffic emissions give approximately the total emissions, regardless for example natural emissions. The difference between run 1 and run 7 is the concentration caused by the total emissions. In considering the contribution of traffic emissions (run 12) and of industry (run 9) it appears that for O_3, O_x and NO_2 the sum of these contributions is less than the difference between the results for run 1 and run 7; for NO this sum is in excess of the difference. It should be mentioned that for the different runs the maxima of O_x, O_3, NO_2 and NO occur at about the same time. It is clear that, caused by the non-linearity of the chemistry, contributions of different source categories are not additive.

Finally, two calculations have been performed to investigate the result of legislation taken in the Netherlands only. Run 11 shows the influence in case there is no emission from local traffic in the Netherlands. As could be expected, there is no influence on O_3- and O_x-levels. For Cabauw and Vlaardingen there is a decrease in NO of about 10%, and only for Cabauw of about 5% in NO_2.

It should be noted here that the model calculates ambient concentrations averaged over the volume of a gridcell, 10 x 10 km^2 x 50 m. It is obvious that this spatial resolution is not sufficient to predict the concentrations in streets close to traffic. In the first place, the proximity of the sources results in considerably higher levels in the streets. Secondly the very short atmospheric transport time of the pollutants and the very fast concentration fluctuations may result in important deviations from the photostationary equilibrium. Qualitively it can be stated that the quencing of O_3 by the freshly emitted NO results in lower O_3-levels while NO and NO_2 are higher at short distances from roadways. A quantative description of these curbside levels however should be given by a different type of model. On the other hand, the formation of O_3 and O_x has a much larger time scale and can be described with a spatial resolution of 10 km.

The effect of lack of emissions from industry in the Netherlands is depicted in run 13. Again there is hardly any influence
on O_3- and O_x-levels, and for Cabauw and Vlaardingen a decrease
in NO of about 10 % and in NO_2 of about 3%.

It is clear that under those kinds of episodic conditions
emission regulations by the Netherlands only have hardly any
influence on concentration levels on a scale of 10 x 10 km^2.
However, this does not imply that there can not be a larger effect on a smaller scale, in particular in an area close to a main
roadway.

Finally, some remarks will be made concerning the O_x-
levels occurring. Given an emission of NO in a volume of air
containing O_3, the NO-molecules will react with the same amount
of O_3-molecules to form NO_2. The total amount of $O_x \equiv NO_2 + O_3$
will remain constant during this process; there is a trade-off
between NO_2 and O_3. Assuming a photo-stationary equilibrium to
a degree of accuracy of the first order, it will depend on the
equilibrium constant $K \equiv (NO) (O_3)/(NO_2)$ how much NO_2 will be
formed, given the NO_x- and O_x-level. The factor K is a function
of the radiationfactor. In accordance with this procedure, and
given an inflow of O_x of about 100 ppb, the maximum value, and
an increase caused by the emissions in the area of up to about
140 ppb, the NO_2- and O_3-levels can be determined without taking
account of the full chemical mechanism. It should be noticed,
however, that the O_x-increase can only be determined by using
the complete dispersion model including the chemical mechanism.

CONCLUSIONS

From the results obtained the following conclusions can be
drawn for the area, covering the Netherlands and its surroundings, and episode considered :

- The agreement between measured and calculated O_3- and NO_2-
  levels for the basic run is remarkably good.

- The sensitivity analysis shows that increasing the boundary conditions for NO_x has a larger effect on O_3-levels
  than decreasing the HC-boundary conditions; an increase
  in the NO_x-boundary conditions leads to a decrease in O_3-
  levels.

- Neither an increase nor a decrease of 30% in both NO_x- and
  HC-emissions has any net effect on O_3- and O_x-levels.

- The emissions outside the area, reflected in the boundary
  conditions, are responsible for, on the average, 85% of
  the O_3-concentration and 50% of the PAN-concentration in-

side the area. For an accurate determination of the O_3-levels occuring and their origin the area considered is too small.

- The influence of the emissions of industry on the O_3- and NO_2-levels occurring is larger than that of the emissions originating from traffic.

- Regulations for emissions inside the Netherlands only will have only a marginal effect on concentration levels of O_3 and NO_2 on the scale of 10 x 10 km^2 considered.

- Concentrations of different source categories are, in general, not additive, because of the non-linearity of the chemical mechanism.

ACKNOWLEDGEMENT

Part of this study has been sponsored by the Dutch Ministry of Health and Environmental Protection, and the presented results have been published with permission of the Ministry.

The authors are indebted to Mr. J. Hulshoff and Ms. J. Kardux for their assistance in carrying out the computer runs.

REFERENCES

1.   L.A.M. Janssen, "System for calculating air pollution I, II and III". CMP-Rep. 78/1, 79/1 and 79/3, TNO, Delft, the Netherlands, 1980.
2.   N.D. van Egmond and C. Huygen, "Evaluation of a meso-scale model of the dispersion of sulphur dioxide", Proc. of the tenth I.T.M. on Air Poll. Modelling and its Applications, Rome, October 1979.
3.   K.D. van den Hout, "Model calculations of ambient SO_2 and NO_x concentration patterns in the Netherlands", Fourth Int. Symp. on Industrial Chimneys; The Hague, the Netherlands, May 1981.
4.   S. Zwerver, "An air quality management system as a tool for establishing a NO_x-poling", To be published in Proc. of the U.S.-Dutch Int. Symp. on Air Pollution by Nitrogen Oxides, Maastricht, The Netherlands, May 1982.
5.   A. Esschenroeder, "Atmospheric dynamics of NO_x- emission controls", Proc. of the 15th Int. Coll. 71-90, Paris, May 1982.
6.   P.D. Gutfreund e.a., "Assessment of NO_x emissions control requirements in the South Coast Air Basin", SAI, Rep. no. 81277, November 1981.

7.   N.D. van Egmond and H. Kesseboom, "Meso-scale transport of
     NO_x, long term and short term concentration levels",
     To be published in Proc. of the U.S.-Dutch Int. Symp.
     on Air Pollution by Nitrogen Oxides, Maastricht, the
     Netherlands, May 1982.
8.   S.D. Reynolds e.a., "An introduction to the SAI-airshed model
     and its usage", SAI-Rep. EF 78-53R4 EF 79-31, March 1979
9.   P.J.H. Builtjes e.a., "Application of a photochemical disper-
     sion model to the Netherlands and its surroundings",
     Proc. of the eleventh I.T.M. on Air Pollution Model-
     ling and its Applications, Amsterdam, November 1980.
10. C. Seigneur e.a., "Sensitivity of a complex urban air quality
     model to input data", J. of Appl. Met. 20, 1020, 1981.
11. K.E. Bencala and J.H. Seinfeld, "An air quality model perfor-
     mance assessment package", Atm. Env. 13, 1181, 1979.
12. R. Stern and B. Scherer, "An application of the empirical
     kinetic modelling approach (EKMA) to the Cologne area",
     Proc. of the eleventh I.T.M. on Air Pollution Model-
     ling and its Applications, Amsterdam, November 1980.

DISCUSSION

G. CLERICI   Have you studied the in-
            fluence of the initial concentrations of O_3,
            NO_x in your runs ?

P.J.H. BUILTJES   No sensitivity analysis
            concerning the initial conditions of O_3 has been
            carried out yet. Both the initial and boundary
            conditions of NO_x have been increased by a factor
            2 in run 2. The total run contains 26 hours,
            which means that the influence of the initial
            conditions is decreased substantially after the
            first hours of the run.

B. RUDOLF   Did you calculate an example
            with north westerly winds ?

P.J.H. BUILTJES   No calculations with north
            westerly winds have been carried out, mainly
            because a photochemical episode in the Nether-
            lands is very unlikely to occur in that case.

A SECOND GENERATION COMBINED TRANSPORT/CHEMISTRY MODEL FOR THE REGIONAL TRANSPORT OF SO_x AND NO_x COMPOUNDS

Gregory R. Carmichael, Toshihiro Kitada
and Leonard K. Peters

Chemical & Materials Engineering Program, University of
Iowa, Iowa City, Iowa 52242 (USA); School of Regional
Planning, Toyohashi University of Technology, Toyohashi
440 (Japan); Department of Chemical Engineering,
University of Kentucky, Lexington, Kentucky 40506 (USA)

INTRODUCTION

The relationships between SO_x and NO_x emissions and the distribution of these compounds are complex and not fully understood. It is known that these trace species may be transported long distances (thousands of kilometers) from source areas, and result in high ambient levels over broad regions. In addition, NO_x and SO_x have been related to a number of environmental problems including acid rain, tropospheric haze and resulting reduced visibility, and have been correlated with a variety of adverse health indicators.

The distribution of SO_x and NO_x in the atmosphere is the result of complex interactions among the emissions (both anthropogenic and natural), the prevailing meteorology, and the chemical transformations and removal mechanisms. Regional scale transport/-chemistry models that describe the circulation of SO_x and NO_x in the troposphere can be extremely beneficial in understanding the physical and chemical processes occuring between the sources and sinks of those pollutants.

In this paper, a combined transport/chemistry model for the regional scale transport of SO_x and NO_x compounds is described. The model is Eulerian and three dimensional and is an extension of an operational SO_x transport model developed by the authors (Carmichael and Peters, 1981). This second-generation model treats 39 species, 18 of which are advected species. The advected species include NO, NO_2, SO_2, SO_4^{2-}, O_3, HNO_3, HC, NH_3, PAN, H_2O_2, and HCHO. Both heterogeneous and homogeneous chemical reactions are modeled, as are both wet and dry removal processes. A finite

element numerical method is used in the model. Model formulation
and initial test results are presented in the following section.

MODEL DESCRIPTION

The regional transport of SO_x and NO_x is modeled within an
Eulerian framework. A block description of the model is presented
in Figure 1. Thirty-nine chemical species are included in the
analysis of which eighteen are sufficiently long lived that they
(under some circumstances) must be treated as advected species.
The remaining species are considered to be short lived and are
modeled by steady state methods.

The mathematical analysis is based on the coupled, three
dimensional advection-diffusion equation:

$$\frac{\partial C_\ell}{\partial t} + \frac{\partial (U_j C_\ell)}{\partial x_j} = \frac{\partial}{\partial x_j} \left[K_{jj} \frac{\partial C_\ell}{\partial x_j} \right] + R_\ell + S_\ell \quad , \tag{1}$$

where C_ℓ is the concentration of species ℓ, U_j is the velocity vector, K_{jj} is the eddy diffusivity tensor ($K_{jj} = 0$ for $i \neq j$ has been
assumed), R_ℓ is the rate of formation or loss by chemical reaction,
and S_ℓ is the emission rate.

The model actually utilizes the surface-following coordinate
system shown in Figure 2. The vertical region, including topography features, is mapped into a dimensionless rectangular region
according to

$$\rho_k = \frac{Z_k - h(x,y)}{H_2 (x,y,t)} \quad h(y) \leqslant Z_k \leqslant h(x,y) + H_2(x,y,t) \quad , \tag{2}$$

and

$$\rho_k = \left(\frac{k-1}{KGRID-1} \right)^\alpha \quad 0 \leqslant \rho_k \leqslant 1 \quad , \quad k = 1, KGRID \tag{3}$$

where the subscript k indicates the vertical grid number, KGRID is
the total number of grids, and α is a parameter which controls the
grid spacing. When $\alpha = 1$ the grid spacing in dimensionless
coordinate is uniform and when $\alpha > 1$ there is higher resolution
near the surface. Our current application uses $\alpha = 2$ and $H_2(x,y,t)$
$= 8$ km. The height of the region is chosen so that boundary layer-
free troposphere processes and interactions can be modeled.

Chemistry

Both gas phase and liquid phase chemical reactions are
treated. The homogeneous gas phase mechanism used is summarized in
Table 1. The mechanism involves 81 reactions and 39 species. NO,
NO_2, HNO_3, NH_3, SO_2, SO_4^{2-}, HCl (alkanes), C_2H_4, HC2 (alkenes), HC3
(aromatics), O_3, PAN, HCHO, RCHO, H_2O_2, HNO_2, $RONO_2$ and RO_2NO_2 are

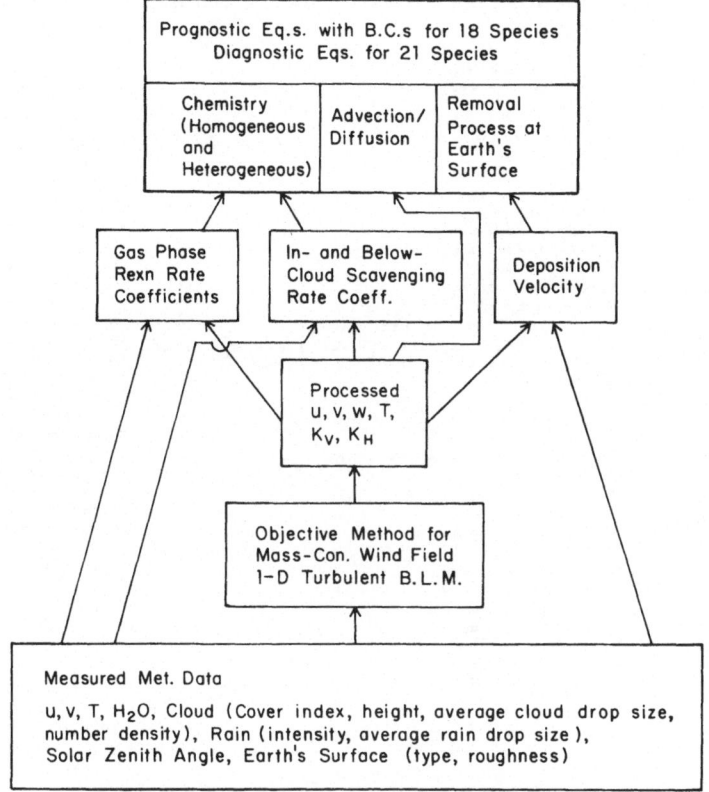

Figure 1. Schematic of model construction

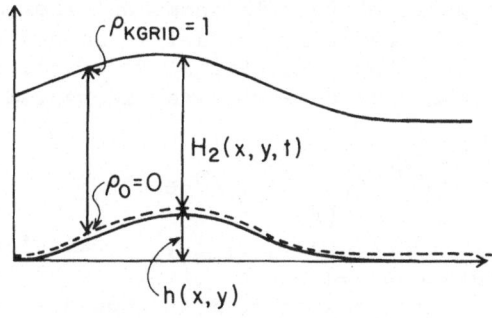

Figure 2. Surface-following coordinate system used in the model.

Table 1. Homogeneous gas phase kinetic mechanism used in the
          model.

1. $NO_2 + hv \rightarrow NO + O(^3P)$
2. $O(^3P) + O_2 + M \rightarrow O_3 + M$
3. $O_3 + NO \rightarrow NO_2 + O_2$
4. $NO_2 + O(^3P) \rightarrow NO + O_2$
5. $NO_2 + O(^3P) \rightarrow NO_3$
6. $NO + O(^3P) \rightarrow NO_2$
7. $NO_2 + O_3 \rightarrow NO_3 + O_2$
8. $NO_3 + NO \rightarrow 2\ NO_2$
9. $NO_3 + NO_2 \rightarrow N_2O_5$
10. $N_2O_5 \rightarrow NO_2 + NO_3$
11. $N_2O_5 + H_2O \rightarrow 2HONO_2$
12. $NO + NO_2 + H_2O \rightarrow 2HONO$
13. $HONO + HONO \rightarrow NO + NO_2 + H_2O$
14. $O_3 + hv \rightarrow O_2 + O(^1D)$
15. $O_3 + hv \rightarrow O_2 + O(^3P)$
16. $O(^1D) + M \rightarrow O(^3P) + M$
17. $O(^1D) + H_2O \rightarrow 2OH$
18. $HO_2 + NO_2 \rightarrow HONO + O_2$
19. $HO_2 + NO_2 \rightarrow HO_2NO_2$
20. $HO_2NO_2 \rightarrow HO_2 + NO_2$
21. $HO_2 + NO \rightarrow NO_2 + OH$
22. $OH + NO \rightarrow HONO$
23. $OH + NO_2 \rightarrow HONO_2$
24. $HONO + hv \rightarrow OH + NO$
25. $CO + OH \rightarrow CO_2 + HO_2$

26. $OH + HONO \rightarrow H_2O + NO_2$
27. $HO_2 + HO_2 \rightarrow H_2O_2 + O_2$
28. $OH + HO_2 \rightarrow H_2O + O_2$
29. $OH + O_3 \rightarrow HO_2 + O_2$
30. $HO_2 + O_3 \rightarrow OH + 2O_2$
31. $HCHO + hv \rightarrow HO_2 + HCO$
32. $HCHO + hv \rightarrow H_2 + CO$
33. $HCHO + OH \rightarrow HCO\ (+H_2O)$
34. $HCO + O_2 \rightarrow HO_2\ (+CO)$
35. $RCHO + hv \rightarrow RO_2 + HCO$
36. $RCHO + OH \rightarrow RCO_3$
37. $Ol\ (olefin) + OH \rightarrow RO_2$
38. $Ol + O \rightarrow RO_2 + RCO_3$
39. $Ol + O_3 \rightarrow 0.5\ \zeta\ HCHO + (1 - 0.5\ \zeta)\ RCHO$
       $+ 0.5\ (\varepsilon)\ (\zeta + 0.5\ \eta)\ RO_2 + 0.25\ (\varepsilon)\ (\zeta + \eta)\ HO_2 + 0.25\ (\varepsilon\zeta)$
       $+ 0.25\ (\varepsilon\eta)\ RO + 0.25\ (\varepsilon\eta)\ HCO + 0.5\ (1 - \varepsilon)\ RCHOO$
40. $Alk + OH \rightarrow RO_2$
41. $Alk + O \rightarrow RO_2 + OH$
42. $C_2H_4 + OH \rightarrow RO_2$
43. $C_2H_4 + O \rightarrow RO_2 + HCO$
44. $RO \rightarrow \alpha\ HO_2 + (1 - \alpha)\ RO_2 + \beta\ HCHO + \gamma\ RCHO$
45. $NO + RO \rightarrow RONO$
46. $RONO + hv \rightarrow RO + NO$
47. $NO_2 + RO \rightarrow RONO_2$
48. $NO_2 + RO \rightarrow RCHO + HONO$
49. $NO_2 + RO_2 \rightarrow RO_2NO_2$
50. $NO_2 + RO_2 \rightarrow RCHO + HONO_2$
51. $RO_2NO_2 \rightarrow NO_2 + RO_2$
52. $NO + RO_2 \rightarrow NO_2 + RO$
53. $NO + RCO_3 \rightarrow NO_2 + RO_2$
54. $NO_2 + RCO_3 \rightarrow PAn$

55. $PAn \rightarrow NO_2 + RCO_3$
56. $Aro + OH \rightarrow RO_2 + RCHO$
57. $SO_2 + hv \rightarrow {}^1SO_2$
58. ${}^1SO_2 + M \rightarrow {}^3SO_2 + M$
59. ${}^3SO_2 + M \rightarrow SO_2 + M$
60. ${}^3SO_2 + O_2 \rightarrow SO_3 + O(^3P)$
61. $SO_2 + HO_2 \rightarrow HO + SO_3$
62. $SO_2 + HO \rightarrow HSO_3$
63. $SO_2 + O(^3P) \rightarrow SO_3$
64. $SO_2\ RO_2 \rightarrow RO + SO_3$
65. $SO_2 + RO \rightarrow R\ SO_3$
66. $SO_2 + RCHOO \rightarrow RCHO + SO_3$
67. $SO_2 \rightarrow Part$
68. $SO_3 + H_2O \rightarrow H_2SO_4(Part)$
69. $HSO_3 + O_2 \rightarrow HSO_5$
70. $HSO_5 + H_2O \rightarrow HSO_5\ (H_2O)\ (Part)$
71. $RSO_3 + O_2 \rightarrow RSO_5$
72. $RSO_5 + H_2O \rightarrow RSO_5\ (H_2O)$
73. $RCHOO + H_2O \rightarrow RCOOH + H_2O$
74. $NH_2 + OH \rightarrow NH_2 + H_2O$
75. $NH_2 + O_2 \overset{M}{\rightarrow} NH_2O_2$
76. $NH_2O_2 + NO \rightarrow NH_2O + NO_2$
77. $NH_2O_2 + OH \rightarrow NH_2OH + O_2$
78. $NH_2O + O_2 \rightarrow HNO + HO_2$
79. $HNO + OH \rightarrow NO + H_2O$
80. $HNO + O_2 \rightarrow NO + HO_2$
81. $NH_3 + HNO_3 \overset{M}{\rightleftharpoons} NH_4NO_3$

treated as transported species and the rest as pseudo steady state
species. The extensiveness of the chemical mechanism enables the
modeling of urban chemistry as well as nonurban tropospheric
chemistry.

    The model's treatment of the chemistry also considers the in-
teractions between the gas phase chemistry and the wet removal processes. The wet removal of soluble gases and the capture of
aerosol particles by hydrometeors acts like an additional first order chemical reaction. Thus, additional reactions of the form
below are added to the gas phase chemical mechanism.

$$C_{i_{(g)}} + H_2O_{(\ell)} \overset{k_{iwet}}{\rightarrow} C_{i_{(\ell)}} + \text{Products} \qquad (4)$$

The wet removal coefficient, k_{iwet}, is calculated in the model
from a parameterization based on the liquid water content of the
cloud, cloud temperature, characteristic drop size and number density of cloud droplets, cloud pH and rainfall intensity, and the
chemical and physical properties of the absorbed species (i.e.,
Henry's Law constant, gas phase diffusivity, and dissociation and

redox reactions in solution). Also included is the liquid phase generation of species like sulfate. The sulfate production rate in clouds is calculated using the above parameters and is based on the $S(IV) + O_3$ and $S(IV) + H_2O_2$ reactions.

The chemical reaction rate constants are calculated in the model and vary with temperature and photon flux. The effects of clouds on the photon flux are included in the calculation through use of an empirical correlation relating photon flux to cloud cover (Kaiser and Hill, 1976); i.e.,

$$G = G_c \, (1 - A \, C^{1.75}) \quad , \tag{5}$$

where G_c is the photon flux for clear sky, C is the cloud cover fraction, and A is a constant dependent on cloud type and has a value of 0.55 for fair weather cumulus or patches of cirrus or cumulus clouds. Furthermore, the enhancement of the photon flux due to cloud albedo is included in the analysis by multiplying the clear sky flux at grid points directly above the cloud level by a factor of 1.3.

Numerics

Simulations of regional transport/chemistry described by Equation (1) requires numerical integration. The method used in the model is a combination of the concept of fractional time steps and one-dimensional finite elements. This is referred to as the Locally-One-Dimensional, Finite-Element Method (LOD-FEM). The LOD procedures (Mitchell, 1969; Yanenko, 1971) split the multi-dimensional partial differential equation into time dependent one-dimensional problems which are solved sequentially.

The time-split equations are of the form:

$$\frac{\partial C_\ell}{\partial t} + L_x \, C_\ell = 0 \tag{6}$$

$$\frac{\partial C_\ell}{\partial t} + L_y \, C_\ell = 0 \quad , \tag{7}$$

$$\frac{\partial C_\ell}{\partial t} + L_z \, C_\ell = S_\ell \quad , \tag{8}$$

and

$$\frac{\partial C_\ell}{\partial t} = R_\ell \quad , \tag{9}$$

where the L's represent the one-dimensional operators (e.g., $L_x C_\ell = \frac{\partial(U \, C_\ell)}{\partial X} - \frac{\partial}{\partial x} (K_{xx} \frac{\partial C_\ell}{\partial x})$). The chemical reaction term is treated separately because many of the reactions have time scales much smaller than that for the transport. Splitting out the reaction term allows the use of different time steps for the transport and the chemistry processes.

The transport equations are solved using a modified Galerkin FEM procedure with asymmetric weighting functions and the Crank-Nicholson approximation for the time derivative. The chemistry equations are solved using a pseudo-linearization procedure which gives analytical approximations to the equations. The solution procedures for the transport and chemistry are discussed in more detail below.

Solution of the Chemistry Equations:

The technique used to integrate the chemistry of the transported species is an adaptation of the semi-implicit Euler method proposed by Preussner and Brand (1981). Equation (9) can be written in the form

$$
\frac{dC_\ell}{dt} = -C_\ell \sum_j d_\ell^j \prod_{k \neq \ell} C_k + \sum_i p_\ell^i \prod_m C_m \quad , \tag{10}
$$

where d_ℓ^j represents the rate constant of the destruction of species ℓ by reaction j, p_ℓ^i represents the rate constant of the production of species ℓ by reaction i, and the subscripts k and m indicate the reactant species involved in the destruction and production reactions.

Under the assumption that all species concentrations except ℓ are known (either at the previous time step or have just been calculated) then Equation (10) is of the form $\frac{dC_\ell}{dt} + D\, C_\ell = P$, which has the analytic solution

$$
C_\ell \Big|_{t_i + \Delta t_i} = \frac{P}{D}\Big|_{t_i} + \left(C_\ell - \frac{P}{D}\right)\Big|_{t_i} e^{-D\Delta t_i} \quad , \tag{11}
$$

where $C_\ell\big|_{t_i}$ at $t = 0$ is the initial concentration of species ℓ. This procedure has the important physical properties that C_ℓ can never become negative provided that the reaction rates are positive and the initial values are non-negative, and that in the limit as $t \to \infty$ the proper equilibrium concentrations are obtained, i.e.,

$$
C_\ell^{eq} = \frac{P}{D} = \frac{\sum_i P_\ell^i \prod_m C_m}{\sum_j d_\ell^j \prod_{k \neq \ell} C_k} \quad . \tag{12}
$$

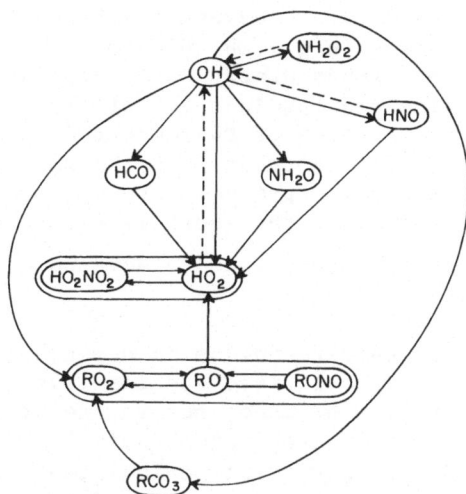

Figure 3. Schematic of free radical interactions

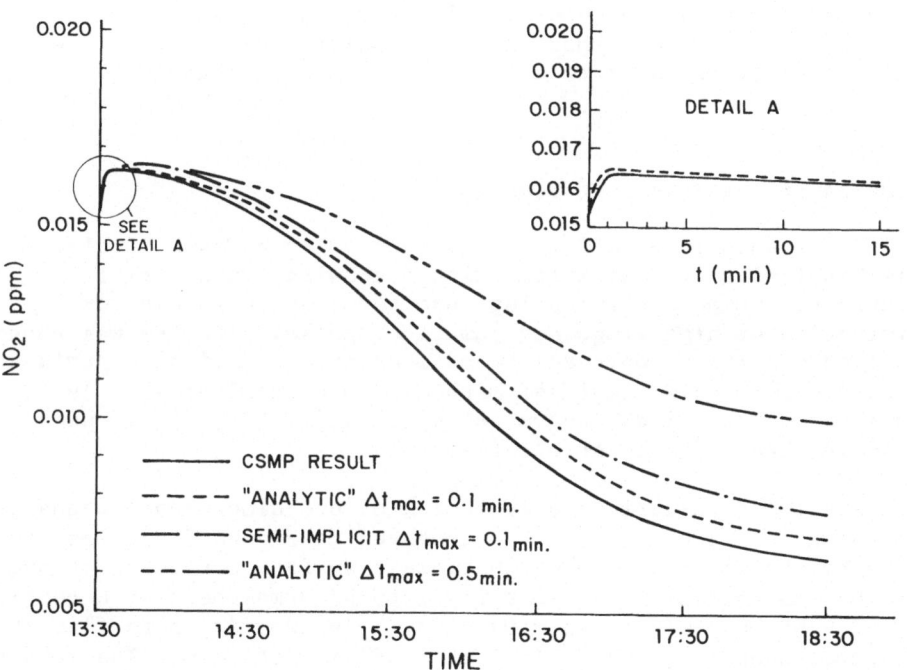

Figure 4. Calculated afternoon NO_2 concentrations using the kinetic mechanism in Table 1.

The above equation is used in the model in the calculation of the advected species only. However, the terms P and D also contain the short lived species (e.g., OH, HO_2, etc.). These species are calculated using the pseudo steady state approximations. The use of this approximation results in a set of algebraic equations which depend on the concentrations of the advected species (a schematic of the radical interactions treated in the solver is presented in Figure 3). The calculation procedure to advance from t_i to $t_i +$ Δt_i uses $C_\ell|_{t_i}$ to calculate the short lived species concentrations and then uses these concentrations to advance $C_\ell|_{t_i}$ to $C_\ell|_{t_i + \Delta t_i}$ using Equation (11).

The accuracy of this technique is dependent on the species of interest and Δt_i. For long lived species (e.g., SO_2, CO, and CH_4) relatively large Δt_i can be used, but for relatively reactive species much smaller time steps are required. Presented in Figure 4 are afternoon NO_2 concentrations calculated using the kinetic mechanism summarized in Table 1. Compared are results obtained using CSMP, the semi-implicit Euler's method (Preussner and Brand, 1981), and the "analytic" method just described. After 5 hours simulation time the "analytic" and the CSMP solutions differ by less than 8%. Over a fifteen minute simulation (fifteen minutes is the step size used for solution of the transport equations) the results differ by less than 1%. The analytic method with the free radical solver can execute a factor of 2 faster than CSMP.

Solution of the Transport Equations:

The transport equations are solved using a Crank-Nicolson Galerkin finite element method with piecewise linear trial functions, asymmetric weighting functions, and a filter for elimination of high frequency numerical noise. The FEM was chosen based on results of one- and two-dimensional numerical experiments comparing available numerical methods (Carmichael et al., 1980, Chock and Dunker, 1982; Pepper et al., 1980). The filter used is one described by McRae et al. (1982).

Results of numerical calculation of one-dimensional transport are shown in Figures 5 and 7. Presented in Figure 5 are results for a hypothetical case when there is a surface source of strength 3×10^{14} molecules/cm^2-s, emitting into an atmosphere with vertical velocity of -0.2 m/s (w ($\rho = 0$) = 0), $K_v = 10$ m^2/s, zero flux at the upper boundary, and an initial condition of zero. The results are after 12 time steps using a 15 minute step size. As shown, the use of symmetric weighting functions can lead to oscillations in the solution. However, the solution using the asymmetric weighting function does not oscillate.

The effect of the filter is to suppress oscillations. This is demonstrated in Figure 6 where a data set which oscillates about the curve $y = (x-1)^2$ is passed through the filter. The effect of the filter is also shown in Figure 5. When filtering is used in conjunction with symmetric weighting functions a marked improvement is shown. However, the results using the asymmetric weighting functions developed by Christie et al. (1976) for problems having locally large peclet numbers show that those weighting functions have a greater effect in suppressing the oscillations than the filter does. This is also demonstrated by the results in Figure 7. Shown are results after 88 time steps for similar conditions as Figure 5 except that the flux at the upper boundary is set at 10% of the surface flux and vertical wind velocity of 0.2 m/s. Very large oscillations occur using symmetric functions (the circled data represent negative predicted concentrations) but no oscillations occur using asymmetric weighting functions.

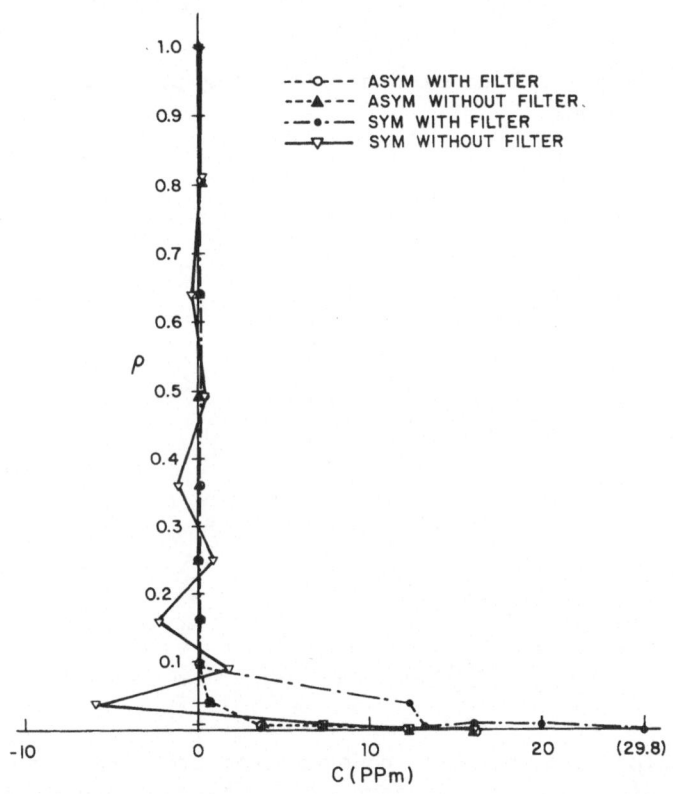

Figure 5. Calculated vertical concentration profiles for one-dimensional transport with a surface source. Results are after 12 time steps.

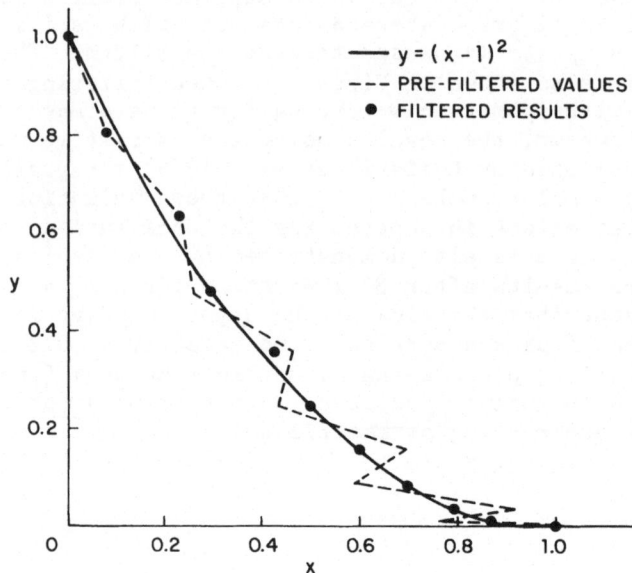

Figure 6. Sample calculation of filtering procedure used in the
model.

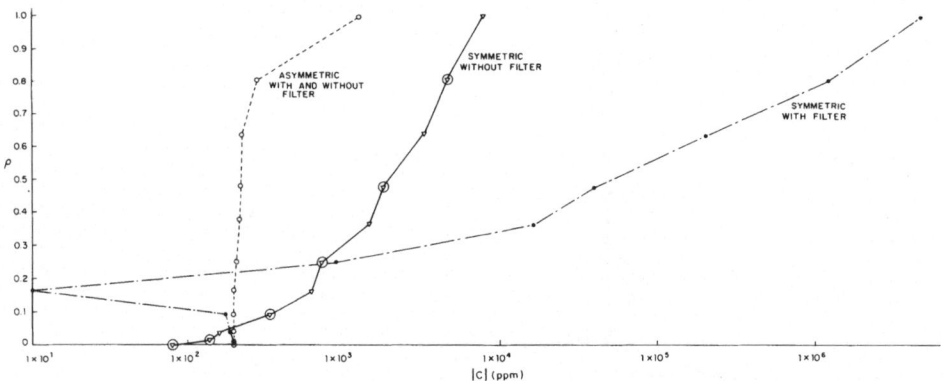

Figure 7. Calculated vertical conentration profiles for one-
dimensional transport with a surface source and nonzero flux at the upper boundary. Results are after 88
time steps.

Model Inputs

The model requires as input source distributions, initial conditions, and meteorological data. The meteorological data required is summarized in Figure 1. Generally, upper air meteorological data are available at 12-hour intervals. However, the boundary layer dynamics, especially over land, have substantial temporal variations and a 24-hour preiodicity. Therefore, the upper air data has to be analyzed to provide the necessary temporal and spatial structure to the input data. This is most important in the calculation of the vertical winds and the eddy diffusivity values.

Objective Analysis of Wind Field:

Typically only the horizontal wind fields are measured and these measurements do not satisfy the continuity equation. A non-mass conservative wind field introduces an artificial pseudo first order loss or generation term into the species mass transport equation (Kitada et al., 1982). In this model, an objective analysis procedure based on variational calculus is being used to obtain a three-dimensional mass consistent wind field (Sasaki, 1970; Peters et al., 1979). Recent studies comparing the output wind field against a numerically-generated land/sea breeze flow show that the procedure can reliably reproduce the vertical motions (Kitada et al., 1982).

One-dimensional Boundary Layer Model:

The distribution of trace gases in the atmosphere is greatly influenced by the dynamic behavior of the boundary layer. In this model, the one-dimensional boundary layer model developed by Yamada and Mellor (1975) is used to describe the diurnal boundary layer (specifically used to calculate K_v profiles). Examples of vertical eddy diffusivity profiles generated by this analysis are shown in Figure 8. This model can include large scale meteorological features (i.e., meso- or synoptic-scale) through a geostrophic wind term. The one-dimensional calculation is performed at each grid point, but is still considerably less expensive than using a three-dimensional boundary layer model.

CLOSURE

A combined transport/chemistry model for the regional scale transport of SO_x and NO_x compounds has been described. The model is Eulerian and three-dimensional and is an extension of an operational SO_x transport model developed by the authors. This second generation model treats 18 species including NO, NO_2, SO_2,

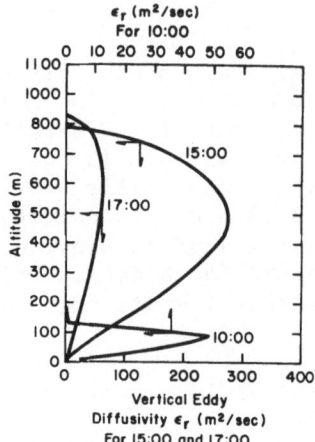

Figure 8. Example vertical eddy diffusivity profiles from the one-
          dimensional turbulent boundary layer model of Yamada and
          Mellor (1975).

SO_4^{2-}, O_3, HNO_3, NH_3, PAN and HC as transported species. Both
heterogeneous and homogeneous chemical reactions are modeled, as
are both wet and dry removal processes. A finite element numerical
method is used in the model and the non-linear chemistry is handled
using a pseudo-linearization scheme which results in virtually
decoupled numerical calculations.

The model has sufficient detail to model urban boundary layer
as well as non-urban free troposphere chemistry, boundary layer-
free troposphere exchange in cloud free and cloudy environments and
incloud and below cloud wet removal and chemistry processes. When
fully developed, the model has the capability and structure to
address leading regional scale scientific as well as regulatory
problems.

ACKNOWLEDGEMENTS

This research was supported in part by the National
Aeronautics and Space Administration under Research Grant NAG 1-
36. Travel for G.R. Carmichael was provided by The University of
Iowa, Faculty Scholars Program. Travel for T. Kitada to U.S.A. was
supported by the Ministry of Education of Japan through the
"Visiting Research Fellowship at Foreign Countries" Program.
Special thanks go to Kay Chambers for making the line drawings and
to Jane Frank for typing the manuscript.

REFERENCES

Carmichael, G.R., Kitada, T. and Peters, L.K., 1980, Application of a Galerkin finite element method to atmospheric transport problems, Computers and Fluids, 8:155.

Carmichael, G.R. and Peters, L.K., 1981, Application of the Sulfur Transported Eulerian Model (STEM) to a SURE data set, in: Proceedings of the 12th International Technical Meeting on Air Pollution Modeling and Its Application, pp. 324-347.

Chock, D.P. and Dunker, A.M., 1982, A comparison of numerical methods for solving the advection equation, Atm. Envirn. (in press).

Kaiser, J.A.C. and Hill, R.H., 1976, Irradiance at sea, JGR, 81:395.

Kitada, T., Kaki, A., Ueda, H. and Peters, L.K., 1982, Estimation of vertical air motion from limited horizontal wind data - a numerical experiment, Atm. Envirn. (in press).

McRae, G., Goodin, W. and Seinfeld, J., 1982, Numerical solution of the atmospheric diffusion equation for chemically reacting flows, J. Comp. Phys., 45:1.

Mitchell, A.R., 1969: "Computational Methods in Partial Differential Equations", John Wiley and Sons.

Pepper, D., Cooper, E., Baker, A., 1980, in "Developments in Theoretical and Applied Mechanics", 10:397.

Peters, L.K., Yamanis, J. and Akhtar, W., 1979, An algorithm to generate input data from meteorological and space shuttle observations to validate a CH_4-CO model. Status report NASA Grant NSG 1501.

Preussner, P.R. and Brand, K.P., 1981, Application of a semi-implicit Euler method to mass aaction kinetics, Chem. Eng. Sci., 10:1633.

Sasaki, Y., 1970, Some basic formulisms in numerical variational analysis, Mon. Wea. Rev., 98:875.

Yamada, T. and Mellor, G., 1975, A simulation of the Wangara atmospheric boundary layer data, J. Atmos. Sci., 32:2309.

Yanenko, N.N., 1971, "The Method of Fractional Steps", Springer-Verlag, New York.

DISCUSSION

J. KNOX   This modeling effort is very
interesting and promising. Hence, is it possible to
add additional species equations of prognostic character to track several conservative tracers through the
region ?

G. CARMICHAEL   Yes, it is quite easy to add
tracer(s) to the analyses. Adding a few (conservative)
tracers to the analysis would not increase significantly the computations (since much of the computations time is tied up in the chemistry.
Indeed adding isolated sources of tracers to the
analysis can provide a trace of the transport of
reactive species from that region and this can provide useful information.

I. TROEN   Basically a dispersion model
such as yours can only give time (or ensemble)
averaged concentrations, and I would therefore
question your conclusion that the model contains the
necessary physics. The addition of quadratic (in
concentration) terms will require a model for the
instantaneous concentration fluctuations. Would you
comment on this problem ?

G. CARMICHAEL   You are worried about whether
the ensemble mean reaction rate can be modeled adequately by the assumption that it is equal to the
reaction rate based on the mean concentrations. My
feeling is that on the time and space scales that
regional models are concerned with, the above assumption is okay (under most circumstances). However one
can think of special cases where it may not be, for
example point sources. Since the model cannot resolve
point sources as such, the problems associated with
segregated reactants are not addressed. This is a
case of a sub grid scale process that requires
further study. In fact, we are currently studying
this problem.

ACID DEPOSITION OF PHOTOCHEMICAL OXIDATION PRODUCTS -

A STUDY USING A LAGRANGIAN TRAJECTORY MODEL

Armistead G. Russell, Gregory J. McRae and Glen R. Cass

Environmental Quality Laboratory 206-40
California Institute of Technology
Pasadena, California, 91125, USA

INTRODUCTION

Photochemical air pollution has long been recognized as one of the major causes of such adverse environmental impacts as: visibility degradation, plant deterioration, eye irritation and lung function impairment. In addition, oxidation of both nitrogenous and sulfurous emissions can lead to acidic deposition products, notably nitric acid (HNO_3) and sulfuric acid (H_2SO_4), that can have major long term impacts on ecosystems. Indeed, the consequences of acid deposition from photochemical reactions may be felt for extended periods after the removal of the source and, in some cases, may be irreversible. This situation, coupled with the need for a methodology for relating changes in precursor emissions to ambient air quality, was one of the major motivations for the research reported in this paper.

Acid deposition generally falls in one of two categories: wet deposition and dry deposition, wet deposition resulting from snow, rain and fog precipitation. In many areas it is believed that the major acid flux to the surface results from dry deposition, the main focus of this work. The impact of nitrogenous oxidation products will be investigated by estimating the magnitude of deposited acids along the path of an air parcel using a vertically resolved, photochemical trajectory model. In a previous study (McRae and Russell, 1982), the fate of total oxides of nitrogen was investigated. In the present study, the fraction of nitrogen oxides that eventually form nitric acid will be estimated using mass balance techniques. A brief discussion of the chemistry leading to nitric acid formation is presented, including an investigation into the relative magnitude of the production of nitric acid by different mechanisms.

MODEL DESCRIPTION

Results obtained in this study were calculated using a model
that describes turbulent transport, gas phase chemical reactions,
ground level deposition and formation of aerosol nitrate (Russell et
al., 1982). Concentrations of gaseous pollutants and ammonium
nitrate aerosol are calculated using a Lagrangian trajectory form of
the atmospheric diffusion equation. The species mass conservation
equation that governs the concentration of species i, $c_i(z,t)$, is
given by (McRae et al. 1982a)

$$\frac{\partial c_i}{\partial t} = \frac{\partial}{\partial z}\left(K_{zz}\frac{\partial c_i}{\partial z}\right) + R_i(c_1, c_2, \ldots, c_n, T) - \frac{\partial v_s c_i}{\partial z} \quad i=1,n \qquad (1)$$

with initial conditions

$$c_i(z,o) = c_i^o \qquad\qquad t = 0 \qquad\qquad (2)$$

and boundary conditions

$$\left(K_{zz}\frac{\partial c_i}{\partial z}\right) = 0 \qquad\qquad z = H \qquad\qquad (3)$$

$$[v_g^i c_i - K_{zz}\frac{\partial c_i}{\partial z}] = E_i \qquad\qquad z = 0 \qquad\qquad (4)$$

where $K_{zz}(z)$ is the vertical turbulent eddy diffusivity,
$R_i(c_1, \ldots, c_n, T)$ is the rate of chemical production of species i at
temperature T, H is the height of the air column and v_s is the settling velocity. A schematic representation of the vertically
resolved trajectory model and computational cells is shown in Figure
1. Implicit in the formulation of (1) are the assumptions that horizontal diffusion and vertical advection are small, and that the
effects of wind shear are negligible. The effects of these simplifications are discussed in Liu and Seinfeld (1975). In (1) the term
v_s is used to describe gravitational settling. For aerosol particles smaller than 1 μm the loss of material by settling is negligible. Since ammonium nitrate is usually associated with micron sized
aerosol particles, the effects of gravitational settling can be
neglected. As a result, the vertical turbulent transport of both
gases and aerosols are treated equivalently in this model.
Since the formulation of the relationship of K_{zz} to atmospheric stability is described in McRae et al. (1982a) it will not be repeated
here.

The parameters associated with the boundary conditions are: v_g^i,
the deposition velocity, and E_i, the species mass emission flux per
unit area. Boundary conditions at the top of the region (z = H)
assume that the mixing depth, Z_i, is less than H, thus turbulent
transport through the top of the air column is negligible. Ground
level (z = 0) boundary conditions are a statement of mass continuity, accounting for surface deposition, diffusive transport, and
emissions. Deposition velocities are used to

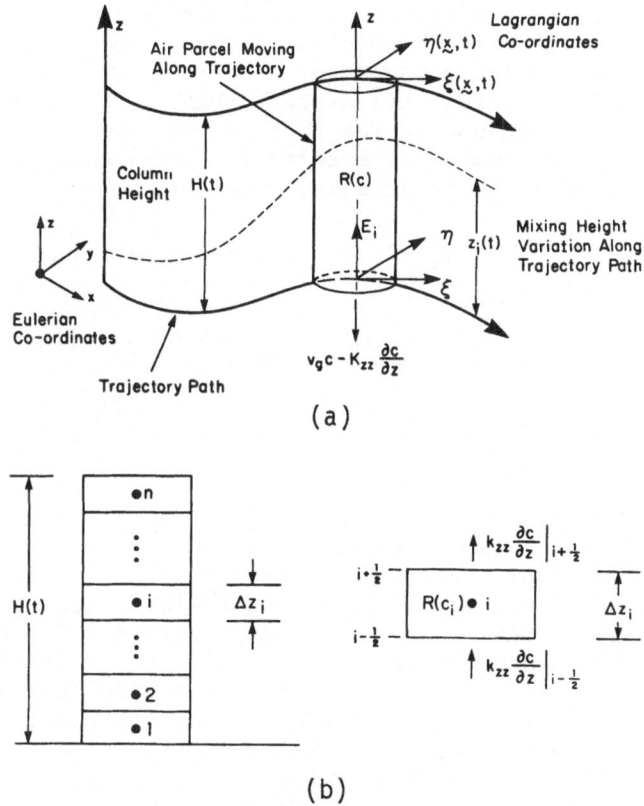

(a)

(b)

Fig. 1.   Schematic representation of (a) vertically resolved
          Lagrangian trajectory model and (b) the computational grid
          cell convention

describe the interaction and reaction of gases and aerosols with
surfaces. In general, v_g^1 is dependent on meteorological conditions
and on the reactivity of species i, with more reactive species having a higher deposition velocity. A more complete description of
the parameterization of surface removal processes is presented in
the next section.

The chemical kinetics associated with the term $R_i(c_1,...,c_n,T)$ in (1), are described using the photochemical reaction mechanism of Falls and Seinfeld (1978), Falls et al. (1979), McRae et al. (1982) and McRae et al. (1982a,b). The chemical interactions between different nitrogen containing species are shown in Figure 2. In addition to the direct emissions of nitric oxide (NO), some of the other more abundant nitrogenous species, produced by photochemical reactions occuring in the atmosphere, are nitric acid (HNO_3), nitrogen dioxide (NO_2) and peroxactyl nitrate (PAN). Nitric acid is particularly important because of its role in acid deposition and in the formation of ammonium nitrate aerosol. The specific reaction pathways that produce nitric acid are given below.

$$NO_2 + OH \xrightarrow{18} HNO_3 \qquad K_{18} = 1.52 \times 10^4 \text{ ppm}^{-1} \text{ min}^{-1} \qquad (5)$$

$$NO_2 + RO_2 \xrightarrow{40} HNO_3 + RCO_3 \quad K_{40} = 5.5 \qquad \text{ppm}^{-1} \text{ min}^{-1} \qquad (6)$$

$$N_2O_5 + H_2O \xrightarrow{46} 2HNO_3 \qquad K_{46} = 1.5 \times 10^{-5} \text{ ppm}^{-1} \text{ min}^{-1} \qquad (7)$$

where T is in $^\circ K$. Other than direct emissions, the only sources of nitric acid are the above reaction steps. During the daytime, reaction 18 is the dominant route for producing gas phase HNO_3. At night the hydrolysis of dinitrogen pentoxide (N_2O_5) by reaction 46 can become important. Some controversy exists about the magnitude of the rate constant for this reaction, an upper limit, due to Morris and Niki (1973), has been used in the following calculations. Because of the uncertainty associated with the kinetics of the mechanism step $N_2O_5 + H_2O \rightarrow 2HNO_3$, a subsequent section will analyze the effects of reducing k_{46}.

In addition to the gas phase chemistry a unique feature of this model is the incorporation of a calculation scheme that describes the formation of ammonium nitrate aerosol. Based on the work of Stelson and Seinfeld (1982ab), the approach uses fundamental thermodynamic principles to find the equilibrium dissociation constant, K, for the ammonium nitrate-nitric acid-ammonia system,

$$NH_4NO_3(\text{aerosol}) \underset{\leftarrow}{\overset{K}{\rightarrow}} NH_3(g) + HNO_3(g) \qquad (8)$$

Given the dissociation constant K,

$$K = [NH_3(g)][HNO_3(g)] \qquad (9)$$

it is a straightforward task to determine the concentration of ammonium nitrate and $[NH_3(g)]$ and $[HNO_3(g)]$, the gaseous concentrations of ammonia and nitric acid, respectively. Further details of this calculation procedure are described in Russell et al. (1982).

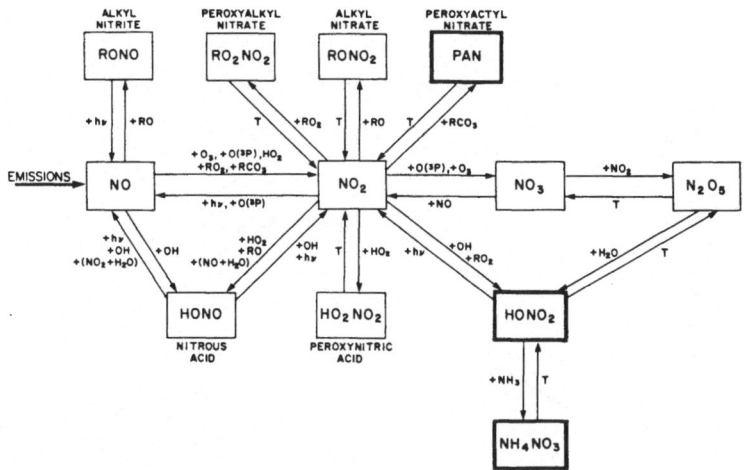

Fig. 2.   Diagram representing the chemistry involving atmospheric nitrogen oxides and the resulting products

Evidence for the ammonium nitrate equilibrium in the atmosphere has been found by making simultaneous measurements of ammonia and nitric acid and comparing the data against calculated dissociation constants (Stelson et al., 1979; Doyle et al., 1979). In the study by Stelson and Seinfeld (1982ab), it was found that both the dissociation constant and the physical state of ammonium nitrate was a function of temperature and relative humidity. Ammonium nitrate is found as a solid if the relative humidity is less than that of deliquescence (RHD) given by

$$\ln RHD = \quad 723.7/T + 1.7037 \qquad\qquad (10)$$

where T is in $^{\circ}$K. Above the relative humidity of deliquescence ammonium nitrate exists in the aqueous phase. Figures 3a and b show the variation of K as a function of temperature and relative humidity. Dissociation constants for typical atmospheric conditions range from 0.04 $(ppb)^2$ at 5°C and 90%RH to 1400 $(ppb)^2$ at 40°C and a RH of 30%.

DRY DEPOSITION OF ACID SPECIES

An important component of any approach to estimating the magnitude of acid deposition (and, similarly, acid neutralizing species) is a procedure for calculating the pollutant flux to the surface. This section provides a brief summary of a technique introduced in McRae et al. (1982a). In most models the deposition rate is described by a single quantity, the pollutant deposition velocity, v_g. The flux of material, F, directed towards the lower boundary

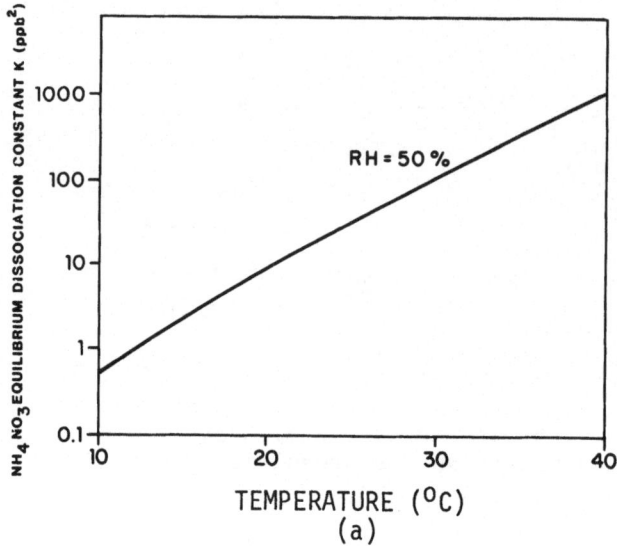

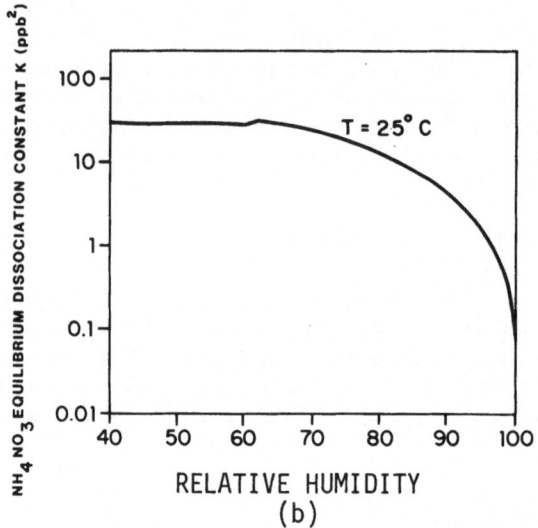

Fig. 3.   Functionality of the equilibrium dissociation constant of
NH_4NO_3 on (a) temperature and (b) relative humidity

surface is defined by $F = v_g c(z_r)$, where $c(z_r)$ is the concentration of the material at some reference height z_r. A basic problem with this expression is that it does not explicitly represent the fact that dry deposition involves a complex linkage between turbulent diffusion in the surface boundary layer, molecular scale motion at the air-ground interface and chemical interaction of the material with the surface.

As a first step towards improving upon the simple model, it is necessary to recognize that there are two basic components associated with pollutant removal, the transport of material to the ground and the interaction of the pollutants with the surface. Unless extensive field experiments have been carried out, it is not possible to accurately characterize the second component of the dry deposition process. An alternative approach, and the focus of this section, is to develop an estimate for the upper limit of v_g in terms of the transport processes, and the concentration at a reference point above the surface. (Typically the height of the lowest computational grid point in the airshed model.) A secondary goal is to identify the significant meteorological variables and surface properties needed either to correlate different measures of v_g or to modify the results for different experimental conditions.

Within the layer $0 \leq z \leq z_r$, deposition is assumed to be a one-dimensional, steady state, constant flux process occurring without reentrainment. With these assumptions the deposition flux of gaseous materials is described by $F = [K_{zz} + D] \partial c / \partial z$ where K_{zz} is the pollutant eddy diffusion coefficient and D the molecular diffusion coefficient of the material in the air. By equating fluxes and assuming that u_* is constant in the surface layer, then it is possible to show, using Monin-Obukov similarity theory, that

$$v_g = \frac{k\left[1 - \dfrac{c(z_d)}{c(z_r)}\right]}{\displaystyle\int_{z_d}^{z_r} \frac{1}{zu_*} \phi_p\left(\frac{z}{L}\right) dz} = \frac{k^2 u(z_r)\left[1 - \dfrac{c(z_d)}{c(z_r)}\right]}{\left[\displaystyle\int_{z_o}^{z_r} \phi_m\left(\frac{z}{L}\right)\frac{dz}{z}\right]\left[\ell n\left(\frac{z_o}{z_d}\right) + \displaystyle\int_{z_o}^{z_r} \phi_p\left(\frac{z}{L}\right)\frac{dz}{z}\right]} \tag{11}$$

The expressions ϕ_m and ϕ_d, used in (7), are experimentally derived functions that account for the influence of atmospheric stability on turbulent transport (Businger et al. 1971). The lower limits of integration, z_d and $c(z_d)$, refer to the elevation and concentration of material at the effective pollutant sink height. If the ratio $c(z_d)/c(z_r)$ is zero then the surface is considered to be a perfect sink and then v_g is simply the reciprocal of the aerodynamic resistance. Variations in chemical or biological resistance at $z = z_d$ can be modeled by changing the concentration ratio. In general z_d is not equal to the surface roughness z_o, a height associated with momentum sink (Brutsaert, 1975). Within the surface layer defined by $z_d \leq z \leq z_r$ the bulk contribution to the diffusive transport from molecular diffusion is negligible. Evaluation of the term

$\ln(z_o/z_d)$ in the denominator of (7) requires a knowledge of z_d and of the transfer processes at the surface. Based on a survey of the heat transfer literature and in particular the work of Brutsaert (1975), Wesely and Hicks (1977) assumed that

$$\ln\left(\frac{z_o}{z_d}\right) = 2\left(\frac{Sc}{Pr}\right)^{2/3} \tag{12}$$

where Sc and Pr are the Schmidt and Prandtl numbers associated with the pollutant material in air. The complete expression is then

$$v_g = \frac{k^2 u(z_r)\left[1 - \dfrac{c(z_d)}{c(z_r)}\right]}{\left[\displaystyle\int_{z_o}^{z_r} \phi_m\left(\frac{z}{L}\right)\frac{dz}{z}\right]\left[2\left(\frac{Sc}{Pr}\right)^{2/3} + \displaystyle\int_{z_o}^{z_r}\phi_p\left(\frac{z}{L}\right)\frac{dz}{z}\right]} \tag{13}$$

The integrals required to evaluate v_g are given in McRae et al. (1982a).

The final result exposes the fact that v_g is directly influenced by the prevailing meteorology and atmospheric stability. The effect of stability is particularly apparent; consider, for example, the conditions shown in Figure 4. With z/L in the range -1.5 to $+1.5$, the deposition velocities vary by almost a factor of five. This result indicates that under typical conditions there could be a significant diurnal variation in the surface removal of pollutant material. The use of a single diurnal average v_g would overpredict the amount of material removed at night time. The functional dependence of v_g on the elevation above the surface highlights the need for reporting the reference height z_r in field or laboratory studies. If v_g, z_r, z_o and $u(z_r)$ are measured, then it is possible to evaluate $c(z_d)/c(z_r)$ and in turn, v_g, for elevations other than the reference height. This is a useful approach for developing the deposition velocities for air quality models in which z_r may be of $O(10-50m)$.

Once the pollutant deposition velocity has been established, either by direct measurement or estimated using the proposed model, the next step is to develop a formal procedure for calculating the amount of material removed at the ground. At the lower surface of the airshed the pollutant removal is typically described by the boundary condition (4) evaluated at $z = z_r$, where z_r is a reference elevation, $v_g(z_r)$ and $c(z_r)$ are the pollutant deposition velocity and concentration at that height.

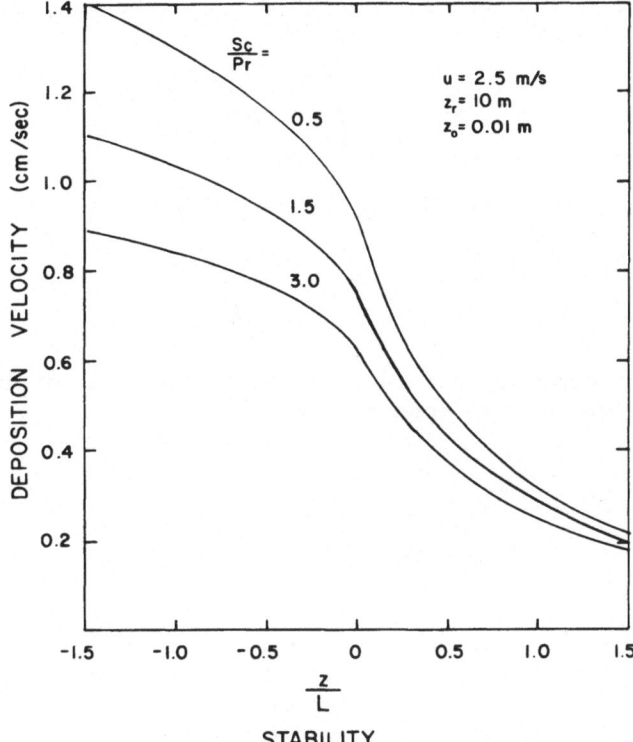

Fig. 4.   Variation of surface deposition velocity as a function of
          atmospheric stability

Note that the elevation of the lowest computation grid point is typically much higher than the reference height, z_r, used to establish
the pollutant deposition velocities, as seen in Figure 5. Because
of the need to approximate the vertical concentration profile in
discrete increments, $c(z_r)$ is not readily available. When coupled
with the observation that v_g varies with height, there is a need to
develop an equivalent deposition velocity v_g that, when applied to
the cell average concentration, c_1, correctly predicts the flux at
the lower boundary. One way to develop such a model is to assume
that most of the lowest cell is within the surface or constant flux
layer. If this is the case then the cell deposition velocity is
given by $v_g = v_g(z_r)c(z_r)/c_1$. If c_1 is to represent the average

value of the actual vertical concentration distribution in the range $z_r \leq z \leq z$ then it must be equivalent to

$$c_1 = \frac{1}{\Delta z - z_r} \int_{z_r}^{\Delta z} c(z) dz \tag{14}$$

Within the constant flux layer $c(z)$ is given by

$$c(z) = c(z_r) \left[1 + v_g(z_r) \int_{z_r}^{z} \frac{dz}{K_{zz}} \right] \tag{15}$$

The equivalent cell deposition velocity can now be determined by combining the above equations to give

$$\overline{v}_g = \frac{v_g(z_r)}{1 + \frac{v_g(z_r)}{ku_*(\Delta z - z_r)} \int_{z_r}^{\Delta z} \int_{z_r}^{z} \phi_p(\frac{x}{L}) \frac{dx}{x} \, dz} \tag{16}$$

An example of the variation of $\overline{v}_g$ with cell size and atmospheric stability is shown in Figure 6. As can be expected, the equivalent deposition velocity becomes smaller as z increases. The variation is most pronounced under stable conditions because of the reduced vertical mixing. One implication of this result is that if $v_g(z_r)$, rather than $\overline{v}_g$, were to be used in a practical calculation then the surface removal flux would be considerably overestimated.

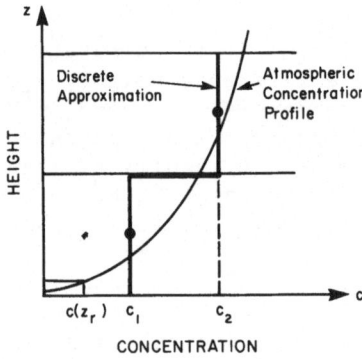

Fig. 5.   Nomenclature for the cell averaged deposition model showing the discrete approximation of the concentration profile

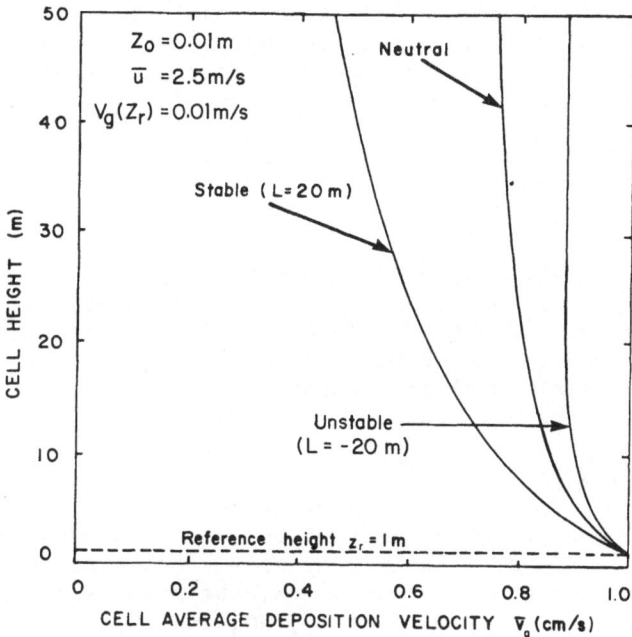

Fig. 6. Variation of cell average deposition velocity as a function of atmospheric stability and computational cell
height

Ammonium nitrate aerosol deposition is treated in the same
manner as gaseous deposition due to the aerosols expected small
size. A value for the approximate deposition velocity was estimated
from the experimentally found deposition velocities of NH_4^+ and NO_3^-
(Slinn and Slinn, 1981).

Calculation of the deposition velocity is a key step in closing
the boundary fluxes into the air parcel. By neglecting horizontal
diffusion, and assuming that transport across the top of the modeling region is negligible, it can be seen that the only bounding
fluxes transpire at the surface, z = 0. Thus a mass balance on
emissions and reaction products, as well as total species deposition, can be found by integrating the surface level flux and the
vertical concentrations. Total deposition of species i is found by
integrating $v_g c_i$ along the trajectory. This should be an upper
estimate.

Results from this study are concerned with the magnitude of
acid deposition resulting from the oxidation of nitrogen oxides
emissions. With this in mind, it is necessary to define what is

meant by the term acid. In this study acid deposition will be
defined as the deposition of species with immediately available pro-
tons minus proton scavenging species. Thus, considering just the
NH_3-HNO_3-NH_4NO_3-HONO system, the net acid deposition is given by

$$\{\text{Acid Deposition}\} = \{\text{Deposition of } HNO_3\} +$$

$$\{\text{Deposition of HONO}\} - \{\text{Deposition of } NH_3\}. \qquad (17)$$

This definition, while adequate for many circumstances, may require
modification to account for reactions in the receiving ecosystem,
such as $NH_4^+ + 2O_2 \rightarrow 2H^+ + NO_3^- + H_2O$ as well as the slightly acidic
character of NH_4NO_3. The deposition of sulfate or the possible acid
producing reactions involving deposited NO_2 and SO_2 have not been
included in the present study.

MODEL EVALUATION

Both the gas phase and aerosol pollutant concentration predictions have been compared against measured concentrations. In
evaluating the gas phase calculations, the model's predictions were
compared against measurements of O_3, NO_2 and NO, three of the most
commonly measured pollutants (McRae et al., 1982b). A later evaluation was performed on the model after it was modified to include an
ammonium nitrate formation mechanism (Russell et al., 1982). In
that study, aerosol and ozone predictions were compared against
ambient measurements. In both studies agreement between the measured and calculated concentrations was seen to be good, and the
results of the latter study are shown in Figures 7a and b.

A detailed inventory of ammonia emissions was assembled for the
evaluation of the aerosol model. This inventory is also quite
important in the study of acid deposition, as NH_3 is one of the
major acid neutralizing species. A comparison of the predicted and
measured concentrations of total ammonia ($NH_3(g) + NH_4^+$) shown in
Figure 8 indicates the spatial accuracy of the inventory. This
inventory was added to pre-existing inventories of reactive hydrocarbons (RHC), nitrogen oxides (NO_x) and carbon monoxide (CO) emissions. Figure 9 illustrates the spatial distribution of emissions.

The ability of a model to describe ambient concentrations of
gases is only one indication of the validity of the treatment for
surface deposition. Extensive experiments in the air basin are
required for definitive evaluation of the performance of the deposition calculations. At present, the needed experimental data do not
exist. A study by Whelpdale and Shaw (1974) can be used for a qualitative evaluation. Their results, shown in Table 1, follow the
same variation with atmospheric stability shown in Figure 4. During
very stable conditions, as might be found at night, the deposition
flux of material to the ground is controlled predominantly by the

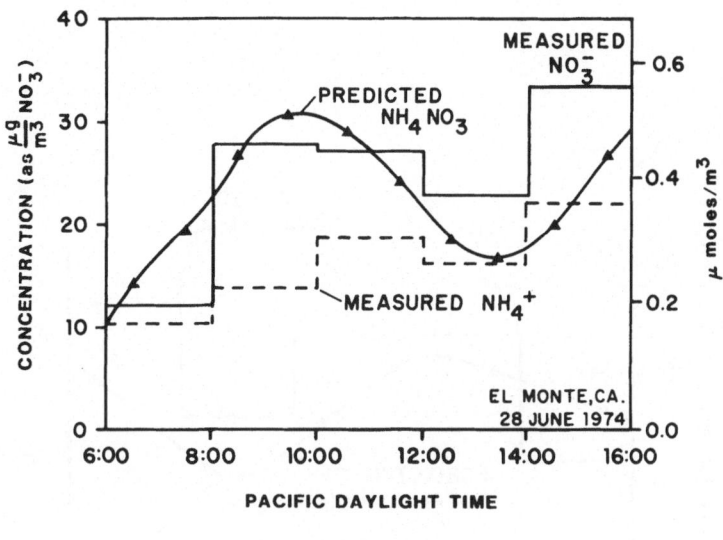

(a)

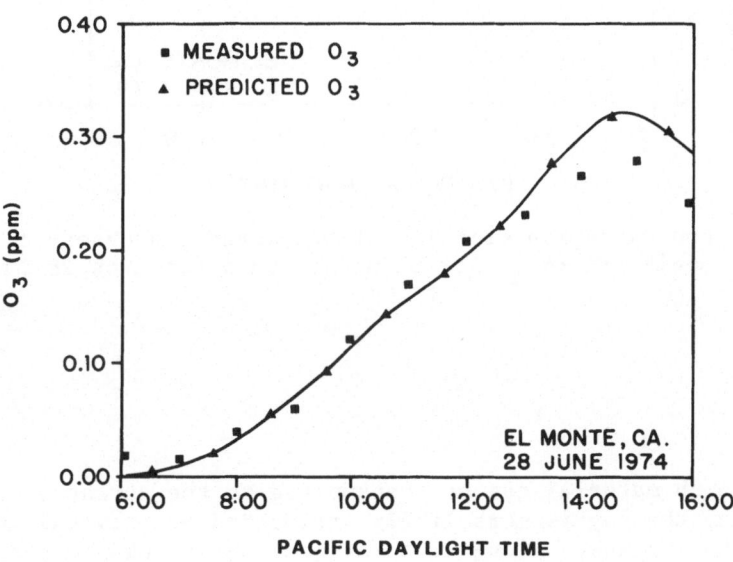

(b)

Fig. 7. Comparison of model predictions against ambient data for
(a) NH_4NO_3 and (b) Ozone on 28 June 1974 at El Monte,
California

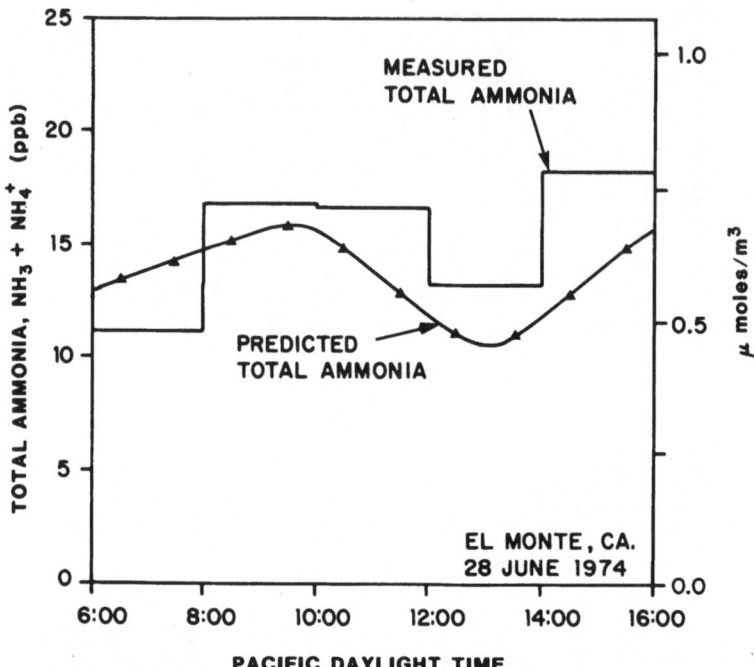

Fig. 8.   Concentration profiles of predicted and observed total am-
          monia ($NH_3 + NH_4^+$) for El Monte, California on 28 July 1974

rate at which material can be transported to the surface. Deposition during the day is more likely influenced by chemical interaction at the surface. Along with the qualitative comparison,
predicted deposition velocities are of the same magnitude as found
by experiment.

DISCUSSION OF RESULTS

     The net flux of acid and basic species to the ground has been
calculated for a major urban area. Available inventories and
meteorological data for 27 June 1974 were used to determine emission
fluxes and the meteorology of the

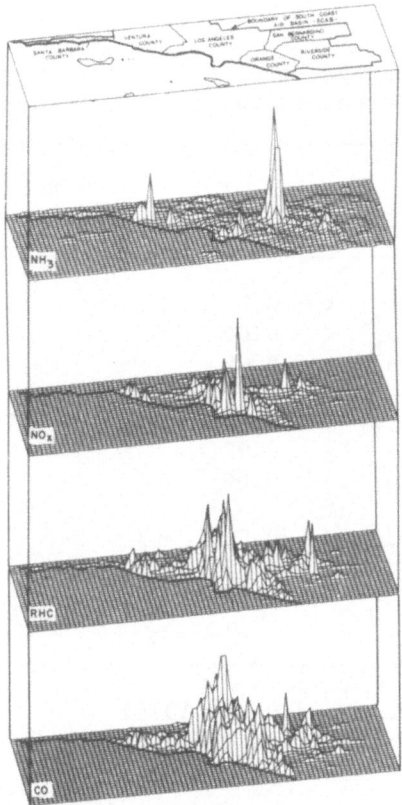

Fig. 9.   Spatial representation of daily emissions of ammonia
(NH$_3$), nitrogen oxides (NO$_x$), reactive hydrocarbons (RHC)
and carbon monoxide (CO) in the Los Angeles, California
basin (Inventory period June 1974)

Los Angeles basin. Ambient conditions on 27 June 1974 were typical
of the severe smog episodes that are experienced in Southern California. These days are characterized by high oxidant levels and
corresponding nitric acid levels. For this study, the air parcel
trajectory was selected so that it crossed the airshed. The particular trajectory started near the ocean at midnight, then proceeded
east over Los Angeles, and finally passed over Upland at 1600
Pacific Standard Time (PST). By starting over the ocean, the dependency of the calculated results on initial conditions is decreased.

TABLE 1

Average Deposition Velocity of SO_2 for
Different Surface and Stability Conditions

SURFACE	STABILITY			NUMBER OF EXPERIMENTS	DEPOSITION VELOCITY v_g (cm/s)
Grass		Ri_b	< −0.02	10	2.4
	−0.02 <	Ri_b	< 0.02	3	2.6
		Ri_b	> 0.02	2	0.5
Snow		Ri_b	< −0.02	1	1.6
	−0.02 <	Ri_b	< 0.02	3	0.52
		Ri_b	> 0.02	8	0.05
Water		Ri_b	< −0.02	7	4.0
	−0.02 <	Ri_b	< 0.02	7	2.2
		Ri_b	> 0.02	4	0.16

a. Source: Whelpdale and Shaw (1974)
b. Stability is defined in terms of the bulk Richardson number Ri_b

$$Ri_b = \frac{g}{T} \, \Delta z \, \frac{\Delta \theta}{(\Delta \overline{u})^2}$$

where T is the ambient temperature, Δz difference in sampling height
$\Delta \theta$ the potential temperature difference and Δu the wind speed

    Deposition of acidic and basic species occurs along the whole
trajectory, as seen in Figure 10. In the early morning near the
coast, the deposition rate of ammonia is higher than that of the
nitrogenous acids. The reason for this is that the nitrogen oxides
emissions have not reacted to form nitric acid. The deposition rate
in the morning is very slow because the stable atmosphere suppresses
rapid vertical mixing. After sunrise the deposition rate increases,
and nitric oxides start to oxidize to nitric acid. This leads to a
net acid flux. This trend continues until sunset when the atmosphere becomes more stable, decreasing the deposition rate. The
nitric acid burden in the air parcel is still much greater than the
ammonia burden. The instantaneous deposition rates shown in Figure
10 demonstrate the pronounced diurnal variation of acid deposition.

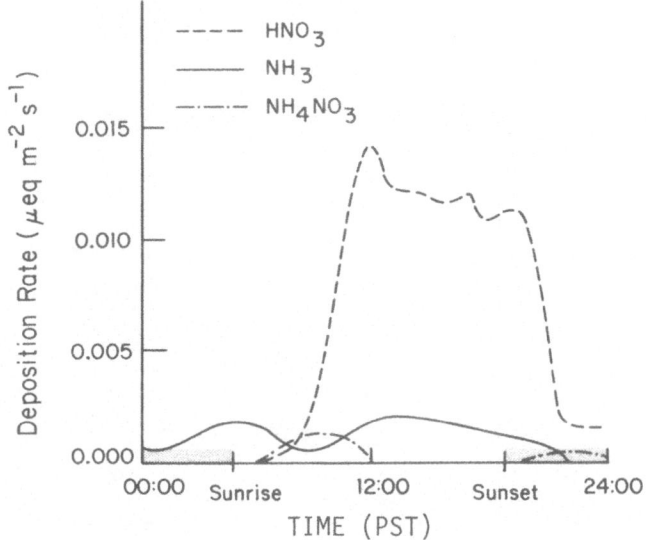

Fig. 10. Instantaneous acid flux to the ground along the trajectory. Acid flux is due almost totally to HNO_3 deposition and ammonia is treated as an acid neutralizer. Ammonium nitrate is treated as neutral

The deposition flux is greatest in the afternoon because of rapid mixing, and much less both in the morning and at night.

Total nitrogeneous species deposition along the trajectory is shown in Figure 11. In comparison to other species, very little nitric acid has been deposited by noon. By the end of the trajectory (2400 PST), a large fraction (31.3%) of the deposited nitrogen was in the form of nitric acid. In Figure 11, other consists primarily of ammonium nitrate. Table 2 summarizes the acid and nitrogen deposition by the end of the trajectory.

The predicted net dry acid deposition, from Table 2, of 422 μeq m^{-2} can be compared to the findings of Liljestrand and Morgan (1980) for wet deposition. Liljestrand and Morgan estimated the wet flux of nitrogenous acids to be 0.016 eq m^{-2} year, compared to 0.19 estimated in this study. Liljestrand (1979) estimated dry deposition to be 0.56 eq m^{-2} year^{-1}, including NO_x deposition. Dry deposition appears to account for a much greater flux of acid than wet deposition. It should be kept in mind that the estimate obtained in this study should be an upper limit due to the deposition model used and trajectory chosen.

DEPOSITION OF NITROGEN SPECIES

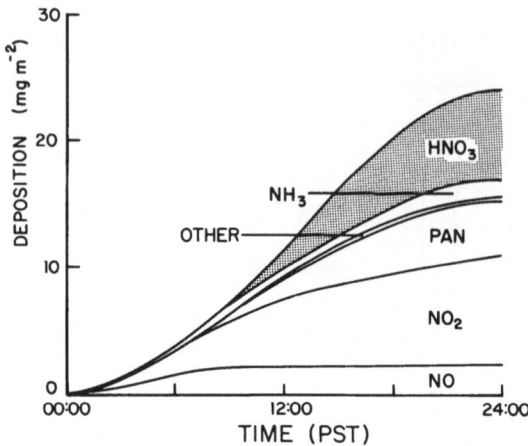

Fig. 11. Total integrated nitrogen deposition by species

A balance may also be performed on all the nitrogen containing
pollutants. Approximately 35% (by mole equivalent) of the nitrogen
oxides emissions are deposited during the day. Recall that 31% is
removed as nitric acid, so about 11% of the nitrogen oxides emissions are deposited as nitric acid during the trajectory. Of the
65% of the nitrogen containing pollutants that remain in the air
parcel, 41% is nitric acid at the end of the trajectory studied.
The remaining nitric acid would be expected to deposit out as the
air parcel continues the following days. Figure 12 depicts the balance between deposited nitrogen oxides emissions and those emissions
remaining in the air column.

Modeling the chemistry of nitric acid involves a study of the
major formation mechanisms. Figure 13 illustrates the diurnal variation in the contributions of the three pathways. Table 3 gives the
percentage of nitric acid formed by each step. As expected, the
NO_2-OH reaction dominates during the day. Immediately after sunset,
however, most of the nitric acid is predicted to be formed by the
the homogeneous hydrolysis of N_2O_5. The rate constant used for that
reaction, $K_{46} = 1.5 \times 10^{-6}$ ppm^{-1} min^{-1}, is, as mentioned previously,
believed to be an upper limit.

TABLE   2

Integrated Fluxes of Nitrogen Containing Species and
Associated Acid Deposition for 24-Hour Period[a]
Along a Representative Trajectory

SPECIES	DEPOSITED NITROGEN $\mu gN\ m^{-2}$	ACID DEPOSITION $\mu eq\ m^{-2}$	AMMONIA RELATED NITROGEN DEPOSITION $\mu gN\ m^{-2}$
NO	2422 (10.4%)		
NO_2	8742 (37.7%)		
PAN	4477 (19.3%)		
HNO_3	7255 (31.3%)	518	
HONO	26.4 (0.1%)	1.9	
NH_4NO_3	277 (1.2%)		277 (16.7%)
NH_3	---	-98 [b]	1377 (83.3%)
TOTAL	23199(100%) [c]	422	1654(100%)

(a) 27 June 1974 Upland trajectory
(b) Negative flux corresponds to neutralization of HNO_3 by NH_3
(c) Total nitrogen deposition (excluding NH_3)

        Because some controversy exists about the magnitude of k_{46}, it
was decided to test the sensitivity of nitric acid formation to k_{46}.
This can be done by investigating the reactions involving N_2O_5
below.

$$NO_2 + O \xrightarrow{6} NO_3 \qquad\qquad K_6 \simeq 3600\ \ ppm^{-1}\ min^{-1} \qquad\qquad (18)$$

$$NO_2 + O_3 \xrightarrow{7} NO_3 \qquad\qquad K_7 \simeq 0.05\ \ ppm^{-1}\ min^{-1} \qquad\qquad (19)$$

$$NO + NO_3 \xrightarrow{8} 2NO_2 \qquad\qquad K_8 \simeq 27200\ \ ppm^{-1}\ min^{-1} \qquad\qquad (20)$$

$$NO_2 + NO_3 \xrightarrow{44} N_2O_5 \qquad\qquad K_{44} \simeq 3900\ \ ppm^{-1}\ min^{-1} \qquad\qquad (21)$$

$$N_2O_5 \xrightarrow{45} NO_2 + NO_3 \qquad\qquad K_{45} \simeq\ 6.9\ \ min^{-1} \qquad\qquad (22)$$

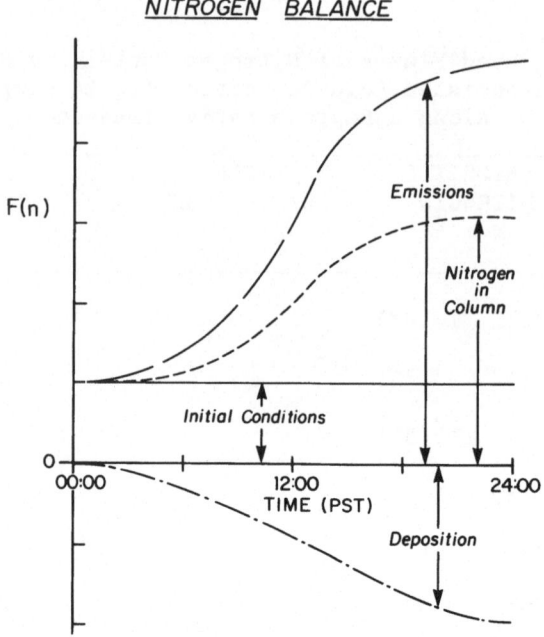

Fig.   Nitrogen balance along the trajectory showing total deposited
       nitrogen, nitrogen emissions, and nitrogen remaining in the air
       parcel.

$$N_2O_5 + H_2O \overset{46}{\to} 2HNO_3 \qquad K_{46} \le 3\times10^{-6} \text{ ppm}^{-1} \text{ min}^{-1} \tag{23}$$

Thus, the rate expressions for NO_3 and N_2O_5 are

$$\frac{d[NO_3]}{dt} = K_6[NO_2][O]+K_7[NO_2][O_3]-K_8[NO][NO_3]-K_{44}[NO_2][NO_3]+K_{45}[N_2O_5] \tag{24}$$

and

$$\frac{d[N_2O_5]}{dt} = K_{44}[NO_2][NO_3]-K_{45}[N_2O_5]-K_{46}[N_2O_5][H_2O] \tag{25}$$

Both N_2O_5 and its precursor NO_3 are short lived species, and the
pseudo-steady state assumption (PSSA) can be used to investigate
their dynamics. From the previous expressions, the PSSA gives:

$$[NO_3] = \frac{(K_6[O][NO_2]+K_7[O_3][NO_2])(K_{45}+K_{46}[H_2O])}{(K_8[NO]+K_{44}[NO_2])(K_{45}+K_{46}[H_2O])-K_{45}K_{44}[NO_2]} \tag{26}$$

$$[N_2O_5] = \frac{(K_6[O][NO_2]+K_7[O_3][NO_2])K_{44}[NO_2]}{(K_8[NO]+K_{44}[NO_2])(K_{45}+K_{46}[H_2O])-K_{45}K_{44}[NO_2]} \tag{27}$$

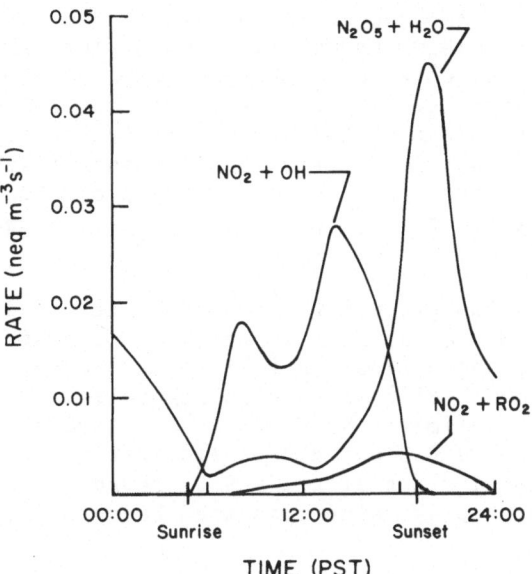

Fig. 13. Diurnal variation in production of HNO_3 by reaction paths

Figure 13 indicated that the homogeneous N_2O_5 hydrolysis is important only during the nighttime when O atom concentrations are negligible. Also the NO concentrations are very low at night, on this
occasion, except near the ground. With this in mind, the PSSA
expression can be simplified, resulting in two equations. Aloft,
where $k_{45}k_8[NO] << k_{46}k_{44}[NO_2][H_2O]$ (because of the low [NO] concentrations), it is found that

$$[N_2O_5] = \frac{K_7[O_3][NO_2]}{K_{46}[H_2O]} \tag{28}$$

Near the ground, where [NO] is large,

$$[N_2O_5] = \frac{K_{44}K_7[O_3][NO_2]^2}{K_{46}K_{44}[NO_2][H_2O]+K_{45}K_8[NO]} \tag{29}$$

Using (28), it is seen that at in the upper parts of the atmosphere
the homogeneous production of HNO_3 hydrolysis at night is given by

$$\frac{d[HNO_3]}{dt} \cong 2K_7[O_3][NO_2] \quad \text{(elevated)} \tag{30}$$

The surprising result is that aloft the production of HNO_3 by N_2O_5
hydrolysis is insensitive to k_{46}, the rate constant for that reaction. Near the ground, the formation of HNO_3 is given by

$$\frac{d[HNO_3]}{dt} \cong \frac{2K_{46}K_{44}K_7[H_2O][O_3][NO_2]^2}{K_{46}K_{44}[NO_2][H_2O]+K_8K_{45}[NO]} \tag{31}$$

which is sensitive to changes in the value used for k_{46}. However, because most of the N_2O_5 hydrolysis occurs aloft where [NO] is low, the column averaged total production of HNO_3 by N_2O_5 is relatively insensitive to small revisions in k_{46}. A test using a hydrolysis rate constant one fifth of the upper limit gave results essentially unchanged to what was found previously, though slightly less nitric acid was produced near the ground. Total production of nitric acid by the different reactions, and with a five-fold decrease in k_{46}, is given in Table 3.

Integral to the preceding discussions is the dynamics of NO_3, and the mechanism's capability to predict NO_3 concentrations. The model predicts a rapid rise and fall in the concentration of NO_3 centered about 1900 (PST) as seen in Figure 14. This is the same trend found in experiments by Platt et al. (1980). The peak predicted value of 0.065 ppb also falls in the range of peak values, <0.006 to 0.35 ppb, Platt et al. (1980) found for Claremont, California, the median peak value presented being 0.036 ppb.

TABLE 3

Nitric Acid Production via Chemical Pathways, and
Sensitivity to Revisions in the N_2O_5 Hydrolysis Rate Constant

REACTION PATHWAY	TOTAL PRODUCTION[a] μ eq m^{-3} day^{-1}
NO_2 + OH	0.76
N_2O_5 + H_2O ($k_{46} = 1.5 \times 10^{-5}$ ppm^{-1} min^{-1})	1.11
NO + RO_2	0.13
N_2O_5 + H_2O ($k_{46} = 3 \times 10^{-6}$ ppm^{-1} min^{-1})	1.08
N_2O_5 + H_2O ($k_{46} = 1.5 \times 10^{-5}$ ppm^{-1} min^{-1}, aerosol scavenging of NO_3 included)	0.51
N_2O_5 + H_2O ($k_{46} = 1.5 \times 10^{-7}$ ppm^{-1} min^{-1}, aerosol scavenging of NO_3 included)	0.01

(a) For 27 July 1974, 00:00 to 23:00, averaged over an air parcel 1000 m high.

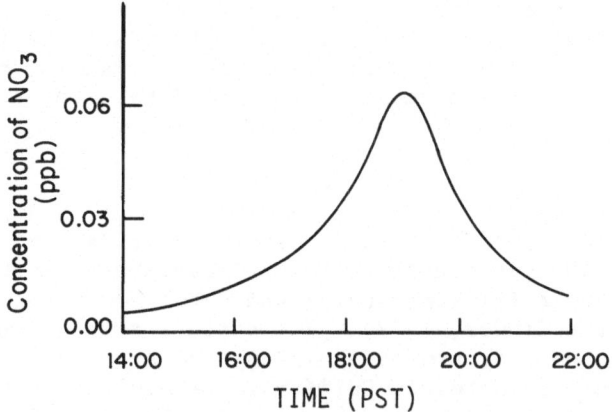

Fig. 14. NO_3 concentration profile

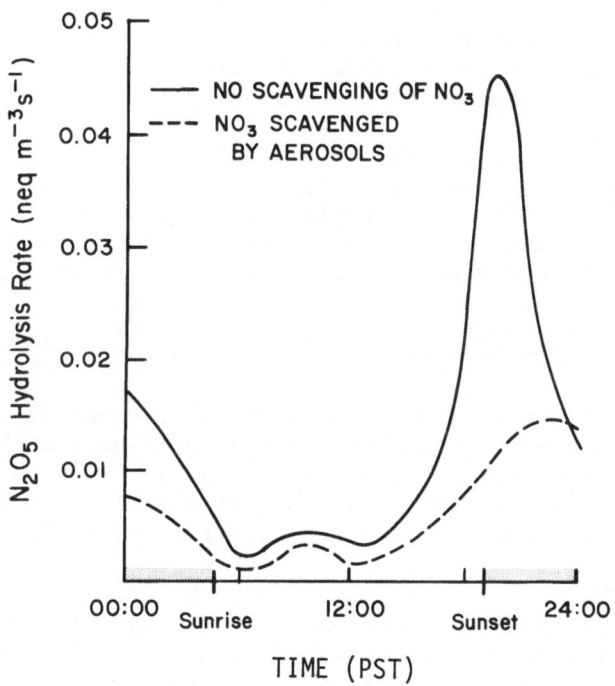

Fig. 15. Diurnal variation in the homogeneous N_2O_5 hydrolysis rate, including aerosol scavenging of NO_3

The mechanism used in the previous analyses, however, did not include the interaction of NO_3 with aerosols. NO_3 is highly reactive, and may decompose when it contacts a surface (Chameides and Davis, 1982). An upper estimate of the scavenging of NO_3 by collisions with aerosol surfaces can be found by using an expression from kinetic theory:

$$\beta = c \ (\frac{RT}{2\pi M})^{1/2} \tag{32}$$

where β is the rate that NO_3 molecules collide with a unit surface area, R is the gas constant, T the temperature, M the molecular weight of the colliding species and c the concentration of colliding species (Friedlander, 1977). The surface area of the aerosol present can be approximated by using the average size distribution of a typical Los Angeles, California aerosol found by Whitby et al. (1972), which has a total surface area of about $900 \ \mu m^2/cm^3$. The nighttime rate expression for NO_3 then becomes

$$\frac{d[NO_3]}{dt} = K_6[NO_2][O_3]+K_7[NO_2][O_3]+K_{45}[N_2O_5]-[NO_3](K_8\cdot[NO]+K_{44}[NO_2]+k^*) \tag{33}$$

where k^* is the rate constant for the destruction of NO_3 by collision with an aerosol particle. For an aerosol with a surface area of $900 \ \mu m^2/cm^3$, k^* is about $3.5 \ min^{-1}$. In deriving this rate constant, it was assumed that the collision efficiency was unity. If aerosol interactions are included the model the peak predicted NO_3 concentration is reduced 50% to 0.033 ppb.

Inclusion of a mechanism to scavenge NO_3 would be expected to influence the N_2O_5 hydrolysis. Figure 15 shows a decrease in N_2O_5 hydrolysis if the effects of aerosol collisions are included. If both aerosol collisions and a two order of magnitude decrease in k_{46} is used, very little nitric acid is produced via this path, as shown in Table 3.

CONCLUSIONS

A vertically resolved, Lagrangian, photochemical trajectory model has been used to study dry acid deposition and the acidification of the atmosphere in an urban air basin. It is seen that the major part of the acid deposition experienced in this area could be due to dry deposition. In the trajectory modeled, 11% of the emitted nitrogen oxides deposit out as nitric acid, and 30% is retained aloft as nitric acid at the end of the day. Nitric acid and PAN are tw of the more abundant oxidation products of NO_x emissions. Ammonia leads to the presence of ammonium nitrate and is available to neutralize deposited acids. Ammonia deposition was only 20% of that of nitric acid on a mole basis along the trajectory studied here.

ACKNOWLEGEMENTS

This work was supported, in part, by a grant from the
Andrew W. Mellon Foundation to the Environmental Quality Laboratory.

REFERENCES

Brutsaert, W., 1975, The roughness length for water vapor, sensible
    heat, and other scalars, J. Atmos. Sci., 32:2028-2031.
Businger, J.A., Wyngaard, J.C., Izumi, Y. and Bradley, E.F., 1971,
    Flux profile relationships in the atmospheric surface layer,
    J. Atmos. Sci., 28:181-189.
Chameides, W.L. and Davis, D.D., 1982, The free radical chemistry
    of cloud droplets and its impact on the composition of rain,
    J. Geophy. Res., 87:4863-4877.
Doyle, G.J., Tuazon, E.C., Graham, R.A., Mischke, T.M., Winer, A.M.
    and Pitts, J.N. Jr., 1979, Simultaneous concentrations of
    ammonia and nitric acid in a polluted atmosphere and their
    equilibrium relationship to particulate ammonium nitrate,
    Env. Sci. & Tech., 13:1416-1419.
Falls, A.H., McRae, G.J. and Seinfeld, J.H., 1979, Sensitivity and
    uncertainty of reaction mechanisms for photochemical air
    pollution, Int. J. Chem. Kin., 11:1137-1162.
Falls, A.H. and Seinfeld, J.H., 1978, Continued development of a
    kinetic mechanism for photochemical smog, Environ. Sci. and
    Tech., 12:1398-1406.
Friedlander, S.K., 1977, "Smoke, Dust and Haze - Fundamentals of
    Aerosol Behavior," John Wiley and Sons, New York, 317 pp.
Liljestrand, H.M., 1979 Atmospheric transport of acidity in Southern
    California by wet and dry mechanisms, PhD Thesis, California
    Institute of Technology, Pasadena, California.
Liu, M-K. and Seinfeld, J.H., 1975 On the validity of grid and trajectory
    models of urban air pollution, Atmos. Environ. 9:555-574.
McRae, G.J. and Russell, A.G., 1982, Dry deposition of nitrogen
    containing species, in Acid Deposition: Wet and Dry, 6,
    ed. B.B. Hicks, American Chemical Society Symposium Volumes
    on Acid Rain (to appear November 1982).
McRae, G.J., Goodin, W.R. and Seinfeld, J.H., 1982a, Development of a
    second generation mathematical model for urban air pollution:
    I model formulation, Atmos. Environ. 16:679-696.
McRae, G.J., Goodin, W.R. and Seinfeld, J.H., 1982b, Numerical solution
    of the atmospheric diffusion equation for chemically reacting
    flows, J. Comp. Phys., 45:1-42.
McRae, G.J. and Seinfeld, J.H., 1982, Development of a second-generation
    mathematical model for urban pollution II: model performance
    evaluation, Atmos. Environ., (in press).
Morgan, J.J. and Liljestrand, H.M., 1980, Spatial variation of acid
    precipitation in southern California, Environ. Sci. and Tech.,
    15:333-339.
Morris, E.D. and Niki, H., 1973, Reaction of dinitrogen pentoxide
    with water, J. Phys. Chem., 77:1929-1932.

Platt, U., Perner, D., Winer, A.M., Harris, G.W. and Pitts, J.N. Jr.,
    1980, Detection of NO_3 in the polluted troposphere by
    differential optical absorption, <u>Geophy</u>. <u>Res</u>. <u>Lett</u>.,
    1:89–92.

Russell, A.G., G.J. McRae and G.R. Cass, 1982, Mathematical modeling
    of the formation and transport of ammonium nitrate aerosol,
    <u>Atmos</u>. <u>Environ</u>., (in press)

Slinn, S.A. and Slinn, W.G.N., 1982, Modeling of atmospheric particulat
    deposition to natural waters, <u>in</u> Atmospheric Pollutants in
    Natural Waters, ed. Steven J. Eisenriech, Ann Arbor Science.

Stelson, A.W., Friedlander, S.K. and Seinfeld, J.H., 1979, A note on
    the equilibrium relationship between ammonia and nitric acid
    and particulate ammonium nitrate, <u>Atmos</u>. <u>Environ</u>., 13:369–371.

Stelson, A.W. and Seinfeld, J.H, 1982a, Relative humidity and
    temperature dependence of the ammonium nitrate dissociation
    constant, <u>Atmos</u>. <u>Environ</u>., 16:983–992.

Stelson, A.W. and Seinfeld, J.H., 1982b, Relative humidity and pH
    dependence of the vapor pressure of ammonium nitrate–nitric
    acid solutions at $25^\circ C$, <u>Atmos</u>. <u>Environ</u>., 16:993–1000.

Wesely, M.L. and Hicks, B.B., 1977, Some factors that affect the
    deposition rates of sulfur dioxide and similar gases on
    vegetation, <u>J</u>. <u>Air</u> <u>Poll</u>. <u>Cont</u>. <u>Assoc</u>., 27:1110–1116.

Whelpdale, D.M. and Shaw, R.W., 1974, Sulfur dioxide removal by
    turbulent transfer over grass, snow and water surfaces,
    <u>Tellus</u>, 26:196–205.

Whitby, K.T., Husar, R.B and Liu, B.Y.H., 1972, The aerosol size
    distribution of Los Angeles smog, <u>J</u>. <u>Coll</u>. <u>Inter</u>. <u>Sci</u>.,
    39:177–204.

SIMULATION OF A PHOTOCHEMICAL SMOG EPISODE IN THE RHINE-RUHR AREA WITH A THREE DIMENSIONAL GRID MODEL

Rainer Stern and Bernhard Scherer

Freie Universität Berlin, Institut für Geophysikalische
Wissenschaften – Fachrichtung Meteorologie –
1000 Berlin 33, Thielallee 50, West Germany

ABSTRACT

An advanced Eulerian photochemical dispersion model is applied
to the Rhine-Ruhr area in the FRG. One of the main features of
this area are the complicated wind conditions induced by the complexity of the terrain. Therefore, special emphasis is given in
the preparation of the three dimensional wind field as input for
the dispersion model. Two methods of wind field generation are
discussed, one based on interpolation of surface winds, the other
on a prognostic mesoscale model. The calculated concentrations of
ozone, NO and NO_2, utilizing both of the wind fields, are compared
with measurements. Preliminary results are presented.

INTRODUCTION

Emission control strategies must be employed to improve ambient air quality under photochemical smog conditions. In order
to assess the value of specific oxidant control strategies, dispersion models are a commonly used tool. Different modeling approaches focus on different aspects of the problem area and hence
incorporate different simplifications of the physical and chemical
processes involved.

One attempt to formulate a relatively simple ozone prediction
relationship, that considers the role of both the precursors hydrocarbons and nitrogen oxides on ozone formation, is the Empirical
Kinetic Modeling Approach (EKMA), proposed by the EPA[1]. EKMA is
based on an isopleth diagram generated by computer simulations of
the chemical reactions among pollutants in a well mixed box. An

application of EKMA to the Cologne area[2] showed that the main
assumption of EKMA -- instantaneous mixing of all pollutants within
the reaction volume -- is strongly violated in this area. The
limited horizontal and vertical resolution of the box model approach underlying EKMA can lead to erroneous control strategies
in regions having spatially inhomogeneous emission patterns.

Therefore it became desirable to investigate the use of a more
physically sophisticated photochemical dispersion model for the
Cologne area. For this purpose an Eulerian grid model the SAI-Airshed model[3],was selected. This model was also recently applied to
the Netherlands[4].

This paper presents some preliminary results of the model
application and, in particular, focuses on differences in oxidant
predictions that result from using two different wind field generators to produce the required SAI-Airshed input. The wind field
generators, one based on interpolation of observations and the
other prognostic, are described. The spatial and temporal variation characteristics of the resulting wind fields are discussed in
the light of the differing oxidant predictions and are compared
with measurements at both rural and urban sites.

DESCRIPTION OF THE AREA AND EPISODE CONSIDERED

The Cologne area is highly industrialized with emissions
mainly from chemical and petrochemical sources. The next adjoining
large urban-industrial area is Düsseldorf, situated about 40 km to
the north and adjacent to the southern edge of the Ruhr area, one
of the largest industrial agglomerations in Europe. Because of
this very proximity of large industrialized areas, the modeling
domain was enlarged to the region shown in Fig. 1 to ensure, that
all major emissions are inside the domain and thus, must not be
handled through boundary conditions. Ozone measurements and NO_x
measurements were available only at a few stations: Two rural
stations, Michelsberg and Oelberg both situated south of Bonn,
and four urban stations,Eifelwall (E) and Godorf (G) in the Cologne
area, and University of Bonn (INST) and Venusberg (V) in and near
Bonn, respectively (see Fig. 1). Flight measurements provide some
additional ozone and NO_x data for several oxidant episodes.

After careful review of the data,the photochemical smog episode
of June 22 to 27,1976 was selected for simulation.During this episode,
a high pressure system was situated over northern Europe.Its center
moved rapidly eastward and a high pressure ridge developed, extending from the Azores to Russia. The wind flow over the Rhine-Ruhr
area changed from westerly to northwesterly on June 22,followed by a
weakening of the winds on June 23 and 24.Finally,a steady easterly
flow with moderate winds developed on June 25.A more detailed description of the selected episode is given elsewhere [5,6]. June 25

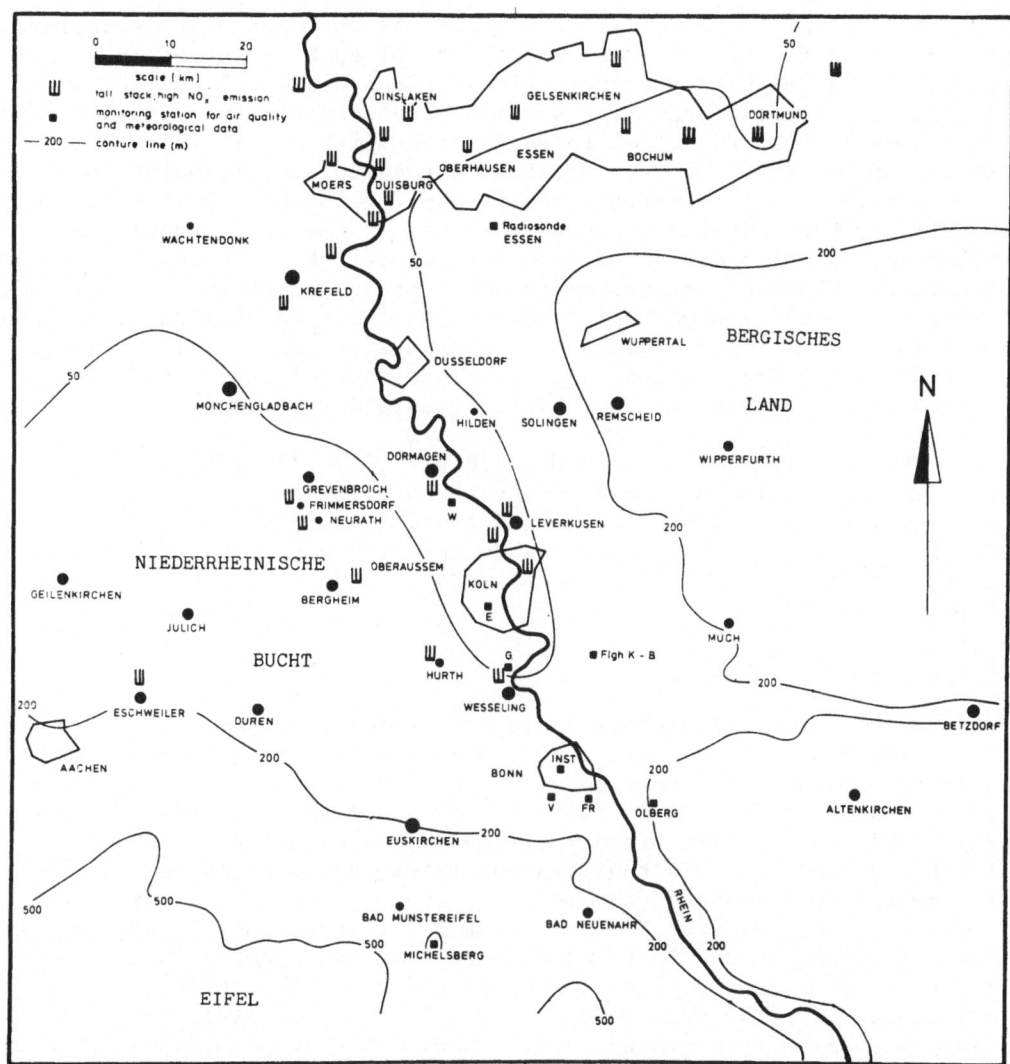

Fig. 1. The modeling region including Bonn, Cologne (Köln) and the Ruhr area in the north. Isolines indicate height of topography above sea level.

was selected for the first model evaluation runs because of the
smoothness and relative simplicity of the flow pattern.

MODEL DESCRIPTION AND INPUT PREPARATION

     The basis of the SAI Airshed model[3] is the continuity equation,
which expresses the conservation of mass of each pollutant in a turbulent fluid in which the chemical reactions occur. Turbulent diffusion is parameterized using gradient-transfer theory. The solution
of the horizontal advection terms is accomplished using a numerical
scheme of the SHASTA type. Vertical advection and diffusion are
treated with an implicit finite difference technique. Pollutant removal processes are incorporated through the use of a deposition
velocity, which is estimated on a cell-by-cell basis using resistance
modeling. Chemical transformations are presented by the carbon-bond
mechanism, a 42-step mechanism which describes the chemical reactions
of organics, nitrogen oxides, ozone,and sulfur oxides. Its special
feature is the splitting of organics into four classes based on
carbon bond characteristics, reactivities,and reaction products.

     The modeling region of 160×176 km^2 (Fig. 1) is divided into
22×20 grid cells of 8×8 km^2 in the horizontal and 5 cells in
the vertical, extending from ground level to a height of 2600 m.
The individual cell height varies dynamically according to the height
of the mixing layer. The temporal resolution of all inputs is one
hour.

Meteorological input

     Most of the meteorological input was derived from routine ob-
servational data. The height of the mixing layer, the so-called
diffusion break, was constructed using temperature soundings from
radiosonde data from Essen (see Fig. 1) and SODAR data from Cologne.
Fig. 2 gives the time history of the estimated diffusion break
height for June 25. In the less populated sections, as well as in
the mountainous areas of the modeling region, the mixing depth was
lowered during the night to account for the increased cooling and
reduced mixing of the lower atmosphere that is expected in rural
regions. Temperature gradients above and below the inversion were
calculated from the radiosonde data. The NO_2-photolysis rate constant was taken from ground level measurements at Bonn. Near ground-
level temperature values were obtained from the Cologne airport data
and assumed constant over the entire modeling region. Atmospheric
pressure, water concentration and stability class of the atmosphere
were also kept spatially constant.

     The generation of the mean horizontal wind fields is discussed
later. The vertical wind component is calculated internally in the
model from the wind continuity relationship.

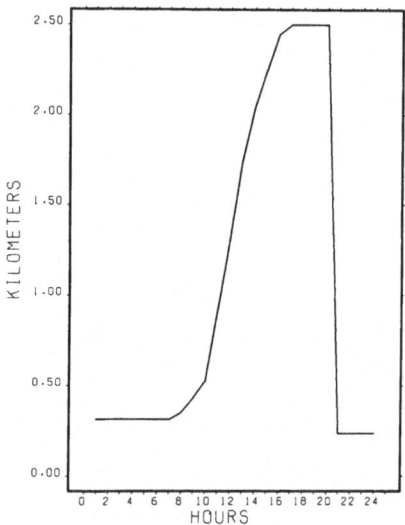

Fig. 2. Height of the mixing layer (diffusion break) for June 25

Emission input

Emissions are treated as volume sources, that is, instantaneous mixing throughout a grid cell is assumed. All emissions of area sources are injected into the lowest layer of grid cells, whereas emissions of point sources are injected into the appropriate grid cell depending on the total effective plume height calculated using the plume rise formulas of Briggs[7] for bent-over plumes.

The emission inventory includes emissions of NO_x, CO, SO_2, and the four carbon bond categories of hydrocarbons. A summary of the emissions is presented in Table 1. To limit the computational burden only 97 effective point sources were considered. This relatively low number was obtained by combining stacks, having similar characteristics and lying within a single grid cell, to form one effective point source. Each of the six source type categories shown in Table 1 are subject to a characteristic diurnal time dependence, which is assumed valid for the entire region.

Inititial and boundary conditions

Initial and boundary conditions had to be prescribed for all species at the ground and aloft. There is a substantial factor of uncertainty in the determination of these values, because of a lack of data, especially for the hydrocarbons.

Table 1. Total emissions in the modeling region and percentage of
         point and area sources. NO_x emissions are divided by weight
         into 90 % NO and 10 % NO_2.

	Total emissions in t/24 h	Point sources	area sources Industry Petrochemical plants Traffic Domestic Natural
		%	%
NO_x	2603	65	35
SO_2	5315	90	10
CO	10065	53	47
Paraffins	710	–	100
Olefins	174	–	100
Aromates	413	–	100
Aldehydes	81	–	100

Initial conditions were estimated from the available data at
0000 CET on June 25, the starting time of the simulation. All
available information about air quality, especially for remote
areas,was used to characterize the conditions at the lateral and
top boundaries of the modeling domain. The right choice of boundary conditions is an important factor for good model performance
since the model is quite sensitive to inflow hydrocarbons concentrations[8]. The fact that the inflow mass of the highly reactive
olefins, based on the chosen boundary values, is comparable with the
mass emitted within the modeling area, dramatizes this problem. For
the less reactive hydrocarbons, the inflow mass is a factor 3 to 5
times greater than the emitted mass.

GENERATION OF THE WIND FIELDS

Due to the complexity of the terrain within the modeling region and the scarcity of information on surface and upper level
winds, one of the most difficult inputs to specify is the wind field.
Two methods were applied to produce gridded wind inputs :
- An interpolation scheme, and
- a dynamical wind prediction model.

The interpolation scheme

The ground level gridded wind components at each grid point
were calculated from winds measured nearby the grid point corrected
by a weighting factor. To accomplish this, different influence areas
were defined for each wind station with the constraint that each
grid point was influenced by at least one of the wind stations. The
size of each influence area around a wind station was subjectively
determined on the basis of the density of stations in its neighbour-
hood and on the representativeness of its measured wind value given
on the surrounding terrain features. The weighting factor had the
following form :

$$ w_{i,j}^{k} = \left[\frac{R^2(k) - r^2(k)_{i,j}}{R^2(k) + r^2(k)_{i,j}} \right]^{1.5} \qquad \text{for } R(k) > r(k) $$

$$ = 0 \qquad\qquad\qquad \text{for } R(k) \leqslant r(k) $$

with k   wind station index
     R(k)   radius of influence for station k
     $r(k)_{i,j}$ distance from station k to gridpoint i,j

The power 1.5 was found to minimize wind field discontinuities introduced by the interpolation procedure. Twenty wind stations, in an
area somewhat larger than the modeling area, were used for the crea-
tion of the gridded hourly surface wind fields.

The 6 hourly radiosonde ascents at the Essen station were the
basis for the calculation of the winds aloft. For the hours between
radiosondes, wind profiles were constructed using linear time interpolation. The spatial variations exhibited in the hourly surface
winds were allowed to decrease with height, resulting in a uniform
wind field above the mixing layer. The shear of the wind profile
served as a weighting factor for the turning of the wind above each
grid point. Wind speed and direction above the mixing layer were
taken from the respective profile values.

The dynamical wind prediction model

To provide more physically consistent descriptions of the complex three-dimensional flow, the University of Virginia mesoscale
model, UVMM, was used. This model, originally developed to simulate
sea breezes, was recently adopted for flow simulations over irregular
terrain[9]. The UVMM is a hydrostatic, prognostic three-dimensional
mesocale model based on the solution of the mass, momentum and energy
balance equations. It employs a terrain following coordinate system
and uses K-theory to parameterize turbulent exchange processes. In
this study, the model has been slightly modified. Surface energy
budget calculations, based on short and long wave radiation contri-

butions, were excluded and surface heat balance was simply prescribed using surface temperature curves for each of five different surface characteristics. In addition, the geostrophic wind was allowed to vary in time. The model was run for June 25 starting in the late afternoon of June 24, with a spatial resolution equal to that of the dispersion model.

Comparison of the calculated wind fields

Figures 3 and 4 show both the interpolated (a) and the prognostic (b) wind fields for two levels and three characteristic hours of the diurnal boundary layer variation : 3 to 4 CET, the stable boundary layer before sunrise, 13 to 14 CET, the well mixed boundary layer and 22 to 23 CET, the stable boundary layer after sunset. Fig. 3 depicts the surface layer windfield (level 1). Fig. 4 presents the flow in level 2, defined as the center height of the second vertical grid layer. The height of this layer varies in time according to the rise and fall of the diffusion break, from about 125 m in the night to 450 m in the afternoon. Fig. 3 also shows the wind observations within the modeling area.

The flow pattern of both ground level wind fields is similiar in the morning and the afternoon (Fig. 3). However, the channeling of the Rhine valley is more pronounced in the prognostic wind field because the interpolated wind field in this area is based on only two wind stations. Also, in the early afternoon, the mesoscale model predicts a rather smooth flow with an average wind direction from the east, whereas the interpolated field shows a shift towards a more north-easterly wind direction in the western part of the model's domain. In the night there are substantial differences in the two flow pattern, especially in the southwest. The terrain effects are much more pronounced in the prognostic wind field whereas the interpolated field has a quite smooth structure with rather high windspeeds. This clearly indicates the weakness of all interpolation techniques when used in sparse data regions. In this case, the wind pattern over the Eifel mountains is determined by only one station, not necessarily representative for the entire region. On the other hand, due to the lack of data, nothing definite can be said about the accuracy of the predicted wind field.

The flow structure of the winds in layer two (Fig. 4) of both interpolated and predicted wind fields is, as expected, much smoother than the structure of the respective ground level fields. Terrain effects have nearly vanished, especially in the prognostic field during the night, where the low, extremely divergent surface winds in the Eifel mountains have turned with height into a nearly homogeneous easterly flow (Fig. 3b and Fig. 4b).

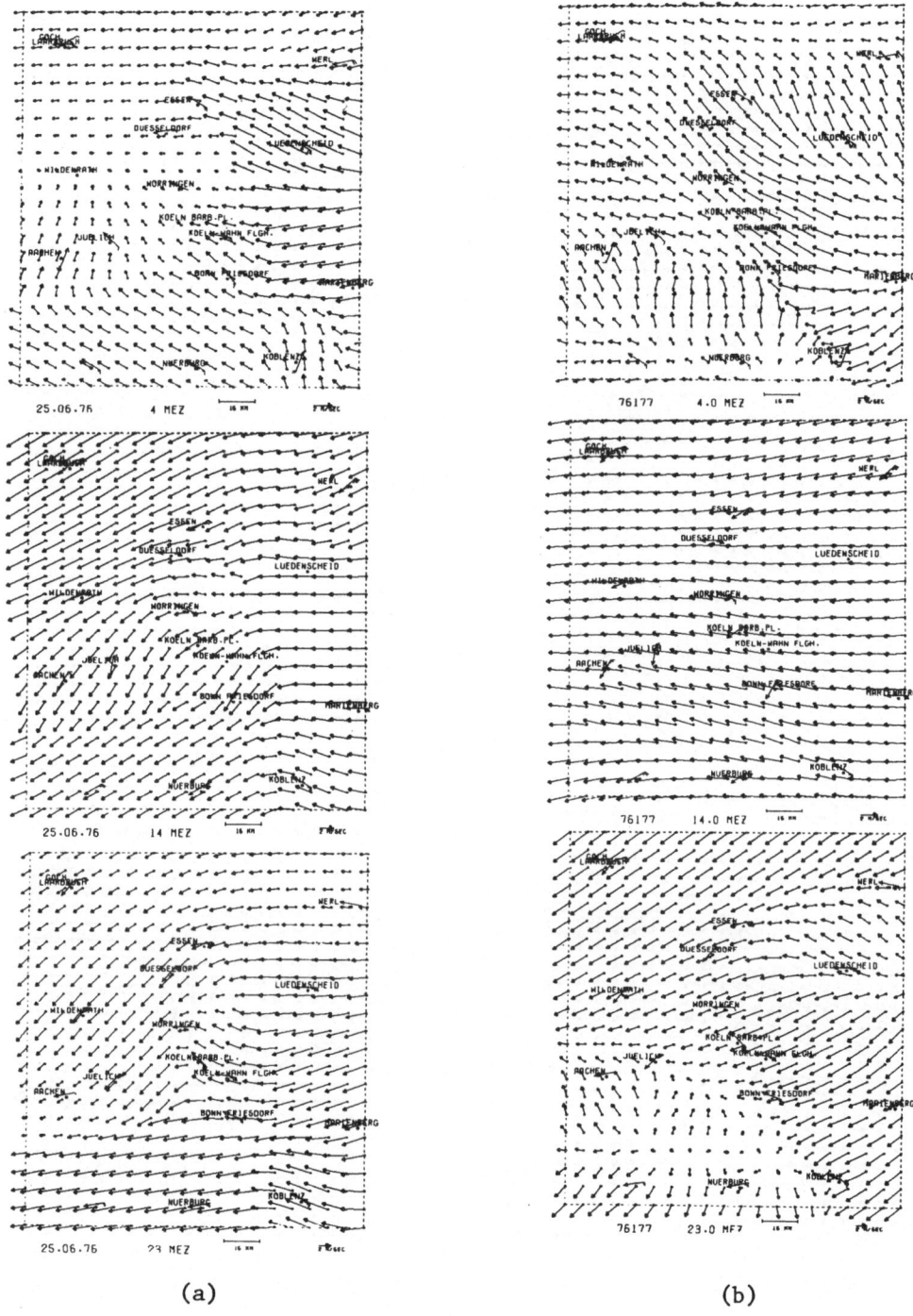

(a)   (b)

Fig. 3. Interpolated (a) and prognostic (b) wind field at ground
          level for three selected times on June 25

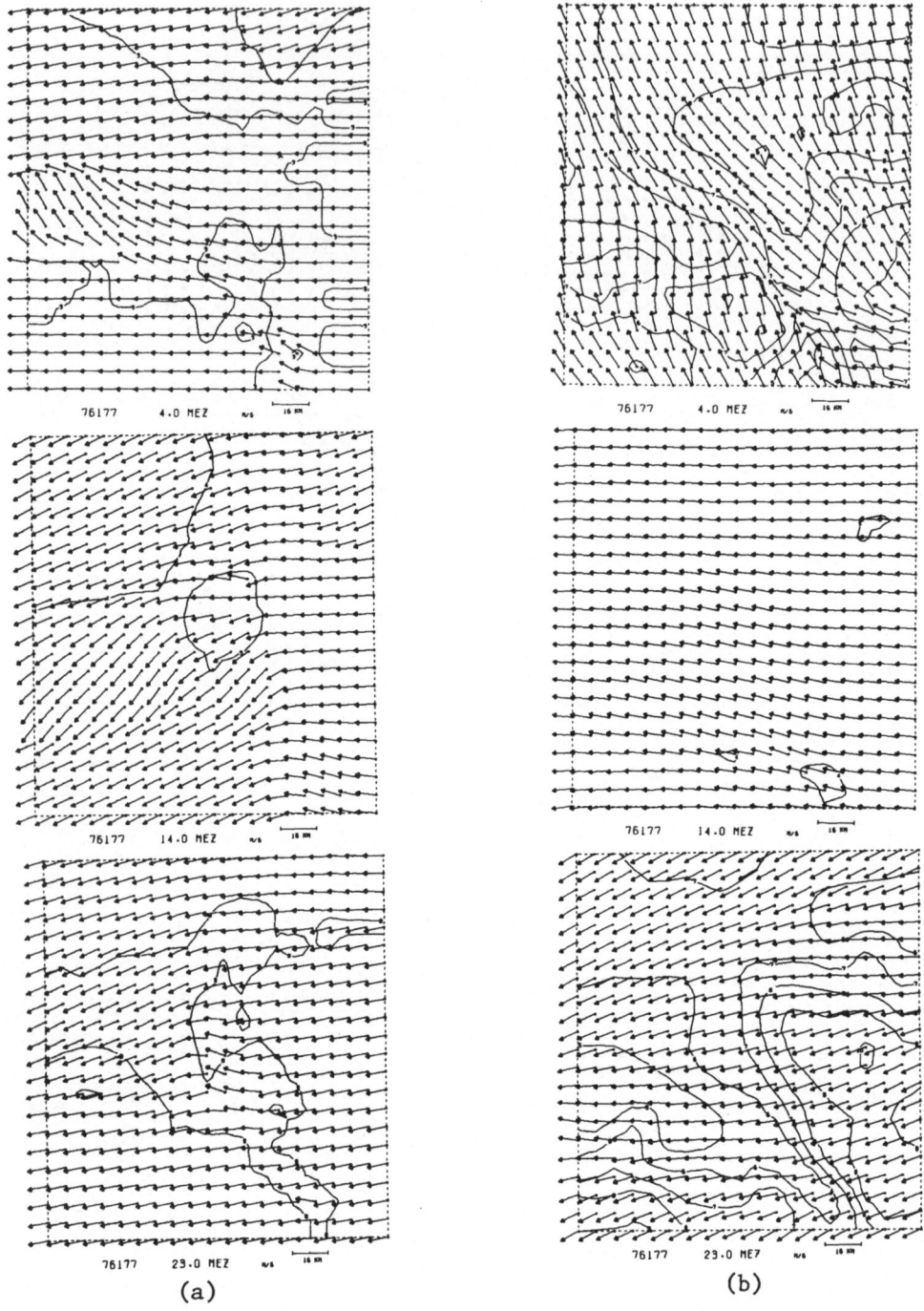

(a)   (b)

Fig. 4. Interpolated (a) and prognostic (b) windfield at level 2
         for three selected times.

In comparing the predicted winds with the observations it should be kept in mind that the mesoscale model calculates volume averaged winds on a large grid size, whereas many of the observations reflect local wind patterns, that cannot be resolved on the 8 x 8 km^2 scale. Averaged over all simulation hours and all vertical levels, the mean wind speed of the predicted flow is considerably higher than the speed of the interpolated fields.

SIMULATION RESULTS

Two simulations were carried out, one utilizing the interpolated wind field (hereafter referred to as run 1), the other the wind field predicted by the mesoscale model (run 2). All the other inputs were the same for both simulations.

Figures 5 and 6 show the predicted ozone concentrations compared with the observations for four urban and two rural stations, respectively. With correlation coefficients of .77 and .82 and standard errors of estimations of 13.2 ppb and 15.0 ppb for all data pairs of run 1 and run 2, respectively, the predicted concentrations are in a rather good agreement with the observations.

At all urban stations (Fig. 5) the ozone peak was observed to last for at least eight hours with concentration fluctuations superimposed on the main peak. While the times of ozone build-up and destruction are similar predicted in both runs, differences appear during the time of maximum ozone levels. In run 1 concentration fluctuations occur similar to those observed. In run 2 only smooth curves are calculated having single maxima. The fluctuations of the observed ozone levels may be explained with precursor transport controlled by the local wind field. Thus, it is more likely that run 1, based on the local wind observations, is able to reproduce these transport related concentration fluctuations. The calculated ozone levels at the Bonn station are too high in both runs during the night and the early morning. This can be attributed to the averaging properties of a grid model. Initial ozone concentrations, calculated as a weighted average of urban and rural land use in each grid cell, are rather high for the cell in which the Bonn station is situated because the small urban area of Bonn is not resolved by the grid. This leads to the overestimation at night and may be also one reason for the excessive peak value. The urban area of Cologne with the stations Godorf and Eifelwall is better resolved. Thus, calculated ozone levels at night are in better agreement with the observations. In Fig. 7 observed and predicted NO and NO$_2$ concentrations are shown for the station in Bonn. Calculated concentrations of both runs fail to reproduce the measured values. This, in fact, reflects again the difficulties of comparing point measurements with calculated box averages. Unfortunately, all available NO$_x$ measurements are influenced by local sources, mainly traffic. Therefore, no model evaluation for the calculated NO$_x$ is attempted.

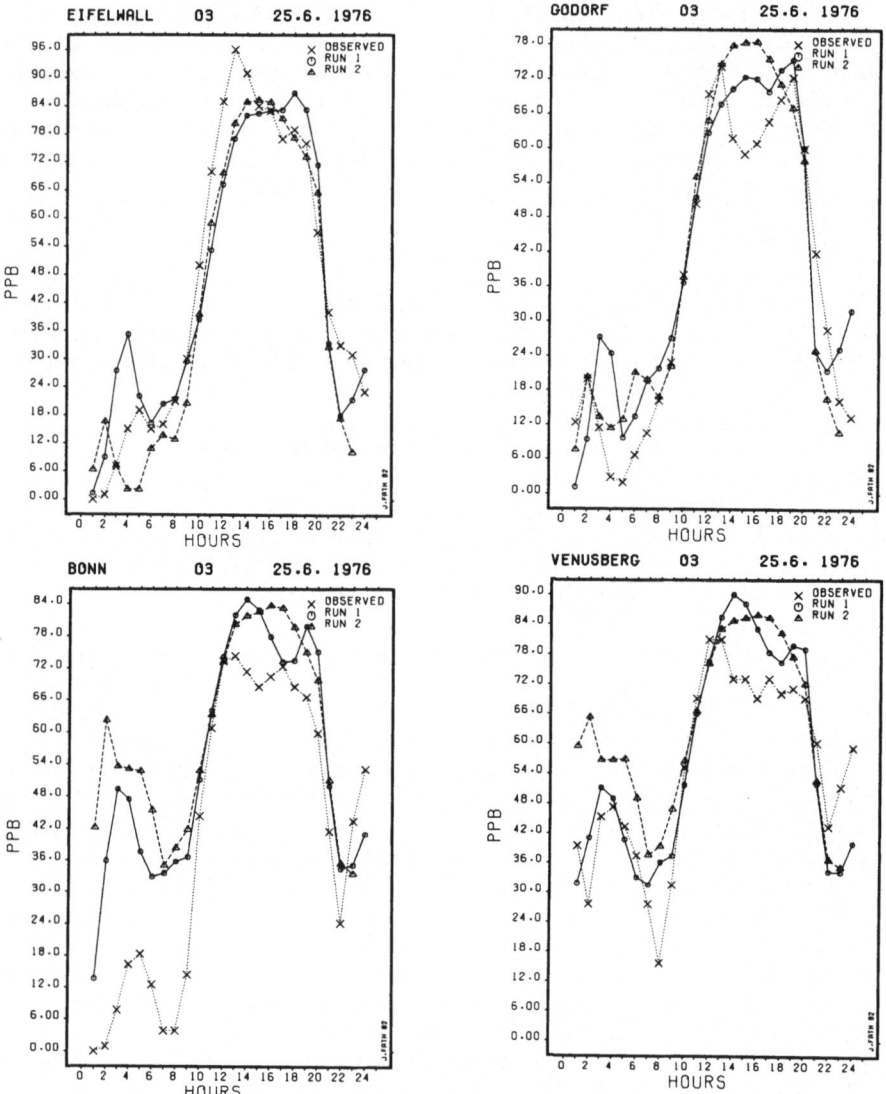

Fig. 5.   Computed and observed ozone concentrations at four urban
          stations. Run 1 is based on the interpolated, run 2 on
          the predicted wind field.

        At the rural stations Ölberg and Michelsberg (Fig. 6), ob-
served ozone concentrations remain at rather high levels (50 to
80 ppb) at night. However, the calculated ozone level in run 1
drops to about 30 ppb during the night, well below the observed
concentrations. In run 2 rural ozone predictions agree more closely with observations.

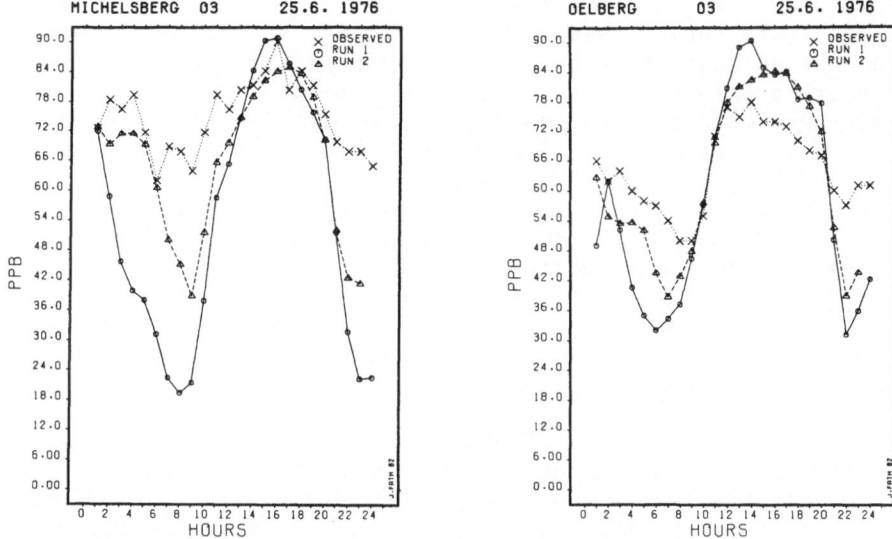

Fig. 6.   Computed and observed ozone concentrations at two rural
          stations. Run 1 is based on the interpolated, run 2 on
          the predicted wind field.

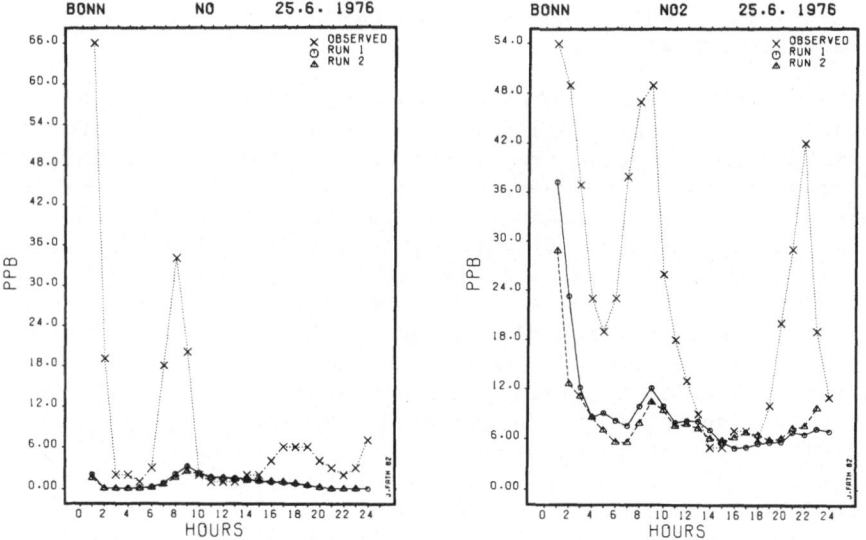

Fig. 7. Computed and observed NO and NO_2 concentrations at the Bonn
        station. Run 1 is based on the interpolated, run 2 on the
        predicted wind field.

The wind fields show remarkable differences both in speed and direction in these rural areas during nighttime, but nothing can be said about the accuracy of either of the flow patterns because no measurements are available in the vicinity of the two rural stations. The closer agreement of run 2 may be the result of a better simulation of the transport or the velocity dependent removal processes. Nevertheless, the underprediction in this area is not completely understood. Also the model's instantaneous spread of emissions, which in this area are completely due to traffic, can lead to an overestimation of the ozone destruction by NO during the night.

Calculated peak ozone values show a tendency toward overestimation at all stations except Eifelwall. However, no final conclusion should be drawn from the results of these first evaluation runs.

One characteristica of photochemical smog formation is the increasing oxidant concentration downwind of the precursor source area. Fig. 8 illustrates the spatial growth level distribution of predicted ozone concentrations for both runs at the time the highest concentrations are calculated. Two ozone plumes are clearly evident about 50 to 70 km to the west and southwest of the maximum hydrocarbon emission area in the center of the modeling domain. Peak ozone concentrations are considerably higher in run 1, which utilizes the interpolated wind field. The prognostic wind field of run 2 is characterized by higher wind speeds and therefore by faster advection of pollutant loaden air parcels. Thus, it is possible that maximum ozone concentrations may form outside the modeling domain. Unfortunately, no ground level measurements are available in the downwind area. Thus, nothing can be said about the accuracy of the predicted concentrations.

Air quality measurements taken aloft using an instrumented aircraft were obtained during the afternoon on June 25. For lack of time no comprehensive comparison can be presented in this paper. A first analysis of the measured ozone concentrations shows ozone levels of 140 ppb with maxima above 150 ppb in 1500 to 2000 m, approximately 50 km downwind of the Cologne area. Fig. 9 illustrates the calculated horizontal ozone distribution for both runs in 2000 m. The structure of the fields is similar to the respective ground level distributions (Fig. 8). Ozone levels above are somewhat higher than at the ground. However, computed ozone maxima in both runs are well below the flight measurements with higher values for run 1. This again can be interpreted as a result of the lower wind speed of the interpolated winds which allow maxima ozone formation inside the modeling region. The flight measurements indicate that the predicted ozone plumes are a few grid cells too far in the south. This discrepancy may indicate that wind directions aloft are actually more directed to the west than those input to the model. However, wind speeds aloft seem better represented in run 1.

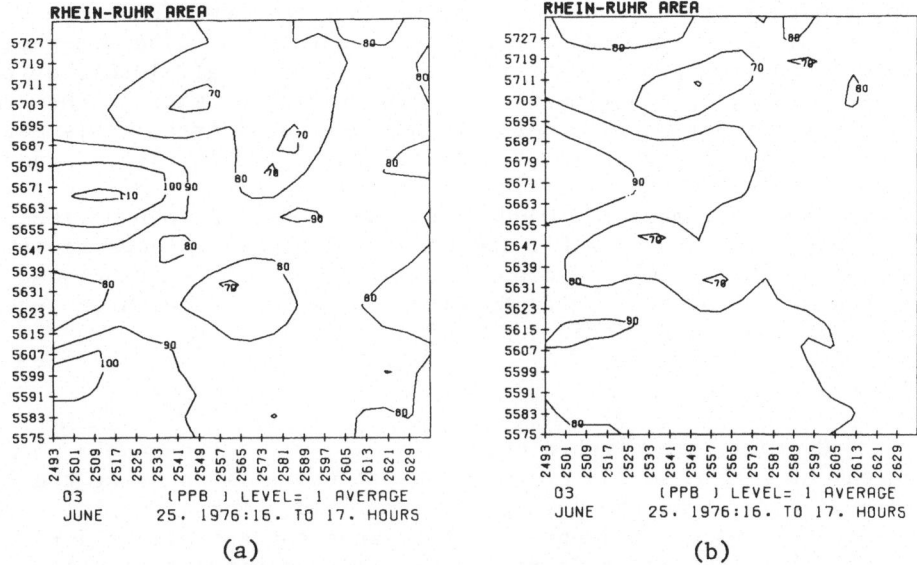

Fig. 8. Calculated horizontal ozone distribution at ground level for run 1 (a), based on the interpolated wind field, and run 2 (b), based on the predicted wind field.

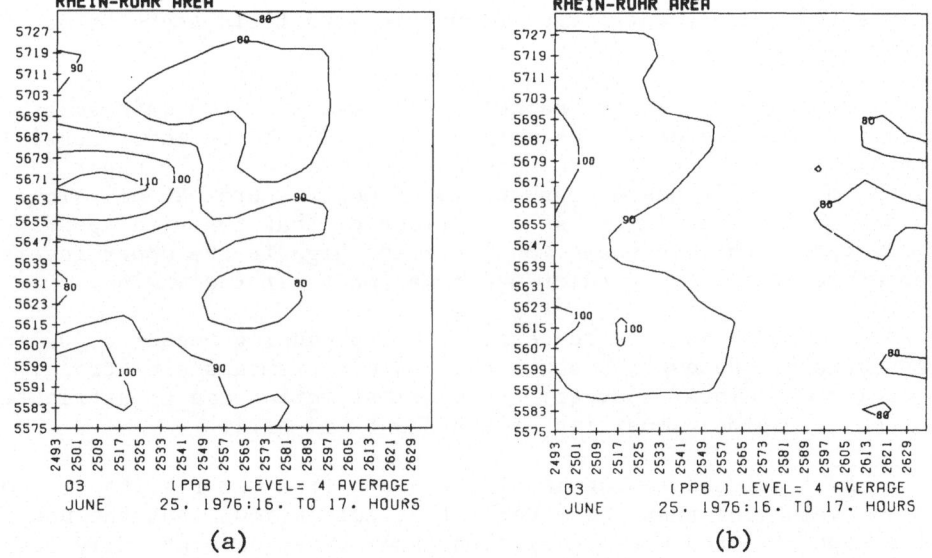

Fig. 9. Calculated horizontal ozone distribution in 2000 m for run 1 (a), based on the interpolated wind field, and run 2 (b), based on the predicted wind field.

The tendency of the model to underpredict the downwind ozone maxima
aloft may be attributed to the following reasons. The NO_2-photolysis
rate taken from ground measurements was used at all levels of the
model.   Thus, attenuation of the solar radiation with height due
to light scattering from aerosols was not considered. This may
lead to a too low photolysis rate for the upper levels. In addition, the instantaneous mixing of emissions into the grid cells
tend to overestimate the O_3-destruction by NO. As can be recognized
from Fig. 1, a considerable number of major power plants is situated
upwind of the calculated ozone maxima. Thus, the instantaneous
spread of the plant's large NO emissions into the cells can create
the erroneous ozone deficit downwind.

SUMMARY

     Overall, the Airshed model results are in reasonable good
agreement with measured ozone values at urban sites using the interpolated as well as the predicted wind field. For the rural sites
the model tends to underpredict the ozone concentrations during the
night. The pattern of diurnal variation was simulated quite well
at all stations. Calculated peak ozone concentrations show a tendency toward overprediction at all stations except the urban site
Eifelwall.

     In the areas where the interpolated wind field is based on
only a few measurements, the prognostic wind field seems to do the
better job.

     Ozone maxima aloft, calculated downwind of the main emission
areas, are predicted approximately 30 to 40 ppb too low. Possible
explanations are a too low NO_2-photolysis rate in the upper levels
or the model's instantaneous spread of point source NO emissions
into a grid cell. There is some evidence, that the wind speeds
predicted by the mesoscale model are too high in the upper levels.
Therefore, run 2 may predict ozone maxima too far downwind.

     The model is less successful in reproducing measured NO_x con-
centrations. However, because the NO_x measurements are strongly
influenced by local sources, the underestimation can be attributed
to the averaging properties of the $8 \times 8 \text{ km}^2$ grid.

     No final conclusion should be drawn from the results of these
first evaluation runs. However, it is not obvious that the use
of a sophisticated wind prediction model gives automatically better
simulation results. This must be seen in view of the uncertainties
related to the other input data, especially boundary conditions and
emissions. In addition, it has to be mentioned that the preparation of the prognostic wind fields required much greater resources
in terms of labor and computer requirements than the generation
of the interpolated winds.

More evaluation and sensitivity studies, especially with regard to the boundary conditions, will be carried out in the future. Once the assessment of the accuracy and precision of the model's predictions is determined, the model will be used for control strategy testing and development.

ACKNOWLEDGEMENT

This study was supported by the Umweltbundesamt.

REFERENCES

1. EPA, Uses, Limitations and Technical Basis of Procedures for Quantifying Relationships between Photochemical Oxidants and Precursors, EPA-450/2-77-21a - November 1977.
2. Stern R., and Scherer B., An Application of the Empirical Kinetic Modeling Approach to the Cologne Area. 11th Technical Meeting on Air Pollution and its Application, Amsterdam 1980.
3. Reynolds S.D., et al., An Introduction to the SAI Airshed Model and Its Usage. Systems Applications Inc. - March 1979.
4. Builtjes P.J., et al., Application of a Photochemical Dispersion Model to the Netherlands and its Surroundings. 11th Technical Meeting on Air Pollution and its Application, Amsterdam 1980.
5. Scherer B., and Stern R., Untersuchung einer Photochemischen Smog-episode in Raume Köln-Bonn, Institut für Geophysikalische Wissenschaften der FU Berlin, November 1980.
6. Scherer B., and Stern R., Analysis of a Photochemical Smog Episode and Preparation of the Meteorological Input Data for a Three-dimensional Air Quality Dispersion Model, Proc. 2nd European Symposium Physico-Chemical Behaviour of Atmospheric Pollutants, 29 september-1 october 1981, Varese, Italy.
7. Briggs G.A., Plume Rise Predictions, in : Lectures on Air Pollution and Environmental Impact Analysis, AMS 1979.
8. Seigneur C., et al., Sensitivity of a Complex Urban Air Quality Model to Input Data. J. of Appl. Met., Vol. 20, September 1981.
9. Mahrer Y., and Pielke R., Numerical Simulation of Air Flow Over Irregular Terrain, Beitr. z. Physik d. Atm. 50, 1977.

OXIDANT FORMATION DESCRIBED IN A THREE LEVEL GRID

Knut Erik Grønskei

Norwegian Institute for Air Research
P.O. Box 130
N-2001 Lillestrøm, Norway

INTRODUCTION

In Grenland, Norway, a combination of industrial emissions
and urban areas may cause high oxidant episodes. To clarify the
local oxidant formation potential within the area, a model
describing both photochemical and dispersion effects was developed. This paper describes the results of a cooperation
between Institute of Geophysics, University of Oslo, and Norwegian
Institute for Air Research (NILU) (1).

Different spatial scales had to be considered, i.e., effects
on the regional scale were taken into consideration by the definition of boundary values of a grid model (each level consisting
of 32x16 gridpoints with grid distance 1 km). Large point-sources
were considered by a puff model in the grid system. Individual
puffs were tracked until their spatial scale became comparable
with the grid distance horizontally or vertically. The mass of
pollutants was then added to the average value for the grid element where the puff was located. To differentiate between high-
level and low-level sources, 3 levels are used for vertical
resolution below the mixing height. A surface layer model are
used to account for the profiles close to the surface and for
dry deposition.

MODEL DESCRIPTION

General description

The dispersion is partly caused by the time variation of
the wind field and partly by turbulent exchange. For each of

the pollution components, the following equation is solved:

$$\frac{\partial c_i}{\partial t} = \vec{v}_h \cdot \nabla_h c_i - \frac{\partial (w'c_i')}{\partial z} - \nabla_h \cdot (\vec{v'c_i'}) + R_i + Q_i \qquad (2.1)$$

$$\quad\text{I} \qquad\qquad \text{III} \qquad\qquad \text{IV} \qquad \text{V} \quad \text{VI}$$

Processes considered in this model are:

I   horizontal advection, $\vec{v}_h \cdot \nabla_h c_i$

II   vertical advection, $w \frac{\partial c_i}{\partial z}$ (assumed to be small and neglected in Eq. (2.1))

III   turbulent exchange vertically
IV   turbulent exchange horizontally
V   chemical reactions (sinks or sources)
VI   emission from natural and anthropogenic sources

The time derivatives caused by advection, turbulent exchange and by emissions (Q_i) are calculated separately. These effects are added to the time variation caused by the chemical reactions and new concentrations are calculated. In this way the time integration follows a method proposed by E. Hesstvedt et al. (2). The calculation of vertical exchange follows a method proposed by Pasquill (3).

Advection

Wind measurements close to the ground are used to define a two-dimensional stream function (ψ) along the lateral boundary of the lowest grid level. A mass-consistent approximation of the horizontal windfield is defined by solving:

$$\nabla^2 \psi = 0 \qquad (2.2)$$

From a dynamical point of view, this windfield will change only as a function of the lateral boundary conditions in the constant flux layer when the vertical exchange of sensible heat and momentum are horizontally homogeneous and quasi-stationary.

The dispersion effect of deviations from this horizontal wind field is taken into account by the description of horizontal turbulent exchange.

Vertical exchange

A finite difference approach to estimate vertical exchange, based on K-theory, requires numerical simulation in many atmospheric layers. To avoid the labour and cost of detailed numerical solutions, Pasquill (3) has proposed a "local similarity" treatment of vertical spread from a ground source.

The basic principle is to express the rate of dispersion, $\frac{d\bar{z}}{dt}$ (where $\bar{z}$ is the mean displacement of an ensemble of particles after a given travel time, t) in terms of basic parameters of the turbulent boundary layer. The assumption is made that the increase in vertical spread, $\bar{z}$, or alternatively z_m (the extreme vertical displacement), is always determined by local properties.

By using detailed numerical dispersion calculations, the horizontal rate of dispersion ($\frac{d\bar{z}}{dt}$) is estimated as simple power function that apply from unstable conditions (Monin Obukhov length: L = -7 m) to moderately stable conditions (L = 4 m) :

$$\frac{d\bar{z}}{dx} = a\left(\frac{K_z}{u\bar{z}}\right)^b \qquad (2.3)$$

where a = 0.95; b = 1.06.
K_z: vertical turbulent diffusivity.

Three layers are considered in the calculations: one layer for the emissions of low level sources, and two upper layers for medium height and high level sources (Table 2.1).

Measurements of temperature stratification with height will be used to determine the mixing height H_{mix}.

Table 2.1 : The vertical structure of the three-level model.
When the mixing height (H_{mix}) is larger than 200 m, this height is used as the upper limit of Layer 3.

z	Turbulent exchange with the background atmosphere
H_{mix} —— (m) ——	
Layer 3	$\Delta z_3 = \frac{H_{mix} - \Delta z_1}{2}$, when $H_{mix} \geq 200$ m
Layer 2	$\Delta z_2 = \frac{H_{mix} - \Delta z_1}{2}$
Layer 1	$\Delta z_1 = 50$ m Emissions from low level sources
	Dry deposition. Emissions from natural sources

It is assumed that the air layer between the first and second level grows according to Equation 2.3, causing an exchange of pollution. The exchange will be proportional to the difference in concentration between the levels. The third layer also exchanges pollution with the background air above the mixing height. The following fluxes are calculated:

F_0 : Dry deposition to the ground

F_{1-2}: $(\frac{d\bar{z}}{dt})_{1-2}$. (C_2-C_1) from Layer 1 to Layer 2

F_{2-3}: $(\frac{d\bar{z}}{dt})_{2-3}$. (C_3-C_2) from Layer 2 to Layer 3

F_{3-b}: $(\frac{d\bar{z}}{dt})_{3-b}$. (C_b-C_3) from Layer 3 to the background atmosphere

The turbulent diffusivity of pollution, K_z, is estimated by using the turbulent diffusivity of heat, K_h, and the following formula is used:

$$ (2.4) $$
$$ K_z = K_h = 0.35\ u_* z/\phi_h(z/L) $$ (Standard meteorological notation is used)

Measurements of wind and temperature at two levels are used as input data, and the caracteristic parameters of the surface layer, u_*, Θ_*, and L, are determined by an iteration process as suggested by Busch et al. (4). Businger's empiracally determined universal functions (5) are used.

Horizontal exchange

Measurements of horizontal wind fluctuations are used to estimate horizontal mixing. The standard deviation of these fluctuations (σ_Θ) is used with the wind speed and the grid distance to define a horizontal turbulent exchange coefficient, D, describing the dispersion effect of turbulence elements smaller than the grid distance, Δx. The following formula is used:

$$ \frac{\partial}{\partial x}(u'c_i') + \frac{\partial}{\partial y}(v'c_i') = D\left(\frac{\partial^2 c_i}{\partial x^2} + \frac{\partial^2 c_i}{\partial y^2}\right) \qquad (2.5) $$

where:
$$ D = \sigma_\Theta^2\ u(z)\Delta x. $$

Near point sources the dispersion is described by following the movements of particles. The horizontal dimension of the puff increases according to the formula :

$$\sigma_y(t + \Delta t) = \sigma_y(t) + \Delta t \ . \ u \ . \ \sigma_\Theta \ . \ B \qquad (2.6)$$

$$B = 1.0, \text{ when } \sigma_y < 100 \text{ m}$$

$$B = 0.5, \text{ when } \sigma_y \geq 100 \text{ m}$$

The particles or puffs are growing until their horizontal (σ_y) or vertical (σ_z) standard deviation is equal to half the grid distance. The mass of pollution that follows the puff is then mixed in the grid element where the puff is located.

Boundary conditions

In calculating vertical exchange, dry deposition is taken into account as a lower boundary condition. At the upper boundary, a flux of pollution out of the system is made proportional to the concentration in the upper layer. The factor of proportionality may change with time of the day.

In the horizontal, the concentrations along the lateral boundaries may be determined by long range transport calculations.

For the horizontal advection upwind finite differences are used.

Dry deposition is a sink of pollution at the ground. However, pine forests covering parts of the area are a source of hydrocarbons that may influence the chemical reactions.

The deposition of gas to the ground, and in similar way the evaporation of a gas from the ground, is often discussed in terms of the resistance of the boundary layer to vertical pollution exchange. The resistance, r, is the inverse of the velocity of deposition Vg:

$$r = V_g^{-1} \ = \frac{\text{concentration difference}}{\text{flux}}$$

The total resistance is further divided into three parts:

$$r = r_a + r_b + r_s \qquad (2.7)$$

where:
    r_a = aerodynamic resistance in free air
    r_b = boundary layer resistance
    r_s = surface resistance

The characteristic turbulence parameters for the atmospheric surface layer may be used to specify the aerodynamic resistance. The surface resistance is dependent on the chemical affinity between the ground and the gas. The boundary resistance is de-

pendent on diffusion processes close to the ground, where the
atmospheric surface layer theory does not apply.

For the constant flux layer of the atmosphere, the following
equation may be written:

$$c_* u_* = K_z \frac{\partial c}{\partial z} \qquad (2.8)$$

where:

  c = pollutant concentration
  c_* = flux of pollution divided by the friction velocity

Equation 2.8 may be used to calculate the variation of concentration with height $c(z)$ in a stable atmosphere using equation (2.4) :

$$c(z) - c_0 = c_* (2.1 \ln \frac{z}{z_0} + 13.4(\frac{z-z_0}{L} + \frac{z_0}{L} \ln \frac{z}{z_0})) \qquad (2.9)$$

  z_0 = roughness heigth where average horizontal wind speed is zero
  $c_0 = c(z_0)$
  $c_* = (c - c_0)/f(z,z_0,L)$

when $f(z,z_0,L) = (2.1 \ln \frac{z}{z_0} + 13.4 (\frac{z-z_0}{L} + \frac{z_0}{L} \ln \frac{z}{z_0}))$

In correspondance with Equation 2.8:

$$c_* u_* = (c - c_0)u_*/f(z,z_0,L) \qquad (2.10)$$

where:
  $r_a = f(z,z_0,L)/u_*$

Further:

$$\frac{c_0}{r_b + r_s} = u_* c_* = (c - c_0)u_*/f(z,z_0,L) \qquad (2.11)$$

Similar equations may be written for an unstable atmosphere.
In this case vertical exchanges in the atmosphere are most effective, and the boundary layer and surface resistance often limit
dry deposition ($r_b + r_s \gg r_a$). Deposition velocity of the order
of 1 cm/s becomes important when the model keeps track of the
pollution during several hours.

In mesoscale modelling dry deposition is considered for SO_2,
NO, NO_2, O_3, and aerosols. The deposition velocities as referred
in the literature are determined by measurements. To estimate emissions of hydrocarbons from forest areas, Equation 2.10 may also be
used. Then it is necessary to measure concentrations at two levels,
as well as parameters determining the aerodynamic resistance.

The following empirically determined values for dry deposition
(6) are used as maximum values for level 1 in the model:

$$V_{max} = \frac{1}{r_b + r_s} \qquad\qquad (2.12)$$

For O_3 : V_{max} = 0.6 cm/s
For NO : V_{max} = 0.1 cm/s
For NO_2 : V_{max} = 0.2 cm/s
For PAN   : V_{max} = 0.2 cm/s

When $r_a \gg r_b + r_s$, it is the turbulent flux in the surface
layer that determines dry deposition.

Before comparing calculated concentrations for the surface
layer (25 m level) with observed values 2 m above the ground,
corrections have to be made for the concentration gradient close
to the ground.

Photochemical reactions

The scheme of reactions is given by Hov et al. (7), employing
the factors of photo-dissociation and of reaction, as given by
Isaksen et al. (8). The transport and diffusion of 29 components
are calculated in the model using as input, data on emission of
6 specific hydrocarbons (C_2H_4, C_3H_6, C_4H_{10}, o-xylene, HCHO, and
CH_3CHO). The characteristic lifetimes of different groups of
hydrocarbons in the model are shown in Table 2.2.

Taking the reaction time into consideration, it is only in
air pollution episodes with stagnant air that oxidants developed as a result of local sources will occur within the area.
Some other reactions are very fast requiring a short time step.
The integration of the combined chemistry-transport model is made
economically feasible by using a simplified calculation procedure,
developed and tested at the University of Oslo (2).

The continuity equation for the concentration C of each component, i, may be written:

$$\frac{dC_i}{dt} = P_i - L_i C_i \qquad i = 1,2,\ldots\ldots,N \qquad\qquad (2.13)$$

where P_i and $L_i C_i$ are the sum of all production (P_i) respective
reduction terms (photochemical reactions, deposition and transport).
The equations may be solved analytically, assuming that P_i and L_i
are constant over the time step Δt:

$$C_{i,t+\Delta t} = C_{i,e} + (C_{i,t} - C_{i,e}) \exp(-L_i \Delta t) \qquad\qquad (2.14)$$

Table 2.2 : Characteristic lifetime $\tau(1/k\ (OH))$
for hydrocarbons in the model.

	$k^{o)}$	(OH)= 5×10^6 [1] τ	(OH)= 1.5×10^6 [2] τ
n-butane	$2.4 \cdot 10^{-12}$	23 hours	77 hours
o-xylene	$1.4 \cdot 10^{-11}$	4 hours	13 hours
ethylene	$7.9 \cdot 10^{-12}$	7 hours	23 hours
propylene	$1.5 \cdot 10^{-11}$	3.7 hours	12 hours

o) k: unit is $cm^3/(molecules \cdot s)$

1) 5×10^6 molecules/cm^3: characteristic daily maximum mean value

2) 1.5×10^6 molecules/cm^3: characteristic daily maximum mean value.

where
$Y_{i,e} = P_i/L_i$ is the concentration at equilibrium.

All components are classified in three categories depending on their lifetime ($\tau_i = 1/L_i$):

1) If $\tau_i < \Delta t/10$,

$$C_{i,t+\Delta t} = P_{i,t+\Delta t}/L_{i,t+\Delta t} \qquad (2.15)$$

2) If $\tau_i > 30\ \Delta t$, the concentrations are calculated by a linear formula:

$$C_{i,t+\Delta t} = (P_i - L_i C_{i,t})\Delta t \qquad (2.16)$$

3) If $\Delta t/10 \leq \tau_i \leq 30\ \Delta t$, equation (2.14) is used.

The classification of a specific component may change with time as a result of changing concentrations.

Some components with short lifetime (class 1 or 2) are strongly interrelated. In order to get a correct interrelation, some iterations are carried out for each time step. Some linear transformations are used to eliminate strong interrelations that may cause numerical instability.

NUMERICAL EXPERIMENTS

Local emissions

The area of calculations is shown in Figure 3.1. A grid system, consisting of 1 km x 1 km squares, covers the area. An emission survey for natural and anthropogenic sources is developed. The total emissions for area sources and for point sources are shown in Table 3.1. A number of sources (between 3-10) are considered separately in 3 industrial areas. The industrial areas are shown in Figure 3.1.

Data on dispersion

Hourly wind measurements from Herøya were used to calculate advection in the surface layer. Wind observations at 25 m level at Aas were used for the two upper levels. The temperature difference between 2 m and 25 m was further used to characterize the turbulence in the surface layer.

Boundary values for the concentrations/long range transport

A two-dimensional model was used to simulate oxidant formation between 50°N and 60°N as a result of precursor emissions from England. The grid system is shown in Figure 3.2. The result of

Table 3.1 : Average total emissions per day (d)
in the area of calculations.

	CH_2O kg/d	CH_3CHO kg/d	C_2H_4 kg/d	C_3H_6 kg/d	C_4H_{10} kg/d	C_8H_{10} kg/d	NO kg/d	NO_2 kg/d	CO kg/d
Area sources[1]	110	115	2265	2045	2260	3110	1520	80	30900
Point sources	-	-	7234[2]	17	34	17	6730	2440	48000
Sum	110	115	9499	2062	2294	3127	8250	2520	78900

[1] The emissions from area sources are mainly caused by car traffic and by the use of solvents. Hydrocarbon emission from forests are calculated as area source emission of ethylen, propylen, n-butane.

[2] The emissions of ethylene include an accidental emission of 300 kg ethylene/h from the petrochemical plants at Rafnes.

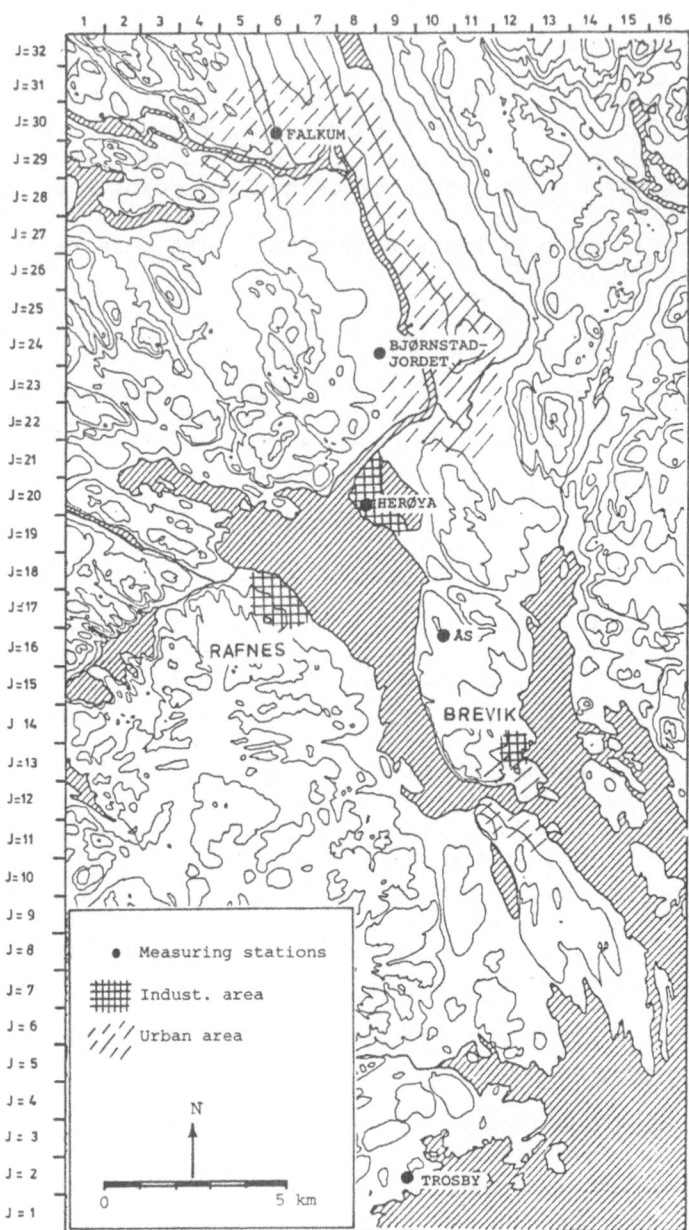

Fig. 3.1 : Area of calculations.
          The main industrial and urban areas are marked.
          Height between contour lines : 50 m.

this calculation for all components is used as a time dependent
boundary values for the local photochemical model. (A complete
description of this model may be found in ref. (10).)

Computer requirements

The programs were run on a CD 74. Integration was developed
for three layers and 29 components in a grid consisting of 512
points (km^2) in each level. The time step was about 5 min. The
program was organized to keep about 52 000 floating numbers in its
internal memory, and about 100 000 numbers as an intermediate
storage area on disk. The forward integration of 1 hour used about
14 minutes of computer time. The results for each hour of integration were stored on magnetic tape.

Design of calculations

Two types of calculations were performed:

1. Simulating an actual situation, using observed upwind
   ozone concentrations at the ground to describe the upwind
   boundary value of the area.
   The results of long range transport calculations could
   explain the measured upwind concentrations of ozone.

2. Calculating the ozone development as a result of local
   sources, assuming clean air along the upwind boundary.
   Low wind and poor dispersion conditions were prescribed
   to study possible local development of oxidants.

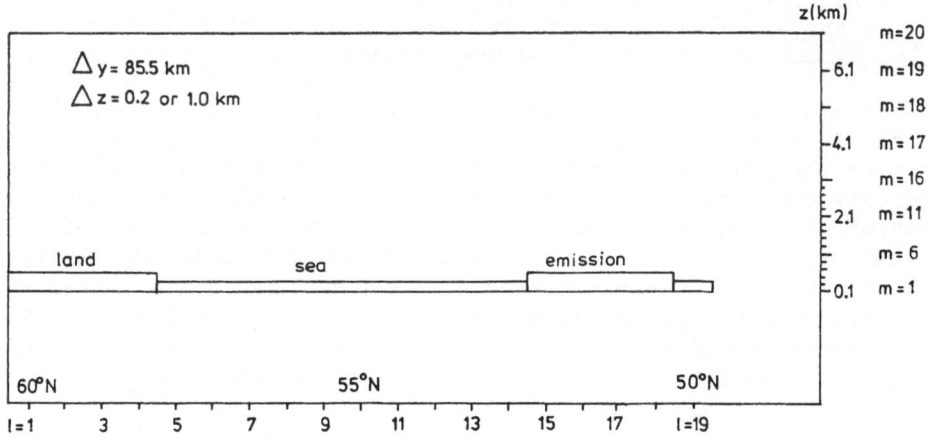

Fig. 3.2 : The two-dimensional long range transport model.

COMPARISON WITH MEASUREMENTS

Simulation of ozone concentrations in episodes

Figure 4.1 shows observed ozone concentration for the period 29-31.7.78. To account for the daily variation close to the ground, it was necessary to include dry deposition, and a time variable mixing in the atmospheric surface boundary layer.

The ozone concentration measurements at Trosby were used as input for the boundary value when the wind was blowing from the south. The distributions of ozone along the ground (level 1) and at Level 3 at noon are shown in Figure 4.2. Measured values at 2 m above the ground are shown in brackets. Measured and calculated concentrations indicate reduced ozone values as the air passes through the area in this episode.

The distributions of NO and O_3 (a), NO_2 and the sum of hydrocarbons (b) in the morning are shown in Figure 4.3. Figure 4.3 shows a decrease in NO and O_3 concentration. Figure 4.3b shows a corresponding increase in the NO_2 values. The Figure further indicates that a zone exists between the industrial areas where the mixture of hydrocarbons and nitrogenoxides is favourable for development of oxidents.

Calculation of local ozone development in an episode with stagnant air

Figure 4.4 shows the ozone concentrations about 100 m above the groud level at 7 a.m. and 2 p.m. in a situation with calm winds (0.5 m/s). It is seen that O_3 is developed as a result of local sources. The ethylene emissions from the petrochemical plant caused the ozone development. If more propylene than ethylene were emitted, more ozone would have been generated in a shorter time.

The highest ozone concentration recorded in this area, was found north of the petrochemical plant at Haukenes in the north-western part of the grid. Measurements north and south of the local sources indicate that local development may occur and an excess ozone concentration of about 80 ppb has been observed (10). Estimation of emissions based on ambient air measurements show that fugitive emissions include ethylene, propylene, ethane and propane (11). Accidental releases may explain that observed ozone concentrations. As a result of emissions from sources in Skien/Porsgrunn area, ozone concentrations often seem to be reduced within the area.

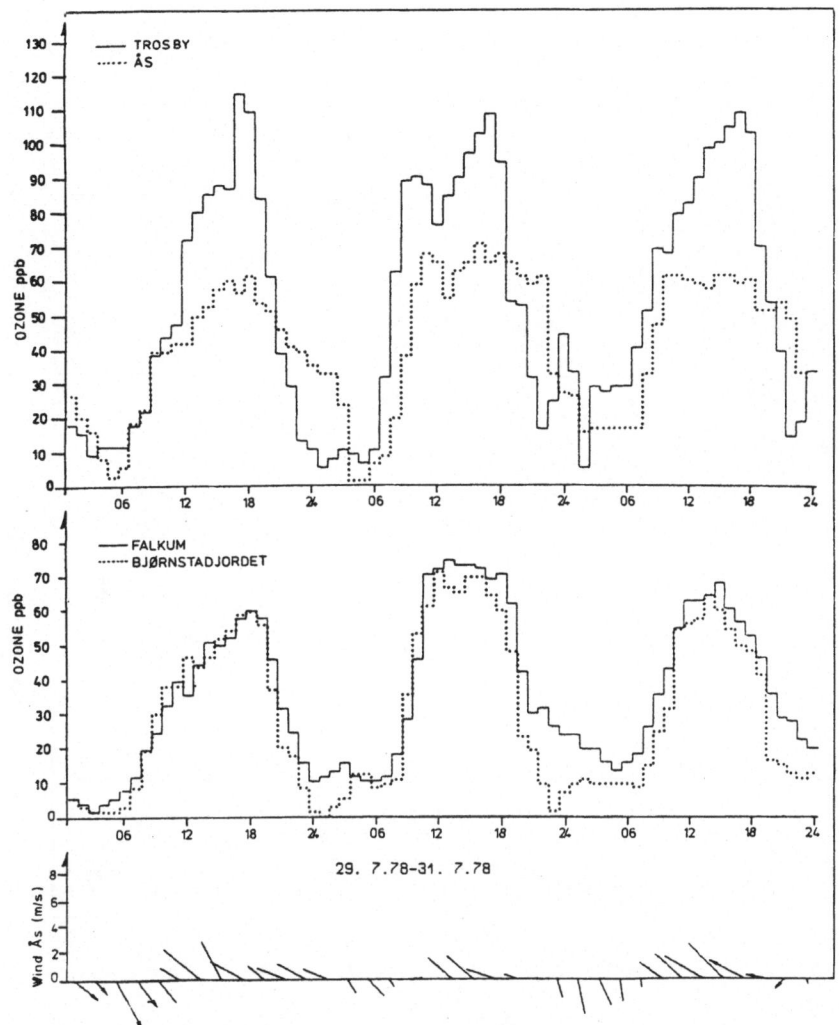

Fig. 4.1 : Ozone episode 29–31.78 in Telemark. Hourly values
are given for ozone (Trosby, Ås, Bjørnstadjordet
and Falkum) and for wind (Ås).
The wind vector refer to a coordinate system with
north along the ordinate and east along the
abscissa.

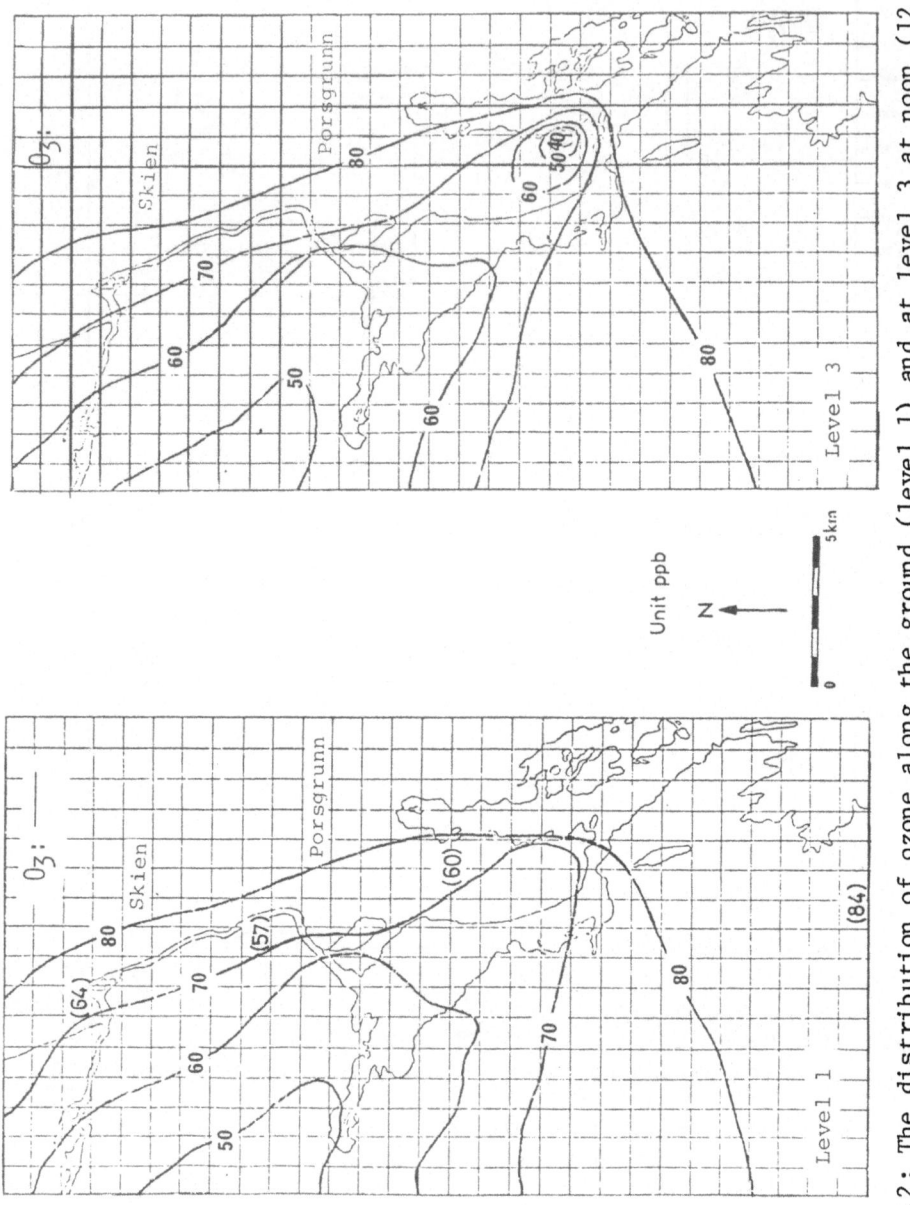

Fig. 4.2: The distribution of ozone along the ground (level 1) and at level 3 at noon (12 a.m.). The measured ozone concentrations at Trosby were used as input for the boundary value. Measured values 2 m above the ground are shown in brackets.

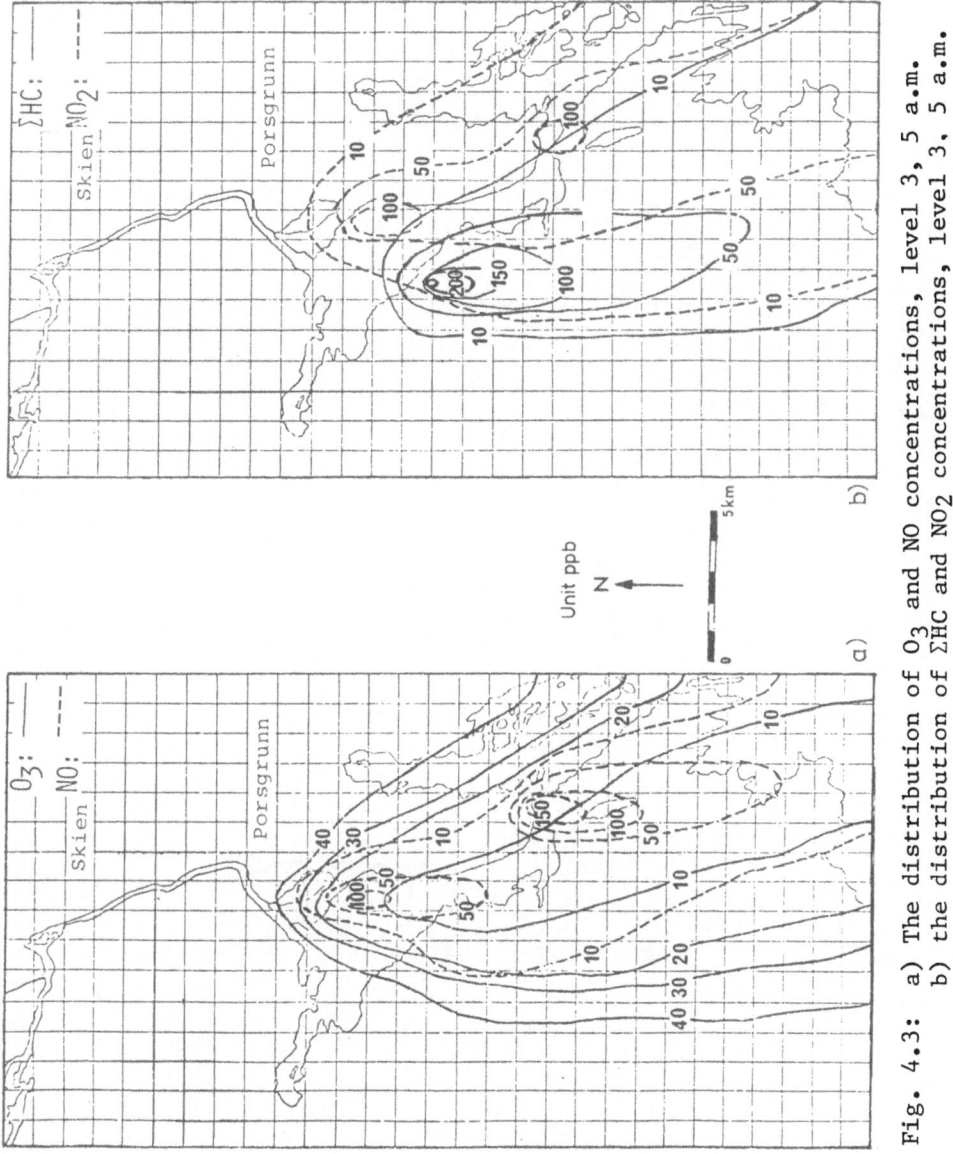

Fig. 4.3:   a) The distribution of O_3 and NO concentrations, level 3, 5 a.m.
             b) the distribution of ΣHC and NO_2 concentrations, level 3, 5 a.m.

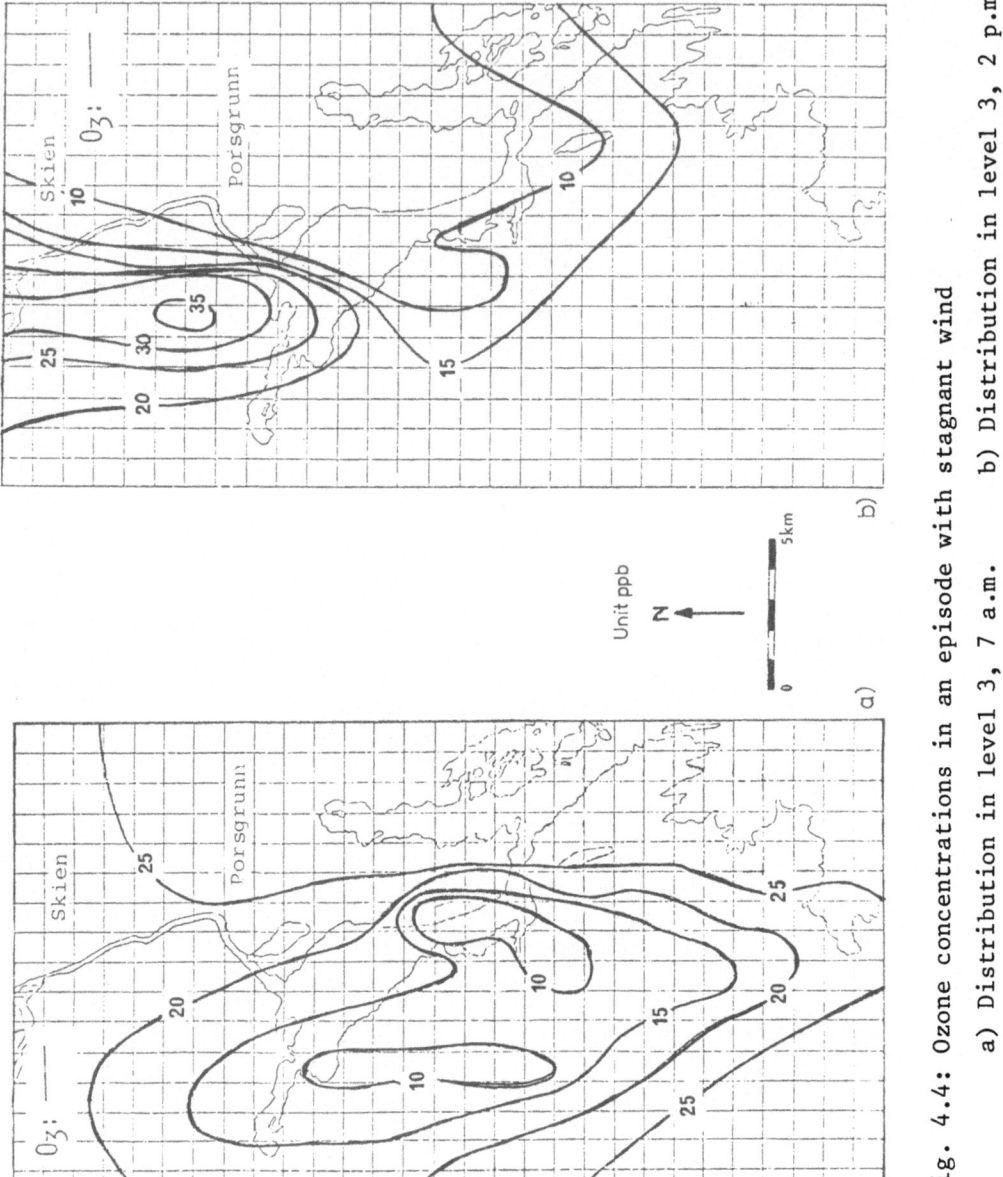

Fig. 4.4: Ozone concentrations in an episode with stagnant wind
      a) Distribution in level 3, 7 a.m.   b) Distribution in level 3, 2 p.m.

Ozone concentrations close to point sources

The dispersion of point source emissions is treated in the puff-model.

Aircraft ozone measurements, recorded high ozone values in one plume 0.5-4 km from one of the industrial plants near Prosgrunn. The ozone values were 30-60 ppb higher than the background values (see Figure 4.5). The instrument was a Dasibi UV absorption monitor.

These values could not be explained by the existing emission survey and the existing models.

CONCLUSIONS

The model calculations indicate the following effects of local air pollution emissions on oxidant formation in the area:

1. The observed ozone concentrations on 31 July 1978 are mainly caused by the concentrations in the air entering the area. To describe this formation of ozone, it is necessary to use a larger scale model.

2. The ozone concentrations on a larger scale entering the area are effectively broken down by local emission of NO. The reactions result in a rapid transformation of NO to NO_2, particularly at night.

3. The local emissions in the area may lead to a moderate increase in ozone concentrations in those areas influenced by the locally polluted air. The potential for increased ozone values may increase with the emission of hydrocarbons from sources outside the area, or if the constituents of the hydrocarbons are changed within the area.

4. If the residence time for the air within the area increases because of weak wind, the calculations indicate a moderate ozone development downwind of the petrochemical plants in the area. A maximum zone, 5 km from the source area, is indicated provided weak wind and average dispersion conditions prevail during a period of more than three hours.

Using the existing emission inventory, natural hydrocarbons do not contribute significantly to the total formation of ozone in the area. The emission may, on the other hand, influence the concentrations near forested areas.

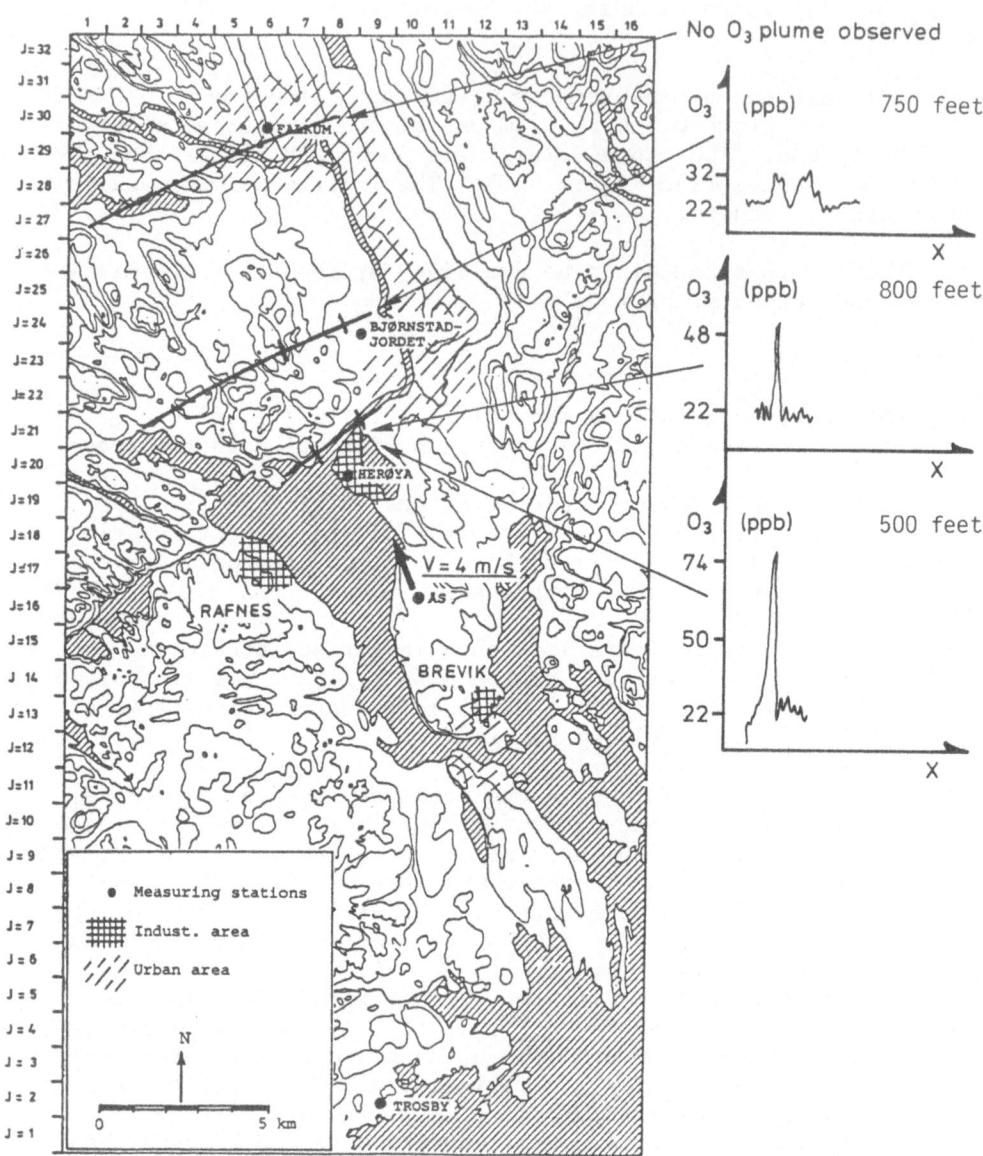

Fig. 4.5: The ozone trace in aircraft traverses through
an industrial plume at different heights and
distances from Herøya.

Long range transport of ozone may account for the values advected into the area.

The model may be further developed to include a telescoping procedure using a Lagrangian procedure close to the large point sources to also account for chemical processes.

A finite-difference approximation is suitable for the meso-scale to account for separation in local emissions. A puff model may be further useful to resolve the effects of emissions from the area on a larger scale.

REFERENCES

1.   F. Gram, K.E. Grønskei, K. Horntveth, K., Ø. Hov, I.S.A. Isaksen and J. Schjoldager, Fotokjemiske oksydanter i Grenland. Modellberegninger. Lillestrøm 1980 (NILU OR 1/80).

2.   F. Hesstvedt, Ø. Hov and I. Isaksen, Quasi-steady state approximations in air pollution modelling: Comparison of two numerical schemes for oxidant prediction. Oslo, Int. J. Chem. Kin., 10, 971-994 (1978).

3.   F. Pasquill, Some topics relating to modelling of dispersion in boundary layer. Washington D.C. 1975. (EPA-650/4-75-015.)

4.   N.E. Busch, S.W. Chang and R.A. Anthes, A multi-level model for the planetary boundary layer suitable for use with mesoscale dynamic models. J. Appl. Meteor., 15, 909-919 (1976).

5.   J.A. Businger, Turbulent transfer in the atmospheric surface layer. In: Workshop on Micrometeorology, D.A. Haugen, ed., Boston Amer. Meteor. Soc., 1973, pp. 67-98.

6.   T.A. McMahon and P.J. Denison, Review paper, Empirical atmospheric deposition parameters. A survey. Atmos. Environ., 13, 571-585 (1979).

7.   Ø. Hov., I.S.A. Isaksen and E. Hesstvedt, Diurnal variation of ozone and other pollutants in an urban area. Atmos. Environ., 12, 2469-2479 (1978).

8.   I.S.A. Isaksen, E. Hesstvedt and Ø. Hov, A chemical model for urban plumes: Test for ozone and particulate sulfur formation in St.Louis urban plume. Atmos. Environ., 12, 599-604 (1978).

9. K. Horntveth, Luftforurensning på regional skala undersøkt i
       en todimensional vertikal transportmodell. Inst. for
       Geofysikk, Univ. i Oslo. Våren 1980.
10. J. Schjoldager, Ambient Ozone measurements in Norway
       1975-1979. Journal APCA, 31, (11), 1187-1191 (1981).
11. B. Sivertsen, Tracer experiments to assess diffuse emissions
       of hydrocarbons from petrochemical factories at Rafnes
       (April 1981). Lillestrøm 1981 (NILU OR 53/81).

DISCUSSION

I. TROEN   Do you employ the Businger
              surface layer formulation for K_z through the whole
              boundary layer ? And if so is it not pushing surface
              layer theory a bit far for example under strongly
              stable conditions ?

K. GRØNSKEI   I agree that the surface layer
              theory may be pushed too far. I will look into the sen-
              sitivity of this approximation. The primary intention
              when estimating L and u_*, was to estimate dry deposi-
              tion and the exchange between layer 1 and layer 3.
              The vertical exchange depends on $\frac{K}{u}$ at the following
              levels:
              z = 50 m and z = 125 m in stable conditions
              z = 50 m and z = $\frac{Hmix}{2}$ + 25 m when the mixing height
              is larger than 200 m.

6: EVALUATION OF MODEL PERFORMANCES IN PRACTICAL APPLICATIONS

Chairmen: K. van Egmond
M. Williams

Rapporteurs: J. van Ham
P. Zannetti
S. E. Gryning

SITE-TO-SITE VARIATION IN PERFORMANCE OF

DISPERSION PARAMETER ESTIMATION SCHEMES

John S. Irwin*

Environmental Sciences Research Laboratory, EPA
Research Triangle Park, N.C. 27711

INTRODUCTION

To promote better practices in air pollution modeling, the
American Meteorological Society sponsored a workshop on stability
classification schemes and sigma curves (Hanna et al., 1977). The
participants at this workshop reviewed the bases and limitations
of the popular dispersion parameter schemes and suggested that
the standard deviation of the crosswind concentration distribution,
σ_y, might be viewed as

$$\sigma_y = \sigma_v \, t \, f_y, \qquad (1)$$

where σ_v is the standard deviation of the horizontal crosswind
component of the wind, t is the downstream travel time and f_y is a
nondimensional function. The standard deviation of the vertical
concentration distribution, σ_z, was viewed as

$$\sigma_z = \sigma_w \, t \, f_z, \qquad (2)$$

where σ_w is the standard deviation of the vertical component of
the wind, and f_z is a nondimensional function. Small-angle
approximations, such as $\sigma_v t = \sigma_a X$ and $\sigma_w t = \sigma_e X$ (where σ_a
is the standard deviation of the horizontal wind angle, σ_e is the
standard deviation of the vertical wind angle, and $X = u \, t$),
allow restatement of (1) and (2) in various forms.

*On assignment from the National Oceanic and Atmospheric
Administration, U.S. Department of Commerce

605

Initial evaluations of some of the currently available schemes for estimating plume dispersion parameters using (1) and (2) revealed variations in the performance of the schemes from one field experiment to the next, for both the lateral and vertical dispersion parameter comparisons (Irwin, 1983). Doran et al. (1978) showed that f_y should be a function of the averaging time, s; the sampling duration, τ, associated with σ_v; and the sampling duration, T, associated with σ_y. In most experiments, the variance of the wind direction is measured during the duration of the tracer measurements, and T equals τ. The Doran et al. (1978) results suggest that the variations in performance of dispersion parameter estimation schemes from site-to-site are in part related to variations in s, τ, and T. The primary purpose of this analysis is to summarize the site-to-site variations in the performance of several dispersion parameter estimation schemes, including schemes presented by Cramer (1976), Draxler (1976), and Pasquill (1976). The secondary purpose of this analysis is to indicate values of s, τ, and T to associate with the dispersion parameter schemes tested, given the available data.

Table 1. Dispersion experiments having near-surface release heights.

Site	Release Height (m)	Receptor Height (m)	Meteorological data	Dispersion data, tracer
Round Hill I (Cramer, Record, and Vaughan, 1958)	0.3	2.0	U, σ_a, σ_e at 2 m ΔT (12 - 1.5 m)	Concentrations along arcs (50, 100, 200 m), 10-min releases, SO_2 gas.
Round Hill II (Cramer, Record, and Vaughan, 1958)	0.5	1.5	U, σ_a at 2 m ΔT (12 - 1.5 m)	Concentrations along arcs (50, 100, 200 m), 10-min releases, SO_2 gas.
Prairie Grass (Barad, 1958) (Haugen, 1959)	0.46	1.5	U, σ_a, σ_e at 2 m ΔT (16 - 2 m)	Concentrations along arcs (50, 100, 200, 400, 800 m), 10 min-releases, SO_2 gas.
N.R.T.S. A&B (Islitzer and Dumbauld, 1963)	~1.0	~1.0	U, σ_a, σ_e at 4 m ΔT (16 - 4 m) ΔT (8 - 1 m)	σ_y and max (100, 200, 400, 800, 1600, 3200 m), 60-min releases, uranine dye.
Green Glow (Fuquay, Simpson, and Hinds, 1964)	2.5	1.5	U, σ_a at 2.1 m Ri (15.2 - 2.1 m)	σ_y (200, 800, 1600, 3200, 12800, 25600 m), 30-min releases, zinc sulfide.
Hanford-30 (Fuquay, Simpson, and Hinds, 1964)	2.5	1.5	U, σ_a at 2.1 m Ri (15.2 - 2.1 m)	σ_y (200, 800, 1600, 3200, 12800 m), 60-min releases, zinc sulfide.
Ocean Breeze (Haugen and Fuquay, 1963)	2 to 3	1.5	U, σ_a at 3.7 m ΔT (16.5 - 1.8 m)	σ_y (1200, 2400, 4800 m) 30-min releases, zinc zinc sulfide.
Dry Gulch (Haugen and Fuquay, 1963)	2 to 3	1.5	U, σ_a at 3.7 m ΔT (16.5 - 1.8 m)	σ_y (853, 1500, 2301, 4715, 5665 m), 30-min releases, zinc sulfide.
Mountain Iron (Hinds and Nickola 1967; Hinds, 1968)	~2.0	1.5	U, σ_a at 3.66 m ΔT (18 - 1.8 m) ΔT (91.4 - 1.8 m)	σ_y (variable, from 260 m to 11400 m) Several 5-min releases, mostly 30 min releases, zinc sulfide.
Hanford-67 (Nickola, 1977)	2.0	1.5	U at 2 m σ_a at 1.5 m ΔT (15.2 - 0.9 m)	σ_y (200, 400, 800, 1200, 1600, 3200, 5000, 7000, 12800 m), 10-to 30-min releases, zinc sulfide, fluoresein (uranine), rhodamine B, krypton-85.

EXPERIMENTAL APPROACH

Tables 1 and 2 summarize the field data used in this analysis. Included in these tables are the sources of the data and a summary of the meteorological and dispersion data available. The details of converting the field data to a similar format are summarized in Irwin (MS). Table 3 summarizes the s and τ associated with the wind turbulence measurements, the T associated with the concentration measurements (and hence with σ_y and σ_z), and the aerodynamic roughness length for each of the field experiments.

The vertical and lateral dispersion parameters were estimated using (1) and (2) for each downwind arc for the models listed in Table 4. Model 1 is Draxler's (1976), and model 2 is Cramer's (1976). Models 3 and 4 are variations of models 1 and 2 in which f_y and f_z do not vary with release height. Model 5 is Pasquill's (1976). For these estimates, the vertical and lateral velocity variances at release height were used.

Table 2. Dispersion experiments having elevated release heights.

Site	Release Height (m)	Receptor Height (m)	Meteorological data	Dispersion data, tracer
Suffield (Walker, 1965)	7.42 and 15	Surface	U, σ_a, σ_e at 8 m and 16 m ΔT (4 - 0.5 m)	σ_y (27.4 to 1097.3 m) 30-to 60-min releases, glass spheres.
N.R.T.S. (Islitzer, 1961) (Bowne, 1960)	46	1	U, σ_a, σ_e at 43 m T (43 - 3 m)	Sutton (n, C_y, C_z) (150 to 3400 m), 30-min releases, uranine dye.
Hanford-67 (Nickola, 1977)	26 and 56	1.5	U, σ_a at 26 and 56 m ΔT (15.2 - 0.9 m)	σ_y (200 to 4000 m), 10-to 30-min releases, zinc sulfide, fluorescent (uranine), rhodamine B, krypton-85.
Ågesta (Högström, 1964),	50	Smoke puff photography	σ_e estimates at 50 m Diffusion time Stability parameter	σ_z (100 to 1500 m) σ_z (20 to 400 s) (Sampling time about 60 min)
Hanford (Elderkin et al., 1963) and (Hilst and Simpson, 1958)	56	1.5 for σ_y Vertical array about release height for σ_z	U, σ_a at 61 m Ri (76 - 15 m) for σ_y ΔT (91 - 30 m) for σ_z	σ_y (100 to 1600 m) σ_z (152 to 1524 m) 15-to 60-min releases, flourescent dye.
Karlsruhe (Thomas et al., 1976) (Thomas and Nester, 1976)	100	Surface	U, σ_a, σ_e at 100 m ΔT (100 - 30 m)	σ_y and σ_z (variable 100 to 3900 m), 20-to 60-min releases, tritiated water vapor, halogenated hydrocarbons.
Porton (Hay and Pasquill, 1957)	140 to 255	Vertical array about release height	U, σ_e at release height ΔT (7 - 1.2m)	σ_z (variable 100 m to 500 m), 30-min releases, lycopodium spores.

Various statistics comparing the estimated dispersion parameters with measured values initially were computed for each possible combination of release height (elevated and near-surface) and for stable and unstable stratification. The fractional error was found to be a useful statistic, as it best summarized, in a quantitative sense, the subjective impressions of model bias and precision of model estimates gained from these analyses. The fractional error

$$e = 2 (P - O) / (P + O), \qquad (3)$$

where P is the model estimate and O is the measured value, has the same numerical value, but different signs, for either an overestimate of n times the measured value or an underestimate of 1/n times the measured value.

Table 3. Concentration sampling durations, wind velocity variance averaging times and sampling durations, and site roughness lengths for the tracer experiments.

	Site roughness (cm)	Concentration sampling duration (min)	Lateral wind fluctuations Averaging time (s)	Sampling duration (min)	Vertical wind fluctuations Averaging time (s)	Sampling duration (min)
Near-surface releases						
Round Hill I	>10	10	1.0[k]	10	1.0	10
Round Hill II	>10	10	2.5	10	n/a	n/a
Prairie Grass	0.6-0.9[a]	10	2.5	10	1.067	10
NRTS A&B	1.5	60	5.0	60	5.0	60
Green Glow	3.0	30	20	30	n/a	n/a
Hanford-30	3.0	20 to 75	20	20 to 75[o]	n/a	n/a
Ocean Breeze	(b)	30	2.0[m]	47	n/a	n/a
Dry Gulch	(b)	30	2.0[m]	47	n/a	n/a
Mountain Iron	(c)	5 to 48[h]	10	5 to 48	n/a	n/a
Hanford-67(2m)	3.0	30	5.0	30	n/a	n/a
Elevated releases						
Suffield	2.0[d]	30 to 60	1.0	30 to 60[p]	n/a	n/a
NRTS	1.5[e]	30	5.0	30	5.0	30
Hanford-67(26, 56m)	3.0	30	5.0	30	n/a	n/a
Ågesta	60	60	n/a	n/a	(s)	60
Hanford	3-5[f]	(i)	5.0	15 to 60[r]	(s)	Unk
Karlsruhe	110	30	Unk[n]	10	Unk	10
Porton	5.0[g]	30	n/a	n/a	2.5	30

n/a = not applicable, Unk = unknown

[a] Högström (1964) suggests 0.9 cm and Horst et al. (1979) suggest 0.6 cm.

[b] Horst et al. (1979). Rolling sand dunes covered with dense palmetto and brushwood (OB). Sloping mesa cut by deep ravines; vegetation mainly grasses with occasional brush and tree lines (DG).

[c] Mountainous terrain in Southern California.

[d] Estimated from site descriptions.

[e] Assuming value suggested by Islitzer and Dumbauld (1963) for the NRTS A&B experiments.

[f] Högström (1964) suggests 5 cm and Horst et al. (1979) suggest 3 cm.

[g] Högström (1964).

[h] 17 cases (5 min), 24 cases (15 to 20 min), 67 cases (25 to 30 min), and 3 cases (44 to 48 min).

[i] None reported; sampling apparently continued as long as conditions were steady-state.

[k] Assuming "movie camera" procedure used in 1952 was also used in 1954-55, Cramer and Record (1953).

[m] Horst et al. (1979) report questionable wind-vane performance occurred that may have led to erroneously high values for the lateral wind fluctuations.

[n] Thomas et al. (1979). Data collected continuously by an automatic data acquisiton system. No information given on sampling rate or averaging time used.

[o] 7 cases (20 to 21 min), 68 cases (30 to 31 min), 10 cases (45 to 51 min), and 28 cases (60 to 75 min).

[p] 32 cases (30 min), 27 cases (44 to 49 min), and 45 cases (56 to 60 min).

[r] Duration for each trial not given.

[s] Velocity variances estimated from dispersion data.

Table 4. Equation used for the functions f_y and f_z to characterize the growth of later and vertical dispersion as a function of travel time, T (in seconds), or of downwind distance, X (in meters). Model 1 is Draxler's (1976) and provides for variation of the functions with release height and stability. Model 2 is Cramer's (1976), where no buoyancy-induced dispersion is assumed. Model 5 is Pasquill's (1976).

	Lateral dispersion		Vertical dispersion	
Model 1	Elevated releases		Elevated releases	
		$1/f_y = 1 + 0.9(T/1000)^{0.5}$	unstable	$1/f_z = 1 + 0.9(T/500)^{0.5}$
			stable	$1/f_z = 1 + 0.945(T/100)^{0.806}$
	Surface releases		Surface releases	
	unstable	$1/f_y = 1 + 0.9(T/300)^{0.5}$	unstable	$f_z = (0.3/0.16)(T/100-0.4)^2 + 0.7$
	stable	$1/f_y = \begin{cases} 1 + 0.9(T/300)^{0.5} & T<550 \\ 1 + 28/(T)^{0.5} & T>550 \end{cases}$	stable	$1/f_z = 1 + 0.9(T/50)^{0.5}$
Model 2		$f_y = (50/X)[(X-5)/45]^{0.9}$		$f_z = 1$
Model 3		$1/f_y = 1 + 0.9(T/1000)^{0.5}$	unstable	$f_z = 1$
			stable	$1/f_z = 1 + 0.9(T/50)^{0.5}$
Model 4		$1/f_y = 1 + 0.9(T/1000)^{0.5}$	unstable	$1/f_z = 1 + 0.9(T/500)^{0.5}$
			stable	$1/f_z = 1 + 0.9(T/50)^{0.5}$
Model 5		$1/f_y = \begin{cases} 1 + (X/2500)^{0.5} & X<10,000 \\ 3(X/10,000)^{0.5} & X>10,000 \end{cases}$		Pasquill-Gifford dispersion scheme

Those cases with downwind travel distances greater than 5.4 km were excluded from analysis. This only involved removing some of the lateral dispersion parameter comparisons, as all the vertical dispersion comparisons were within 5.4 km. The measurements of the lateral dispersion at distances beyond 5.4 km were considered most suspect, due to limitations in the design of the field experiments. Restricting the analysis to distances less than 5.4 km would result in negligible contributions to the lateral dispersion due to wind direction shear in the vertical. Of the schemes tested, Cramer's (1976) and Draxler's (1976) characterizations of f_y implicitly account for the effects of wind-direction shear in the vertical on σ_y. Pasquill's (1976) characterization of the lateral dispersion accounts for the effects of wind-direction shear on σ_y separately from f_y. By limiting the downwind distances employed in this analysis, the lateral dispersion parameter schemes are on a more equal basis for comparison.

ANALYSIS

Lateral Dispersion

The values of the median of the residuals presented in Table 5 for the σ_y comparisons suggest that models 1, 3, 4, and 5 have a bias to underestimate σ_y, while model 2 tends to overestimate σ_y.

In Table 6, the mean and the standard deviation of the fractional errors (in percent) and the percent of the model estimates within a factor of two of the measured values are given for each model in each field experiment.

No trend, as a function of concentration sampling duration, could be seen in the variation of the performance of the models in estimating the lateral dispersion parameters.

The trend towards underestimates of the lateral dispersion parameters as averaging time increases is what would be expected, but the scatter and uncertainty of the results, allow only qualitative judgements to be made. This tendency towards smaller estimates of σ_y as s increases is due to a reduction in the computed

Table 5. Lateral dispersion parameter estimates compared with the measured values. The three statistics listed for each are the mean, the median, and the standard deviation of the residuals (estimated values minus measured values) in meters. N is the number of cases. Values less than ±0.5 are shown as ±0.

	N	Model 1	Model 2	Model 3 Model 4	Model 5
Stable stratification					
Near-surface releases					
Round Hill I	45	-0, +0, 6	4, 3, 7	2, 1, 6	2, 1, 6
Round Hill II[a]	18	5, 4, 4	10, 8, 8	7, 6, 6	7, 5, 6
Prairie Grass	160	-9, 1, 69	-1, 3, 67	-5, 2, 68	-6, 1, 67
NRTS A&B	35	-3, -4, 32	26, +0, 56	11, -2, 40	-1, -3, 30
Green Glow	60	-41, -23, 84	-5, -0, 81	-36, -10, 83	-44, -17, 84
Hanford-30	75	-55, -23, 108	-6, 2, 98	-58, -11, 121	-61, -15, 118
Ocean Breeze[a,b]	50	83, 50, 120	158, 121, 115	67, 61, 80	59, 59, 72
Dry Gulch[a,b]	17	116, 86, 250	212, 193, 271	97, 72, 209	80, 67, 203
Mountain Iron[b]	19	-45, -25, 85	11, 17, 71	-51, -25, 93	-68, -29, 103
Hanford-67(2m)	87	1, -8, 114	72, 30, 144	-5, -4, 76	-9, -4, 84
Elevated releases					
Suffield	55	-1, +0, 20	1, 1, 21	-1, +0, 20	-2, -0, 21
Hanford-67(26, 56m)	208	-31, -22, 69	-6, -9, 76	-31, -22, 69	-51, -37, 67
Karlsruhe[c]	6	-528, -516, 420	-516, -510, 410	-528, -516, 420	-541, -526, 429
Unstable stratification					
Near-surface releases					
Round Hill I	27	-4, -1, 10	2, 2, 9	-1, 1, 10	-3, -1, 9
Round Hill II[a]	12	-1, -1, 3	6, 4, 4	2, 1, 3	+0, -0, 3
Prairie Grass	155	-8, -3, 19	4, 1, 22	-1, -1, 20	-6, -3, 19
NRTS A&B	93	-6, -5, 50	37, 4, 76	22, +0, 59	-10, -5, 49
Hanford-30	38	-39, -20, 75	24, 7, 65	-7, 2, 62	-28, -7, 70
Ocean Breeze[a,b]	122	28, 32, 70	193, 159, 117	101, 87, 77	72, 63, 65
Dry Gulch[a,b]	190	7, 8, 125	202, 162, 181	86, 61, 136	65, 46, 115
Mountain Iron[b]	92	-42, -25, 139	105, 44, 220	22, 9, 164	-7, -8, 146
Hanford-67(2m)[c]	7	11, 10, 7	74, 49, 59	45, 31, 32	18, 17, 11
Elevated releases					
Suffield	49	-3, 1, 24	1, 2, 22	-3, 1, 24	-9, -1, 28
NRTS	96	-7, -4, 42	7, -1, 55	-7, -4, 42	-34, -14, 50
Hanford-67(26, 56m)	46	-11, -8, 41	24, 8, 59	-11, -8, 41	-40, -34, 46
Hanford	57	-13, -11, 32	+0, -2, 38	-13, -11, 32	-28, -22, 34
Karlsruhe	33	-129, -52, 204	-115, -36, 193	-129, -52, 204	-145, -74, 202

[a] Lateral turbulence measurements questionable.
[b] Complex terrain.
[c] Fewer than 10 cases.

values of σ_a as s increases. As discussed in Pasquill (1974),
the effect of s on the computed variance is to smooth out the fine
structure of the velocity variations. The contributions to the total
variance becomes very small for fluctuations of frequency n > 1/s.

As site roughness increases, the frequency spectrum of wind-
velocity fluctuations is expected to reflect an increase in the
small-scale mechanical turbulence. Such an increase would be most
noticeable for measurements near the surface during stable stratification. All other factors being equal, it is anticipated that the
effects of s on the computed values of σ_a, and hence on the estimates of σ_y, may be more dramatic at sites having large surface
roughness compared to the effects seen at sites having small surface

Table 6. Lateral dispersion parameter compared with the measured
values. The three statistics listed for each are the mean fractional
error (percent), the standard deviation of the fractional errors
(percent), and the percent of the estimates within a factor of two
of the measured values. N is the number of cases.

	N	Model 1			Model 2			Model 3 Model 4			Model 5		
Stable stratification													
Near-surface releases													
Round Hill I	45	-4,	35,	91	22,	35,	91	8,	35,	93	7,	34,	93
Round Hill II[a]	18	30,	21,	94	54,	23,	72	42,	22,	83	40,	22,	83
Prairie Grass	160	-4,	56,	85	21,	58,	74	9,	58,	80	4,	55,	82
NRTS A&B	35	-13,	36,	94	16,	39,	86	3,	38,	91	-8,	34,	97
Green Glow	60	-37,	29,	87	-3,	29,	98	-25,	31,	87	-32,	28,	87
Hanford-30	75	-32,	32,	92	4,	31,	93	-22,	33,	91	-25,	31,	89
Ocean Breeze[a,b]	50	42,	42,	64	73,	36,	38	44,	41,	62	41,	39,	74
Dry Gulch[a,b]	17	48,	48,	76	79,	46,	41	55,	53,	53	48,	53,	59
Mountain Iron[b]	19	-23,	29,	*	12,	23,	*	18,	22,	*	28,	23,	89
Hanford-67(2m)	87	-12,	37,	93	26,	35,	86	-2,	36,	92	-4,	36,	92
Elevated releases													
Suffield	55	-2,	41,	85	7,	42,	84	-2,	41,	85	-8,	42,	87
Hanford-67(26, 56m)	208	-26,	41,	84	-11,	41,	88	-26,	41,	84	-42,	39,	75
Karlsruhe[c]	6	-161,	11,	0	-155,	13,	0	-161,	13,	0	-167,	8,	0
Unstable stratification													
Near-surface releases													
Round Hill I	27	-10,	21,	*	8,	19,	*	-1,	20,	*	-8,	19,	*
Round Hill II[a]	12	-2,	9,	*	18,	9,	*	8,	9,	*	2,	9,	*
Prairie Grass	155	-18,	24,	97	3,	25,	99	-6,	25,	98	-16,	24,	98
NRTS A&B	93	-8,	27,	*	15,	29,	92	7,	29,	95	-10,	26,	*
Hanford-30	38	-20,	23,	95	17,	23,	97	1,	22,	*	-12,	24,	97
Ocean Breeze[a,b]	122	23,	33,	89	74,	28,	41	51,	31,	70	41,	31,	74
Dry Gulch[a,b]	190	7,	41,	89	62,	33,	58	35,	39,	78	28,	34,	85
Mountain Iron[b]	92	-21,	50,	82	28,	52,	75	6,	50,	87	-7,	52,	86
Hanford-67(2m)[c]	7	12,	6,	*	52,	11,	*	36,	8,	*	18,	9,	*
Elevated releases													
Suffield	49	4,	20,	*	12,	20,	*	4,	20,	*	-5,	22,	98
NRTS	96	-10,	20,	*	-2,	21,	*	-10,	20,	*	-28,	22,	98
Hanford-67(26, 56m)	46	-6,	13,	*	10,	13,	*	-6,	13,	*	-22,	13,	*
Hanford	57	-12,	18,	*	-3,	20,	*	-12,	18,	*	-28,	21,	98
Karlsruhe	33	-47,	56,	64	-38,	54,	67	-47,	56,	64	-60,	50,	58

[a] Lateral turbulence measurements questionable.
[b] Complex terrain.
[c] Fewer than 10 cases.
* All estimates within a factor of two.

roughness. Analyses of the results presented in Tables 5 and 6 are not conclusive, as there are so few cases for comparison.

The results having 1-s averaging time seem to be unaltered in going from 2- to 10-cm roughness. It is best not to compare results from surface and elevated releases, as the effects of s on σ_a for the elevated releases would be expected to be less than for a surface releases. However, the elevated releases at 2-cm roughness, with a 1-s averaging time, are the Suffield experiments, which were the lowest of the elevated releases having only 7.4- and 15-m release heights. The comparisons having 5-s averaging times suffer, as there is only a small change in roughness (from 1.5 to 3 cm). From these results, it is not apparent that σ_a is more sensitive to averaging time effects as roughness increases.

Earlier in the discussion, it was mentioned that differences between the sampling durations used to compute σ_v and σ_y might affect the dispersion parameter estimates. Differences between τ and T occured at three of the experiments: Karlsruhe, Ocean Breeze, and Dry Gulch. The effect of differences between τ and T on the σ_y estimates for the Ocean Breeze and Dry Gulch experiments, where τ was 47 min and T was 30 min, could not be computed from the original data listings. The effect of differences in τ and T can be computed for the Karlsruhe experiments. A reanalysis to compute 30-min values of σ_a from the 10-min summary data, indicates that the 30-min values of σ_a are on average 28 % larger than the 10-min values of σ_a used in the original estimates of the lateral dispersion parameters.

Vertical Dispersion

Tables similar to Tables 5 and 6 were developed for the σ_z comparisons. No trends as a function of concentration sampling duration or site roughness length could be seen in the variation of the performance of the models in estimating the vertical disperison paraemters. Excluding those experiments where the wind turbulence was questionable, the terrain was complex, or there were fewer than 10 cases for analysis, shifts in the median of from 3 to -20 m are typical from one field experiment to the next.

There was a strong tendency for model 2 (Cramer, 1976) to overestimate the σ_z values during stable stratification. This was discounted, as it resulted from not including the depth of the nocturnal mixed surface layer in the characterization of the vertical dispersion parameter. Sufficient data, such as detailed profiles of the variance of the vertical velocity fluctuations, were not available for the depth of the nocturnal mixed surface layer to be specified confidently. Limiting of the vertical dispersion during stable stratification is an important consideration in Cramer's model.

The analyses of the mean fractional errors of the vertical dispersion parameter estimates suggest that models 1, 3, and 4 underestimate the σ_z values for the NRTS A&B experiments, Islitzer and Dumbauld (1963), especially for the cases during stable stratification. Whether this is due to the 5-s averaging time or due to some other effect such as deposition is not clear.

CONCLUSIONS

Dispersion parameter estimates were made using 5 models for comparison with parameters determined during 17 field tracer experiments conducted at 11 sites to investigate whether the variations in averaging time, s; the sampling duration, τ, associated with the turbulence measurements; and the sampling duration, T, associated with the dispersion measurements, accounted for observed site-to-site variations in the performance of the schemes as suggested by Doran et al. (1970).

The variation in the performance of the lateral dispersion parameter estimates with respect to s suggest that underestimates of the dispersion parameters occur if the averaging time is too long. The decrease in σ_y is related to the decrease in σ_a as averaging time, s, increases. The variations in performance of the dispersion parameter schemes from one experiment to the next are quite large. In this analysis, it was not clear whether all or most of these variations could be related to variations in s, τ, and T. The relationships between the performance of the dispersion parameter estimation schemes and s, τ, and T may well have been obscured by the fact that the data compared was collected at different sites using different experimental designs. Even with the scatter in the results, it would seem that for the models tested, s should be 5 s or less for the lateral dispersion parameter estimates and perhaps less than 2 s for the vertical dispersion parameter estimates. The underestimates of the lateral dispersion parameters for the Karlsruhe experiments suggest that the sampling duration for the computations of σ_v should be similar to sampling duration associated with the σ_y values. More comparisons are needed in which the values of s, τ, and T are systematically varied to better characterize the sensitivity of the dispersion parameter estimates to s, τ, and T.

REFERENCES

Barad, M. L. Ed., 1958: Project Prairie Grass, a field program in diffusion. Geophysical Research Papers, No. 59, Vols. I and II. Air Force Cambridge Research Center Report AFCRC-TR-58-235, 479 pp. [NTIS PB 151 425 and PB 151 424]

Bowne, N., 1960: Measurements of atmospheric diffusion from an elevated source. Proceedings of 6th A.E.C. Air Cleaning Conference, Atomic Energy Commission, 76-88. [NTIS TID-7593].

Cramer, H. E., 1976: Improved techniques for modeling the dispersion of tall stack plumes. Proceedings of the 7th International Technical Meeting on Air Pollution Modeling and its Application, No. 51, NATO/CCMS, 731-780. [NTIS PB 270 799]

_____, F. A. Record and H. C. Vaughan, 1958: The Study of the Diffusion of Gases or Aerosols in the Lower Atmosphere. AFCRL-TR-58-239, The MIT Press, 133 pp. [NTIS AD 152 582]

_____, and F. A. Record, 1953: The variation with height of the vertical flux of heat and momentum. J. Appl. Meteor., 10, 219-226.

Doran, J. C., T. W. Horst and P. W. Nickola, 1978: Variations in measured values of lateral diffusion parameters. J. Appl. Meteor., 17, 825-831.

Draxler, R. R., 1976: Determination of atmospheric diffusion parameters. Atmos. Environ., 10, 99-105.

Elderkin, C. E., W. T. Hinds and N. E. Nutley, 1963: Dispersion from elevated sources. Hanford Radiological Sciences Research and Development Annual Report for 1963, 1.29-1.36. [NTIS HW-81746]

Fuquay, J. J., C. L. Simpson, and W. T. Hinds, 1964: Prediction of environmental exposures from sources near the ground based on Hanford experimental data. J. Appl. Meteor., 3, 761-770

Hanna, S. R., G. A. Briggs, J. Deardorff, B. A. Egan, F. A. Gifford, and F. Pasquill, 1977: AMS-workshop on stability classification schemes and sigma curves - summary of recommendations. Bull. Amer. Meteor. Soc., 58, 1305-1309

Haugen, D. A. Ed., 1959: Project Prairie Grass, a field program in diffusion. Geophysical Research Papers. No. 59, Vol. III. Air Force Cambridge Research Center Report AFCRC-TR-58-235, 673 pp. [NTIS PB 161 101]

_____, and J. J. Fuquay, Eds., 1963: The Ocean Breeze and Dry Gulch diffusion programs, Vol. I. Air Force Cambridge Research Laboratories and Hanford Atomic Products Operations, Report HW-78435, 240 pp. [Available from Technical Information Center, P.O. Box 62, Oak Ridge, TN]

Hay, J. S., and F. Pasquill, 1957: Diffusion from a fixed source at a height of a few hundred feet in the atmosphere. J. Fluid Mech., 2, 299-310

Hilst, G. R., and C. L. Simpson, 1958: Observations of vertical diffusion rates in stable atmospheres. J. Meteor., 15, 125-126

Hinds, W. T., 1968: Diffusion over coastal mountains of Southern California. Pacific Northwest Laboratories Annual Report for 1967, Vol. II: Physical Sciences, Part 3. Atmospheric Sciences, D. W. Pearce, Ed., Pacific Northwest Laboratories Report BNWL-715-3, 19-53. [NTIS BNWL 7-5-3]

_____, and P. W. Nickola, 1967: The Mountain Iron diffusion program: Phase I, South Vandenberg, Vol. I, Pacific Northwest Laboratories Report BNWL-572, 220 pp. [NTIS AD 721 858]

Högström, U., 1964: An experimental study on atmospheric diffusion Tellus., 16, 205-251

Horst, T. W., J. C. Doran, and P. W. Nickola, 1979: Evaluation of empirical atmospheric diffusion data. Pacific Northwest Laboratories Report NUREG/CR-0798, 137 pp. [NTIS PNL-2599]

Irwin, J.S., (1983): Estimating plume dispersion - a comparison of several sigma schemes. q. Appl. Meteor.

Islitzer, N. F., 1961: Short-range atmospheric dispersion measurements from an elevated source. J. Meteor., 18, 443-450.

_____, and R. K. Dumbauld, 1963: Atmospheric diffusion-deposition studies over flat terrain. Int. J. Air Water Pollut., 7, 999-1022.

Nickola, P. W., 1977: The Hanford 67-series: a volume of atmospheric field diffusion measurements. Battelle Pacific Northwest Laboratories, 454 pp. [NTIS PNL-2433]

Pasquill, F., 1974: Atmospheric Diffusion. John Wiley & Sons, 429 pp.

_____, 1976: Atmospheric dispersion parameters in Gaussian plume modeling, Part II: possible requirements for change in the Turner workbook values. U.S. Environmental Protection Agency Report EPA-600/4-76-030b, 53 pp. [NTIS PB 258 036]

Thomas, P. W., W. Hubschmann, L. A. Konig, H. Schuttelkopf, S.
     Vogt and M. Winter, 1976: Experimental determination of the
     atmospheric dispersion parameters over rough terrain, Part
     I: measurements at the Karlsruhe Nuclear Research Center.
     Federal Republic of Germany, Central Division for Nuclear
     Research, M.B.H., Report KFK2285, 132 pp.

_____, and K. Nester, 1976: Experimental determination of
     the atmospheric dispersion parameters over rough terrain,
     Part II: evaluation of measurements. Federal Republic of
     Germany, Central Division for Nuclear Research, M.B.H.,
     Report KFK2286, 116 pp.

Walker, E. R., 1965: A particulate diffusion experiment.
     J. Appl. Meteor., 4, 614-621.

DISCUSSION

R. STEENKIST   Some authors give dispersion
               as a function of time, others as a function of x or
               u x t. So then the dispersion in the first scheme is
               also a function of wind velocity.Can you comment on
               that ?

J.S. IRWIN   I looked forward to a day when
               the universal functions handle release height effects,
               stability effects, wind speed effects etc. in a con-
               sistent manner without having to "switch schemes".
               At present, the comparisons I have seen, suggest that
               even very simple schemes (that, for instance, allow
               no variation in F_y and F_z with release height) work
               well. This suggests to me, that the other effects will
               be best parameterized using new experiments construc-
               ted for the expressed purpose of developing new impro-
               ved formulas for F_y and F_z.

B. SIVERTSEN   If any of the σ_v data were taken
               at different heights above the surface for the cases
               of elevated releases, did you see any effects on the
               F-functions on measuring height for σ_v.

J.S. IRWIN   Yes, Cramer's model worked best
               for elevated releases and Pasquill's model served best
               for surface releases. However, the evaluation of both
               models suffered from large scatter in comparison re-
               sults. I am inclined to advise the use of simple mo-
               dels, for now, and worry about release height effects
               later, based on new and better controlled experiments
               designed to study such effects in particular.

S.E. GRYNING   Were you always able to estimate
a value of σ_z from the data that were obtained from
elevated source experiments ?

J.S. IRWIN   No. In some cases, the Gaussain-
plume model (assuming conservation of mass) was not
useful. The problem may have resulted from poor
sampling, the plume may have travelled completely
below the sampling array, as the problem may have
resulted from a significant violation of the assumption of conservation of mass.

F.B. SMITH   You suggest an upper limit to
the averaging time of the basic v and w data of 5 and
2 seconds respectively. However, except at very short
range, the σ_y or σ_z do not depend on the high frequency ends of the spectra and larger averaging times
could be tolerated in principle, although I appreciate
the schemes incorporate these effects in a different
way.

J.S. IRWIN   The upper limits are suggested
to minimize error in estimates using the schemes
tested. As you said, the schemes allow for the variations that occur during transport downwind using means
other than averaging time. Use of turbulence data
appreciably smoothed will degrade performance at all
distances. Whether smoothing effects at very large
distances (beyond 20 Km) are "significant" in practice
is difficult to tell and will depend on the purpose
of the estimates.

ESTIMATES OF VERTICAL DIFFUSION FROM SOURCES

NEAR THE GROUND IN STRONGLY UNSTABLE CONDITIONS

A.A.M. Holtslag

Royal Netherlands Meteorological Institute

P.O. Box 201, 3730 AE De Bilt

INTRODUCTION

This paper deals with estimates of vertical diffusion from sources near the ground in strongly unstable, or convective conditions. We will compare the ground level concentrations predicted by three diffusion models. For the comparison the observations of the Prairie grass experiment are used (Barad, 1958). The models describe the dispersion of a passive contaminant released from a continuous crosswind line source.

The first diffusion model is the well-known Gaussian plume model. In this case the vertical dispersion is described by a dispersion coefficient (σ_z) related to stability classes and the surface roughness length (e.g. Pasquill, 1974).
The second model is based on Monin-Obukhov scaling applied to the diffusion equation by Van Ulden (1978). Monin-Obukhov scaling is in principle only permitted in the surface layer. We will investigate the results of the surface layer diffusion model also for plumes with greater depth than the surface layer. The third model is based on so-called free convection scaling in the convective boundary layer (Nieuwstadt, 1980). This means that the turbulence and also the diffusion of contaminants is dominated by buoyancy. In this study the free convection model is extended to dispersion within the surface layer.

The models in this paper use characteristic boundary layer parameters as recommended by Hanna et al. (1977). These parameters are the friction velocity u_*, the buoyancy flux $(g/T)\overline{wT}$, the surface roughness length z_0 and the Obukhov stability parameter L, defined by

$$L = - \frac{u_*^3}{k \frac{g}{T} \overline{WT}} \quad , \qquad\qquad (1)$$

where k is the Von Karman constant. At present reliable estimates of the mentioned boundary layer parameters can be made from standard meteorological data (Holtslag et al, 1981; Holtslag and Van Ulden, 1982). Therefore the described diffusion models can be used in practice for air pollution studies.

In the next section the data is described which is used for the comparison of the diffusion models in this paper. In the three sections thereafter the diffusion models are discussed in more detail and compared with observations. Finally the relative skill of the models is investigated.

EXPERIMENTAL DATA

During the classic Prairie grass experiment data was collected on the dispersion from a point source at a height z_s = 0.46 m (Barad, 1958). Nieuwstadt (1980) and Van Ulden (1978) computed the cross-wind integrated concentrations from measurements along an arc at a height of 1.5 m. Because of the limit of resolution of the sample technique we will separate runs with a maximum concentration less than 0.5 mgm^{-3} (Van Ulden, 1978). We will use the results of 20 convective runs for three downwind distances (50, 200 and 800 m). For these runs the condition $-h/L > 10$ is satisfied (h is the boundary layer height). The boundary layer height is obtained from radiosonde data and aircraft soundings by Nieuwstadt (1980).

The characteristic boundary layer parameters u_*, $\overline{WT}$ and L are computed with semi-emperical relations (Dyer, 1974; Paulson, 1970). In this calculation the temperature difference between two heights (0.5 m and 8 m) is used together with the windspeed at 8 m (U_8). The surface roughness length (z_0) was taken equal to z_0 = 0.008 m (Pasquill, 1974; Van Ulden, 1978). This procedure is described in more detail by Holtslag and Van Ulden (1982).In table 1 results for the 20 convective runs are summarized. The values of u_* and L differ from those reported by Nieuwstadt (1980) and Van Ulden (1978) because they used the semi-empirical relations of Businger (1973) which probably underestimate the eddy stress (see Wieringa, 1980). This results in smaller values of u_* of ~ 15 % and L of ~ 25 %.

In table 1 also the stability class is given which is required for the application of the Gaussian plume model. These classes are obtained from L and z_0 with a method reported by

Table 1 The derived characteristic boundary layer parameters
        u_*, H (= $\rho c_p \overline{WT}$) and L together with the stability class
        for each run of the Prairie grass diffusion experiment in
        convective conditions.

Run	$u_*(\text{ms}^{-1})$	$H(\text{Wm}^{-2})$	$-L(\text{m})$	stability class
1	0.22	93	10	B
5	0.45	239	33	C
7	0.35	331	11	B
8	0.35	168	22	C
9	0.53	298	44	C
10	0.37	363	12	B
15	0.27	202	8	B
16	0.27	360	5	A
19	0.46	277	31	C
20	0.71	648	49	C
25	0.23	171	6	A
26	0.49	284	37	C
27	0.48	208	46	C
30	0.54	268	50	D
43	0.40	311	18	C
44	0.46	278	31	C
49	0.52	389	31	C
50	0.51	269	42	C
51	0.50	207	54	D
61	0.58	353	49	C

Golder (1972). This determination of stability class is for these
data in agreement with the method of Pasquill (1974), which uses
sensible heat flux and 10 m windspeed.

THE GAUSSIAN PLUME DIFFUSION MODEL.

      The well known Gaussian plume diffusion model for a cross-
wind ground level line source reads (Pasquill, 1974)

$$\frac{C_y}{Q} = \frac{(2/\pi)^{\frac{1}{2}}}{\overline{U}_p \, \sigma_z} \tag{2}$$

where C_y is the cross wind integrated concentration at the
surface, Q is the source strength, σ_z is the vertical dispersion
coefficient and $\overline{U}_p$ is the mean transport velocity of the plume.

      The vertical dispersion coefficient σ_z is related to a
stability class, the surface roughness length z_0 and the travel

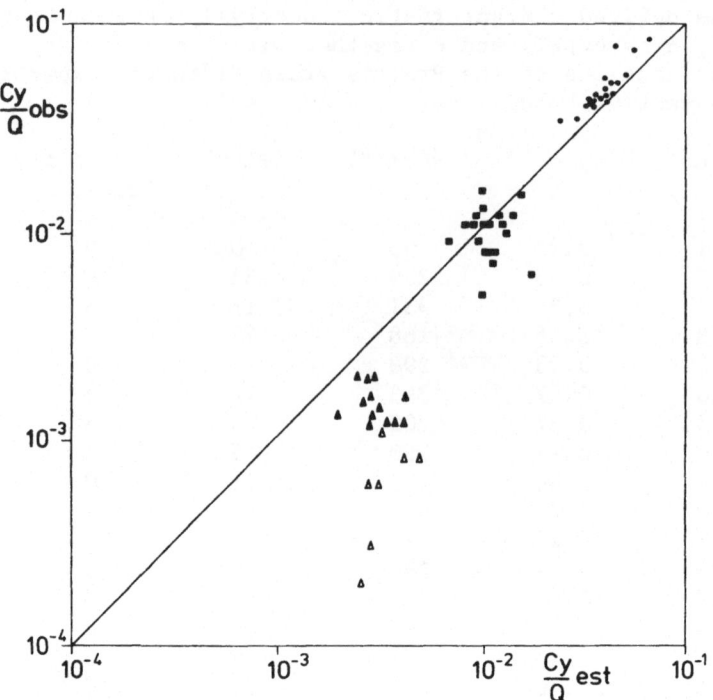

Fig. 1 A comparison between observed normalised cross wind
concentrations (C_y/Q obs) with calculated values (C_y/Q est)
of Eq (2) at three distances (x). Here, dots refer to x =
50 m, squares to x = 200 m and triangles to x = 800 m. At x
= 800 m the open triangles refer to less reliable
concentrations (runs 1, 5, 7, 10, 16, 19, 25 and 51).

distance x by (Pasquill, 1974)

$$\sigma_z = a \, x^b \, (10 \, z_0)^c \quad , \tag{3}$$

where

$$c = 0.53 \, x^{-0.22} \quad . \tag{4}$$

Here z_0, x and σ_z are in metres. The stability class dependent
coefficients a and b are tabulated by Pasquill (1974). The
coefficient c gives a correction of σ_z for different values of z_0.

     In Fig. 1 results are shown between calculated and observed
values of C_y/Q. We have used the observed 8 m windspeed U_8 as a
first guess for U_p. This choice is discussed below. It is seen
that the agreement is satisfactory for x = 50 m and x = 200 m, but

for x = 800 m the calculated values are about a factor two larger than the most reliable concentrations. Apparently the calculated values of σ_z are too small on larger distances.

THE SURFACE LAYER DIFFUSION MODEL

    In the atmospheric surface layer the characteristic parameters are the friction velocity u_*, the surface roughness length z_o and the Obukhov stability parameter L (Tennekes, 1982). According to Monin-Obukhov similarity theory the profiles of wind speed and eddy diffusivity can be written as universal functions of z_o, u_* and L.

    Van Ulden (1978) derived similarity expressions of the concentration profile using profiles of wind speed and eddy diffusivity with the diffusion equation for a continuous cross-wind line source. Then the normalised crosswind integrated concentration at height z $C_y(z)/Q$ at travel distance x is

$$\frac{C_y(z)}{Q} = \frac{0.73}{\overline{U}_p \cdot \overline{z}} \exp\left[-(0.66\, z/\overline{z})^{1.5}\right] \qquad 9 \qquad (5)$$

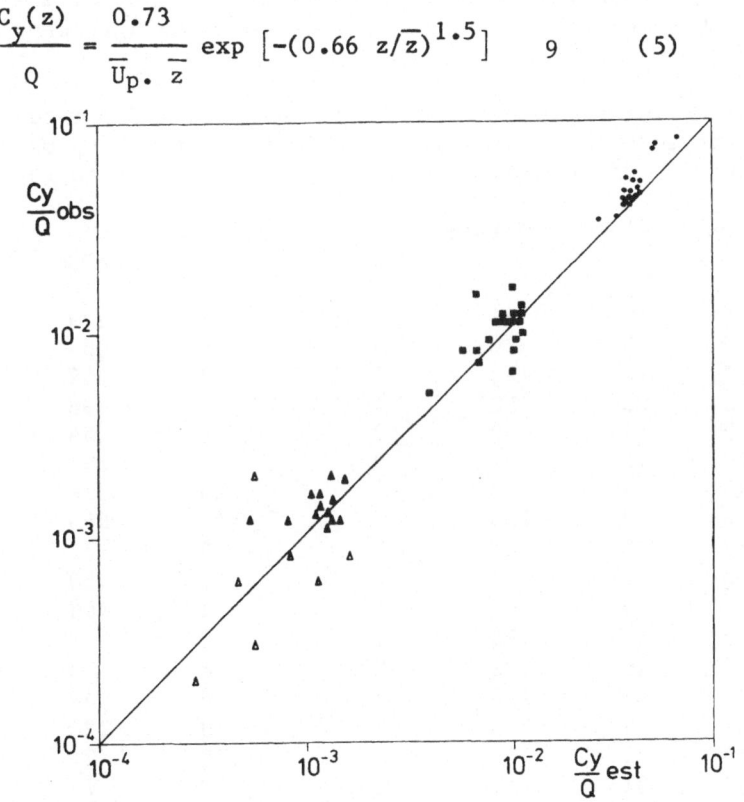

Fig. 2 As Fig. 1 but here (C_y/Q) est is calculated with Eq (5).

where

$$\bar{U}_p = \frac{u_*}{k} [\ln(0.6 \frac{\bar{z}}{z_o}) - \psi(0.6 \frac{\bar{z}}{L})] \quad . \tag{6}$$

In (6) ψ is a stability function defined by Paulson (1970).
Further $\bar{z}$ is the mean height of the particles, which is a solution
of

$$x + x_o = \frac{\bar{z}}{k^2} [\ln(0.6 \frac{\bar{z}}{z_o}) - \psi(0.6 \frac{\bar{z}}{L})](1 - d \frac{\bar{z}}{L})^{-\frac{1}{2}}, \tag{7}$$

for given x, x_o, z_o and L.
Here $d = 6.2$ and $k = 0.41$ when the empirical similarity relations
of Dyer (1974) are used. In (7) x_o is the virtual distance
determined by the source height z_s. The value of x_o can be
obtained by the initial condition $\bar{z} = z_s$ at $x = 0$ in Eq. (7).

Fig. 2 give results of calculated values of C_y/Q at $z = 1.5$ m
compared with observed values. It is seen that the agreement is
good for all three distances.

Table 2 The calculated height of the plumecentre $\bar{z}$ for three
distances (x) compared with the Obukhov stability parameter
L and 10% of the boundary layer height h for each run.

Run	−L(m)	0.1h(m)	$\bar{z}$(m) x=50m	$\bar{z}$(m) x=200m	$\bar{z}$(m) x=800m
1	10	26	3	21	229
5	33	78	2	9	69
7	11	134	3	19	208
8	22	138	2	11	98
9	44	55	2	8	54
10	12	95	3	19	195
15	8	−	4	26	296
16	5	106	5	45	560
19	31	65	2	9	73
20	49	71	2	8	50
25	6	65	4	34	406
26	37	90	2	9	63
27	46	128	2	8	53
30	50	156	2	8	49
43	18	60	3	13	124
44	31	145	2	10	74
49	31	55	2	10	74
50	42	75	2	9	57
51	54	188	2	8	47
61	49	45	2	8	50

   Table 2 gives calculated values for the averaged height of the particles $\bar{z}$ and the Obukhov stability parameter of table 1. It is seen that at x =50 m the plume is within the atmospheric surface layer ($\bar{z} < -L$). This is also true for 14 runs at x = 200 m. At x = 800 m the plume extends well above the surface layer ($\bar{z} > -L$). Then so-called free convection scaling is appropriate. But from Fig. 2 it is seen that the predictions of the surface layer model are still acceptable at x = 800 m.

   Finally, in table 3 calculated values of $\bar{U}_p$ are given with (6). Also the observed 8 m wind speed ($\bar{U}_8$) is listed. It is seen that $\bar{U}_8$ overestimates U_p for x = 50 m, U_8 compares reasonable with $\bar{U}_p^8$ for x = 200 m and $\bar{U}_8$ underestimates $\bar{U}_p$ for x = 800 m.

However, the differences are within 25%. Thus $\bar{U}_8$ is a reasonable good estimate for $\bar{U}_p$ in our data set.

Table 3 The calculated windspeed of the plume ($\bar{U}_p$) at three distances (x) together with the observed wind speed at 8 m($\bar{U}_8$) for each run.

Run	$\bar{U}_8$ (ms^{-1})	$\bar{U}_p$(ms^{-1}) x= 50 m	$\bar{U}_p$(ms^{-1}) x= 200 m	$\bar{U}_p$(ms^{-1}) x= 800 m
1	3.2	2.8	3.3	3.8
5	7.0	5.5	6.8	8.1
7	5.1	4.3	5.3	6.0
8	5.4	4.4	5.3	6.3
9	8.4	6.5	8.0	9.6
10	5.4	4.5	5.5	6.3
15	3.8	3.3	4.0	4.5
16	3.6	3.3	3.9	4.3
19	7.2	5.7	7.0	8.3
20	11.3	8.7	10.7	12.8
25	3.2	2.9	3.4	3.8
26	7.8	6.1	7.4	8.9
27	7.6	5.9	7.2	8.6
30	8.5	6.6	8.0	9.6
43	6.1	5.0	6.1	7.1
44	7.2	5.7	7.0	8.3
49	8.0	6.4	7.8	9.2
50	8.0	6.2	7.6	9.1
51	8.0	6.2	7.6	9.1
61	9.3	7.2	8.7	10.5

THE FREE CONVECTION DIFFUSION MODEL

Free convection takes place in the convective boundary layer in a region between the surface layer and the mixed layer. In the free convection region it appears that only $\frac{g}{t}\overline{WT}$ is the characteristic parameter for homgeneous and stationary turbulence (Nieuwstadt,1980). Panofsky (1978) states that the free convection region occurs for $-L < z < 0.1\ h$.

In the free convection region the vertical dispersion must follow free convection scaling. Then the normalised crosswind integrated concentration C_y/Q is given by (Nieuwstadt, 1980)

$$\frac{C_y}{Q} = 0.9\ \overline{U}_p^{\frac{1}{2}}(\frac{g}{T}\overline{WT})^{-\frac{1}{2}}\ x^{-3/2}\ . \tag{8}$$

It is seen that C_y/Q not depends on the boundary layer height h, nor on the surface roughness length z_0.

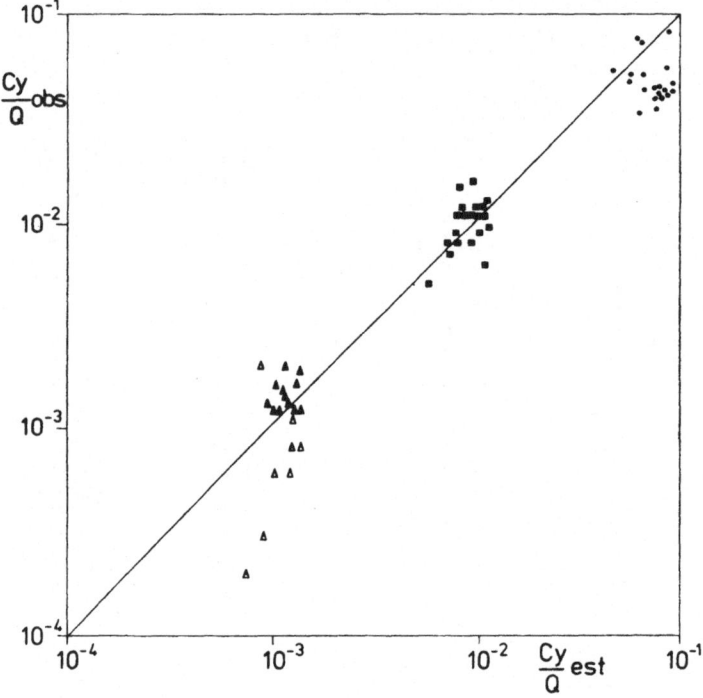

Fig. 3. As Fig. 1 but here (C_y/Q) est is calculated with Eq (8).

Fig. 3 shows a comparison of estimates with (8) and observations for C_y/Q. For the calculations we have used $\overline{U}_8$ for the velocity $\overline{U}_p$ in (8). It is seen that the agreement is good at x = 200 m and x $\overset{\text{p}}{=}$ 800 m. At x = 50 m the agreement is less. This is not surprisingly because the plume is still within the atmospheric surface layer and then z < -L (see table 2). At x = 800 the plume is sometimes outside the free convection layer because z > 0.1 h (see table 2). In the latter conditions still a reasonable agreement is found which indicates that free convection scaling also applies for z > 0.1 h (Nieuwstadt, 1980).

COMPARISON OF THE DIFFUSION MODELS

In this section we will compare the relative skill of the three models of this paper for three distances (50, 200 and 800 m). Table 4 shows observed and computed averages of the normalised concentration C_y/Q together with root mean square errors between computed and observed values.

At x = 50 m it is seen that the predictions of the Gaussian plume model and the surface layer model are similar. Both models furnish concentrations which are within 25% of the observed average. The free convection model predictions are about 50% too large. This can be explained by the fact that at x = 50 m the plume is still within the surface layer. Then free convection scaling is not appropriate.

Table 4 Comparison of observed averages of the observed normalised concentration C_y/Q with the three models for C_y/Q at three distances. Here n is number of runs. Indices 0, 1, 2, 3 refer to observed value, Gaussian plume model, surface layer model and free convection model, respectively. σ_{ij} is the root mean square error between $(C_y/Q)_i$ and $(C_y/Q)_j$.

	x=50m	x=200m	x=800m
n	20	20	12
C_y/Q 0	499	103	14
C_y/Q 1	416	115	34
C_y/Q 2	422	93	12
C_y/Q 3	752	94	12
σ_{01}	107	39	20
σ_{02}	105	27	4
σ_{03}	318	27	3
σ_{12}	35	37	22
σ_{13}	365	33	22
σ_{23}	356	8	2

At x = 200 m it is seen that the predictions of the surface layer model and the free convection model are similar. Apparently both scaling principles can be applied to this distance. The same is found for x = 800 m with the Prairie grass data.

At x = 200 m the agreement of the Gaussian plume model with observations is still satisfactory, but at x = 800 m the estimate is about a factor 2 larger (only the most reliable concentrations are taken into account). This result partly can be explained by the fact that the observed 8 m wind speed is used, which is on the average 25% smaller than the transport speed $\overline{U}_p$ at x = 800 m (see table 3). After correction the values are still overpredicting a factor of 1.75.

CONCLUSIONS

In this study the skill of three diffusion models is compared with observations at three distances (50, 200 and 800 m). These models use characteristics boundary layer parameters as recommended by Hanna et al. (1977).

It is found that the surface layer diffusion model predicts the concentrations very well on all the three distances. For larger distances the agreement of the free convection model is also satisfactory. At shorter distances free convection is not appropriate because the plume is still within the surface layer.

The Gaussian plume model overestimates on larger distances. This can be explained by underestimates of the vertical dispersion coefficient given by Pasquill (1974). In practice the other models have a better overall skill. The basic parameters needed for practical application of the models can be obtained from standard meteorological data (Holtslag and Van Ulden, 1982).

ACKNOWLEDGEMENTS

The author thanks Dr. H. van Dop and Mr. A.P. van Ulden for their comments on the draft of this paper.

REFERENCES

Barad, M.L., Ed., 1958, Project Prairie grass, a Field Program in diffusion. Geophys.Res. Rap. no. 59, Vols. 1 and 2, Geophysics Research Directorate. Air Force Cambridge Research Center, Bedford.

Businger, J.A., 1973, Turbulent transfer in the atmospheric surface layer, in: "Workshop on Micrometeorology", D.A. Haugen, ed., A.M.S., Boston.

Dyer, A.J., 1974, A review of flux-profile relationships. Boundary Layer Meteorol. 7, 363:372.

Golder, D., 1972, Relations among Stability parameters in the surface layer, Boundary Layer Meteorol., 3, 47:58.

Hanna, S.R., G.A. Briggs, J. Deardorff, B.A. Egan, F.A. Gifford F. Pasquill, 1977, AMS Workshop on stability classification schemes and sigma curves-Summary of recommendations, Bull. Amer. Meteor. Soc., 58, 1305:1309.

Holtslag, A.A.M., H.A.R. de Bruin and A.P. van Ulden, 1981, Estimation of the sensible heat flux from standard meteorological data for stability calculations during daytime, in: "Air Pollution Modeling and its Application I, C. De Wispelaere, ed., Plenum, London.

Holtslag, A.A.M. and A.P. van Ulden, 1982, A simple scheme for daytime estimates of the surface fluxes from routine weather data, Submitted to J. Appl. Meteor. (Institute report FM-82-12 available on request).

Nieuwstadt, F.T.M., 1980, Application of mixed-layer similarity to the observed dispersion from a ground-level source, J. Appl. Meteor. 19, 157:162.

Panofsky, H.A., 1978, Matching in the convective planetary boundary layer, J. Atmos. Sci., 35, 272:276.

Pasquill, F., 1974, "Atmospheric Diffusion", Wiley, London.

Paulson, C.A., 1970, The mathematical representation of wind speed and temperature profiles in the unstable atmospheric surface layer, J. Appl. Meteor., 9, 856:861.

Tennekes, H., 1982, Similarity relations, scaling laws and spectral dynamics, in: "Atmospheric Turbulence and Air Pollution Modelling", F.T.M. Nieuwstadt and H. van Dop, eds., Reidel, London.

Van Ulden, A.P., 1978, Simple estimates for vertical diffusion from sources near the ground, Atm. Env., 12, 2125:2129.

Wieringa, J.,1980, A revaluation of the Kansas mast influence on measurements of stress and cup anemometer overspeeding, Boundary Layer Meteorol., 18, 411:430.

DISCUSSION

Remarks from the author :   Please note that in the above
                    models no deposition is taken into account. We have
                    just taken the three models as they are published
                    in literature to investigate the skill on the same
                    data set. When deposition is taken into account the
                    skill of the Gaussian plume model might be better
                    at 800 m (see the paper of S.E. Gryning and S.E.
                    Larsen in this book for the influence of deposition)

                    In the paper we have illustrated that the skills of
                    the surface layer model and the free convection model
                    are at least comparable to the skill of the Gaussian
                    plume model. Besides of this the first two models are
                    based directly on boundary layer parameters which
                    rule the turbulence. Therefore we prefer the first
                    two models.
                    It should be noted that the constant in Eq. (8) is
                    obtained by Nieuwstadt (1980) partly from the same
                    data set as we have used here.
                    Finally we pay your attention on the fact that at
                    larger distance (beyond 1 Km in this data set) the
                    plume will be in the well-mixed boundary layer. Then
                    so-called mixed layer scaling is appropriate (see
                    Nieuwstadt (1980) and the paper of Venkatrom in this
                    book).

N.D. VAN EGMOND   Can the underestimation of σ_z
                    in the Gaussian model result from truncation or
                    systematic discrepancies in the Golder stability
                    conversion scheme which was used to derive Pasquill
                    stability classes ?

A.A.M. HOLTSLAG   Indeed systematic discrepancies
                    in the Golder stability conversion scheme can exist.
                    But it is likely unprobably that this results in more
                    than a factor of 2 systematic underestimation of σ_z
                    as we have found in Fig. 1 for the travel distance
                    x = 800 m. Besides of this the Golder scheme was found
                    in agreement with the method of Pasquill (1974) for
                    the Prairie Grass data. Therefore it is more likely
                    that the σ_z values are doubtfull in combination with
                    the skill of the Gaussian plume model.

VALIDATION OF A SINGLE SOURCE SIMULATION BY MEANS OF PAST AND

ACTUAL AIR QUALITY MEASUREMENTS

J.G. Kretzschmar, G. De Baere and J. Vandervee

SCK/CEN
B-2400 Mol
Belgium

INTRODUCTION

When the construction of the new 600 MWe fossil fuel power plant of Langerlo, in the northern part of Belgium, started, the electric power company initiated a SO_2 monitoring programme in order to determine the already existing SO_2-pollution in the vicinity of the planned power plant. Daily measurements in eight monitoring stations (Fig. 1) were started in 1973 by the N.V. Vinçotte. These measurements continued once the plant became operational. The statistical interpretation of the data collected during several successive years, respectively without and with the power plant emitting sulphur dioxide, made it possible to define the measured impact of the new source as a function of time and space.

Once the plant was operational ambient SO_2-levels in the eight monitoring stations were also calculated by means of the SCK/CEN's bi-Gaussian Immission Frequency Distribution Model (IFDM), using actual emission data and meteorological data from the meteorological tower at the SCK/CEN in Mol, some 35 km towards the northwest of the power plant.

As the SO_2-pollution in the region under investigation was already clearly above what could be named general background levels, even before the installation of the power plant, and taking into account the simple fact that a single source only influences a specific site when the wind is blowing in the right direction, the best way to define the impact of the new source seemed to be by means of the following wind direction dependent analysis of measured ambient levels respectively before and after the construction of the power plant.

631

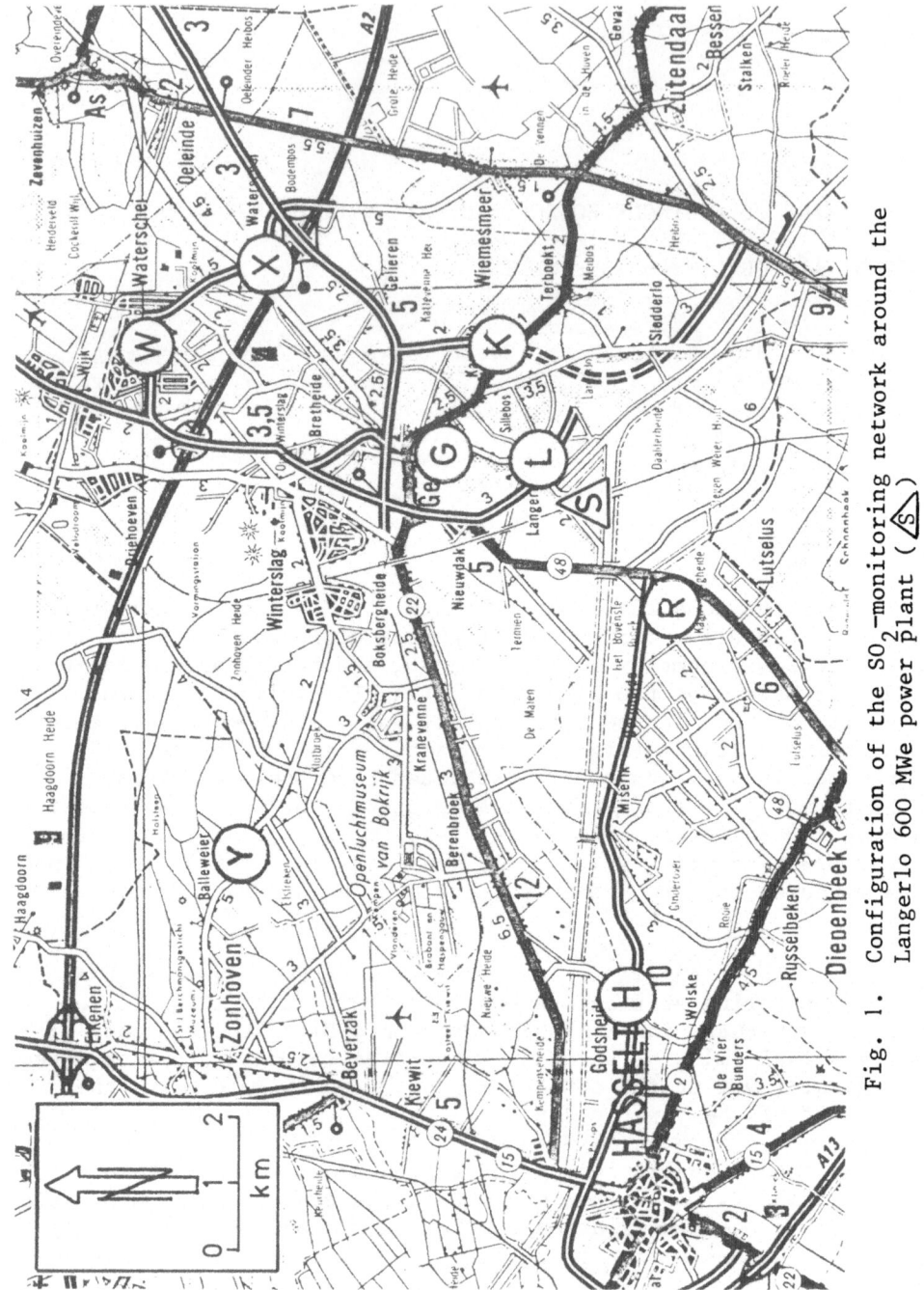

Fig. 1. Configuration of the SO_2-monitoring network around the Langerlo 600 MWe power plant (Ⓢ)

Measured daily SO_2-levels and hourly wind directions during a reference year (1974) before the start of the power plant are used to determine average SO_2-levels as a function of wind direction in each of the eight monitoring sites. The spatial homogeneity of the SO_2-pattern with respect to the location of the planned power plant is controlled by means of the ratios between downwind and upwind SO_2-levels for representative receptor-wind direction combinations. Assuming that, apart from the new power plant, all other sources remained constant or changed in the same way as a function of time, the increase in the corresponding ratios for a second reference year (1979), once the power plant was emitting SO_2, defines the impact of that source as a function of wind direction.
Subtracting the calculated SO_2-levels, solely due to the power plant in the said reference year 1979, from the corresponding measured ones must reduce the ratios to their original values of the first reference year 1974 if the simulation is realistic.

Once the model has been verified by this procedure it is used to calculate statistics of short-term levels, e.g. half-hourly or hourly SO_2 ground-level concentrations for actually occurring meteorological situations. The former are checked then by means of corresponding semi-mobile measurements up- and downwind the power plant, in order to verify the model with respect to its ability to simulate the maximum impact of the source under investigation.

ANALYSIS OF THE MEASURED SITUATION IN 1974 AND 1979

As well for 1974, reference year before the start of the power plant, as for 1979, reference year with the power plant in operation, daily average SO_2-levels are available for the eight monitoring stations given in figure 1 (Vinçotte, 1975 and 1981). These data are combined with the corresponding hourly wind direction measurements at the 69 m-level on the meteorological tower in Mol to obtain SO_2 pollution roses for each of the monitoring sites in the following way (J. Kretzschmar et al., 1979):
    - only days with the 24 successive wind direction measurements within a sector not larger than 80 degrees are retained for further analysis,
    - the daily average SO_2-level measured during such a day is taken as being representative for each individual hourly wind direction (discretisation angle 10 degrees) observed during that day,
    - over a period of one year the arithmetic average of all the in this way attributed SO_2-levels to a specific wind direction sector of 10 degrees is the average yearly level for that direction.

It is obvious that this procedure to obtain pollution roses
from daily averages is only a first order approximation. Several
tests on data sets of half-hourly or hourly SO_2-measurements have
nevertheless proved that pronounced peaks in the corresponding
pollution roses are well reproduced when all SO_2-measurements are
first of all combined into daily averages and treated afterwards
with the previously described procedure to derive pollution roses
from daily SO_2-averages and hourly wind direction observations.
The major shortcoming of the method is the obvious decrease in the
peak values occurring when the receptor is under the plume axis
and the increase of the levels in the adjacent wind direction
sectors. Peak values in pollution roses based on daily average
ground-level concentration measurements are as a consequence
always lower limits for the actual peak values that would occur
when the wind is steadily blowing from the source to the measuring
site, while the levels in adjacent wind direction sectors tend to
be increased by the smearing out effect of the procedure. As in what
follows measured and calculated daily averages will be treated in
exactly the same way, and as the results will be solely used to
follow changes in ratios between levels obtained for the same wind
direction in different monitoring sites, these shortcomings are of
less importance anyway.

The resulting pollution roses for 1974 and 1979 are given in
figure 2. Note that the previously mentioned selection criterion for
the days with less or more persistent wind direction led to 158
acceptable days for 1974 and 121 for 1979. The 1974 situation in
figure 2 reveals clearly a wind direction dependent SO_2 pattern in
the region around the planned power plant S. Wind direction
dependency is similar in the different monitoring sites. Within the
sector of the dominant south to southwesterly winds (see windrose)
ratios between average levels in (future downwind) sites such as
L, G, K, X and the (upwind) site R are varying between 1.1 and 1.5
as can be noted in the upper part of Table 1. In 1979, once the
power plant is operating at 60 % of its full capacity on the average,
at all these sites relevant increases are noted (Table 1, lower part)
except for site L being too close ($\sim$ 800 m) to the two 140 m stacks
of the power plant. Note also that from 1974 to 1979 the absolute
levels in sites R and L increased proportionally so that their ratios
didn't change. As pointed out before, the impact of the new source
is clearly marked in some sites despite the unavoidable smearing out
effect of the applied procedure. In each of the sites K, G, X and W
the wind direction, wherefore the maximum increase in the SO_2
concentration (Table 4) occurs with respect to the background level
in the advected air (site R), corresponds very well with what's
expected taking into account the position of the different sites
with respect to the source (fig. 1 and 2).

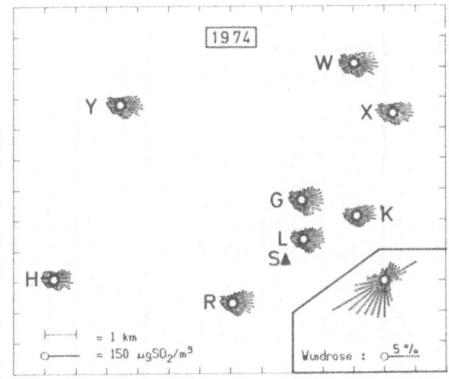

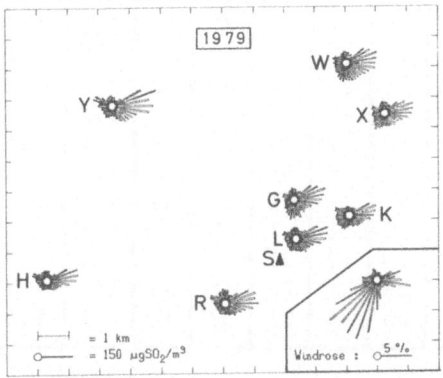

Fig. 2. Average measured SO₂-levels as a function of wind direction
respectively before (1974) and after (1979) the start of
the power plant S.

Table 1.   Average levels and their ratios with respect to the
            average levels in site R as a funtion of wind direction
            in the sector 160 to 260 degrees for respectively 1974
            and 1979

Year	dd	average levels (µg/m^3)						ratios				
		R	L	K	G	W	X	L/R	K/R	G/R	W/R	X/R
	160°	50	57	39	64	56	45	1.1	0.8	1.3	1.1	0.9
	170°	41	49	38	59	51	43	1.2	0.9	1.4	1.2	1.0
	180°	30	34	34	43	40	36	1.1	1.1	1.4	1.3	1.2
	190°	25	30	31	38	37	32	1.2	1.2	1.5	1.5	1.3
	200°	21	25	25	32	34	27	1.2	1.2	1.5	1.6	1.3
1974	210°	23	29	29	35	40	31	1.3	1.3	1.5	1.7	1.4
	220°	27	36	31	36	49	38	1.3	1.2	1.3	1.8	1.4
	230°	26	33	29	33	48	37	1.3	1.1	1.3	1.8	1.4
	240°	27	32	31	32	49	37	1.2	1.2	1.2	1.8	1.4
	250°	29	33	33	33	47	42	1.2	1.2	1.2	1.7	1.5
	260°	35	39	41	39	56	47	1.1	1.2	1.1	1.6	1.3
	dd	R	L	K	G	W	X	L/R	K/R	G/R	W/R	X/R
	160°	41	49	39	73	71	49	1.2	1.0	1.8	1.7	1.2
	170°	36	42	30	62	68	46	1.2	0.8	1.7	1.9	1.3
	180°	35	41	33	72	65	43	1.2	0.9	2.1	1.9	1.2
	190°	32	38	41	80	74	50	1.2	1.3	2.5	2.3	1.8
	200°	36	44	52	62	70	59	1.2	1.4	1.7	1.9	1.6
1979	210°	37	45	55	55	60	72	1.2	1.5	1.5	1.6	2.0
	220°	29	38	48	40	46	58	1.3	1.7	1.4	1.6	2.0
	230°	30	38	53	41	47	50	1.3	1.8	1.4	1.6	1.7
	240°	26	30	47	31	41	39	1.2	1.8	1.2	1.6	1.5
	250°	27	32	46	27	38	36	1.2	1.7	1.0	1.4	1.3
	260°	30	35	46	27	37	42	1.2	1.5	0.9	1.2	1.4

Figure 2 also illustrates why it is very difficult, and quite
often risky or even scientifically unsound, to compare air pollution
data from two different years without taking into account the
meteorological data and hoping nevertheless to determine the impact
of a known or unknown change in the source configuration.
Although in 1979 the influence of the new source is readily noted
for certain "site-wind direction" combinations, it is also obvious
from figure 2 that 1974 was in general quite different from 1979.

During the latter much more SO_2-pollution was noted for north-easterly to easterly winds, and this has nothing to see with the new industrial SO_2-source. The real reason is the exceptionally cold January and February months of 1979, increasing the overall SO_2-pollution in Belgium in general, due to increased emissions (space heating) as well as persistent limited dispersion conditions almost all over Europe. A somewhat unfortunate consequence of the specific characteristics of 1979 is that the eventual impact of the new source in sites R and H is masked by the high SO_2-levels already present in the advected air upwind the power plant.

From the previous analysis it can be concluded that once the power plant was in operation its SO_2-impact upon the ambient SO_2-levels, as measured in the monitoring network around the power plant, was detectable when the data were analysed as function of wind direction.

CALCULATED IMPACT OF THE POWER PLANT

By means of the bi-Gaussian Immission Frequency Distribution Model (J.G. Kretzschmar et al., 1980 and 1982) the impact of the power plant's emissions upon the SO_2-levels in the eight monitoring stations has been calculated for 1979 with the following input data :
- power plant at 60 % of its maximal capacity of 600 MWe
- two 140 m chimneys releasing each some 600 g SO_2/s (423 K, 126 Nm^3/s)
- hourly meteorological measurements at the 69 m-level in Mol
- plume rise formula of Stümke 2
- mixing height unlimited

Calculated hourly SO_2-averages are grouped to determine calculated daily average values, that are synchronous with the measured values. After subtraction of the former from the latter new daily averages are obtained that should be representative for measured values in the absence of the power plant, if the simulation is realistic. Pollution roses of these simulated 1979 measurements (actual measurements minus calculated fraction due to the power plant) are then computed as before and the results are given in figure 3. This situation must be comparable to the 1974 situation. Upwind-downwind ratios for the dominant wind directions, summarized in Table 2a, show that these approach unity after subtraction of the calculated contribution from the power plant. In comparison with the corresponding ratios in Table 1 for 1974, year during which there wasn't any power plant, the values of Table 2a are somewhat low. It has not been possible to determine if this is due to an overestimation in the simulations or if it is part of the variability of these ratios from one year to another.

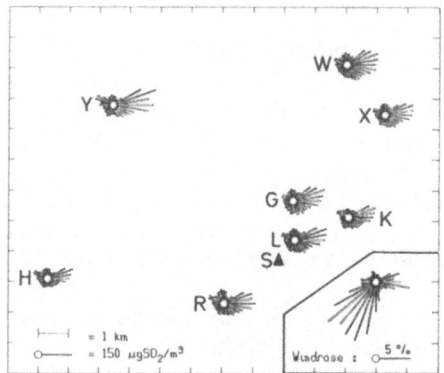

Fig. 3.   Average measured daily SO_2-levels as a function of
          wind direction after subtraction of the calculated
          daily averages due to the emissions of the power plant S

Table 2a. Average levels and their ratios for the simulated 1979
           measurements (measured values minus calculated
           contribution of the power plant)

dd	average levels ($\mu g/m^3$)						ratios				
	R	L	K	G	W	X	L/R	K/R	G/R	W/R	X/R
160°	41	49	39	53	55	45	1.2	1.0	1.3	1.3	1.1
170°	36	42	30	36	46	40	1.2	0.8	1.0	1.3	1.1
180°	35	40	29	42	43	33	1.1	0.8	1.2	1.2	0.9
190°	32	35	31	47	43	32	1.1	1.0	1.5	1.3	1.0
200°	36	41	38	33	36	34	1.1	1.0	0.9	1.0	0.9
210°	37	41	36	35	35	35	1.1	1.0	1.0	1.0	1.0
220°	29	32	26	27	29	23	1.1	0.9	0.9	1.0	0.8
230°	30	33	25	31	35	20	1.1	0.8	1.0	1.2	0.7
240°	26	27	17	24	30	20	1.0	0.6	0.9	1.2	0.8
250°	27	29	20	25	32	24	1.1	0.7	0.9	1.2	0.9
260°	30	30	21	25	32	28	1.0	0.7	0.8	1.1	0.9

Results of the same procedure applied on two other wind
directions are summarized in Table 2b. Here too good agreement is
obtained between the 1974 measured situation and the 1979 one
after correction for the calculated contributions of the power
plant. Note that the observed overestimation (or yearly fluctuations)
for the dominant southwesterly winds is not seen for east-
northeasterly to southeasterly wind directions.

A second independent check of the model was possible by means
of a limited set of measured half-hourly SO_2-levels obtained in a
minimax-approach (J. Kretzschmar et al., 1981) around the power plant.
Since the start-up of the power plant the N.V. Vinçotte (Vinçotte,
1982) systematically searched and measured in a mobile way the
maximum half-hourly SO_2 concentrations due to the power plant's
emissions. The collected data showed that the highest values in the
measured half-hourly levels ranged from 320 µg SO_2/m^3 (98th-
percentile) to 477 µg SO_2/m^3 (max.max.). The corresponding
calculated highest hourly levels by means of the IFDM for the
reference year 1979 ranged from 310 µg SO_2/m^3 (98th-percentile
in site K) to 440 µg SO_2/m^3 (max.max. in sites K and R).
The excellent agreement between measured and calculated upper values
is obvious.

Table 2b. Downwind and upwind average levels and their ratios
           for east-northeasterly to southeasterly winds.

dd	1974 (measured)					1979 (measured)					1979 (meas.-calc.)				
	H	L	G	K	$\frac{3H}{L+G+K}$	H	L	G	K	$\frac{3H}{L+G+K}$	H	L	G	K	$\frac{3H}{L+G+K}$
60°	65	78	85	77	0.8	95	103	115	96	0.9	83	103	115	96	0.8
70°	66	79	85	80	0.8	137	146	143	119	1.0	116	146	143	119	0.9
80°	60	75	81	67	0.8	134	136	129	111	1.1	100	136	129	111	0.8
90°	58	79	85	75	0.7	121	126	116	102	1.1	87	126	116	102	0.8
100°	73	84	88	76	0.9	91	105	100	87	0.9	71	105	100	87	0.7

dd	1974 (measured)					1979 (measured)					1979 (meas.-calc.)				
	Y	L	G	R	$\frac{3Y}{L+G+R}$	Y	L	G	R	$\frac{3Y}{L+G+R}$	Y	L	G	R	$\frac{3Y}{L+G+R}$
110°	68	67	73	70	1.0	102	71	71	81	1.4	89	71	71	80	1.2
120°	85	80	88	79	1.0	94	61	58	67	1.5	76	61	58	67	1.2
130°	52	55	58	48	1.0	82	53	52	52	1.6	62	53	52	52	1.2
140°	54	61	71	47	0.9	78	61	59	55	1.3	59	61	54	55	1.0
150°	58	61	66	51	1.0	64	47	45	41	1.4	45	47	42	41	1.0

CONCLUSIONS

By means of a practical example it has been shown how 24 h-
average SO_2-levels, measured over a sufficiently long period of time
in a number of sites before and after the start of a new 600 MWe
power plant, can be combined with hourly wind direction measurements
to determine the contribution of that source to the overall SO_2-
pollution already present in the region under investigation.

It has alo been illustrated that a dispersion model can be
validated in such a situation by subtracting calculated daily
averages levels (solely due to the new source) from measured ones,
followed by an appropriate analysis of the data as a function of the
wind direction and the source-monitoring network configuration.

Short-term levels calculated by means of the same model showed
to be in very good agreement with the results of an independent
mobile minimax-campaign measuring the maximum half-hourly levels
due to the source under investigation.

ACKNOWLEDGEMENTS

This research has been carried out in close collaboration with the power company EBES and the control agency N.V. Vinçotte making the emission and immission data available for further analysis and use in the IFDM code. The stimulating discussions with J. Decoussemaeker (Ebes, Langerlo), Deprez (Ebes, Antwerpen) F. Gheeraert (Ebes, Kallo) and M. Warzée (Vinçotte) were very helpful indeed.

REFERENCES

Cosemans, G., Kretzschmar, J.G., De Baere, G. and Vandervee, J., 1982, Large scale validation of a bi-Gaussian dispersion model in a multiple source urban and industrial area, Proc. of the 12th NATO/CCMS, San Francisco 1981.

Kretzschmar, J.G. and Cosemans, G., 1979, A five year survey of some heavy metal levels at the Belgian North Sea coast, Atm. Env., 13:267.

Kretzschmar, J.G., De Baere, G. and Vandervee, J., 1980, Dispersion estimates for short or long-term releases based upon the Immission Frequency Distribution Model (IFDM) of the SCK/CEN, Mol, Belgium, Proc. of the Seminar on Radioactive Releases and their Dispersion in the Atmosphere following a hypothetical Reactor Accident, CEC, Luxembourg.

Vereniging Vinçotte, 1974, Eerste metingsprogramma van de lucht-verontreiniging in de omgeving van de centrale van Langerlo. Toestand voor de ingangzetting van de centrale, N71304E GMW.

Ibidem, 1975, N17304E NMW.

Ibidem, 1981, Meetgegevens meetnet Langerlo voor 1979, personal communication.

Ibidem, 1982, N73072B, Vergelijking van de door het SCK berekende niveaus met de door Vinçotte gemeten niveaus, personal communication.

Kretzschmar, J.G. and Cosemans, G., 1981, Random- and minimax-campaigns for the determination of the actual air pollution levels in an unknown region, Atm. Env., 15:1047.

APPLICATION OF A THREE-DIMENSIONAL MODEL TO THE CALCULATION OF TRAJECTORIES: A CASE STUDY

Thomas T. Warner, Rosa G. de Pena, John F. Takacs and
Roderick R. Fizz

The Pennsylvania State University
Department of Meteorology
503 Walker Bldg.
University Park, PA 16802

INTRODUCTION

A complex winter storm occurred in eastern United States on
14-15 January 1980. During its passage, five sequential samples of
precipitation were collected at State College, Pennsylvania as part
of an intensive study of the storm by the Dept. of Meteorology at
the Penn State University. Twenty four-hour back trajectories from
State College were calculated for each precipitation sample, using
the horizontal and vertical winds predicted by a 3D numerical primitive equation model. The source regions of the precipitating atmosphere were then compared with some chemical properties of the rainfall. The rainfall samples were analyzed for the following ions:
hydronium(H_3O^+), sulfate ($SO_4^=$), nitrate (NO_3^-), chloride (Cl^-) and
sodium (Na^+) among others. Some of the concentrations changed significantly during the period of the precipitation event. Of perhaps
greatest interest is the ratio of Cl^- to Na^+ (Cl^-/Na^+) which decreased from values of 0.9 and 1.15 near the beginning of the event
to less than 0.6 during the sampling period of about 12 h. Because
the Cl^-/Na^+ ratio in the sea water is 1.18, the data suggest a source
region which changes from maritime to continental during the observation period. The transport model results support this hypothesis.
Trajectories that arrive over State College at 900 mb during the
early part of the sampling period show that the transport was from
a maritime region over the Atlantic Ocean to the southeast. As time
progressed in the sampling period, however, the source regions for
this low-level air became more continental. The predicted 24 h
transport that occurred prior to the last rainfall sample showed a
source region to the west of the Appalachian Mountains.

These calculations illustrate how a 3D dynamic model can be used to predict transport within a meteorological situation that is sufficiently complicated by vertical motions and small scale effects so that conventional transport calculations would be rendered largely ineffective.

METEOROLOGICAL SITUATION

This case was studied for two reasons. First, the presence of copious precipitation throughout the experiment, associated with a widespread winter storm, provided a difficult transport modeling environment. Second, storm analyses and data acquisition performed at Penn State during the period of interest aided in the calculation and verification of air parcel trajectories.

During the period under study, 0000 GMT 14 January - 0000 GMT 15 January 1980, most of the eastern United States received an onslaught of bad weather. Precipitation amounts of over one inch were recorded from Florida to Pennsylvania. Coastal regions received rain, while the Appalachian Mountain regions experienced a mixture of snow, rain, and freezing rain. State College had a 2.11 cm accumulation of mixed precipitation. The Penn State radar received precipitation echoes between the surface and 600 mb for most of the period. At times, echoes returned from as high as 450 mb. Surface temperatures at State College remained between -2 and +0.5 degrees C throughout the study.

A large high pressure system prevailed over northeast United States early in the period and was characterized by strong cross-isobaric flow. The high moved northeastward out of the domain while a ridge persisted east of the Appalachians. From Virginia southward, north-northeasterly ageostrophic flow continued into the middle of the period. Cyclogenesis was occurring off the South Carolina coast, as a 1010 mb low intensified while moving off of Cape Hatteras. By the end of the period, this cyclone dominated the coastal wind and pressure fields. Frontogenesis took place in a confluence zone off of the North Carolina coast. At upper levels, a trough was situated in the western half of the domain, resulting in south-southwesterly flow over the eastern portion of the region. In the northwestern sector of the domain, an intense low weakened as it progressed eastward. Later in the period, a high pressure system swept into the northwestern portion of the region.

PRECIPITATION SAMPLING PROCEDURE

Sequential precipitation samples for the event were collected with a modified version of an ERN1 721 automatic wet deposition only collector.

The modification consisted of adding a motor powered turntable which supports a tray with twelve sample containers. A chrome-steel collecting funnel is connected to a steel spout that is suspended over the collecting containers.

The time each container spends under the collecting spout can be controlled with a digital timer. This time can be varied according to the precipitation rate, from one minute to ninety-nine hours.

For the event under study, five samples were collected. The collection times, the precipitation type, the surface air temperature and the relative humidity for the five samples are given in Table 1.

Table 1   Collection times and precipitation
type for the sequential samples.

Sample	Time (GMT)	Precipitation Type	Temperature °C	Relative Humidity %
1	0800-1330	Snow-Freezing Rain	-2.0	98
2	1330-1830	Rain	0	95
3	1830-1950	Rain	0	95
4	1950-2115	Rain	0	95
5	2115-0230	Rain	0.5	95

Each sample was analyzed by the Battelle P.N. Laboratories according to the procedure established by the MAP3S program.

The chemical analysis results indicate that the concentration of sulfate was practically the same during the whole event, exhibiting the low-value characteristic of the winter samples.

The concentration of hydronium and nitrate ions changed during the event in a similar way. These concentrations decreased from the first to the fourth sample, but increased again in the last sample. The precipitation rate changed in an opposite way. This inverse correlation between concentration and precipitation intensity has been observed for these ions in every event at the Penn State site.

Chloride and sodium concentrations followed a different pattern. Both ion concentrations decreased from the first to the second sample, but from then on while chloride concentration increased slightly in the third sample and remained constant for the rest of the event, the sodium concentration increased continuously up to the fifth sample with the result that the ratio of chloride to sodium changed constantly. These ratios are shown in Fig. 1.

Whereas for samples one and two the ratio is near one, it decreases appreciably for the other three samples. We note that this ratio is about equal to the value 1.18 for sea water and very near t that for the maritime aerosol.

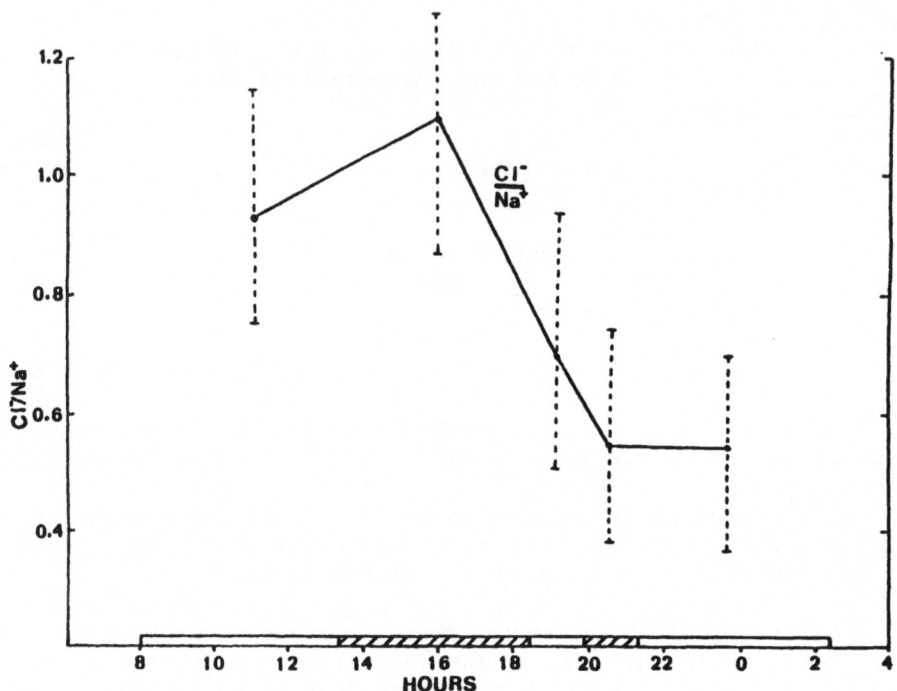

Fig. 1.   Ratio of Cl^-/Na^+ in five precipitation samples from State
College, PA, 0800 GMT 14 January to 0215 GMT 15 January.
Length of sampling times is indicated by alternating hatched
and unhatched areas.

TRAJECTORY MODEL

The mesoscale dynamic model which supplied wind data for the transport calculations was developed, tested, and refined at Penn State over the past 8 years. The model, as described by Anthes and Warner (1978), is hydrostatic and applicable over a wide range of horizontal space scales. This version of the mesoscale model is applied on a Lambert conformal map projection. The vertical coordinate is sigma defined by

$$\sigma = \frac{p_s - p_t}{p_*}$$

where $p_* = p_s - p_t$, p_s is the surface pressure, and p_t is the pressure at the top of the model. The major structural characteristics of the three model versions are listed in Table 2.

Table 2 Model Structural Characteristics

Structural Characteristic	Value
Number of model levels (σ values)	11: (1.0, .94, .87, .79, .71, .61, .50, .39, .28, .17, 0.0)
grid spacing	70 km
pressure at top of model	100 mb
time step	140 s
forecast grid mesh size	41 x 41

An initialization procedure creates the necessary initial conditions of temperature, horizontal wind, moisture, and surface pressure for the model, using archived analyses and rawinsonde data. Temperatures are balanced with the nondivergent wind field to reduce the intensity of spurious gravity waves which can mask important meteorological features.

The planetary boundary layer was represented by the lowest model layer and the vertical fluxes of momentum were treated by a version of the bulk parameterization of Deardorff (1972). No surface heat fluxes existed.

The lateral boundary conditions for the meteorological variables were obtained through a linear interpolation in time between the 12-hourly observations of those variables.

Kinematic trajectories are produced by interpolating horizontal and vertical winds in three dimensions from the model grid points to the trajectory endpoints. The endpoint is displaced in the

direction of the wind with its speed for a small time interval (say
5 to 15 min.).   This process is repeated using new wind information
at the new endpoint of the trajectory.   Updated wind information is
available from the dynamic model at every time step (5 min. in this
case).

RESULTS

     Figure 2 displays backward-in-time trajectories computed using
the previously described technique.   All trajectories begin at a

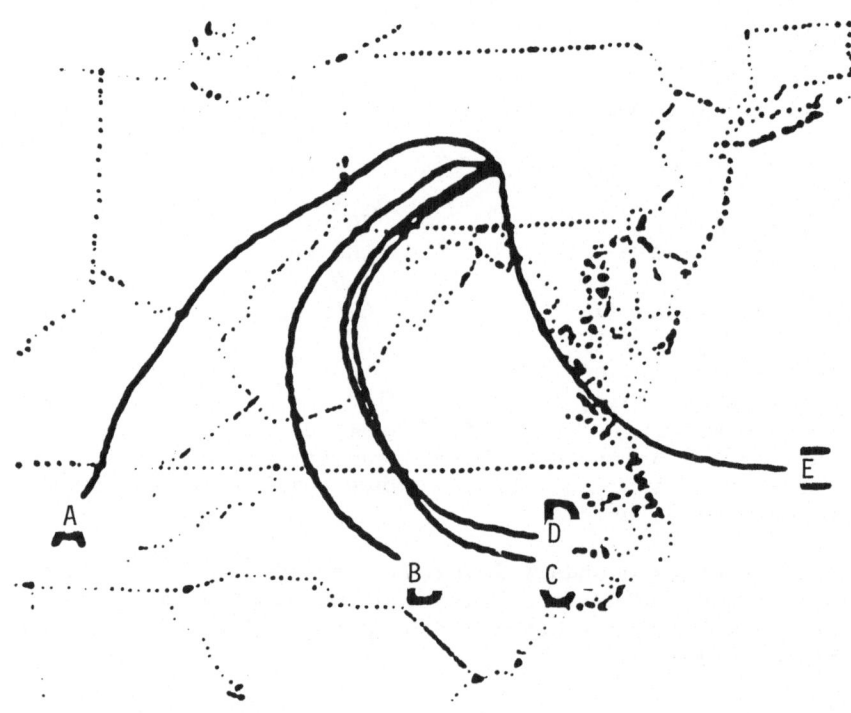

Fig. 2.   Backward-in-time trajectories with a common starting point
          at 900 mb over State College, PA.   Starting times over
          State College are 1100 GMT (E), 1600 GMT (D), 1900 GMT (C),
          2100 GMT (B) 14 January 1980; and 0000 GMT 15 January 1980
          (A).   All back trajectories terminate at 0000 GMT 14 Jan-
          uary 1980.

pressure level of 900 mb over State College, PA. The back trajectories end at 0000 GMT 14 January and begin over State College at
0000 GMT 15 January (A), 2100 GMT 14 January (B), 1900 GMT 14 January
(C), 1600 GMT 14 January (D), and 1100 GMT 14 January (E). The source
region of the low-level air arriving over State College, PA is diag-
nosed by the model as clearly changing from one that is maritime in
character to one that is continental. The source region for the air
arriving over State College at 1100 GMT 14 January (trajectory E in
Figure 2), the midpoint of the first precipitation sampling period,
was over the Atlantic Ocean. The observed ratio Cl^-/Na^+ is consistent with this analysis. During the next three sampling period, the
transport is from more continental regions to the south of State
College (trajectories B, C, D) and by the time of the last sampling
period the air originates to the west of the Appalachian Mountain
Range, a continental region (trajectory A). Again, the observed
trend in the ratio Cl^-/Na^+ reflects this predicted change in the
source region from a maritime one to a continental one.

REFERENCES

Anthes, R. A., and T. T. Warner, 1978: Development of hydrodynamic
    models suitable for air pollution and other mesometeorological
    studies. Mon. Wea. Rev., 106, 1045-1078.
Deardorff, J. W., 1972: Parameterization of the planetary boundary
    layer for use in general circulation models. Mon. Wea. Rev.,
    100, 93-106.

DISCUSSION

A. VENKATRAM   When you march forward in time
            in your dynamic model, do you attempt to make your
            predictions compatible with available observations ?

T. WARNER   Not in this case. However, we
            are now testing a procedure that will incorporate
            radiosonde observations into the model solution du-
            ring the forecast to limit the error growth in the
            large scale solution.

N.D. VAN EGMOND   Is the model already used for
            frontal precipitations after episodes of high sul-
            fate levels ?

T. WARNER   No, but I agree that it would
            be an interesting application of the model.

ESTIMATION OF UNCERTAINTIES IN TRANSFER MATRIX

ELEMENTS IN LONG RANGE TRANSPORT MODELING

P. K. Misra

Air Resources Branch, Ministry of the Environment
880 Bay Street, 4th Floor
Toronto, Ontario M5S 1Z8 Canada

INTRODUCTION

In long range transport problems the source receptor relationships are conveniently studied by the use of a transfer matrix when the model uses linear chemistry and scavenging phenomena. The size of the transfer matrix depends on the number of sources and receptors chosen. It is expected that each element of the transfer matrix will have a degree of uncertainty associated with it, which is attributed to the modelling errors and the variability of the atmospheric variables. The latter source of uncertainty is best understood if one considers the fact that models are designed to give ensemble average values which are difficult to obtain in the atmosphere because of the large number of scales which govern atmospheric flows.

It would be desirable to obtain quantitative estimates of the uncertainty of the transfer matrix elements in order to accurately determine the relationship between sources and receptors. There is no direct method of quantifying this uncertainty from modeling results alone. In this paper we present a method to obtain this uncertainty quantitatively by making suitable assumptions on the nature of atmospheric motions and using observed data. The paper also discusses the implications of the spatial and temporal variability of mixed layer heights in the long range transport of air pollutants.

VARIABILITY OF MIXED LAYER HEIGHT IN LONG RANGE TRANSPORT MODELLING

Most of the pollutant interaction with the surface occurs in the mixed layer. The mixed layer is the lowest layer of the

651

atmosphere where three dimensional turbulence of convective and
mechanical origin prevails. The mixed layer exhibits a diurnal
cycle at a fixed location, being of the order of 1 km during the
day time convective conditions and of a few meters during the
night time stable conditions. The mixed layer also exhibits
spatial variability due to topographical and weather variability.
We can therefore imagine the top of the mixed layer on a continent
scale to be a two dimensional surface which changes in a random
fashion.

Pollutants emitted from a tall stack can initially be within
or above the mixed layer depending on the height of the mixed
layer. Further, the trajectory of the pollutants in the long range
intersects the mixed layer in a random fashion being randomly
within or above the mixed layer. When averaged over a large number
of such pollutant trajectories one finds a fraction of the emitted
pollutants to reside above the mixed layer and the remaining fraction to reside within the mixed layer. There is of course a continuous interchange of the vertical position of the pollutants.

The implication of the above phenomenon is that only the mass
which resides within the mixed layer undergoes dry deposition,
whereas the mass above the mixed layer undergoes no dry deposition.
For those models which ignore the variability of the mixed layer
height the pollutant mass is assumed to experience dry deposition
always which may lead to over prediction of dry deposition.
Venkatram et al (1981) and Fisher (1978) account for this by
assuming smaller values for the dry deposition velocities.

DISTRIBUTION OF MASS INTO AND ABOVE THE MIXED LAYER

Let us define the mass which resides above the mixed layer by
a subscript a and the mass inside the mixed layer by a subscript
b. If M denotes the mass originating at the source, and M_a and M_b
the corresponding average masses above and inside the mixed
layer, then it follows that;

$$\frac{dM_a}{dt} = -\frac{M_a}{\tau_a} + \frac{M_b}{\tau_b} \tag{1}$$

and

$$\frac{dM_b}{dt} = -\frac{M_b}{\tau_b} + \frac{M_a}{\tau_a} \tag{2}$$

where τ_a and τ_b are the average periods for which the parcel is above and within the mixed layer.

Denoting the fraction of mass emitted above the mixed layer initially by f, the solutions to (1) and (2) can be given by the following:

$$\frac{Ma}{M} = \left[\frac{f\tau_b-(1-f)\tau_a}{\tau_a+\tau_b}\right] \left[e^{-\frac{(\tau a+\tau b)t}{\tau a \tau b}} -1\right] + f \tag{3}$$

$$\frac{Mb}{M} = \left[\frac{-f\tau_b+(1-f)\tau_a}{\tau_a+\tau_b}\right] \left[e^{-\frac{(\tau a+\tau b)t}{\tau a \tau b}} -1\right] +1-f \tag{4}$$

The right hand sides of (3) and (4) denote essentially the density functions for the probability with which the parcel is above or inside the mixed layer.

APPLICATION

The above concepts are best used in a statistical model. For this analysis we shall use the model developed by Venkatram et al. (1981). In this model the concentration field is determined from a knowledge of the probability density function and is given by the following equation.

$$C(\vec{r}) = \int_0^\infty Q\ P\ (\vec{r}|\vec{r}_s,t)dt \tag{5}$$

where Q is the emission rate, C the concentration field and P the density function of the probability that a particle released as $\vec{r}_s$ would be found at $\vec{r}$.

Atmospheric transport, wet and dry scavenging, and chemical transformation deterine P. As in Venkatram et al. (1981) we shall assume that all these processes are independent of each other. Thus the right hand sides of (3) and (4), which denote the probability of the parcel position in the vertical, are simply multiplied to the density function of Venkatram et al. (1981).

Several points are noteworthy . First, a mixed layer, however small, always exists close to the surface. However, during the night time and stable atmospheric conditions the amount of mass inside this shallow mixed layer may be assumed to be negligibly

small. Thus most of the mass exists above the mixed layer, so that
for Ma dry deposition can be assumed to be zero at night and during
stable atmospheric conditions. Second, it is assumed that the mass
is well distributed inside a depth Z_i, which is the average mixed
layer height taken as 1000m in Venkatram et al. (1981). This is
regardless of whether the pollutants are above or inside the mixed
layer. This arises from the consideration of continuity.

CASE STUDY

With the above modifications, the model of Venkatram et al.
(1981) is applied to the NE US/CANADA. We assume that f is 0.5 for
the top 55 sources in this region. f is set to be zero for the rest
of the sources. The value of f is expected to vary between sources.
It will depend on the actual plume height in relation to the mixed
layer height. The value of 0.5 for f for the top 55 sources is
based on our experience with Sudbury's INCO stack and other power
plant stacks in Canada. For details on source configuration and
agreegation for use in the model, the reader is referred to the
paper of Venkatram et al. (1981). The dry deposition velocity for
SO_2 is assumed to be 1 cm/sec and for SO_4 it is 0.1 cm/sec.

Table 1 shows the model prediction against observed wet
sulphur deposition at several receptors in NE America. It is difficult to standardise criteria for the validation of air quality
models. However, if the air quality variables are assumed to be
log-normally distributed, then the geometric mean of the ratio of
the observed and predicted values and the geometric standard deviation of the observed with respect of the predicted values can
be used as measures of the accuracy of the model. It is noted that
the geometric mean and the geometric standard deviations should be
ideally equal to one.

Table 1 shows the statistics for the predictions by the old
and the new versions of the models. The statistics are similar and
slightly improved for the new version of the model. It is noted
that in the old version of the model dry deposition velocities for
SO_2 and SO_4 are assumed to be 0.5 cm/sec and 0.05 cm/sec respectively. This is equivalent to the assumption of, f=0.5, in the new
version. Thus the similarity between the two versions is not
surprising. The new version, however, is more realistic.

UNCERTAINTY IN TRANSFER MATRIX ELEMENTS

The source receptor relationship for linear long range
transport models can be expressed in a matrix form as follows:

$$E_i T_{ij} = D_j \qquad\qquad\qquad (6)$$

Table 1

RECEPTOR NAME	PRED OBS $(gm^{-2}a^{-1})$	PRED (OLD)	(NEW)	WET SULPHUR DEPOSITION OBS/PRED (OLD)	OBS/PRED (NEW)
KINGSTON, ONT.	1.26	0.91	1.03	1.38	1.22
MOOSONEE, ONT.	0.58	0.33	0.38	1.76	1.51
MOUNTFOREST, ONT.	2.32	0.97	1.07	2.40	2.16
PETERBOROUGH, ONT.	1.81	0.93	1.34	1.95	1.74
PICKEL LAKE, ONT.	0.39	0.28	0.32	1.39	1.23
SIMCOE, ONT.	2.34	1.47	1.58	1.59	1.48
WAWA, ONT.	0.91	0.52	0.57	1.75	1.58
WINDSOR, ONT.	2.98	3.98	4.06	0.75	0.73
CHIBOUGAMAN, QUE.	1.06	0.43	0.49	2.47	2.15
MANIWAKI, QUE.	0.71	0.78	0.87	0.91	0.81
MONTREAL, QUE.	2.35	2.22	2.33	1.06	1.01
MERRIMACH CNTY, N.Y.	0.91	0.94	1.07	0.97	0.85
ALBANY CNTY, N.Y.	1.20	1.28	1.42	0.94	0.85
ALLENGANY CNTY, N.Y.	2.20	1.55	1.68	1.42	1.31
DUTCHNESS CNTY, N.Y.	1.20	2.67	2.81	0.45	0.43
ESSEX CNTY, N.Y.	0.84	0.84	0.96	1.00	0.88
ONEDIA CNTY, N.Y.	1.70	1.08	1.21	1.57	1.40
ONONDAGA CNTY, N.Y.	0.79	1.17	1.33	0.68	0.59
ONTARIO CNTY, N.Y.	1.20	1.34	1.47	0.90	0.82
ST. LAW. CNTY, N.Y.	1.00	0.89	1.00	1.12	1.00
OAK RIDGE, TENN.	1.30	1.05	1.19	1.18	1.09
CHARLO TESBILLE, VIR.	0.90	1.37	1.51	0.66	0.60
TUCKER CNTY, W.V.	2.00	2.07	2.21	0.97	0.91
WASHINGTON, D.C.	1.00	1.92	2.06	0.52	0.49
LEWISTOWN, PENN.	0.98	2.27	2.41	0.43	0.41
PADUCAH, KENTUCKY	0.57	1.99	2.05	0.2944	0.28

LINEAR ANALYSIS : OBSERVED DEPOSITION + b * PREDICTED DEPOSITION

RECEPTOR	R SQUARE	a	b	s	
ALL PT (NEW PREDICTED)	0.32	0.63	0.46	0.57	–
24, 25, 26 Removed	0.48	0.61	0.56	0.51	

GEOMETRIC MEAN AND STANDARD DEVIATION

RECEPTOR	GEOM. MEAN	GEOM. STANDARD DEV.
ALL PT. (OLD PREDICTED)	1.03	1.69
ALL PT. (NEW PREDICTED)	0.94	1.67

Where E_i is the pollutant emission rate of the ith source, D_j the deposition rate at the jth receptor, and T_{ij} the transfer matrix representing the mechanism of transport dispersion, scavenging, and chemical transformation. The summation is over repeated indices.

T_{ij} is a random variable which can be expressed as follows:

$$T_{ij} = \hat{T}_{ij} + T_{ij}' \tag{7}$$

where a 'hat' represents ensemble average and the prime the departure form the ensemble mean. The variance of T_{ij} about the ensemble average is given as follows:

$$\hat{T}_{ij}'^2 = \sigma_{ij}^2 \tag{8}$$

$\hat{T}_{ij}$ denotes the model prediction and T_{ij} the 'observed' values, if T_{ij} can be measured. The departure T_{ij}' is a consequence of two processes; the first is the uncertainty in model parameterisation, and the second is associated with the variability of large scale atmospheric flows. Assuming that these two processes are independent we can write σ_{ij}^2 as $m\sigma_{ij}^2$ corresponding to model parameterisation and $a\sigma_{ij}^2$ corresponding to the atmospheric variability.

Since we are mostly concerned with large scale flows, we can assume that $a\sigma_{ij}^2$ are independent of sources. Thus we can drop the subscript i form $a\sigma_{ij}^2$. Therefore, we get,

$$\sigma_{ij}^2 = a\sigma_j^2 + m\sigma_{ij}^2 \tag{9}$$

$m\sigma_{ij}^2$ can be obtained from distribution of the model parameters. To determine $a\sigma_i^2$ let us divide D_j into two components as in equation (7); i.e.

$$D_j = \hat{D}_j + D_j' \tag{10}$$

$\hat{D}_j$ is the ensemble average value and D_j is the value which is measured. $\hat{D}_j$ is the same as the model prediction at the j^h receptor. Thus comparing the model prediction to the observed deposition rates we can determine $\hat{D}_j^2 = \sigma D_j^2$.

Combining equations (6), (7), (8), (9) and (10) we get,

$$E_i \left(\hat{T}_{ij} + T_{ij}' \right) = \hat{D}_j + D_j' \tag{11}$$

Squaring both sides of (11) and assuming that the transfer matrix elements are uncorrelated (i.e. source areas are sufficiently removed from each other), we get the following:

$$a\sigma_j^2 = \left[\sigma D_j^2 - \sum_j E_j^2 m\sigma_{ij}^2 \right] / \sum_j E_j^2 \tag{12}$$

σ_{ij}^2 is a measure of the uncertainty in transfer matrix elements. An application of this in a source control scenario can be explained by the following inequality:

$$E_i \left[\hat{T}_{ij} + a\sigma_{ij} \right] \leq D_{Tj} \tag{13}$$

where D_{Tj} is the target deposition at the jth receptor and α is a factor denoting the probability of exceedance of T_{ij} over $\hat{T}_{ij}$.

CONCLUSIONS

In this paper we have presented a methodology to quantitatively obtain the uncertainties in transfer matrix elements of linear long range transport models. We have also presented a simple method to incorporate the effects of mixed layer variability in statistical long range transport models. Further research is necessary to accurately determine the average periods which the pollutants spend above and within the mixed layer. This requires simultaneous analyses of air parcel trajectories and mixed layer heights along the trajectories. Improved statistics of the mixed layer heights at fixed points in space would also be necessary to improve the model further. The preliminary results are very encouraging.

As the data base for observed deposition rates at receptors improves we can apply the method outlined in Section-3 to quantitatively obtain uncertainties in transfer matrix elements.

REFERENCES

Fisher, B.E.A., 1978. The Calculation of Long Term Sulphur Deposition in Europe, Sulfur in the Atmosphere, Proceedings of the International Symposium held in Dubrovnik, Yugoslavia, 489-509.

Venkatram, A., B.E. Ley, and S.Y. Wong; 1982, A Statistical Model to Estimate Long-Term Concentrative of Pollutants Associated with Long-Range Transport, Atmospheric Environment, Vol. 16, No. 2. pp. 249-257.

DISCUSSION

F.B. SMITH   I would question your assumption that the Transfer Matrix elements T_{ij} should be mutually independent. Common meteorological characteristics would suggest significant correlations between neighbouring T_{ij}.

P.K. MISRA   I agree with you and this concerns me as I pointed out in my presentation. However, if the source regions are large enough then this may be a reasonable assumption.

F.B. SMITH   Would you consider that your model which allows some of the pollution to travel some of the time above the boundary layer, would overcome the tendency of most current LRT models to underpredict depositions at long range from the major source areas ?

P.K. MISRA   I agree with you. I am currently working on a model of allowing the mass above the mixed layer to be embedded into clouds when in the wet phase. Chemical transformation within clouds being faster, this will also enhance wet deposition. This should overcome the tendency for the models to underpredict wet depositions.

EVALUATION OF A K-MODEL FORMULATED IN TERMS OF MONIN-OBUKHOV

SIMILARITY WITH THE RESULTS FROM THE PRAIRIE GRASS EXPERIMENTS

Sven-Erik Gryning and Søren E. Larsen

Physics Department
Risø National Laboratory
DK-4000 Roskilde, Denmark

ABSTRACT

A K-model describing the dispersion of a passive substance
was formulated in terms of Monin-Obukhov similarity theory, and
then solved numerically for the case of the release of a passive
substance from a point-source at ground-level. An extensive
analysis was undertaken to evaluate this numerical model. To do
this, we applied the widely used set of dispersion data for ground-
level sources that was obtained during the Prairie Grass experiments.
We simulated these dispersion experiments. Since SO_2, which was
used as tracer, is known to deposit on the ground, also the effect
of deposition of the tracer was investigated. The simulations
showed that the vertical concentration profile is greatly affected
by deposition. At a height of 1.5 m, where the measurements were
made, the effect is still weak 50 m from the source, but 800 m
downwind the deposition causes a decrease in the concentration by
about a factor of 2. When consideration is given to the effect
of deposition, the numerical solution of the diffusion equation
yields excellent agreement with the measurements of the crosswind-
integrated ground-level concentrations for both the 200 m and 800 m
distance. Indeed, at 800 m the predictions are consistent with
measurements even into the convective regime, where the use of
K-models becomes dubious.

INTRODUCTION

A cross-wind integrated K-model with wind- and K-profiles
described by Monin-Obukhov similarity relations was solved
numerically for the case of the release of a passive substance
from a point-source at ground-level.

An extensive analysis was undertaken to compare this numerical model with experimental results from the Prairie Grass dispersion experiments. The Prairie Grass experiments are of particular interest. They have formed the experimental basis for the evaluation of many semi-empirical models, e.g. Horst (1979), Nieuwstadt (1980), Briggs and McDonald (1978). Also these experiments form part of the experimental basis for the widely used curves for σ_y and σ_z by Turner (1970).

In these model evaluations the effect of deposition on the measured tracer concentrations has been neglected. The present model suggests that the measured tracer concentrations may be significantly influenced by deposition of the tracer.

THE DISPERSION MODEL

The modelling approach is based on a numerical solution of the diffusion equation for a continuous point-source

$$u(z) \frac{\partial \chi(x,z)}{\partial x} = \frac{\partial}{\partial z} \left[K(z) \frac{\partial \chi(x,z)}{\partial z} \right] \tag{1}$$

where χ is the crosswind-integrated concentration, u the horizontal wind speed, K the eddy diffusivity of matter in the vertical direction, x the downwind distance from the source, and z the height above ground. Diffusion in the x-direction is assumed to be negligible compared with the advection. The wind-profile and eddy diffusivity are expressed in terms of Monin-Obukhov similarity relations, which restrict the validity of this model to the surface boundary layer. Herein, the profiles can be described in terms of the surface roughness z_0, the friction velocity u , and the Monin-Obukhov stability length L.

Monin-Obukhov similarity theory predicts that the vertical eddy diffusivity can be expressed generally as

$$K = \kappa u_* z / \phi(z/L)$$

where κ is von Kármán's constant taken here to be 0.35, and $\phi(z/L)$ is an empirical function that describes the dependence on atmospheric stability. In the case of diffusion of heat, $\phi = \phi_h$ where ϕ_h is the dimensionless temperature gradient. Here, we assumed that the diffusion of heat and matter is identical. Several forms for ϕ_h have been suggested in the literature. The empirical formulas of Businger (1973) obtained from the Kansas experiment (Haugen et al., 1971) have been used here. They read, for L<0,

$$\phi_h = 0.74 \ (1-9z/L)^{-0.5}$$

and for L>0

$$\phi_h = 0.74 + 4.7 \ z/L \ .$$

The wind-profile used here reads

$$u(z) = \frac{u_*}{\kappa}(\ln(z/z_0) - \psi(z/L) + \psi(z_0/L)) \ ,$$

where the $\psi(z_0/L)$ term makes $u(z_0) = 0$. The stability correction in unstable conditions of the neutral wind profile, $\psi(z/L)$, was obtained (Paulson, 1970) by integrating the expression for the dimensionless wind shear. The expression reads

$$\psi(z/L) = 2 \ \ln((1+x)/2)+\ln((1+x^2)/2)-2\tan^{-1}(x)+\pi/2 \ \text{for} \ L<0,$$

where
$$x = (1-15 \ z/L)^{0.25} \ .$$

For stable conditions, it reads (Businger, 1973)

$$\psi(z/L) = - \ 4.7 \ z/L \ .$$

The boundary conditions applied to Eq. (1) are taken as

$$K\frac{\partial \chi}{\partial z} = v_g \chi \qquad \qquad \text{for} \ z = z_0$$

$$K\frac{\partial \chi}{\partial z} = 0 \qquad \qquad \text{for} \ z \rightarrow \infty \ ,$$

where v_g is the characteristic deposition velocity to be used close to the ground. It incorporates both the resistance to transport of the material through the laminar sublayer adjacent to the surface, and the resistance of the surface itself (Garland, 1977). The point-source was approximated by a Gaussian concentration profile in the vertical direction with maximum concentration at the release height and a standard deviation of $2z_0$.

The diffusion equation with boundary and initial conditions as described in the foregoing paragraph, was solved numerically by a finite-difference scheme proposed by Keller (1971). The scheme has second-order accuracy and is unique in the sense that $\chi(x,z)$ and $\partial\chi(x,z)/\partial z$ are approximated with the same accuracy.

The lower boundary condition is applied at the height z_0, which is taken as the ground-level. The upper boundary condition is applied at the height of the top of the grid. The vertical spacing of the grid, which is linear, is adjusted after each integration step in a way such that the concentration at the top of the grid is always kept below 0.1% of the maximum concentration.

SIMULATION OF THE PRAIRIE GRASS EXPERIMENTS

An extensive and widely used set of dispersion data for ground-
level sources was obtained during the Prairie Grass experiments,
(Barad, 1958), conducted in the summer of 1956 at O'Neill, Nebraska.

Sulphur dioxide, released at a height of 0.46 m, was used as
a tracer. Lateral concentration profiles (averaged over 10 min)
were measured at a height of 1.5 m along arcs situated 50, 100,
200, 400, and 800 m from the source. Vertical concentration profiles were measured at towers 100 m from the source at measuring
levels between 0.5 and 17.5 above the ground.

Here we used the values of the crosswind-integrated concentrations normalized with the tracer release rate that has been
calculated by van Ulden (1978) for the arcs 50, 200, and 800 m
from the source; estimates of the surface layer parameters u_* and
L were also made, as derived from the wind- and temperature-
profiles from the various experiments.

We simulated these dispersion experiments and then compared
the results from the numerical modelling with this data set. The
numerical calculations were performed for a source height of 0.46 m.

Since SO_2, which was used as tracer, is known to deposit on
the ground, the effect of deposition of the tracer was also investigated. The numerical computations were made without deposition
and with $v_g = 0.05\ u_*$, the latter corresponds crudely to a deposition velocity of 0.01 m/s at a height of 1 m.

Two examples of vertical concentration profiles under neutral
conditions are shown in Fig. 1. It can be seen that the concentration profile is greatly affected by deposition. At a height of
1.5 m, where the measurements were made, the effect is still weak
50 m from the source, but 800 m downwind the deposition causes a
decrease in the concentration by about a factor of 2.

In Fig. 2 the computed concentrations are compared with those
measured at a height of 1.5 m. It is seen that the effect of deposition is very weak at a downwind distance of 50 m. At a distance
of 200 m from the source most of the measured concentrations are
smaller than those predicted by the model with no deposition, and
this is even more pronounced at a distance of 800 m. However, incorporation of the deposition velocity into the model results in
a general reduction of computed concentrations, and excellent
agreement with measurements over the stability range $-0.05 < 1/L < 0.05$
for both the 200- and 800 m distances. Indeed, at 800 m the predictions are consistent with measurements even into the convective
regime, where the use of K-models becomes dubious.

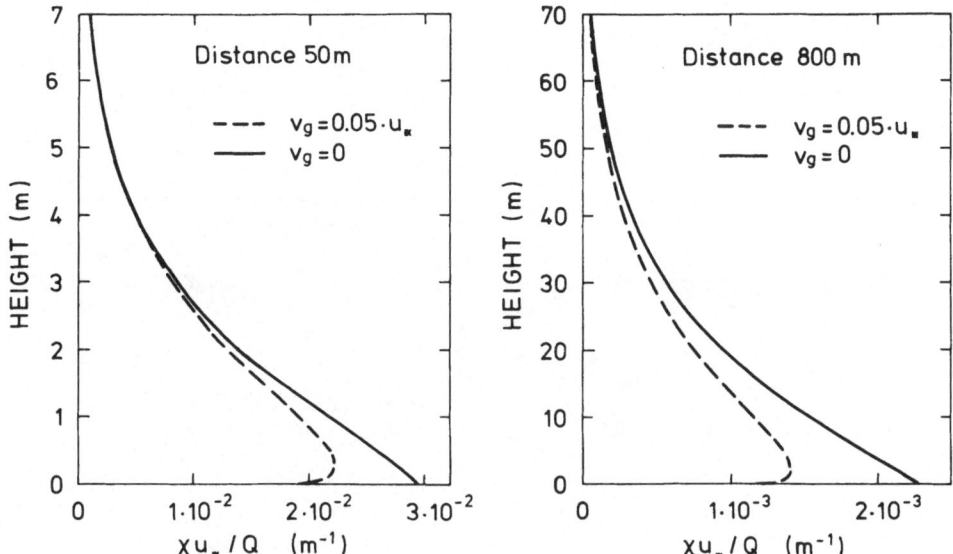

Fig. 1. Predicted vertical concentration profiles, calculated for neutral stability conditions 50 and
800 m from the source, without deposition of the
tracer as well as by assuming $v_g = 0.05u_*$.
The roughness length and source height were equal
to that of Project Prairie Grass, which is
reported to be $z_0 = 0.008$ m (Pasquill, 1974) and
0.46 m respectively.

The tendency in the convective regime is for the measured
concentrations at both 200 and 800 m to be slightly lower than
those predicted by the model. The diffusion process in this
stability regime has been modelled by Deardorff and Willis (1975)
making use of mixed-layer similarity. Nieuwstadt (1980) compared this model with the Prairie Grass data. It is interesting
to note that the performance of Deardorff and Willis's (1975)
model, shown in Nieuwstadt's paper in Fig. 1, is poorer than the
predictions by the K-model used here. However, the agreement
with the measured data will be improved by taking into account
the effect of deposition.

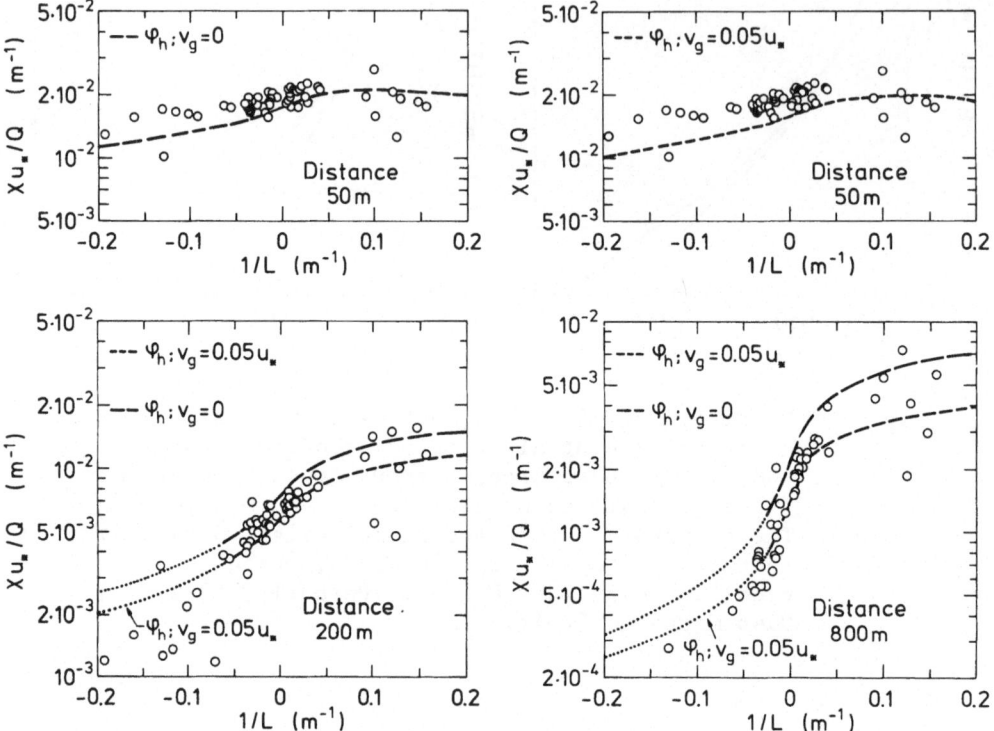

Fig. 2.   Predicted and measured crosswind-integrated
tracer concentrations 50, 200, and 800 m downwind
from the source as a function of stability for
Project Prairie Grass. Predicted concentrations
are calculated without deposition of the tracer
as well as by assuming $v_g = 0.05\ u_*$. Those
portions of the curves in which $\bar{z}/L < -1$ are dotted.

REFERENCES

Barad, M.L. 1958. "Project Prairie Grass, a field program in diffusion", Report AFCRL-TR-235 (2 volumes).

Briggs, G.A., and McDonald, K.R. (1978). "Prairie grass revisited: Optimum indicators of vertical spread". Environmental Research Laboratories. Air Resources, Atmospheric Turbulence and Diffusion Laboratory, Oak Ridge, Tennessee, ATDL contribution No. 78/8, 31 pp.

Businger, J.A. 1973. Turbulent transfer in the atmospheric surface layer, in "Workshop in Micrometeorology". pp. 67 - 100, Am. Met. Soc.

Deardorff, J.W. and Willis, G.E. 1975. A parameterization of diffusion into the mixed layer. J.Appl. Met., 14, 1451-1458.

Garland, J.A. 1977. The dry deposition of sulphur dioxide to land and water surfaces, Proc. R. Soc. Lond. A., 354, 245-268.

Haugen, D.A., Kaimal, J.C. and Bradley, E.F. 1971. An experimental study of Reynolds stress and the heat flux in the surface boundary layer, Quart. J. Met. Soc., 97, 168-180.

Horst, T.W. 1979. Lagrangian similarity modelling of vertical diffusion from a ground-level source, J. Appl. Met., 18, 733-740.

Keller, H.B. 1971. A new difference scheme for parabolic problems, in "Numerical Solutions of Partial Differential Equations 2", pp. 327-350, Academic Press, New York.

Nieuwstadt, F.T.M. 1980. Application of mixed-layer similarity to the observed dispersion from a ground-level source, J. Appl. Met., 19, 157-162.

Nieuwstadt, F.T.M. and van Ulden, A.P. 1978. A numerical study on the vertical dispersion of passive contaminants from a continuous source in the atmospheric boundary layer, Atmos. Envir., 12, 2119-2124.

Pasquill, F. 1974. "Atmospheric diffusion", 2nd ed. Chichester, Ellis Horwood Ltd., New York, John Wiley and Sons.

Paulson, C.A. 1970. The mathematical representation of wind speed and temperature profiles in the unstable atmospheric surface layer, J. Appl. Met., 9, 857-861.

Turner, D.B. 1970. "Workbook of atmospheric dispersion estimates" National Air Pollution Control Administration, Cincinnati, Ohio HPS Pub. 999-AP-26. 88 pp.

van Ulden, A.P. 1978. Simple estimates for vertical diffusion from sources near the ground, Atmos. Envir., 12, 2155-2129.

DISCUSSION

A. VENKATRAM   How sensitive are your results
to your assumptions about the deposition velocity ?

S.E. GRYNING   This has not yet been investigated, but I expect that the sensitivity is rather
weak. Also I expect that physical realistic variations in the deposition velocity will not change the
results more than the scatter in the Prairie Grass
data.

A. VENKATRAM   Why do you expect your results
to be better for unstable conditions than for stable
conditions ?

S.E. GRYNING   Basically the model is valid
in near neutral conditions only ; that it works so
well in unstable/convective conditions is rather
surprising as use of K-theory becomes dubious in
convective conditions. It is characteristic that
there is much scatter in the measured concentrations
from the Prairie Grass experiments in stable conditions, I do not know why, but it could be that
Monin-Obukhov theory only works at heights actually
lower than that of the plume and therefore is used
beyond its range of validity.

ATMOSPHERIC DIFFUSION MODELING AND

OPTIMAL AIR QUALITY CONTROL STRATEGIES x

M. Posch and E. Runca

International Institute for
Applied Systems Analysis
A-2361 Laxenburg, Austria

INTRODUCTION

Atmospheric diffusion models are used in a rapidly increasing number of studies for controlling and/or managing air quality. In general these studies try to find the least-cost control strategy by solving the following mathematical program:

$$\text{select } \underline{x} \tag{1}$$

$$\text{that minimizes } \gamma\,(\underline{x}) \tag{2}$$

$$\text{subject to } \underline{C}(\underline{x},\underline{q}) \leq \underline{C}^{\mathbf{x}} \tag{3}$$

A solution to this problem selects the control measures x_i ($i=1,\ldots,L$) that, when applied to an emission pattern, $\underline{q}$, minimize the cost of control, γ, subject to concentration, $\underline{C}$, remaining below air quality standards $\underline{C}^{\mathbf{x}}$ at all receptor points.

The degree of sophistication of the mathematical programming method for solving (1)-(3) depends on the type of dispersion model, the type of cost function as well as the control variables chosen. (For a review see Cass and Mc Rae, 1981 and the literature quoted therein). In most of the existing programs of the above type (piecewise) linear cost functions are used and the control variables are the efficiencies of abatement technologies, which lead to a linear relationship between emissions and concentrations – altogether resulting in a linear programming problem.

xResearch partly supported by ICSAR grant (International Cooperation for Systems Analysis Research) and the Municipality of Vienna (MA22)

In part A of this paper we investigate problem (1)–(3) for a single point source emitting M different pollutants, where the control variables are not only the efficiencies of specific abatement technologies but also physical parameters (such as stack height, stack diameter, stack exit velocity, etc.) which enter in (3) in a nonlinear manner. We also allow the cost-functions to be nonlinear

Reduction of pollutants concentration can be also achieved by modifying the spatial emission pattern. For example, the provision of a combined heat and electrical power generation plant in an urban area does not only improve the efficiency with which primary energy is consumed but also improves the ambient air quality in a city by replacing individual residential heating by district heating.

In part B of this paper we describe a program which identifies the minimum cost district heating scheme necessary to reduce the ambient air pollution below a given level.

Before moving to the outline of problem A and B started above, we give a brief summary of the dispersion model used.

THE DISPERSION MODEL

In the present study we focused on the assessment of long term average concentrations on a local scale (City of Vienna). The analytical structure of the Gaussian diffusion model used draws on Martin (1971) with modifications by Runca et al. (1976) : wind direction, wind speed, atmospheric stability and temperature are divided into N_d, N_w, N_s and N_t classes, respectively. These classes are used to build the climatological frequency matrix F (id, iw, is, it) for a selected period. Making use of this matrix and assuming that the source is located at the origin of the reference frame, the ground level concentration, averaged over the selected period, at distance x along the centerline of the id - th wind sector is given by

$$<C(x;id)> = \frac{QN_d}{2\pi x} \sum_{iw=1}^{N_w} \sum_{is=1}^{N_s} \sum_{it=1}^{N_t} \frac{F(id,iw,is,it)}{u(id,iw,is,it)} G(x;id,iw,is,it)$$

$$(4)$$

where Q is the source strength, $\bar{u}$ is the mean wind speed for the meteorological conditions (id,iw,is,it) and G is the unit concentration for these conditions.

For points not falling in the centerline, the concentration value is determined by interpolation along the arc of the radius x between two adjacent sectors.

In the Gaussian model for inert pollutants the function G is given by

$$G(x;id,iw,is,it) = \frac{2}{\sqrt{2\pi}\sigma_z(x;is)} \exp\left[-\frac{(h+\Delta h(x;it)^2}{2\sigma_z^2(x;is)}\right] \quad (5)$$

In (5) h is the stack height, Δh is the plume rise (we used the plume rise formulae by Briggs, 1971 and 1975) and σ_z is the vertical dispersion coefficient (we used the curves from Pasquill, 1974, and Gifford, 1961 and 1976, interpolated by a power law expression).

Equation (5) has been deduced for gaseous pollutants; for the simulation of dispersion of particulates or droplets with significant gravitational settling velocities (5) must be modified. Following Dumbauld and Bjorklund (1975), G is given by

$$G(x;id,iw,is,it) = \frac{1+\beta}{\sqrt{2\pi}\sigma_z(x;is)} \exp\left[-\frac{(h+\Delta h(x;it)-\dfrac{xV_s}{\bar{u}(id,iw,is,it)})^2}{2\sigma_z^2(x;is)}\right]$$

$$(6)$$

where V_s is the gravitational settling velocity and β is the reflection coefficient for the particulates ($\beta = 0$: no reflection from ground, $\beta = 1$: total reflection). Equations (5) and (6) are a simplified description of the atmospheric diffusion processes for gaseous and particulate matter, respectively. For example, they do not include the effect of the inversion height.

However, we would like to mention that the optimization algorithms outlined below do not depend on the adopted dispersion model. The model for computing the ground level concentration is an independent module of the whole algorithm and therefore any suitable dispersion model can be used.

MINIMUM COST SOLUTIONS

A. Abatement and Stack Height

Let us assume that we have chosen a certain control strategy, depending on L control parameters x_i (i=1,2,...,L), with which concentration patterns of N pollutants can be influenced. If γ denotes the total cost due to a given set of the L parameters $x_1, x_2, ..., x_L$, the general optimization problem can be stated as follows:

$$\min_{x_1,...,x_L} \gamma(x_1,...,x_L) \quad (7)$$

subject to

$$C_j(x,y;x_1,\ldots x_L) \leq C_j^* \text{ for all } (x,y) \in A \text{ and } j = 1,\ldots,N \qquad (8)$$

$$x_i^{min} \leq x_i \leq x_i^{max} , \quad i = 1, \ldots, L$$

where A is the geographical area under consideration and C_j^* is the standard for the j-th pollutant (at ground level).

To tackle this constrained nonlinear optimization problem we will make the following simplifications:

(i)   The overall cost function is separable, i.e., the total costs are the sum of the costs for each technology

$$\gamma(x_1,\ldots,x_L) = \sum_{i=1}^{L} K_i(x_i) , \qquad (9)$$

(ii) The set of control parameters $x_1,\ldots,x_L$ can be devided into N+1 groups:

$$\underline{x}^{(1)} = (x_1^{(1)},\ldots x_{n_1}^{(1)}) \text{ affects only concentration } C_1$$
$$\vdots$$
$$\underline{x}^{(N)} = (x_1^{(N)},\ldots,x_{n_N}^{(N)}) \text{ affects only concentration } C_N$$

and finally $\underline{x}^{(0)} = (x_1^{(0)},\ldots,x_{n_0}^{(0)})$ affects all concentrations

as the stack height of a plant. To differentiate it from the other groups we use the following notation $\underline{x}^{(0)} = \underline{h}$. (Note that $n_0+n_1+\ldots+n_N = L$).

To proceed further we assume the principle of independency that means if C_j is the initial concentration and $x_1(j),\ldots,x_n(j)$ are the (normalized to unity) control parameters (efficiencies) of the applied abatement techniques, then the concentration is given by

$$C_j(x,y;\underline{x}^{(j)},\underline{h}) = C_j(x,y;\underline{h}) \prod_{k=1}^{n_j} (1-x_k^{(j)}) \quad j = 1,\ldots,N \qquad (10)$$

The optimization problem now reads:

$$\min_{\underline{x}^{(1)},\ldots,\underline{x}^{(N)},\underline{h}} \left\{ \sum_{j=1}^{N} \sum_{k=1}^{n_j} K_{kj}(x_k^{(j)}) + \sum_{k=1}^{n_0} K_k(h_k) \right\} \qquad (11)$$

subject to

$$C(x,y;\underline{h}) \prod_{k=1}^{n_j}(1-x_k^{(j)}) \leq C_j^{\underline{x}} \text{ for all } (x,y)\epsilon A \text{ and } j=1,\ldots,N \quad (12a)$$

$$0 \leq x_{min}^{(j)} \leq x^{(j)} \leq x_{max}^{(j)} \leq 1$$

$$\underline{h}_{min} \leq \underline{h} \leq \underline{h}_{max} \tag{12b}$$

The dependency of the concentration-functions on the parameters $h_1,\ldots,h_{n_0}$ might be very complicated and even not differentiable.

Therefore, we will proceed in two steps:

(1) we keep the parameter $\underline{h}$ constant; then the optimization problem (11)-(12) splits in N subproblems:

$$\min_{\underline{x}^{(j)}} \sum_{k=1}^{n_j} K_{kj}(x_k^{(j)}) \tag{13}$$

subject to

$$\prod_{k=1}^{n_j}(1-x_k^{(j)}) \leq B_j(\underline{h}) \text{ where } B_j(\underline{h}) = \min_{(x,y)\epsilon A} \frac{C_j^{\underline{x}}}{C_j(x,y;\underline{h})} \tag{14a}$$

$$x_{min}^{(j)} \leq x^{(j)} \leq x_{max}^{(j)}$$

(2) with a sequence of N sub-optimal solutions $\underline{x}^{(j)}(\underline{h})$ (j=1,...,N) from step (1) we compute by a suitable search algorithm the minimum of the function:

$$Z(\underline{h}) = \sum_{j=1}^{N} \sum_{k=1}^{n_j} K_{kj}(x_k^{(j)}(\underline{h}) + \sum_{k=1}^{n_0} K_k(h_k) \tag{15}$$

In order to be able to perform this in a reasonable amount of computer time, a fast algorithm has been developed for solving the subproblems (13)-(14). (For a description of this algorithm, see Posch and Runca, 1982).

To illustrate results achievable by solving (12)--(13),an application has been done to the case of two pollutants, one gaseous and the other composed of particulates, released by the same stack.

Consistently with the above formulation it was assumed that the
emission rate of the two pollutants could be controlled independently
by two adequate abatement techniques.

Concentration at the ground for the two pollutants were provided
by (4) with the joint probability frequency matrix F(id,iw,is,it)
computed by the meteorological data recorded in Vienna for the period
October 1977–April 1978.

The results reported below were achieved by assuming that both
costs of stack and abatement techniques were growing with the square
of the stack height and abatement techniques efficiencies respectively.

Figure 1 gives for each pair of concentration values (in
arbitrary units), to be interpreted as prescribed standards for
pollutants 1 and 2, the minimum cost γ for the adopted control
strategy which guarantees that the prescribed standard will not be
exceeded in the area of interest. The stack height and the optimal
efficiencies corresponding to the minimum cost are reported in
Figure 1b, c and d, respectively.

Graphics of the type of Figure 1a–d provide a practical way to
analyze alternative control strategies. The mathematical optimization program has been conceived in a modular way and can be used
interactively. Cost functions are an input to the program. Analysis
on temporal and spatial scales different from the seasonal and local
scales treated by the diffusion model used in this study can be done
by implementing a suitable diffusion model.

B. Urban Centralized Heating System

Centralization of the heat supply in a densely populated area
is not only increasing the efficiency of the use of primary energy,
but provides also a tool to reduce ambient air pollution. Any design of a district heating network has an impact on the air quality
because of the reduction of emissions due to individual heating.
In the following we discuss a model which allows to assess the air
quality impact of a given design of a network; furthermore we
developed an algorithm to compute the minimum cost network to
achieve a prescribed reduction of the existing pollution level.
Although the methods are general, we use the City of Vienna as a
reference case.

In mathematical terms the optimization problem can be stated
as follows:

Let S_k, k = 1,...,N be the nonintersecting subregions of the
urban area, which can be potentially supplied with heat; E_k the
thermal heat, which is needed to supply S_k, and E_T the total amount

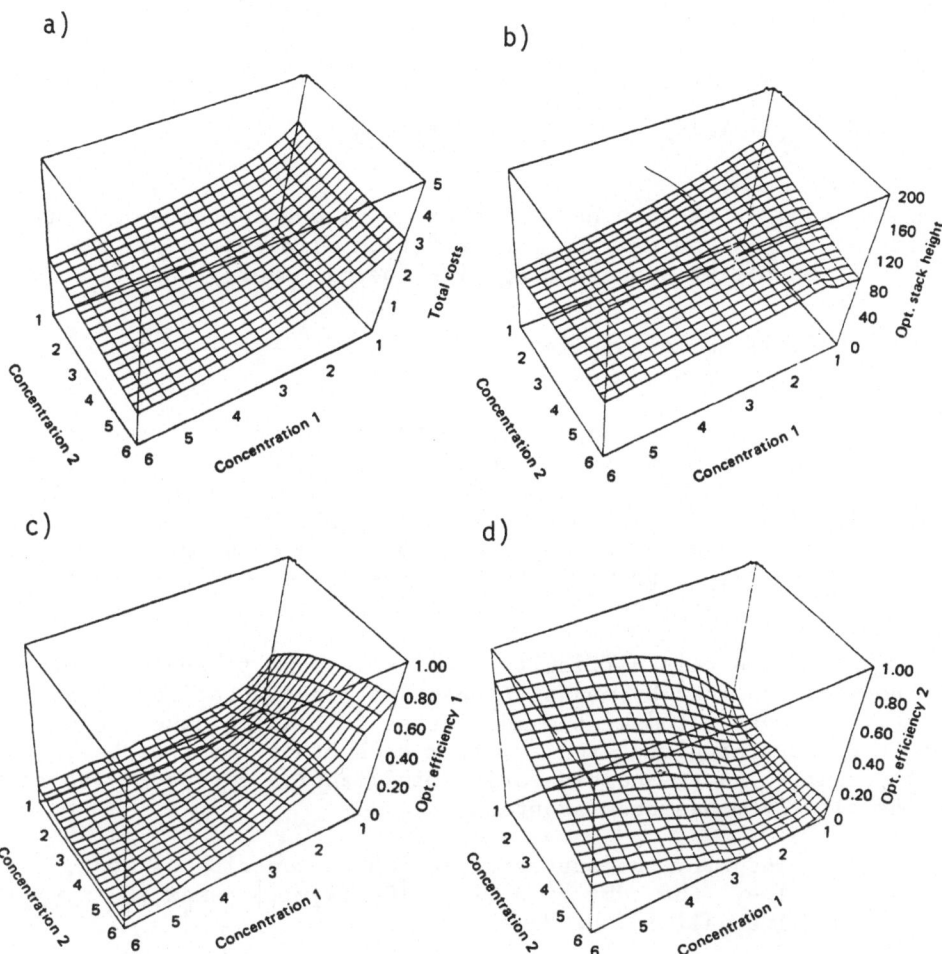

Figure 1. Minimum cost (a), optimal stack height (b), optimal
efficiencies for pollutant 1 and 2 (c,d) as functions
of a possible set of concentration values, to be interpreted as standards for pollutants 1 and 2 (see text).

of thermal energy available ; then we want to solve :

$$\min_{k=1}^{N} \sum \varepsilon_k \gamma_k \tag{16}$$

$$\text{subject to} \quad \sum_{k=1}^{N} \varepsilon_k \gamma_k \leq E_T \tag{17a}$$

$$\text{and} \quad \sum_{k=1}^{N} (1-\varepsilon_k) C_k(x,y) \leq C^* \quad \text{for all } (x,y) \tag{17b}$$

where $C_k(x,y)$ is the concentration at (x,y) due to the emissions within S_k, C^* is the maximum concentration allowed, γ_k are the costs for supplying subregion S_k with thermal energy and

$$\varepsilon_k = \begin{cases} 1, & \text{if } S_k \text{ is supplied} \\ 0 & \text{otherwise} \end{cases}$$

The straightforward approach to the solution of this combinatorial minimization problem is to identify out of the 2^N possible combinations of the N subregions those, which verify constraints (17a) and (17b) and then to select the one which gives the minimum costs. This approach is not implementable even on a large computer because of the rapid growth with N of the number of possible combinations.

To overcome this difficulty it is necessary to reduce a priori the possible choices by means of practical considerations. For example, it does not make too much sense to analyze cases in which the selected subregions are far from each other. On this basis, as an alternative to the combinatorial minimization approach the following algorithm has been adopted :

(i)   Identify the subregion S_{k_0} which contributes most, i.e., the subregion with $||C_{k_0}(x,y)|| = \max_k ||C_k(x,y)||$ (we call it the "core").

(ii)   Check (17b), if verified go to step (v), otherwise

(iii)   Identify the subregions surrounding the "core" (we call it the "belt") and select from it the subregion which contributes most. Removes this subregion from the "belt" and add it to the "core".

(iv)   Check (17b), if verified to step (v), otherwise go to (iii).

(v)   Check (17a), if verified, a feasible solution to (16)-(17) has been found.

To minimize the costs, steps (i) to (v) are repeated N times taking at each time a new subregion as initial "core" (not considering the previous initial "core), and out of the identified feasible solutions the one for which the costs are minimal is selected.

In the above formulation a receptor can be located at any point (x,y) ; for computational reasons, however, we have to discretize : let $C_{k,ij}$ be the concentration at receptor point (i,j) $i=1,\ldots,n_i$, $j=1,\ldots,n_j$, due to emissions in S_k. In the case of Vienna we have $n_i = 30$ and $n_j = 24$, altogether 720 grid points, where the concentration values due to every single source have to be computed.

Also the norm used in the optimization procedure (i)-(v) has to be specified: to check whether the results depend on the norm chosen, we used the tow "extreme cases", i.e. the 1-norm and the max-norm :

$$||C_{k,ij}||_1 = \sum_{i=1}^{n_i} \sum_{j=1}^{n_j} C_{k,ij} \qquad (18a)$$

$$||C_{k,ij}||_\infty = \max_{i,j} C_{k,ij} \qquad (18b)$$

To compute the norm of the long-term average concentration for a single source requires the evaluation of (4) at the 720 grid points, i.e. the evaluation of $2 \times N_w \times N_s \times N_t$ (in the case of Vienna $2 \times 6 \times 6 \times 4 = 288$) exponential functions, altogether $720 \times 288 \times 207360$ function calls. (This number is reduced by the fact that not all elements of the climatic frequency matrix are nonzero: for the period October 1977-April 1978 which has been used in the Vienna case, there are 21 to 46 nonzero elements for each wind direction which yields $720 \times 21 = 15120$ to $720 \times 46 = 33120$ function calls, nevertheless the order of magnitude is some tens of thousands.) But in an urban area one has not only to deal with one single source, but with many hundreds or even thousands (1232 in Vienna) of them. To compute the total concentration field only once requires several millions of function calls. Under these circumstances it is impossible to execute algorithm (i)-(v).

The straightforward way to overcome this difficulty would be the computation of the "transmission matrices" $(C_{k,ij})_{i=1,\ldots,n_i,}$ $j=1,\ldots,n_j$ for every source k, to store it and to utilize it during the runs of the algorithm. This is the way as it is down in most of the optimization programs containing air quality objectives or constraints.

Between these two "extremes" of recomputing the concentration field at every step of the algorithm on the one hand (too much CPU-time) and the preprocessing and storing of all the transmission matrices in advance on the other hand (too much core memory) we made the following compromise:

The 1232 sources recorded in the emission inventory of Vienna are the sum of 140 point sources and 1092 area sources. The point are the power plants, incinerators, and industrial, stacks which are not influenced by district heating; their concentration field is computed in advance and can be considered as a "background". For the area sources, however, we proceeded as follows: the concentration field due to an area source is computed using the point source formula (4), but with a suitable emission height h instead of the stack height plus plume rise. Therefore the unit concentration of an area source depends only on the location of the source and its emission height (all area sources are ax-parallel and have the same dimensions - 1000 m square). According to equation (4) the concentration is computed only along the $N_d = 8$ radials of the wind-rose in the distance r desired. Discretizing also the radial distance into N_r intervals of length $d_r = 250$ m we computed a matrix $R(ir,id)$, $ir = 1,...,N_r, id = 1,...,N_d$ (we call it a "rose"). Seven such "roses" have been prepared for the emission heights of 40, 50,..., 100 m respectively.

The concentration field of a specific area source located at (x_s,y_s) with source strength Q and emission height h is now easily determined. Let h_1 and h_2 ($h_1 \le h \le h_2$) be the emission heights for which "roses" R_1 and R_2 have been prepared. Place their origin in (x_s,y_s) and compute the concentration value at receptor (i.j) by interpolation. Multiplication with Q and interpolation between R_1 and R_2 yields the actual concentration value.

In a first application of the algorithm developed, we focused on the thermal energy supply of the private households. The emission data of these nonpoint sources have a resolution of 1 km^2 (see Figure 2; in this figure there are also given the isolines of the SO_2-concentration field before introducing district heating). The same areas (a total of 265) have been taken as units of heat supply, S_k.

The amount of thermal energy, E_k, necessary to supply S_k was computed from the amount of coal and oil consumed by the private households in S_k and the average number of hours per year during which the heating systems were in operation.

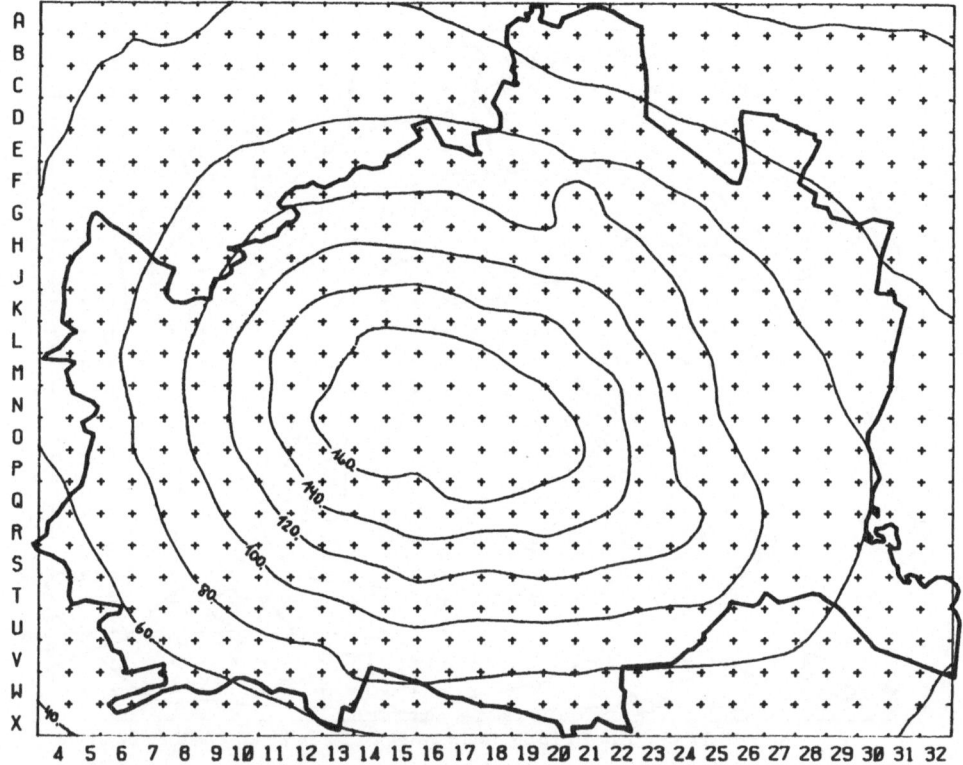

Figure 2. Isolines of the SO_2-concentration ($\mu g/m^3$) in the city
of Vienna averaged over the heating period October '77–
April '78 (the crosses (+) indicate the corners of the
1 km^2 – subareas S_k (see text); thick line: city border).

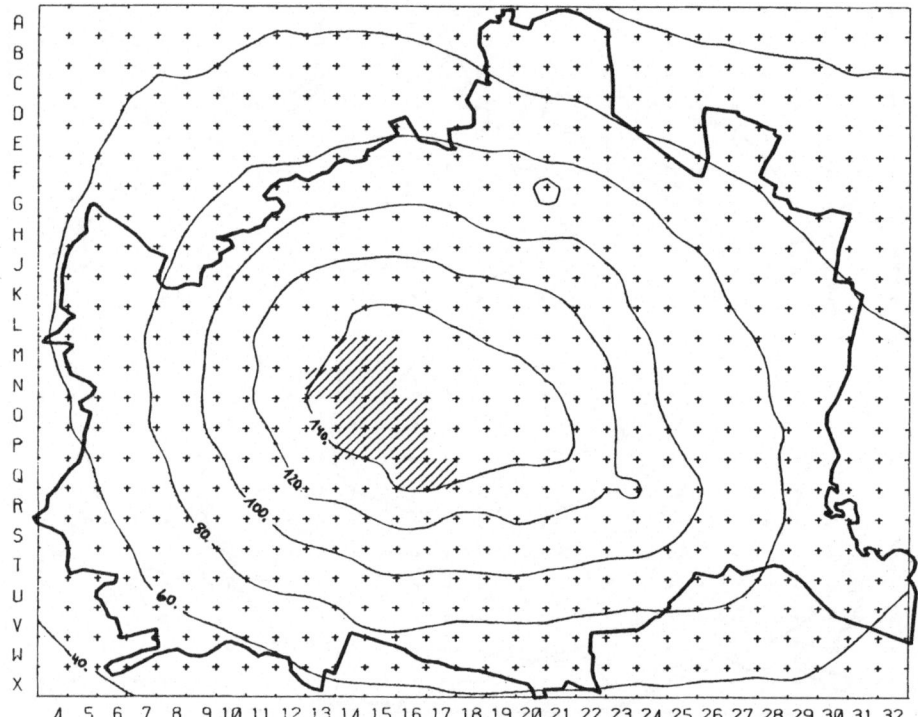

Figure 3.   Isolines of the SO$_2$-concentration (μg/m^3) in the City
            of Vienna after introducing district heating in the
            shaded area (compare with Figure 2).

The costs γ_k for the heat conveyance network supplying S_k were represented by the length of this network. To determine this length the total area of the city has been partitioned according to the structure of the build-up area: large, mostly old buildings in the center of the city, one- and two-family houses in the outer districts, and a mixture of both in the "transition" area between.

For each of these 3 zones an average length for the connection of the buildings of subarea S_k to the main network and an average value for the main network itself has been assigned. The total costs for the optimal heat distribution network gained by algorithm (i)-(v) is the weighted sum of these two contributions.

The shaded area in Figure 3 indicates the "optimal" region which has to be supplied with thermal power in order to reduce the pollution level below 160 $\mu g/m^3$; also the isolines of the resulting concentration field are given in Figure 3.

CONCLUSIONS

It was shown that information on atmosphere as well as control strategies can be integrated in a mathematical program which determines under given constraints, an optimal configuration of a selected control strategy. Specifically this was done in two cases: (a) the control of air pollution from a large point source (thermal power plant) and (b) reduction of pollution in an urban area by means of a centralized heating system.

Although simplifications were introduced in the description of both the atmospheric system and adopted control strategy, numerical experiments conducted with emission and meteorological data of the city of Vienna provided results which could not have been achieved on a purely "good sense" basis, due to the complex relationship between emissions, meteorology, concentration patterns and the possible control strategies.

REFERENCES

Briggs, G.A., 1971, Some Recent Analyses of Plume Rise Observations. 2nd International Clean Air Congress, AP, New York, USA.

Briggs, G.A., 1975, Plume Rise Predictions. Lectures on Air Pollution and Environmental Impact Analysis. Workshop Proceedings, Boston, Mass., September 29-Ocotober 3, 1975. pp. 59-111, American Meteorological Society, Boston, Mass., USA.

Cass, G.R., and McRae, G.J., 1981, Minimizing the Cost of Air Pollution Control. Environmental Science & Technology, 15, 7, pp. 748-757.

Dumbauld, R.K., and Bjorklund, J.R., 1975. NASA/MSFC Multilayer
    Diffusion Models and Computer Programs--Version 5. NASA
    Contactor Rep. No. NASA CR-2631, NASA,G.C. Marshall Space
    Center, Alabama.

Gifford, F.A., 1961, Use of Routine Meteorological Observation for
    Estimating Atmospheric Dispersion. Nuclear Safety $\underline{2}$,
    pp. 47-51.

Gifford, F.A., 1976, Turbulent Diffusion Tuping Schemes: A Review.
    Nuclear Safety, $\underline{17}$, pp. 68-86.

Martin, D.O., 1971, An Urban Diffusion Model for estimating Long--term
    Average Values of Air Quality. Socio-Economic Plann. Sci., $\underline{3}$,
    pp. 329-349.

Pasquill, F., 1974, Atmospheric Diffusion, Halsted Press of John
    Wiley & Sons, New York.

Posch, M., and Runca, E., 1982, Abatement of Air Pollutants and
    Cogeneration: Search for an Optimal Solution. WP-82-34,
    International Institute for Applied Systems Analysis, Laxenburg,
    Austria.

Runca, E., Melli, P., and Zannetti, P., 1976, Computation of Long-
    term Average SO_2 Concentration in the Venetian Area.
    Appl. Math. Mod., 1, pp. 9-15.

DISCUSSION

M.L. WILLIAMS   In view of the complexity of
                    the application of your procedure in the multiple-
                    source situation in Vienna, does your method give
                    a significant improvement over or give different
                    results from a simple first estimate from the basic
                    Gaussian model analysis of the contributions from
                    different grid squares.

M. POSCH   In fact, due to the long
                    averaging period,the computed "solution" is similar
                    to a "solution" one would estimate by a quick ana-
                    lysis. Nevertheless the method outlined above can
                    help to rationalize the decision-making process
                    concerning the design of a district heating scheme.

THE APPLICATIONS OF AN INTEGRAL PLUME RISE MODEL

M.C. Underwood and S. Buchner*

Applied Physics Branch, BP Research Centre
Chertsey Road, Sunbury-on-Thames
Middlesex, England

INTRODUCTION

The mathematical description of plume rise is well established and has been reviewed by Briggs[1]. As the ground level concentration of pollutants is a strong function of the effective rise of the plume above the chimney, Δh, it is important that this quantity can be calculated over a wide range of meteorological conditions. However, as noted by Briggs[1] various formulae give a range of values of Δh that differ by more than an order of magnitude. In order to clarify the situation equations have been used that are closely based upon the physics of plume rise. Schatzmann[2,3,4] has presented a description of plume rise based upon the equations of conservation of mass, momentum, concentration and thermal energy. To close the system of equations an entrainment function is used and this paper discusses the application of this model to a variety of situations.

THEORY

The following equations are used

Continuity of mass

$$\frac{\partial \rho}{\partial t} + \nabla \cdot (\rho V) = 0 \qquad (1)$$

---

*Temporary visitor from University of Hamburg

Continuity of momentum

$$\rho \left[\frac{\partial V}{\partial t} + \nabla \left(\frac{V^2}{2} \right) - V \times (\nabla \times V) \right] = \nabla \rho + \rho g \tag{2}$$

Continuity of a passive scalar

$$\frac{\partial (\rho c)}{\partial t} + V \cdot \nabla (\rho c) + \rho c \nabla \cdot V = 0 \tag{3}$$

Continuity of heat energy

$$\frac{\partial (\rho T)}{\partial t} + V \cdot \nabla (\rho T) + \rho T \nabla \cdot V = 0 \tag{4}$$

where,   V = velocity vector
            T = temperature
            ρ = density
            g = gravity vector
            c = concentration

Using a set of coordinates as defined by Schatzmann, shown in Fig. 1, and the assumptions listed below the equations can be integrated with respect to the cross-sectional coordinates.

The assumptions are that

   (i)   the ambient wind velocity is approximately constant
          in magnitude and direction.

   (ii)   the mean excess and turbulent quantities are
          axisymmetric.

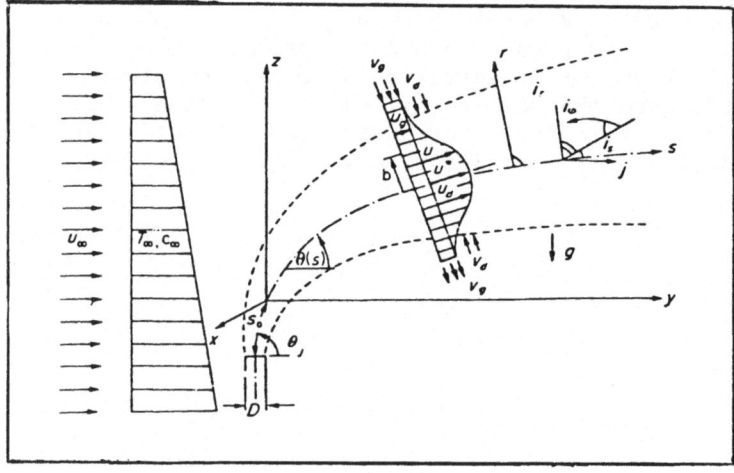

Fig. 1   Definition of coordinates

(iii) the flow is assumed to be fully turbulent a short distance after the vent.

(iv) Gaussian profiles are assumed.

Integration with respect to ϕ and r give five ordinary differential equations and seven unknowns, ie $u^*(s)$, $b(s)$, $\Theta(s)$, $\rho^*(s)$, $T^*(s)$, $c^*(s)$ and $E(s)$. In order to close the system of equations an equation of state is required

$$\rho = \rho\,(c,\ T) \tag{5}$$

and an entrainment function

$$E = E\,(u^*,\ b,\ \Theta,\ \rho^*,\ T^*,\ c^*) \tag{6}$$

The solution is applicable from the point at which the flow is assumed to be fully turbulent, ie when $s = s_o$ and up to where the excess momentum flow is zero.

The Entrainment Function

The entrainment function used in plume rise modelling is a crucial parameter in producing a physically plausible description of the process occurring. Davidson and Slawson[5] have discussed the application of various entrainment hypotheses in assessing the performance of plume rise models based upon the solution of the conservation equations. For the data modelled by Davidson and Slawson the entrainment hypothesis suggested by Ooms[6] was found to be adequate, ie

$$V_e = \alpha\ \left|V - \frac{u_\infty V_x}{V}\right| + \beta\ \frac{V_x}{V}\ \left|\frac{u_\infty W}{V}\right| \tag{7}$$

Eq. (7) reduces to $V_e = \alpha|V|$ near the source and $V_e = \beta|W|$ for a bent over plume. The optimum values of α and β were found to be 0.15 and 0.68 respectively. Henderson-Sellers[7] has shown that for emissions that have both jet and plume like characteristics the value of α is given by

$$\alpha = \left\{I_1/2^{2-a}\right\}\ C_p\ (a + 1)\ \left\{1 + \left(\frac{a + 3}{3a + 3}\right)\ \frac{F_{D_p}^2}{F_D^2}\right\} \tag{8}$$

In this work a dimensionless entrainment function, ϵ, is used as defined by

$$\epsilon = \frac{E}{u^*b} \tag{9}$$

For a momentum jet in a stagnant environment

$$\varepsilon = A_1 = 0.057$$

For a simple buoyant jet

$$\varepsilon = A_1 + A_2 \cdot \frac{1}{F^2}$$

where $A_2 = -0.67$ giving

$$\varepsilon = 0.057 - \frac{0.67}{F^2}$$

For a jet in a cross-flow

$$\varepsilon = \frac{2 A_1}{2 + A_3 \cdot \dfrac{u_\infty}{u^*}}$$

where $A_3 = 10$, giving

$$\varepsilon = \frac{0.14}{2 + 10 \cdot \dfrac{u_\infty}{u^*}}$$

For a momentum jet in a cross-flow

$$\varepsilon = \frac{2 A_1 + 2 A_2 \cdot \dfrac{\sin \Theta}{F^2}}{2 + A_3 \cdot \dfrac{u_\infty}{u^*} \cdot \cos \Theta} \left(1 + A_4 \cdot \frac{u_\infty}{u^*} \cdot \sin \Theta \right)$$

where $A_4 = 2$

RESULTS AND DISCUSSION

The equations were solved numerically using a Runge–Kutta procedure. The data of Slawson and Csanady[8] for a buoyant plume in a neutral atmosphere is shown in Fig. 2. The predictions of the model are shown and slightly underestimate the measured values. This could be due to deviation from the dry adiabatic lapse rate, which was used for the calculation as a fit to the measured temperature gradient shown in Fig. 2. At a downwind distance of 1 km, the difference between the measured and predicted height was about 15 m, implying an error of about 5% in the predicted total plume rise above the ground.

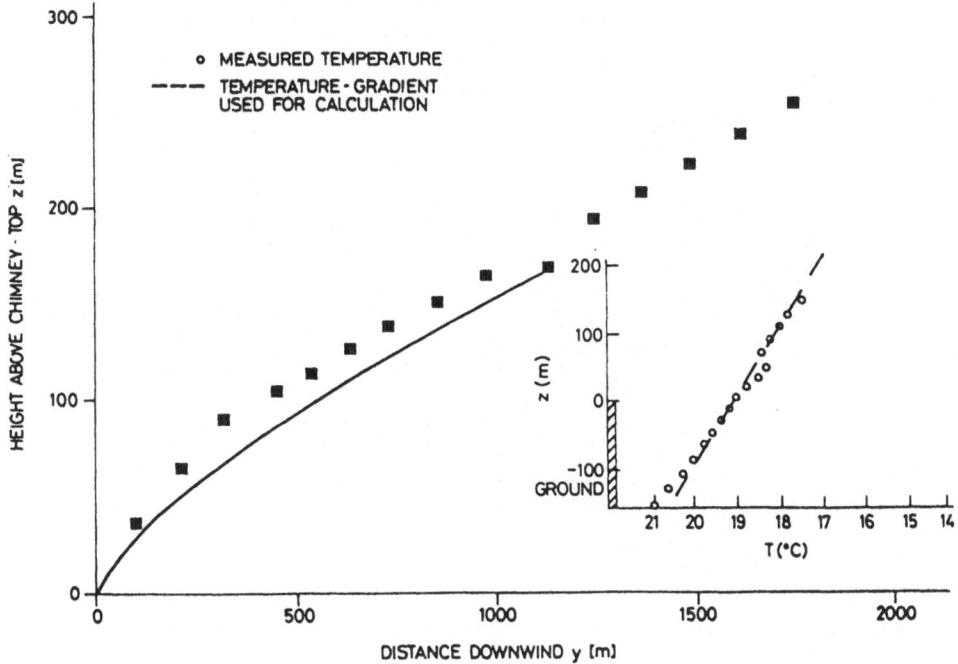

Fig. 2. Predictions and data for a buoyant plume in a
        neutral atmosphere.

Data of Slawson and Csanady[8] for an atmosphere with an
inversion layer is shown in Fig. 3. The measured temperature
lapse rate was assumed to be given by two linear lapse rates as
shown in Fig. 3, with an inversion occurring 170 m above the
stack. The worst agreement between predictions and measurements
occurred at a downwind distance of approximately 600 m, where the
predicted value (above ground) was 14% above the measured value.
At a height of 250 m above the stack, the measured plume centre-
line cannot legitimately be compared with theory, as no
atmospheric temperature measurements were presented above this
height.

For conditions with a strong positive temperature gradient
the data of Slawson and Csanady[8] for the measured temperature
gradient was treated in two ways, as shown in Fig. 4. Case 1
shows the linear fit used for calculation purposes from the top of
the stack to the inversion layer, then from the inversion layer
onwards. The predicted plume rise is shown in Fig. 5. However,
the temperature data are only for heights up to 220 m above the
stack, and the assumption of a positive temperature gradient

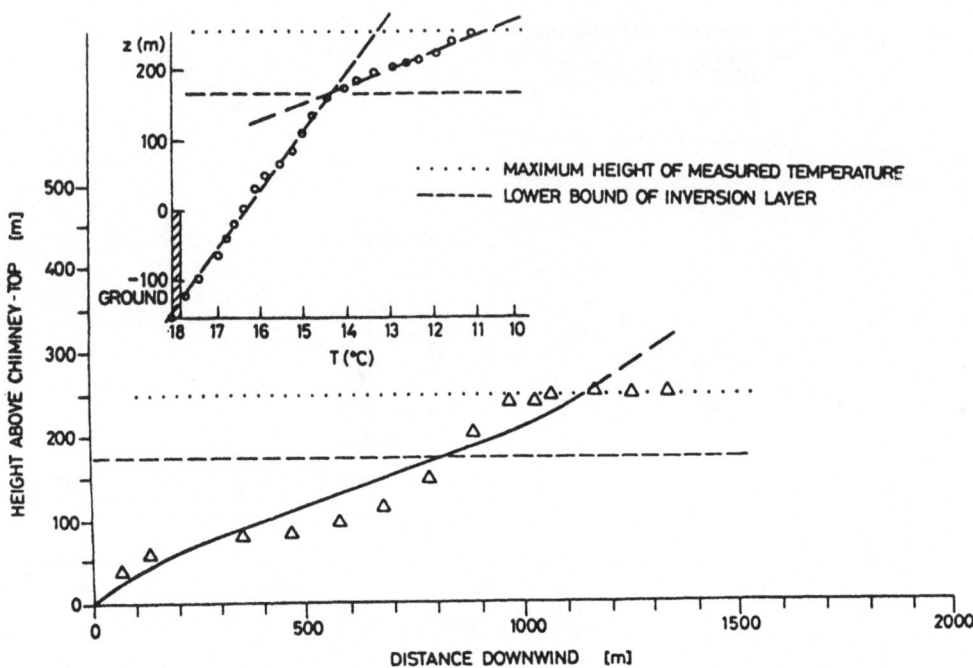

Fig. 3   Predictions and data for a plume in an atmosphere
         with an inversion

remaining unchanged above this height was made.   This assumption
may be invalid, and so to try and obtain a good fit to data a dry
adiabatic lapse rate was assumed above 220 m, as shown in Fig. 4,
Case 2.   The results are shown in Fig. 6.

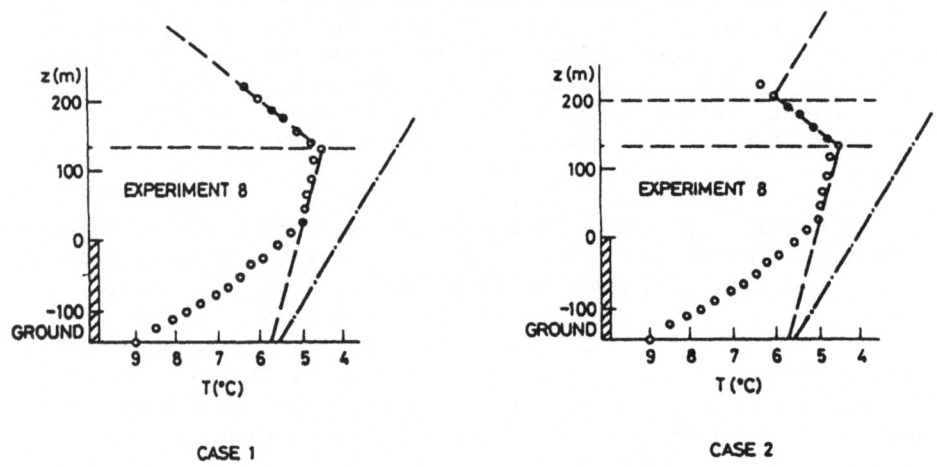

Fig. 4   Temperature lapse rate with strong positive
         temperature gradient

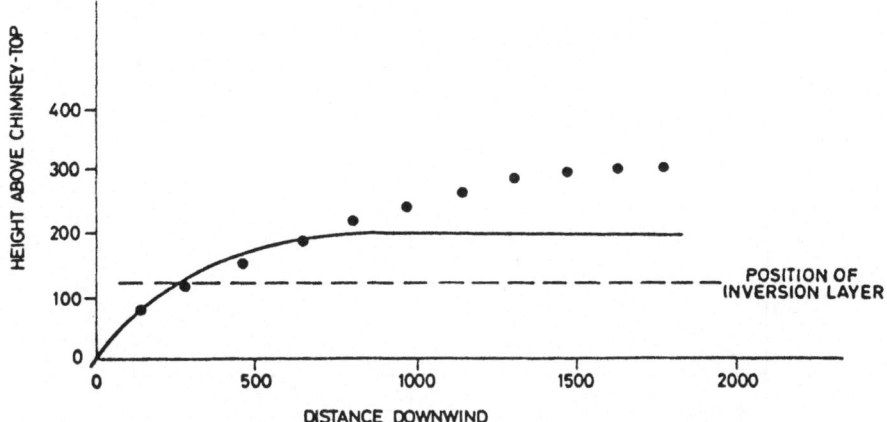

Fig. 5   Predictions and data for positive temperature
          gradient – Case 1

This is instructive as it is clear that to obtain good agreement
between prediction and experiment good measurements of the
atmospheric temperature gradient are required.

The case of a very hot plume, from the data of Bringfeld[9],
has been considered. Predictions are shown in Fig. 7 for a
plume 301K above ambient temperature. The agreement between the

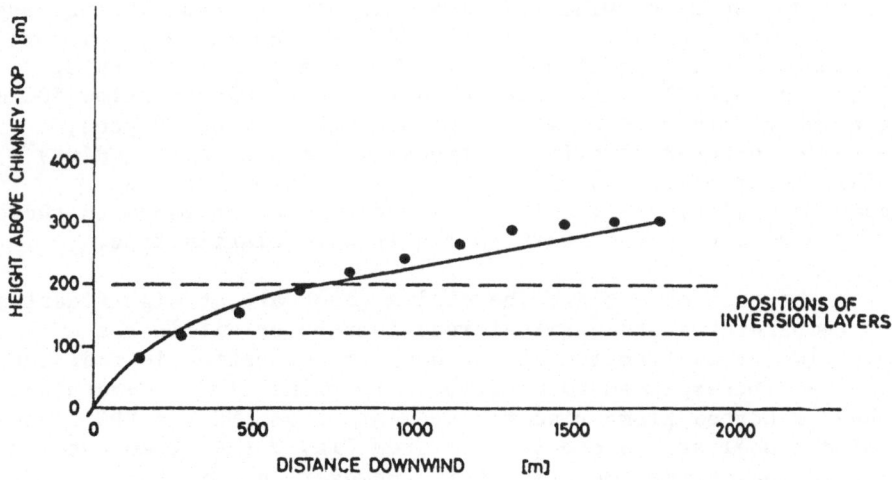

Fig. 6   Predictions and data for positive temperature
          gradient. – Case 2

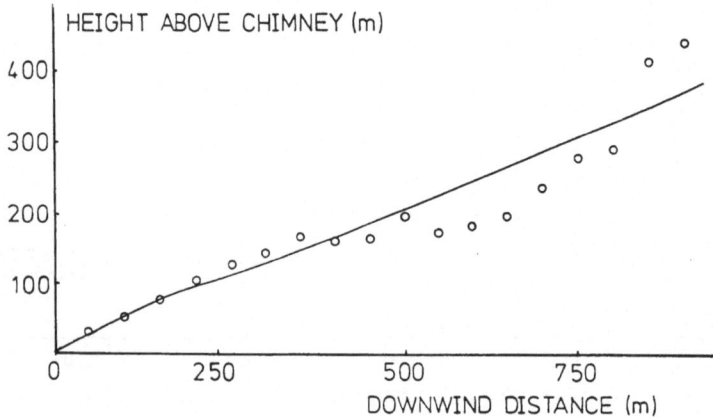

Fig. 7 Predictions and data for a hot plume, ie 301K
         above ambient

predicted position of the plume centre-line and measurement was
good, indicating that significant temperature differences between
the efflux gases and the surrounding atmosphere can be modelled.

For the case of gas turbines the efflux gases are generally
hot and have significant velocities, ie temperatures several
hundred degrees (K) above ambient and velocities significantly
greater than the local wind velocity. In Fig. 8 is shown the case
of a gas turbine operating at over 400K above ambient temperature
and at an exit velocity of 45 ms^{-1}. The data from the work of
Hamilton and Moore[10]. Agreement at downwind distances below 500 m
between the predictions of the model and measurement is good,
however, for greater downwind distances the model underpredicts
the plume centre-line position. This could be because the
atmospheric temperature gradient was not well established in these
measurements, and further work is required to clarify this.

For the case of a flare the efflux gases are at significantly
higher temperatures and chemical reactions occur within the
flame. Also, some fraction of the heat of combustion is radiated
so would not be expected to contribute to plume rise. Fumarola et
al[11] have reported plume rise measurements from LNG and LPG flares
in a wind tunnel and as can be seen from Fig. 9 the agreement with
the prediction of the model is poor. However, a suitable
entrainment function can probably be formulated to include eg
chemical reactions in the flare, and this is being investigated.

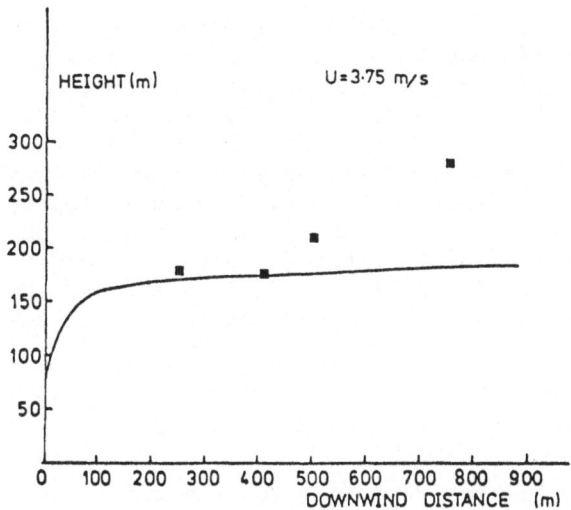

Fig. 8   Predictions and data for gas turbine efflux

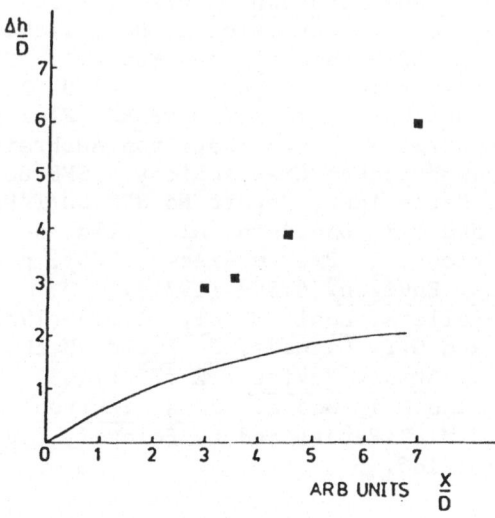

Fig. 9   Predictions and data for LPG flare (data normalised
to flare stack diameter)

CONCLUSIONS

The integral plume rise model of Schatzmann gives excellent
agreement with measurement for the case of buoyant plume centre-
line position, provided that the atmospheric temperature gradient
is known. For hot emissions, where significant density
differences between the efflux gases and the ambient fluid would
be expected, the predictions are in good agreement with
measurement as the commonly employed Bousinnesq approximation was
not used. The case of hot, fast emissions, ie gas turbines, again
can successfully be modelled, but for a flare a modified
entrainment function is required. Such an approach to plume rise
will provide a model applicable to most conditions encountered in
practice.

ACKNOWLEDGEMENTS

I thank The British Petroleum Company plc for permission to
publish this paper. It is a pleasure to record the interest of
Mr J. McKay and Mr R.P.M. Dawson in this work.

REFERENCES

1. G.A. Briggs, Plume rise predictions, AMS Workshop in
        Meteorology and Environmental Assessment, Boston, 1975.
2. M. Schatzmann, Atmos. Environ. 13:721 (1979).
3. M. Schatzmann, Auftriebsstrahlen in Natürlichen Stromüngen-
        Entwicklung EINES Mathematischen Modells. SFB 80,
        Universität Karlsriche, Report No. SFB 80/T/86 (1976).
4. M. Schatzmann and W. Flick, "ATMOSPHÄRE" Fluiddynamisches
        Simulationsmodell zur Vorhersage von Ausbreitungsvorgängen
        in der Atmosphärischen Grenzschicht. SFB 80,
        Universität Karlsriche, Report No SFB 80/T/90.
5. G.A. Slawson and P.R. Davidson, Air pollution modelling and
        its applications, I, Plenum Press (1980), p 417.
6. G. Ooms, Atmos. Environ. 6:899 (1972).
7. B. Henderson-Sellers, Ecol. Model. 11:253 (1982).
8. P.R. Slawson and G.T. Csanady, J. Fluid. Mech. 47:33 (1971).
9. B.J. Bringfeld, Atmos. Environ. 2:575 (1968).
10. P.M. Hamilton and D.J. Moore, Atmos. Environ. 991 (1973).
11. G. Fumarola, D.M. DeFaveri and E. Polazzi, Hydrocarb. Proc.
        (Jan. 1982) p 165.

APPENDIX: NOMENCLATURE

ρ density
V velocity vector
c concentration
T absolute temperature
g gravity vector
u^* excess component of velocity at plume centre-line
b nominal jet radius
Θ angle of plume trajectory
ρ^* excess component of density at plume centre-line
T^* excess component of absolute temperature at plume centre-line
c^* excess component of concentration at plume centre-line
V_e entrainment velocity
V_x downward component of velocity
W vertical component of velocity
I_1 shape constant ($\frac{1}{2}$ for 3-D Gaussian plume)
a constant to indicate number of dimensions (1 for 3-D point
        source)
C_p constant (0.102 for 3-D Gaussian plume)

$F_{D_p}{}^2$ Froude number for pure plume (16.5 for 3-D Gaussiann plume)

$F_D{}^2$ Froude number

F local densimetric Froude number $F = \dfrac{u^*}{\sqrt{g \cdot b \cdot \rho^*/\rho_\infty}}$

E entrainment rate

DISCUSSION

J. KNOX   What range of the parameters
        of orifice diameter and exit velocity have been used
        in your calculations ?

M.C. UNDERWOOD   The agreement between experi-
        ments and the model predictions has been tested for
        small orifices i.e. of the order of a few millimetres,
        and the agreement is good. For the case of larger
        scale releases e.g. gas turbines, agreement between
        the model predictions and measurement has been investi-
        gated for exit velocities of up to about 50 ms^{-1}.
        Agreement is reasonable between measurement and pre-
        diction.

G. RESELE   I would like to make a comment.
One should mention that this sort of modelling has
been used for 10 years for the prediction of the rise
of cooling tower plumes.

M.C. UNDERWOOD   The use of integral plume rise
models for the rise of cooling tower plumes is well
established. However, the detailed comparison of
predictions of plume behaviour for buoyant plumes,
gas turbines etc. has not been fully investigated.
An integral plume rise model should be generally
applicable for most sources, but problems such as
the entrainment function are not yet resolved.

F.B. SMITH   Since prediction of plume rise
is only a step on the road to calculating ground level
concentrations, does it really make sense to ignore
time-variations in the wind field (& in particular
the vertical velocity) arising from large-scale
boundary layer eddies that have a non-linear effect
on the concentrations at the ground ?

M.C. UNDERWOOD   You are of course correct
that time variations in the wind field should be
accounted for. At present the model cannot do this,
but it is in principle possible. The input data
often is not available, but refinements of the type
you mention will give a physically more realistic
model.

ESTIMATION OF DIFFUSE HYDROCARBON LEAKAGES FROM

PETROCHEMICAL FACTORIES

Bjarne Sivertsen

Norwegian Institute for Air Research

P.O.Box 130, N-2001 Lillestrøm, Norway

ABSTRACT

Tracer experiments were carried out to quantify the diffuse leakages of hydrocarbons at two petrochemical complexes in Norway. Single tracer (SF_6) and dual tracer ($SF_6/CBrF_3$) experiments were performed for different test designs and different meteorological conditions.

A simple proportionality model was applied to estimate leakage rates of ethylene, propylene, ethane, propane and isobutane from different parts of the plants. A discussion of uncertainties in the release rate estimates is also included.

Dispersion models were applied to verify concentration profiles and to identify leakage areas. A model using K_z for estimating vertical spread and σ_Θ for estimating horizontal spread, including building size parameters as initial dilution factors, was best correlated to measured tracer concentrations.

INTRODUCTION

From thousands of fittings, such as valves, flanges pumps and compressors, at a petrochemical factory, there is a potential for hydrocarbon leakages. These diffuse leakages can normally not be quantified, and can be prevented only by systematic maintenance. Some of the lower hydrocarbons as ethylene and propylene are active in the formation of photochemical oxidants and their release rates should be kept as low as possible.

To establish a tool for quantifying total diffuse leakages of hydrocarbons (HC) from these sources, the use of tracer techniques was proposed. The first experiments were carried out in 1979 using sulphur hexafluoride (SF_6) as the tracer. Experiments were also undertaken at different parts of the petrochemical factories in the Telemark area of Southern Norway in 1980 and 1981, during the latter applying a dual tracer ($SF_6/CBrF_3$) technique.

The dispersion of gas releases in the vicinity of buildings and other structures of petrochemical complexes is influenced by complex flow fields. Several studies have been performed to modify the dispersion estimates in such flows (1,2). The most often applied approach has been to adjust the standard Gaussian plume dispersion parameters σ_y and σ_z (3,4) to reflect the enhanced dispersion in the turbulent cavity and wake zones downwind from the structures. In this paper a comparison of two methods of estimating concentrations downwind of the petrochemical complexes is presented.

ESTIMATES OF DIFFUSE HC LEAKAGES

The tracer technique

Known amounts of tracers; sulphur hexafluoride (SF_6) and a freon ($CBrF_3$) were released in those parts of the petrochemical factories where leakages were expected. These areas were detected prior to the experiments with non-quantitative portable gas chromatographs. The tracers were collected in 100 cm^3 plastic syringes downwind from the release areas as 15-min samples or as instantaneous samples. Up to fifty 15-min average automatic samplers with electronic timers were used in the experiments. The samplers were analyzed immediately after each experiment with two portable gas chromatographs (capacity $\simeq$ 100 SF_6 samples/hr). SF_6 was most often used as the tracer gas ($CBrF_3$ only when dual tracer technique was needed). The lower detection limit for SF_6 was $\simeq$ 2 ppt (10^{-12} parts SF_6/part air), with useful range from $\simeq$ 2-10^6 ppt. The background level of SF_6 in Norway is expected to be less than 0.4 ppt (5). The detection limit for $CBrF_3$ was $\simeq$ 50 ppt. The tracer techniques and experimental design have been described earlier (6,7).

In order to use the simple proportionality model presented below simultaneous samples of hydrocarbons were collected at selected sampling points during the same tracer test periods. These samples were collected in aluminium-plastic laminated bags by means of electric pumps, and analyzed at the NILU laboratory for ethylene, propylene, ethane, propane, iso-butane etc.

Tracer experiments

A total of 16 experiments have been carried out at different petrochemical factories in the Telemark area of Southern Norway. The experiments were undertaken during different meteorological conditions and with various release configurations. During a few test periods it was possible to check the leakage rate estimates from single test results under opposite wind directions.

Normally tracer experiments of this type can be conducted when the following conditions are fulfilled :
- wind speed should be > 3 m/s to create mechanically induced turbulence for tracer and hydrocarbons to be sufficiently well mixed at the sampling points;
- periods with precipitation should be avoided;
- release time must adjust to wind speed and sampling point distances;
- sampling time must be long enough to avoid short period fluctuations due to atmospheric turbulence;
- sampling point location should assure at least one horizontal cross-section of the concentration distribution;
- tracer release points and release rates relative to each other should be located and proportioned to match actual leakages of HCs;
- meteorological variables should be continuously recorded.

As an example of the experiments carried out, Figure 1 shows the 15-min average SF_6 plume due to releases from the ethylene cracker area at the Rafnes complex. SF_6 was released at different heights above the ground in the cracker areas and at different release rates proportional to the expected leakage rates.

Figure 2 shows that the ethylene, propylene and ethane leakages in the ethylene cracker plant, were adequately simulated with the SF_6 tracer releases, while propane seemed to have its largest source in the eastern part of the factory where the SF_6 release rates were low. The total propane leakage rate estimated from this experiment is therefore more uncertain than the extimates for ethylene, propylene and ethane.

In Figure 3 the 15-min average concentration distribution for the two tracers SF_6 and $CBrF_3$ is presented for releases at the Saga factories at Rønningen. SF_6 was released from 7 locations within the low density polyethylene factory (PEL) and 3 locations in the high density factory (PEH), while $CBrF_3$ was released from 3 locations in the polypropylene factory (PP). The wind was blowing at 4.5 m/s from SSE in near neutral to unstable atmospheric conditions. Most of the $CBrF_3$ cloud was transported to the west of the PEL-factory structures, while SF_6 followed along the larger scale trajectories from SSE.

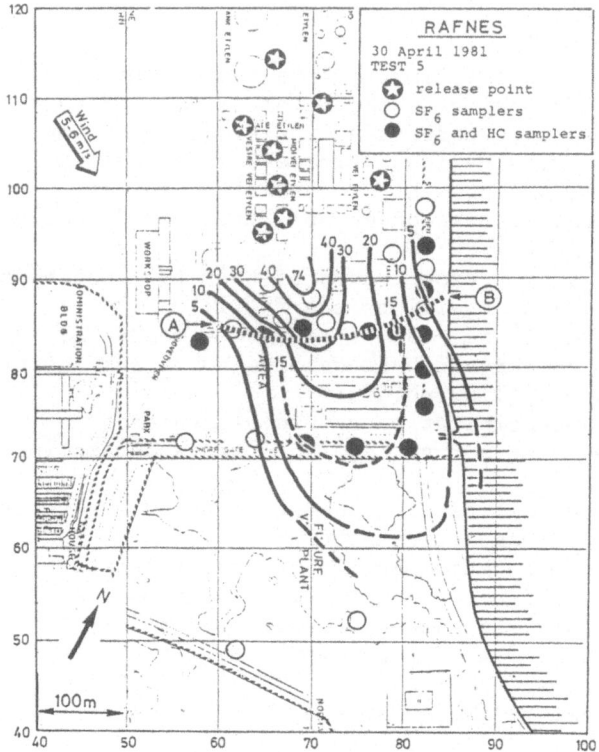

Fig. 1. Test 30 April 1981: 15-min average SF$_6$ concentrations
        (µg/m^3) measured downwind from the ethylene cracker area
        at Rafnes.

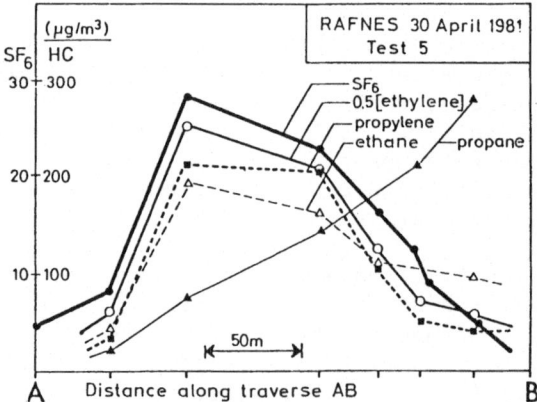

Fig. 2. Test 30 April 1981, Rafnes: 15-min average concentration
        profiles of SF$_6$, ethylene, propylene, ethane and propane
        along traverse AB (see Fig. 1) downwind from the ethylene
        cracker area.

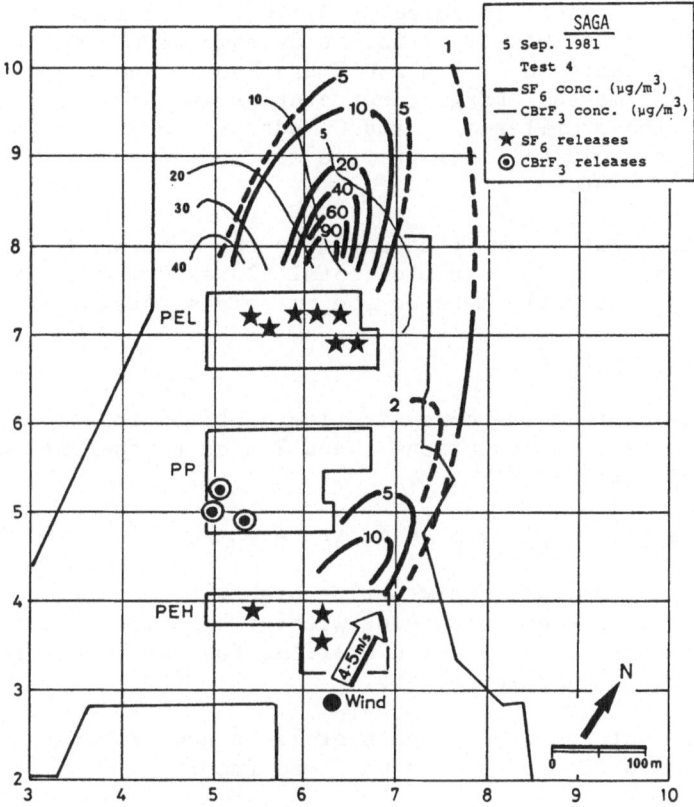

Fig. 3. Test 4,5 Sep. 1981 Saga: 15-min average SF_6 and $CBrF_3$-
concentrations ($\mu g/m^3$) downwind from the low density (PEL),
high density (PEH) polyethylene and polypropylene (PP)
factories at Røningen.

Emission estimates

During near quasi stationary wind conditions, assuming that
there are no atmospheric chemical reactions or deposition of the
hydrocarbons of interest between the leakage points in the
factory release areas and the sampling points, the flux of hydrocarbons (F_{HC}) through an infinite cross section perpendicular to
the wind direction at a distance x_i downwind is equal to the total
release rate (Q_{HC}):

$$F_{HC} = \int_{\infty}^{\infty} \int_{\infty}^{\infty} u(x_i,y,z) \cdot C_{HC}(x_i,y,z)dydz = Q_{HC} \qquad (1)$$

C are concentrations measured at different points within the cross section at x_i and $u(x_i,y,z)$ is the average wind speed perpendicular to the cross section. If u and C ere known, the total release rate could be estimated. Large uncertainties are, however, usually linked to the profiles of u and C, but by the use of tracers it is possible to avoid vertical profile measurements and thus eliminate these uncertainties.

If the tracers are released from the same point (or area) as the hydrocarbons, the ratio of total fluxes downwind (F) will be the same as the ratio between the emissions rates (Q):

$$Q_{HC}/Q_{tr} = F_{HC}/F_{tr} \tag{2}$$

Assuming that the concentration distributions of the two gases are the same, Equations 1 and 2 lead to the following simple expression:

$$C_{HC}(x,y,z)/C_{tr}(x,y,z) = const. = Q_{HC}/Q_{tr} \tag{3}$$

Equation 3 would apply during high wind conditions, when the mechanically induced turbulence within the building structures produce similar concentration profiles for the tracer gases and the HCs at the sampling points.

Note that the wind speed profile is now eliminated and the total release rate of HC can be found from:

$$Q_{HC} = Q_{tr} \cdot (C_{HC}/C_{tr}) \tag{4}$$

The accuracy in the determination of the total HC release rate now depends upon the accuracy of the concentration measurements and the tracer release rate. These concentration measurements must be taken at exactly the same sampling points using the same sampling time to avoid differences due to concentration fluctuations.

To reduce some of the uncertainties in the assumption of equal concentration distributions of HC and tracer, thus taking into account the fact that the HC leakages might not be correctly simulated, Equation 2 was used more directly to estimate the total leakage rate of HC:

$$Q_{HC} = Q_{tr} \frac{F_{HC}}{F_{tr}} \approx Q_{tr} \underset{xy}{\Sigma\Sigma} \left\{ C_{HC}(x,y,z_1) / \underset{xy}{\Sigma\Sigma} \ C_{tr}(x,y,z_1) \right\} \tag{5}$$

Assuming that the wind speed was the same at all sampling
points, (all were located one metre above the ground) and that
the sampling points were equally spaced in the xy-directions.
From this equation the average release rates of HC during each
tracer experiment were estimated.

Uncertainties

The experiments undertaken at the different factories have
revealed a rather large variation of estimated HC leakage rates
from one experiment to another, and also from one sampling point
to another when single sampling points were used for estimating
leakages.

In addition to the variation that is actually present in the
leakage rate of HC from one experiment to another, the measured
variations can be explained by the following main arguments:

1) the goodness of the HC leakage simulation with tracers,
   (unfortunate simulation cause the HC/tracer concentration
   ratio at individual sampling points to vary due to changes
   in wind and turbulence pattern from one sample to another).

2) the overall uncertainty in the determination of the parameters entering eq. 5 (accuracy, precision, systematic
   errors).

The first part of the uncertainties can not easily be quantified. These uncertainties can, however, be reduced by the following precautions:

- increase the number of sampling points to ensure that the
  width of "plumes" from single tracer release points is
  larger than the distance between the sampling points (no
  plumes should escape between samplers).

- perform experiments during high wind conditions (> 3 m/s)
  (high winds increase the mechanically induced turbulence
  to ensure a good mixture of tracer releases and HC leakages, so that the excact location of tracer release points
  becomes less important).

- locate samplers sufficiently far away from the source area
  so that the samplers do not "see" the inhomogeneties in the
  release patterns. (However, the measured concentrations
  would then be lower, and ucertainties might increase for
  this reason).

The second part of the uncertainty evaluation can be quantified and has also been discussed in earlier papers (8,9). The total uncertainty in the HC leakage rate estimate; $Q = f(x_1, x_2 \ldots x_n)$
is a function of the uncertainties in each of the independent
variables; x_i

$$\sigma_Q^2 = \sum_{i=1}^{n} \sigma_{x_i}^2 \left(\frac{\partial f}{\partial x}\right)^2 \tag{6}$$

when the total relative uncertainty (s_i) of each independent variable entering the $_{HC}$ estimate (eq. 5) is known, the following
simple expression for the overall uncertainty in $_{HC}$ can be used:

$$s_{Q_{HC}} = (s_{\Sigma Q_{tr}}^2 + s_{\Sigma C_{HC}}^2 + s_{\Sigma C_{tr}}^2)^{1/2} \tag{7}$$

The tracer gas release uncertainty $(s_{\Sigma Q_{tr}})$ is dependent upon
systematic errors (flowmetre calibration etc) and precision (flow
meter scale reading). The uncertainty in determining the total
release rate from several release points will also be inverse proportional to the square root of the number of release points n:

$$s_{\Sigma Q_{tr}}^2 = s_i^2/n \tag{8}$$

The HC and tracer gas concentration uncertainties involve
errors in sampling, sample handling and precision or standard
error in the analysis of individual samples. The total uncertainty in the summed concentrations used in eq. 5 is also reduced with
an increasing number of samples.

The best estimate for input variable uncertainties for the
HC leakage rate experiments are summarized in Table 1:

Table 1: Relative uncertainties (%) affecting the total
HC leakage estimates.

	Tracers		HC-measurements
	% Releases	Concen-trations	Concentrations
Systematic errors	$\approx$ 5		
Precision	$\approx$ 10	6	5
Accuracy		7	7

This part of the uncertainties of the HC leakage rates were estimated to vary between 6 and 10% for the different experimental designs. In addition, the unquantified uncertainties due to the leakage simulation have to be taken into account.

Mass balance studies including flux estimates through the sampling cross sections, and plume size estimates (plume width and "mixing height") were performed to test the quality of each experiment. Typically 50 to 95% of the tracers released were found when integrated fluxes were estimated.

DISPERSION MODELING

The aim of this part was to study the influence of building structures on the near field dispersion of gases released within these structures. To establish a future tool for estimating downwind hydrocarbon concentrations at such plants, two dispersion modeling approaches were tested on the tracer gas data.

The standard Gaussian model with building dilution factors

The concentration downwind from a single source was assumed to be influenced by both "homogeneous flat terrain" type atmospheric turbulence and building induced turbulence:

$$C = Q \left(\pi \, \Sigma_y \Sigma_z \, \bar{u} \right)^{-1} \exp \left[-(y^2/2\Sigma_y^{\;2} + h^2/2\Sigma_z^2) \right] \tag{9}$$

$$\text{where} \quad \Sigma_y = (\sigma_y^{\;2} + c \cdot B \cdot H/\pi)^{1/2} \tag{10}$$

$$\Sigma_z = (\sigma_z^{\;2} + c \cdot B \cdot H/\pi)^{1/2} \tag{11}$$

B and H is the average building height and width respectively, c is a constant factor usually equal 0.5, and h is the plume centre line height above ground. The dispersion parameters σ_y and σ_z were taken from the Brookhaven data for roughness lengths representative of our test site (4,10). This model is later named the Gifford model.

Vertical K_z, horizontal σ_θ model with initial dilution

The horizontal diffusion was treated according to the recommendations given by Pasquill (11) based upon Taylors statistical treatment of diffusion

$$\sigma_y = \sigma_\theta \cdot x \circ f(t/t_L) \tag{12}$$

where σ_θ is the standard deviation of horizontal wind direction fluctuations. The function f of travel time t and the Lagrangian integral scale t_L has been investigated in earlier tracer experiments (12), and the empirical derived expression was used:

$$f = (1 + 0.055 \cdot t^{0.5})^{-1} \qquad\qquad (13)$$

The 5 min. and 1 hour average σ_θ-values were routinely estimated from data recorded by the NILU automatic weather station. The total horizontal spread was estimated from eq. 10 with the expression for σ_y as given by eqs. 12 and 13.

The vertical diffusion was also treated according to the Pasquill hypothesis, where the mean displacement ($\bar{z}$) of an ensemble of particles after a travel distance x is given by:

$$\frac{d\bar{z}}{dx} = a \left(\frac{K}{u\bar{z}}\right)^b \qquad\qquad (14)$$

where a = 0.95; b = 1.06, and K is the eddy diffusivity.

This equation is valid from unstable to moderately stable conditions. Our experiments are well within these limits. Assuming that the ratio K/u is constant with height, which is a reasonable assumption in the surface boundary layer, eq. 14 can be integrated to become:

$$\bar{z} = \left[a(b+1) (K/u)^b \cdot x + \bar{z}_{x=0}^{b+1}\right]^{(1/(1+b))} \qquad\qquad (15)$$

Assuming that the vertical distribution of gases is Gaussian $\sigma_z = \bar{z} \cdot (\pi/2)^{0.5}$, and that the vertical extension of the plume at $x = 0$ is equal to the average building height, we get:

$$\sigma_z = \left[(\pi/2) (1.96(K/u)^{1.06} x + H)\right]^{0.5} \qquad\qquad (16)$$

The vertical eddy diffusivity K is assumed to be strongly related to the eddy conductivity K_h:

$$K \simeq K_h = X \cdot u_* \cdot z/\phi(z/L) \qquad\qquad (17)$$

where the friction velocity is given by:

$$u_* = X \cdot u \cdot (\ln(z/z_o) - \psi_m (z/L))^{-1} \qquad\qquad (18)$$

L is the Monin Obukhov length and z_o is the surface roughness.

The temperature difference between 36 m and 10 m, DT, and the temperature at 10 m, T, was measured at a meteorological tower. From these data the bulk Richardson number, RB was estimated:

$$RB = g(DT + 0.3)/(T \cdot u^2) \qquad\qquad (19)$$

The Monin-Obukhov length L is assumed to be inverse proportional to the bulk Richardson number RB:

$$L \simeq k(z_o)/RB \tag{20}$$

For our site where the surface roughness length $z_o \simeq 0.5$ m, $k(z_o) \simeq 1$. The value of $k(z_o)$ decreases with decreasing z_o. The von Karman constant; $\chi = 0.35$ and the functions ψ_m are taken from Businger [13] in the following way:

Unstable	RB ≤ -0.01	$\phi = 0.74 (1-9 \ z/L)^{-0.5}$ $\psi_m = 2 \ln 0.5 \cdot (1+x)$ $\quad + \ln(0.5(1+x^2))$ $\quad - 2 \ Atan \ (x+\pi/2)$ $x = (1-15 \ z/L)^{0.25}$
Near neutral	$-0.01 < RB < 0.01$	$\phi = 1$ $\psi_m = 0$
Stable	RB ≥ 0.01	$\phi = 0.75 (1+6.3 \ z/L)$ $\psi_m = 4.7 \ z/L$

(21)

Model evaluations

Tracer data from several experiments were applied in testing the two models. Figure 4 shows typical crosswind concentration profiles of SF_6, measured and estimated using data from two experiments.

Both models adequately estimate the SF_6 concentrations downwind from the source areas. The average deviation between estimated and measured concentrations in all sampling points was 56 and 33% for test 4 and 5 respectively, using the K-sigθ-model, and 83 and 48% using the Gifford model. The standard Gaussian model with building dilution factors (named: Gifford model) apparently give too narrow plumes. The estimates are slightly better with the "K-sigθ"-model.

A linear regression analysis was performed on data from five experiments at the Saga and Rafnes factories. The coefficients a_1 and a_0 of the linear least square fit line :

$$C_{est} = a_1 \cdot C_{meas} + a_0 \tag{22}$$

and the correlation coefficients r between estimated (C_{est}) and measured concentrations (C_{meas}) are given in Table 2 for the application of the Gifford model and the K-sig θ model on five experiments.

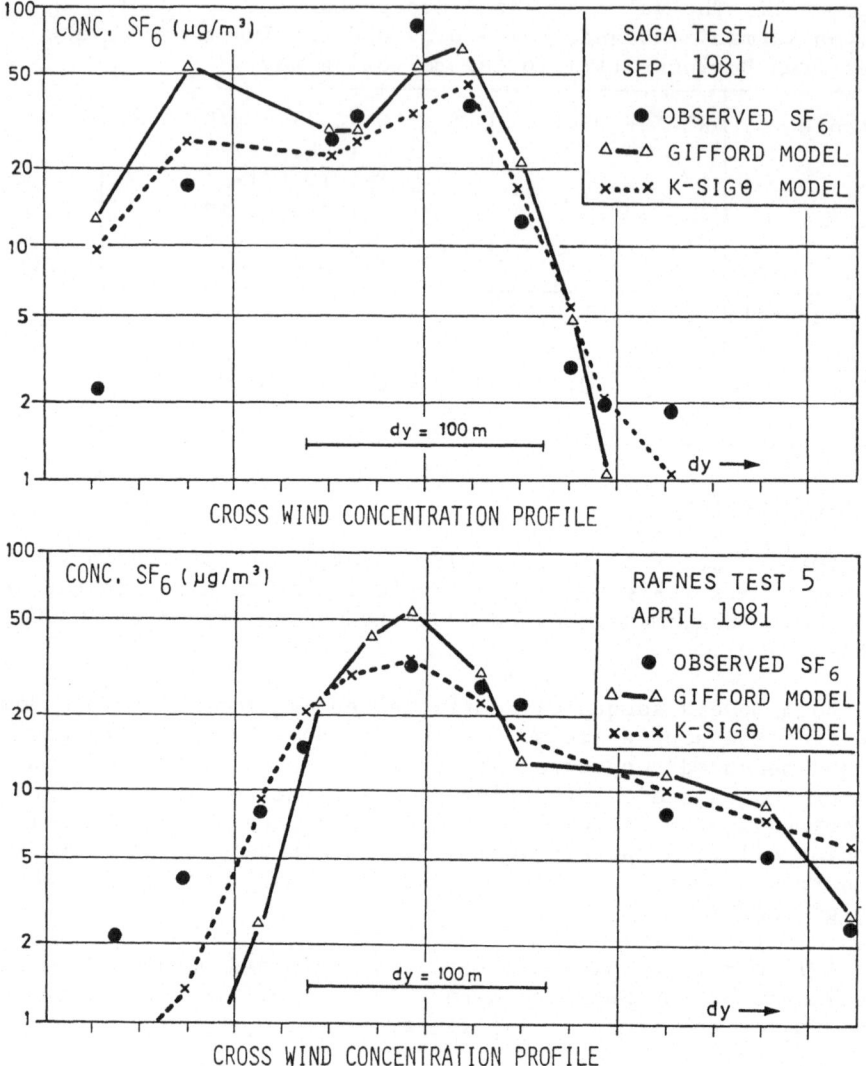

Fig. 4. Downwind cross sections of tracer concentration
         profiles observed and estimated at the Saga
         factories (Test 4) and Rafnes factories (Test 5).

Table 2 : Linear regression coefficients for estimated vs. measured tracer concentrations applying two dispersion models on five experiments.

		Gifford model			K-sig θ model		
		a_1	a_0	r	a_1	a_0	r
Saga	Test 1	0.67	14.4	0.56	0.59	7.7	0.72
	Test 2	1.8	-3.1	0.80	1.2	-2.0	0.90
	Test 3	2.0	-12.9	0.81	1.0	-5.2	0.86
	Test 4	0.7	6.4	0.78	0.55	5.3	0.80
Rafnes	Test 5	1.44	-3.3	0.92	0.97	1.08	0.96

The Gifford model had a tendency of over-estimating concentrations while the opposite applied for the K-sig θ model. The results of the latter model correlated better with the measured concentrations in all experiments.

A summary of estimated versus measured concentrations is presented in Figure 5, which shows that the points are grouped more closely around the "aquality-line" when the K-sig θ model was applied.

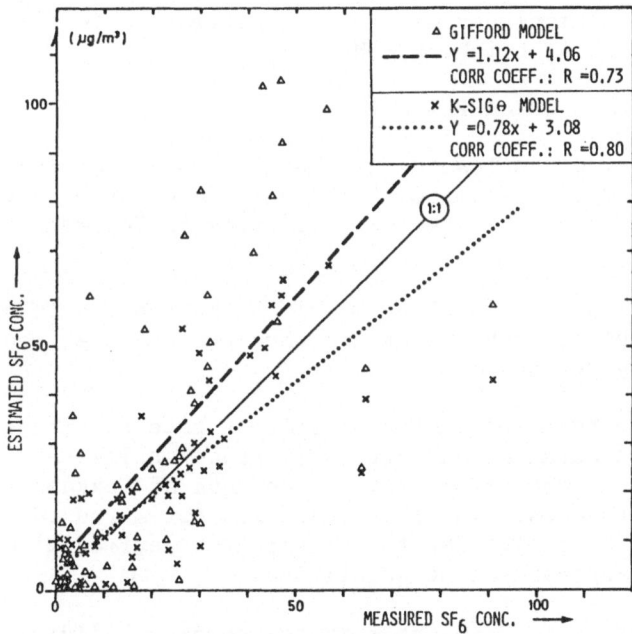

Fig. 5. Estimated versus measured tracer concentrations based upon five tracer experiments at the Saga and Rafnes petrochemical factories.

Table 3 : The linear regression coefficients (see eq. 22) for
three sets of input variables using the Gifford
model, four sets of input variable using the K-sig θ
model on data from Test 5.

		Gifford model			K-sig θ model			
	Run no.	1	2	3	1	2	3	4
In: Wind speed (m/s)		5	6	6	5	6	6	6
σ_θ (deg)		N	N	N	15	15	7	7
Build.height (m)		10	10	25	10	10	10	25
Build.width (m)		50	50	50	50	50	50	50
Out: a_1		1.44	1.2	0.88	0.97	0.8	1.05	0.51
a_0		-3.3	-2.6	0.07	1.08	1.0	-1.6	0.16
r		0.92	0.92	0.92	0.96	0.96	0.89	0.88

A sensitivity test of the two models was also done. Table 3
briefly presents some of the runs using date from Test 5.

Table 3 demonstrates first of all that the K-sig θ model was
more sensitive to the influence of average building height on the
spread in the vicinity of building structures. The reason is that
the value of H is included also in the estimate of K (in eq. 16),
as K is estimated at height z = H.

CONCLUSIONS AND FUTURE APPLICATIONS

The following conclusions can be drawn from the tracer experiments designed to testimate total fugitive emission rates of
hydrocarbons at the petrochemical factories in Telemark Southern
Norway :

- a simple proportionality model can be applied to estimate
  the leakages of hydrocarbons when tracer experiments are
  properly designed;

- the uncertainties in the HC release rate estimates are due
  to: 1) accuracies and errors in tracer releases and concentration determination, 2) goodness of leakage simulation
  by the tracers. The first part was estimated to vary
  between 6 and 10% for the different experimental designs,
  the latter can not be quantified;

- concentrations downwind from the complex building structures at petrochemical factories can best be estimated by a
  dispersion model using: a) K_z for vertical spread, b) wind
  direction fluctuations σ_θ for horizontal spread, c) building

size parameters to include initial dilution near building structures.

The results from the tracer experiments have provided the basis for designing future routine experiments at the factories for control purposes. The diffuse leakage rate estimates will specify whether the fugitive emissions are within the operating limits for the plant, and will also provide information about the location of leakage areas, as a basis for systematic maintenance in the various process areas.

This being known, the "K-sig0" dispersion model will be applied to estimate the concentration distribution of the HCs in question downwind from the factory.

ACKNOWLEDGEMENTS

This study has been supported by Norsk Hydro, Rafnes and Saga Petrochemi A.S. & Co.

REFERENCES

1. A.H. Huber and W.H. Snyder, Building wake effekcts on short stack effluents. In: Third Symp. on Atm. Turb. Diff. and Air Quality. Oct 1976. Raleigh N.C. Am. Met. Soc., Boston, Mass., (1977).
2. R.H. Thuillier, Dispersion characteristics in the lee of complex structures. J. Air Poll. Contr. Ass., 32: 526 (1982).
3. F.A. Gifford, Atmospheric diffusion models and applications. In: Meteorology and atomic energy, D.H. Slade Ed. Atomic Energy Con., Springfield, VA, 1969, Chapter 3.3.
4. M. Smith, Recommended guide for the prediction of the dispersion of airborne effluents. ASME, New York (1968).
5. M. De Bortoli and E. Peechio, Measurements of some halogenated compounds in air over Europe. Atmos. Environ. 10: 921 (1976).
6. B. Lamb and B. Sivertsen, Atmospheric dispersion experiments using the NILU automatic weather station and SF_6 tracer techniques. (NILU OR 12/78). Lillestrøm (1981).
7. B. Sivertsen, Tracer experiments to assess diffuse emissions of hydrocarbons from the petrochemical factories at Saga. (NILU OR 33/81). Lillestrøm (1981).
8. B. Sivertsen, Tracer experiments to assess diffuse emission of hydrocarbons from the petrochemical factories at Rafnes. (NILU OR 27/80). Lillestrøm (1981).
9. P.A. Sackinger, D.D. Reible and F.H. Shair, Uncertainties associated with the estimation of mass balances and Gaussian parameters from atmospheric tracer studies. J. Air Poll. Contr. Ass. 32: 720 (1982).

10. B. Sivertsen, The application of Gaussian dispersion models
       at NILU. (NILU TN 11/80). Lillestrøm (1980).
11. F. Pasquill, Some topics relating to modelling of dispersion
       in the boundary layer. (EPA 650/4-75-015).
       Washington D.C. (1975).
12. B. Sivertsen, Dispersion parameters determined from measure-
       ments of wind fluctuations, temperature and wind pro-
       files. In: Proceedings of the ninth international
       technical meeting on air pollution modelling and its
       application. (NATO/CCMS no. 103). Toronto (1978).
13. J.A. Businger, Turbulent transfer in the atmospheric sur-
       face layer. In: Workshop on micrometeorology, D.A.
       Haugen, Ed., Amer. Meteor. Soc. Boston, Mass. (1973),
       67-98.

DISCUSSION

D.R. MIDDLETON   How important is it to have
       the tracer relase from the same height as indivi-
       dual process plant structures ?

B. SIVERTSEN   As long as you have strong
       wind through the petrochemical structures the relaese
       height is not very critical. However, we always try
       to release tracer at the same height as we suspect
       the leakages

D.R. MIDDLETON   How sensitive are the model-
       ling and the results to the surface roughness esti-
       mate since it takes time for a flow to become fully
       developed overchanges in roughness ?

B. SIVERTSEN   Within the turbulent wake
       zone of the structures, surface roughness is not very
       critical, but as soon as you leave the wake zone
       the surface roughness starts to play a role.
       It is included in our model. To enable concentration
       estimates also outside the wake zone.

PARTICIPANTS

The 13th NATO/CCMS Internation Technical Meeting on Air Pollution
Modeling and its Application
Ile des Embiez, September 14-17, 1982   France

AUSTRIA

Pillmann W.   Technical University of Vienna
                                      and Osterreischisches Bundes-
                                      institut für Gesundheitswesen
                                      Kegelgasse 19-4
                                      A-1030   Vienna

Piringer M.   Zentralanstalt für Meteorologie
                                      Hohe Warte 38
                                      A-1190 Wien

Posch M.   IIASA
                                      Schlossplatz 1
                                      A-2361 Laxenburg

Runca E.   IIASA
                                      Schlossplatz 1
                                      A-2361 Laxenburg

BELGIUM

Coppieters C.   Municipal Laboratory of Antwerp
                                      Slachthuislaan 68
                                      B-2000 Antwerp

Kretzschmar J.   SCK/CEN - Studiecentrum
                                      voor Kernenergie
                                      Boeretang 200
                                      B-2400   Mol

Schayes G.   UCL - Institut d'Astronomie
                                      et de Géophysique
                                      Chemin du Cyclotron 2
                                      B-1348   Louvain-la-Neuve

Schreurs P.   KUL - C.I.T.
                                         De Croylaan 2
                                         B-3030   Leuven-Heverlee

Stief-Tauch H.   Commission of the
                                         European Communities
                                         Rue de la Loi 200
                                         B-1049   Bruxelles

Vergison E.   Solvay S.A.
                                         Rue de Ransbeek 310
                                         B-1120   Bruxelles

Wispelaere C., De   Diensten van de Eerste Minister
                                         Programmatie van het Weten-
                                         schapsbeleid
                                         Wetenschapsstraat 8
                                         B-1040   Bruxelles

CANADA

Davidson A.   Mechanical Engineering
                                         University of Waterloo
                                         C-Waterloo, Ontario N2L 3G1

Fanaki F.   Atmospheric Environment Service
                                         Rue Dufferin Street 4905
                                         C-Downsview, Ontario M3H 5T4

Misra P.   Ministry of Environment
                                         Bay Street 880, 4th Floor
                                         C-Toronto, Ontario

Slawson P.   Mechanical Engineering Dpt.
                                         University of Waterloo
                                         C-Waterloo, Ontario N2L 3G1

DENMARK

Baerentsen J.   National Agency of Environmental
                                         Protection - Air Pollution Lab.
                                         Riso National Laboratory
                                         D-4000   Roskilde

Berkowicz R.   National Agency of Environmental
                                         Protection - Air Pollution Lab.
                                         Riso National Laboratory
                                         D-4000   Roskilde

Gryning S.   Riso National Laboratory
                                    Meteorology Section
                                    D-4000   Roskilde

Markvorsen J.   Cowiconsult
                                    Teknikerbyen 45
                                    D-2380   Virum

Mikkelsen T.   Meteorology Sect., Physics Dept.
                                    Riso National Laboratory
                                    D-4000   Roskilde

Troen I.   Riso National Laboratory
                                    Meteorology Section
                                    D-4000   Roskilde

FRANCE

Boudet   Université de Provence
                                    Place Victor Hugo
                                    F-13331 Marseille Cedex 3

Blondin C.   Etablissement d'Etudes et
                                    Recherches Météorologiques
                                    Direction de la Météorologie
                                    rue de Sèvres 73-77
                                    F-92106   Boulogne-Billancourt Cedex

Caneill J.   Electricité de France
                                    Direction des Etudes et Recherches
                                    Service AEE
                                    Quai Watier 6
                                    F-78400   Chatou

Clement J.   ENRS, Shell Française
                                    Rue du Berri 29
                                    F-75397 Paris Cedex 08

Doury A.   Commissariat à l'Energie Atomique
                                    IPSN/DSN - BP n° 6
                                    F-92260   Fontenay-aux-Roses

Geneve C.   Association Airmaraix
                                    Rue Paradis 307
                                    F-13008 Marseille

Gland H.   Electricité de France
                                    Service AEE
                                    Quai Watier 6
                                    F-78400   Chatou

Grollier R.   Institut Français du Pétrole
                                         BP n° 3
                                         F-69390 Vernaison

Hodin A.   Electricité de France
                                         Direction des Etudes et Recherches
                                         Quai Watier 6
                                         F-78400 Chatou

Klein F.   L.E.C.E.S.
                                         Voie Romaine
                                         F-5710 Maizieres-les-Metz

Lombard E.   Centre de Recherches ESSO
                                         BP n° 6
                                         F-76130 Mont Saint Aignan

Medard T.   Institut National de Recherche
                                         Appliquée
                                         Centre de Recherches
                                         BP n° 1
                                         F-91710 Vert-le-Petit

Méry M.   Electricité de France
                                         Direction des Etudes et Recherches
                                         Service AEE
                                         Quai Watier 6
                                         F-78400 Chatou

Muller M.   Ministère de l'Environnement
                                         Mission des Etudes et de la
                                         Recherche
                                         Bd. du Général Leclerc 14
                                         F-92524 Neuilly s/Seine Cedex

Nadal R.   Direction Interdépartementale
                                         de l'Industrie
                                         Bd. Périer 37
                                         F-13008 Marseille

Oppeneau J.   Ministère de l'Environnement
                                         Mission des Etudes et de la
                                         Recherche
                                         Bd. du Général Leclerc 14
                                         F-92524 Neuilly s/Seine Cedex

Pentel R.   SNPE CE
                                         Rue de Bercy 209-211
                                         F-75012 Paris

Quinault J.

Commissariat à l'Energie
Atomique
Tour d'Aygosi 19
F-13100 Aix-en-Provence

Roulet J.

Elf-Aquitaine
Direction GAN
Cedex
F-92082 Paris-la-Défense

Saab A.

Electricité de France
Direction des Etudes et Recherches
Service AEE
Quai Watier 6
F-78400 Chatou

Sauvageot H.

Observatoire du Puy de Dôme
Centre de Recherches Atmosphériques
F-65300 Lannemezan

FEDERAL REPUBLIC OF GERMANY

Brose G.

Rheinisch-Westfälischer Technischer
Uberwachungs-Verein e.V.
Postfach 10 21 61
D-4300   Essen 1

Eppel D.

Institut für Physik
GKSS - Forschungs Zentrum
D-2054   Geesthacht

Geiss H.

Kernforschungsanlage Jülich
Abteilung Sicherheit und
Strahlenschutz
Postfach 1913
D-5170   Jülich

Glaab H.

Institut für Meteorologie
Hochschulstrasse 1
D-6100   Darmstadt

Janicke L.

Dornier-System
Postfach 1360
D-7990   Friedrichshafen

Klug W.

Institut für Meteorologie
Technische Hochschule
D-6100   Darmstadt

Lehmann K.

DIN - Deutsches Institut für
Normung e.V.
Burggrafenstrasse 4-10
D-1000   Berlin 30

Ludwig C.

Umweltbundesamt
Bismarckplatz 1
D-1000   Berlin 33

Nester K.

Kernforschungszentrum Karlsruhe GmBH
Hauptabteilung sicherheit
Postfach 3640
D-7500   Karlsruhe

Pankrath J.

Umweltbundesamt
Bismarckplatz 1
D-1000   Berlin 33

Reichenbacher W.

Senatsverwaltung für Stadtent-
wicklung und Umweltschutz - VC 6
Lentzeallee 12-14
D-1000   Berlin 33

Rockle R.

Institut für Meteorologie
Hochschulstrasse 1
D-6100   Darmstadt

Roeckner E.

Meteorologische Institut
Bundesstrasse 55
D-2000   Hamburg 13

Rudolf B.

Deutscher Wetterdienst - Zentral-
amt
Frankfurter Strasse 135
D-6050   Offenbach a.M.

Scherer B.

Institut für Geophysikalische
Wissenschaften der Freien Uni-
versität Berlin
Thielallee 50
D-1000   Berlin 33

Schorling M.

Industrieanlagen-Betriebsgesell-
schaft mbH
Einsteinstrasse
D-8012   Ottobrunn

Schultz H.

Universität Hannover
Zentrum I. Strahlenschutz
Appelstrasse 9 A
D-3000   Hannover 1

Thomas P.   Kernforschungszentrum Karlsruhe GmbH
                                   Hauptabteilung Sicherheit
                                   Postfach 3640
                                   D-7500   Karlsruhe

Verenkotte H.   Hauptabteilung Sicherheit/Umwelt-
                                   meteorologie
                                   Kernforschungszentrum Karlsruhe GmbH
                                   Postfach 3640
                                   D-7500   Karlsruhe

Wilcke F.   Freie Universität Berlin
                                   Institut für Geophysikalische
                                   Wissenschaften
                                   Thielallee 50
                                   D-1000   Berlin 33

Zimmermann P.   Max Planck Institut für Chemie
                                   P.B. 3060
                                   D-6500   Mainz

ISRAEL

Yoram Nir-El   Soreq Nuclear Research Center
                                   Yavne 70600
                                   Israel

ITALY

Bonino G.   Instituto di Cosmo-geofisica-CNR
                                   Corso Fiume 4
                                   I-10133   Torino

Borghi S.   Osservatorio Meteorologica di Brera
                                   Via Brera 28
                                   I-20121   Milano

Clerici G.   ARS spa
                                   Viale Maino 36
                                   I-20121   Milano

Hasenjaeger H.   Commission of the European
                                   Community
                                   I-21010 JRC Ispra, Varese

Latini A.   Air Force Meteorological Service
                                   Viale Ungheria 17/2
                                   I-20138   Milano

Levy A.   Snamprogetti S.P.A.
                                           Via Toniolo 1
                                           I-61032 Fano

Longhetto A.   Universita di Torino
                                           Instituto di Cosmogeofisica
                                           CNR - Corso Fiume 4
                                           I-10133 Torino

Melli P.   Centro Scientifico IBM
                                           Via Giorgione 129
                                           I-00147 Roma

Santomauro L.   Osservatorio Meteorologica di Brera
                                           Via Brera 28
                                           I-20121 Milano

Santomauro G.   Osservatorio Meteorologico di Brera
                                           Via Brera 28
                                           I-20121 Milano

Sandroni S.   Joint Research Center
                                           I-21020 Ispra, Varese

Stingele A.   C.C.R. - Program Protection of
                                           the Environment
                                           I-21020 Ispra, Varese

JAPAN

Ide Y.   Mitsubishi Heavy Industries
                                           Nagasaki Technical Institute
                                           I-I Akunoura-machi
                                           Nagasaki
                                           Japan

KUWAIT

Zannetti P.   KISR-EES
                                           P.O. Box 24885
                                           Safat
                                           Kuwait

THE NETHERLANDS

Builtjes P.   MT - TNO
                                           P.O. Box 342
                                           NL-7300 Apeldoorn

Burgers J.

Openbaar Lichaam Rijnmond
Vasteland 96-104
P.O.Box 23073
NL-3001 KB Rotterdam

Colenbrander I.

Koninkl. Shell Laboratorium
Badhuisweg 3
NL-1031 CM Amsterdam

Dop H., Van

Koninklijk Nederlands Meteoro-
logisch Instituut
P.O.Box 201
NL-3730 AE De Bilt

Egmond N., Van

National Institute of Public
Health
Postbus 1
NL-3720 BA Bilthoven

Ham J., Van

SCMO-TNO
Postbox 186
NL-2600 AD Delft

Holtslag A.

Koninklijk Nederlands Meteoro-
logisch Instituut
P.O.Box 201
NL-3730 AE De Bilt

Hout K., Van Den

Research Institute for Environ-
mental Hygiene
TNO
P.O.Box 214
NL-2600 AE Delft

Steenkist R.

KEMA
Utrechtseweg 310
NL-6812 AR Arnhem

Zonneveld S.

Zonnegge 18-23
NL-6903 GS Zevenaar

NORWAY

Eliassen A.

Norwegian Meteorological Institute
P.O.Box 320, Blindern
N-Oslo 3

Grøskei K.   Norsk Institute for Luftforskning
                                             Postbox 130
                                             N-2001   Lillestrøm

Sivertsen B.   Norwegian Institute for Air
                                             P.O.Box 130
                                             N-2001   Lillestrøm

PORTUGAL

Carvaliso   Instituto National de Meteoro-
                                             logica e Geofisica
                                             Rua C Aeroporto de Lisboa
                                             P-Lisbon 5

SPAIN

Manuel de Castro Munoz de Lucas   Catedra de Fisica del Aire
                                             Fac. Fisicas
                                             Madrid 3
                                             Spain

SWITZERLAND

Resele G.   Motor Columbus Ing. AG
                                             Parhstrasse 27
                                             5401 Baden
                                             Switzerland

TURQUIE

Ulug S.E.   Middle East Technical University
                                             Environmental Engineering Dpt.
                                             Ankara
                                             Turkey

UNITED KINGDOM

Apsimon H.   Nuclear Power Section
                                             Imperial College
                                             Mech. Eng. Department
                                             South Kensington
                                             UK-London SW 7

Blackmore D.R.   Shell Research Ltd
                                             Thornton Research Centre
                                             P.O.Box 1
                                             UK-Chester CHI 3SH

Crabtree J.   Meteorological Office (Met. 014)
                                     London Road
                                     UK-Bracknell, Berkshire RG 12 2SZ

Middleton D.   The British Petroleum Co.
                                     Applied Physics Branch, Bldg 100
                                     BP Research Centre
                                     Chertsey Road
                                     UK-Sunbury-on-Thames, Middlesex

Puttock J.S.   Shell Research Ltd
                                     Tornton Research Centre
                                     P.O.Box 1
                                     UK-Chester CHI 3SH

Smith F.   Meteorological Office (Met. 014)
                                     London Road
                                     UK-Bracknell, Berkshire RG 12 2SZ

Underwood M.   BP Research Centre
                                     Applied Physics Branch
                                     Chertsey Road
                                     UK-Sunbury-on-Thames, Middlesex

Williams M.   Warren Spring Laboratory
                                     Gunnels Wood Road
                                     UK-Stevnage, Herts

UNITED STATES OF AMERICA

Bhumralkar C.   National Oceanic and Atmospheric
                                     Administration (NOAA)
                                     6010 Executive Boulevard
                                     W5C - RD2, Romm 805
                                     Rockville, Maryland 20852
                                     U.S.A.

Carmichael G.   University of Iowa
                                     129 Chem Bld
                                     Iowa City IA 52240
                                     U.S.A.

Chan M.   AeroVironment Inc.
                                     145 Vista Avenue
                                     Pasadena, CA 91107
                                     U.S.A.

Chan S.   Lawrence Livermore National Lab.
                                        P.O.Box 808, L-262
                                        Livermore, CA 94550
                                        U.S.A.

Demerjian K.   U.S. Environmental Protection
                                        Agency
                                        Mail Drop 80
                                        Research Triangle Park
                                        North Carolina 27711
                                        U.S.A.

Irwin J.   Environmental Protection Agency
                                        Mail Drop 80
                                        Research Triangle Park
                                        North Carolina 27711
                                        U.S.A.

Knox J.   Lawrence Livermore National Lab.
                                        G-Division, L-262
                                        Livermore, CA 95440
                                        U.S.A.

Ludwig F.   SRI International
                                        333 Ravenswood
                                        Menlo Park, CA 94025
                                        U.S.A.

Mei Kao Liu P.   Systems Applications Inc.
                                        101 Lucas Valley Road
                                        San Rafael, CA 94903
                                        U.S.A.

Russel A.   Environmental Quality Laboratory
                                        California Institute of Technology
                                        Mail Code 206-40
                                        Pasadena, CA 91125
                                        U.S.A.

Taft. J.   Deygon-RA Inc.
                                        P.O.Box 3227
                                        La Jolla, CA 92038
                                        U.S.A.

Venkatram M.   Environmental Research and
                                        Technology Inc.
                                        696 Virginia Road
                                        Concord, Massachusetts 01742
                                        U.S.A.

Warner T.   The Pennsylvania State University
                                        Department of Meteorology
                                        503 Walker Bldg.
                                        University Park, PA 16802
                                        U.S.A.

Yamartino R.   ERT Inc.
                                        696 Virginia Road
                                        Concord, Massachusetts 01742
                                        U.S.A.

YOUGSLAVIA

Loncar E.   Republicki Hidrometeoroloski
                                        Zavod SRH
                                        41001 Zagreb, Gric 3
                                        Yougoslavia

Sinik N.   Republicki Hidrometeoroloski
                                        Zavod SRH
                                        41001 Zagreb, Gric 3
                                        Yougoslavia

Vidic S.   Republicki Hidrometeoroloski
                                        Zavod SRH
                                        41001 Zagreb, Gric 3
                                        Yougoslavia

AUTHOR INDEX

SUBJECT INDEX